Mobile Ad Hoc Networks

Current Status and Future Trends

Mobile Ad Hoc Networks

Current Status and Future Trends

Edited by
Jonathan Loo, Jaime Lloret Mauri, and Jesús Hamilton Ortiz

CRC Press
Taylor & Francis Group
Boca Raton   London   New York

CRC Press is an imprint of the
Taylor & Francis Group, an **informa** business
AN AUERBACH BOOK

CRC Press
Taylor & Francis Group
6000 Broken Sound Parkway NW, Suite 300
Boca Raton, FL 33487-2742

© 2012 by Taylor & Francis Group, LLC
CRC Press is an imprint of Taylor & Francis Group, an Informa business

No claim to original U.S. Government works

Library of Congress Cataloging-in-Publication Data

Mobile ad hoc networks : current status and future trends / editors, Jonathan Loo, Jaime Lloret Mauri, Jesus Hamilton Ortiz.
   p. cm.
Includes bibliographical references and index.
ISBN 978-1-4398-5650-5 (hardback)
   1. Ad hoc networks (Computer networks) 2. Mobile communication systems. I. Loo, Jonathan. II. Lloret Mauri, Jaime. III. Ortiz, Jesus Hamilton.

TK5105.77.M63 2011
004.6'167--dc23   2011038265

Visit the Taylor & Francis Web site at
http://www.taylorandfrancis.com

and the CRC Press Web site at
http://www.crcpress.com

Contents

SECTION III   FUTURE NETWORKS INSPIRED BY MANET

Contributors

Ali Naser Al-Khwildi
Commission of Media and
    Communications (CMC)
Jadreiah, Iraq

Marcelo Atenas
Universidad Politécnica de
    Valencia
València, Spain

Luigi Atzori
Department of Electrical and Electronic
    Engineering (DIEE)
University of Cagliari
Cagliari, Italy

Saeed Bastani
School of Information Technologies
University of Sydney
New South Wales, Australia

Deniz Cokuslu
Ege University International
    Computer Institute
Bornova, Izmir, Turkey

and

Computer Engineering Department
Izmir Institute of Technology
Urla, Izmir, Turkey

Jose Costa-Requena
School of Electrical Engineering
Department of Communications and
    Networking
Aalto University
Aalto, Finland

Orhan Dagdeviren
Computer Engineering Department
Izmir University
Uckuyular, Izmir, Turkey

and

Ege University International
    Computer Institute
Bornova, Izmir, Turkey

Kayhan Erciyes
Computer Engineering Department
Izmir University
Uckuyular, Izmir, Turkey

Mahmood Fathy
Computer Engineering Department
Iran University of Science and Technology
Narmak, Tehran, Iran

Dominique Gaiti
Autonomic Networking Environment
ICD/ERA CNRS UMR-STMR 6279
University of Technology of Troyes
Troyes, France

Miguel García
Universidad Politécnica
de Valencia
València, Spain

Hasan Gumus
Aselsan Ltd. Corporation
Ataturk Organize Sanayi Bolgesi
Cigli, Izmir, Turkey

Julie Hsieh
Department of Computer Science
San José State University
San José, California

Katherine Isaacs
Department of Computer Science
San José State University
San José, California

Bego Blanco Jauregui
Department of Computer Languages
and Systems
University of the Basque Country

and

EUITI Bilbao
Plaza de la Casilla
Bilbao, Spain

Shafiullah Khan
School of Engineering and Information
Sciences,
Computer Communications Department,
Middlesex University,
London, United Kingdom

and

Kohat University of Science and Technology
(KUST)
Kohat, Pakistan

Lyes Khoukhi
Autonomic Networking Environment
ICD/ERA CNRS UMR-STMR 6279
University of Technology of Troyes
Troyes, France

Ilias Kiourktsidis
School of Engineering and Design
Brunel University
London, United Kingdom

Panagiotis Kokkinos
Research Academic Computer Technology
Institute,
University of Patras
Patras, Greece

Grigorios Koulouras
Department of Electronics
Technological Education Institute (T.E.I) of
Athens
Athens, Greece

Raquel Lacuesta
Computer Science and Systems Engineering
University of Zaragoza
Teruel, Spain

Bernardo Leal
Department of Industrial Technology,
Simón Bolivar University
Miranda State, Venezuela

Jaime Lloret
Universidad Politécnica de
Valencia
València, Spain

Jonathan Loo
School of Engineering and Information
Sciences
Computer Communications
Department
Middlesex University
London, United Kingdom

Juan Carlos López
Computer Architecture and Networks,
School of Computer Science,
University of Castilla-La Mancha
Ciudad Real, Spain

Fidel Liberal Malaina
Department of Electronics and
    Telecommunications
University of the Basque Country

and

ETSI
Alameda de Urkijo S/N
Bilbao, Spain

Ali El Masri
Autonomic Networking Environment
ICD/ERA CNRS UMR-STMR 6279
University of Technology of Troyes
Troyes, France

Mohammad Reza Meybodi
Department of Computer Engineering and IT
Amirkabir University of Technology
Tehran, Iran

Melody Moh
Department of Computer Science,
San José State University
San José, California

Jesús Hamilton Ortiz
Computer Architecture and Networks,
School of Computer Science,
University of Castilla-La Mancha
Ciudad Real, Spain

Guillermo Palacios
Computer Science and Systems Engineering
University of Zaragoza
Teruel, Spain

Christos Papageorgiou
Research Academic Computer Technology
    Institute
University of Patras
Patras, Greece

Jorge Luís Perea Ramos
System Engineering
University of Cartagena
Cartagena, Colombia

Julio Cesar Rodríguez Ribons
System Engineering
University of Cartagena
Cartagena, Colombia

Sandra Sendra
Universidad Politécnica de
    Valencia
València, Spain

Javad Akbari Torkestani
Department of Computer Engineering
Islamic Azad University
Arak Branch, Arak, Iran

Emmanouel Varvarigos
Research Academic Computer Technology
    Institute
University of Patras
Patras, Greece

Luiz Filipe M. Vieira
Computer Science Department
Federal University of
    Minas Gerais
Belo Horizonte, Brazil

Onur Yılmaz
Ege University International
    Computer Institute
Bornova, Izmir, Turkey

and

Computer Engineering Department
Izmir University of Economics
Balcova, Izmir, Turkey

Saleh Yousefi
Computer Department
Faculty of Engineering
Urmia University
Urmia, Azarbaijan, Iran

Christos K. Zachos
IP Partners
Thessaloniki, Greece

FUNDAMENTAL OF MANET—MODELING AND SIMULATION

Chapter 1

Mobile Ad Hoc Network

Jonathan Loo, Shafiullah Khan, and Ali Naser Al-Khwildi

Contents

1.1 Introduction

Wireless industry has seen exponential growth in the last few years. The advancement in growing availability of wireless networks and the emergence of handheld computers, personal digital assistants (PDAs), and cell phones is now playing a very important role in our daily routines. Surfing Internet from railway stations, airports, cafes, public locations, Internet browsing on cell phones, and Information in file exchange between devices without wired connectivity are just a few examples. All this ease is the result of mobility of wireless devices while being connected to a gateway to access the Internet or information from fixed or wired infrastructure (called *infrastructure-based wireless network*) or ability to develop an on-demand, self-organizing wireless network without relying on any available fixed infrastructure (called *ad hoc networks*). A typical example of the first type of network is office wireless local area networks (WLANs), where a wireless access point serves all wireless devices within the radius. An example of mobile ad hoc networks (MANETs) [1] can be described as a group of soldiers in a war zone, wirelessly connected to each other with the help of limited battery-powered devices and efficient ad hoc routing protocols that help them to maintain quality of communication while they are changing their positions rapidly. Therefore, routing in ad hoc wireless networks plays an important role of a data forwarder, where each mobile node can act as a relay in addition to being a source or destination node.

1.2 Wireless Networks

Wireless networks can be broadly categorized into two classes: infrastructure-based wireless networks and infrastructure-less wireless networks (ad hoc wireless networks). Infrastructure-based wireless networks rely on an access point, which is a device that acts as a bridge between the wired and wireless networks. With the help of such an access point, wireless nodes can be connected to the existing wired networks. Examples of infrastructure-based wireless networks are wireless networks set up in airports, offices, homes, and hospitals, where clients connect to the Internet with the help of an access point. Figure 1.1 shows an infrastructure mode wireless network.

The other type of wireless networks does not rely on fixed infrastructure, and it is more commonly called an *ad hoc wireless network*. The word *ad hoc* can be translated as "improvised" or "not organized," which often has a negative meaning; however, in this context the sense is not negative, but it only describes the dynamic network situation. An ad hoc mode is used to connect wireless clients directly together, without the need for a wireless access point or a connection to an existing wired network. There are different example of MANET in ad hoc mode such as building-to-building, vehicle-to-vehicle, ship-to-ship etc.; they communicate with each other by relying on peer-to-peer routing. A typical ad hoc mode wireless network is shown in Figure 1.2.

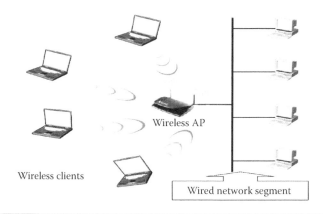

Figure 1.1   Infrastructure mode wireless network.

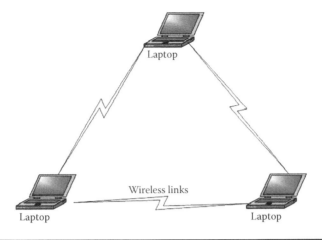

Figure 1.2 Ad hoc mode wireless network.

In wireless network communication, nodes communicate with other nodes via wireless channels. There are two important metrics that are used in the wireless networks: spectrum ranges and different radio frequencies. For example, IEEE 802.11a [2], IEEE 802.11b [3], and IEEE 802.11g [4] use a radio frequency of 5.15–5.35, 2.4–2.58, and 2.4–2.58 GHz, respectively. The signal strength in a wireless medium decreases when the signal travels further beyond a certain distance, and it reduces to the point where reception is not possible [5]. Several medium access (MAC) layers are used in wireless networks to control the use of the wireless medium: Bluetooth MAC layer 802.15 [6] and WLAN MAC layer 802.11 [3]. The topology of the wireless network can be different with time because of the mobility feature. Besides the concept of mobility, another type of mobility is defined and well studied. For example, in wireless networks, the hosts or subnets may be moved from one place to another. Traditional networks require reconfiguration of the IP address used by these hosts or subnets at the new place. A network enabled with mobile IP [7] allows these hosts or subnets to move without any manual IP address reconfiguration. The hosts can remain connected while they are moving around.

1.3 Mobile Ad Hoc Network

A *wireless ad hoc network* is a collection of two or more wireless devices having the capability to communicate with each other without the aid of any centralized administrator. Each node in a wireless ad hoc network functions as both a host and a router. The network topology is in general dynamic because the connectivity among nodes may vary with time due to node mobility, node departures, and new node arrivals. Hence, there is a need for efficient routing protocols to allow the nodes to communicate.

Ad hoc nodes or devices should be able to detect the presence of other such devices so as to allow communication and information sharing. Besides that, it should also be able to identify types of services and corresponding attributes. Since the number of wireless nodes changes on the fly, the routing information also changes to reflect changes in link connectivity. Hence, the topology of the network is much more dynamic and the changes are often unpredictable as compared to the fixed nature of existing wired networks.

The dynamic nature of the wireless medium, fast and unpredictable topological changes, limited battery power, and mobility raise many challenges for designing a routing protocol. Due to the immense challenge in designing a routing protocol for MANETs, a number of recent developments focus on providing an optimum solution for routing. However, a majority of these solutions attain a specific goal (e.g., minimizing delay and overhead) while compromising other factors (e.g., scalability and route reliability). Thus, an optimum routing protocol that can cover most of the applications or user requirements as well as cope up with the stringent behavior of the wireless medium is always desirable.

However, there is another kind of MANET nodes called the *fixed network*, in which the connection between the components is relatively static; the sensor network is the main example for this type of fixed network [8]. All components used in the sensor network are wireless and deployed in a large area. The sensors can collect the information and route data back to a central processor or monitor. The topology for the sensor network may be changed if the sensors lose power. Therefore, the sensors network is considered to be a fixed ad hoc network.

Each of the nodes has a wireless interface and communicates with each other over either radio or infrared frequency. Laptop computers and PDAs that communicate directly with each other are some examples of nodes in an ad hoc network. Nodes in the ad hoc network are often mobile, but can also consist of stationary nodes, such as access points to the Internet. Semi-mobile nodes can be used to deploy relay points in areas where relay points might be needed temporarily. Figure 1.3 shows a simple ad hoc network with three nodes. The outermost nodes are not within the transmitter range of each other. However, the middle node can be used to forward packets between the outermost nodes. Node B is acting as a router and nodes A, B, and C have formed an ad hoc network.

An ad hoc network uses no centralized administration. This ensures that the network would not collapse just because one of the mobile nodes moves out of the transmitter range of the other nodes. Nodes should be able to enter or leave the network as they wish. Because of the limited transmitter range of the nodes, multihops may be needed to reach other nodes. Every node wishing to participate in an ad hoc network must be willing to forward packets to other nodes. Thus, every node acts both as a host and as a router. A node can be viewed as an abstract entity consisting of a router and a set of affiliated mobile hosts. A router is an entity that, among other things, runs a routing protocol. A mobile host is simply an IP-addressable host or entity in the traditional sense.

Ad hoc networks are also capable of handling topology changes and malfunctions in nodes. They are fixed through network reconfiguration. For instance, if a node leaves the network and

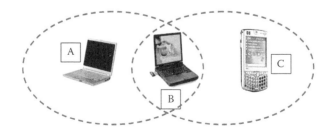

Figure 1.3   Connectivity between nodes A, B, and C.

causes link breakages, affected nodes can easily request new routes and the problem will be solved. This will slightly increase the delay, but the network will still be operational.

1.4 Mobile Ad Hoc Network History

The history of wireless networks dates back to 1970s, and the interest has been growing ever since. During the last decade, the interest has almost exploded, probably because of the fast-growing Internet. The tremendous growth of personal computers and the handy usage of mobile devices necessitate the need for ad-hoc connectivity.

The first generation goes back to 1972. At the time they were called PRNET (packet radio network). In conjunction with ALOHA (areal locations of hazardous atmospheres) [1], approaches for MAC control and a type of distance vector routing PRNET were used on a trial basis to provide different networking capabilities in a combat environment.

The second generation of ad hoc networks emerged in 1980s, when the ad hoc network was further enhanced and implemented as a part of the SURAN (Survivable Adaptive Radio Networks) project that aimed at providing ad hoc networking with small, low-cost, low-power devices with efficient protocols for improved scalability and survivability [9]. This provided a packet-switched network to the mobile battlefield in an environment without infrastructure.

In the 1990s, the concept of commercial ad hoc networks arrived with notebook computers and other viable communications equipment. At the same time, the idea of a collection of mobile nodes was proposed at several research conferences.

The IEEE 802.11 subcommittee had adopted the term "ad hoc networks" and the research community had started to look into the possibility of deploying ad hoc networks in other areas of application. Meanwhile, work was going on to advance the previously built ad hoc networks. GloMo (global mobile information systems) and the NTDR (near-term digital radio) are some of the results of these efforts [10]. GloMo was designed to provide an office environment with Ethernet-type multimedia connectivity anywhere and anytime in handheld devices.

1.5 Mobile Ad Hoc Network Definition

A clear definition of precisely what is meant by an ad hoc network is difficult to identify. In today's scientific literature, the term "ad hoc network" is used in many different ways. There are many different definitions that describe ad hoc networks, but only three are presented here. The first one is given by the Internet Engineering Task Force group [11], the second one is given by National Institute of Standard and Technology [12], and the final definition is given by the INTEC Research group [13].

In MANETs, the wireless nodes are free to move and still connected using the multihop with no infrastructure support. The goal of mobile ad hoc networking is to support robust and efficient operation in mobile wireless networks by incorporating routing functionality into mobile nodes. Ad hoc networks have no fixed routers; all nodes are capable of movement and can be connected dynamically in an arbitrary manner. Nodes of these networks function as routers, which discover and maintain routes to other nodes in the network. Example applications of ad hoc networks are emergency search and rescue operations, meetings, and conventions in which a person wishes to make a quick connection for sharing information.

1.6 MANET Applications and Scenarios

With the increase of portable devices as well as progress in wireless communication, ad hoc networking is gaining importance because of its increasing number of widespread applications. Ad hoc networking can be applied anywhere at anytime without infrastructure and its flexible networks. Ad hoc networking allows the devices to maintain connections to the network as well as easily adds and removes devices to and from the network. The set of applications of MANETs is diverse, ranging from large-scale, mobile, highly dynamic networks to small and static networks that are constrained by limited power. Besides the legacy applications that move from traditional infrastructure environment to the ad hoc context, a great deal of new services can and will be generated for the new environment. Typical applications include the following:

- Military battlefield: Military equipment now routinely contains some sort of computer equipment. Ad hoc networking can be very useful in establishing communication among a group of soldiers for tactical operations and also for the military to take advantage of commonplace network technology to maintain an information network between the soldiers, vehicles, and military information headquarters. Ad hoc networks also fulfill the requirements of communication mechanism very quickly because ad hoc network can be set up without planning and infrastructure, which makes it easy for the military troops to communicate with each other via the wireless link. The other important factor that makes MANET very useful and let it fit in the military base is the fact that the military objects, such as airplanes, tanks, and warships, move at high speeds, and this application requires MANET's quick and reliable communication. Because of the information that transfers between the troops, it is very critical that the other side receives secure communication, which can be found through ad hoc networks. At the end, the primary nature of the communication required in a military environment enforced certain important requirements on ad hoc networks, such as reliability, efficiency, secure, and support for multicast routing. Figure 1.4 shows an example of the military ad hoc network.
- Commercial sector: The other kind of environment that uses an ad hoc network is emergency rescue operation. The ad hoc form of communications is especially useful in public-safety and search-and-rescue applications. Medical teams require fast and effective communications when they rush to a disaster area to treat victims. They cannot afford the time to run cabling and install networking hardware. The medical team can employ ad hoc networks (mobile nodes) such as laptops and PDAs and can communicate via the wireless

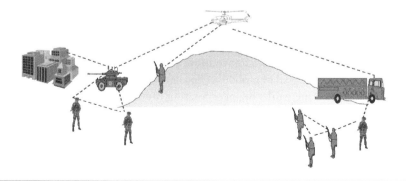

Figure 1.4   Military application.

link with the hospital and the medical team on-site. For example, a user on one side of the building can send a packet destined for another user on the far side of the facility, well beyond the point-to-point range of WLAN, by having the data routed from client device to client device until it gets to its destination. This can extend the range of the WLAN from hundreds of feet to miles, depending on the concentration of wireless users. Real-time communication is also important since the voice communication predominates data communication in such scenarios. Figure 1.5 shows the ad hoc search-and-rescue application.

■ Local level: Ad hoc networks can autonomously link an instant and temporary multimedia network using notebook computers or palmtop computers to spread and share information among participants at conferences, at meetings, or in classrooms. Another appropriate local level application might be in home networks, where devices can communicate directly to exchange information. Similarly, in other civilian environments such as taxicab, sports stadium, boat, and small aircraft, mobile ad hoc communications will have many applications.

■ Personal area network (PAN): It is the interconnection of information technology devices within the range of an individual person, typically within a range of 10 m. For example, a person traveling with a laptop, a PDA, and a portable printer could interconnect them without having to plug anything in by using some form of wireless technology. Typically, this type of PAN could also be interconnected without wires to the Internet or other networks. A wireless personal area network (WPAN) is virtually a synonym of PAN since almost any PAN would need to function wirelessly. Conceptually, the difference between a PAN and a WLAN is that the former tends to be centered around one person while the latter is a local area network (LAN) that is connected without wires and serve multiple users.

Bluetooth is an industrial specification for WPANs. A Bluetooth PAN is also called a *piconet* and is composed of up to eight active devices in a master–slave relationship (up to 255 devices can be connected in the "parked" mode). The first Bluetooth device in the piconet is the master, and all other devices are slaves that communicate with the master. A piconet has a range of 10 m that can reach up to 100 m under ideal circumstances, as shown in Figure 1.6.

The other usage of the PAN technology is that it could enable wearable computer devices to communicate with nearby computers and exchange digital information using the electrical conductivity of the human body as a data network. Some concepts that belong to the PAN technology are considered in research papers, which present the reasons why those concepts might be useful:

Figure 1.5   Search-and-rescue application.

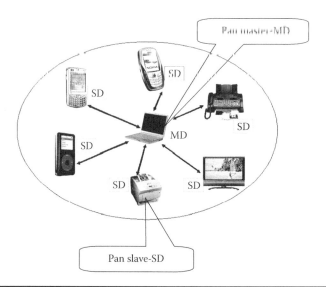

Figure 1.6 Personal area network.

- Small size of the device
- No need for huge power (lower power requirements)
- Not expensive
- Used specially for bodies and for sensitive information
- No methods for sharing data
- Networking can reduce function of input/output
- Allow new conveniences and services

1.7 Ad Hoc Network Characteristics

MANETs have the following features that are necessary to consider while suggesting or designing solutions for these types of networks:

- MANET has a feature of distributed operation because in MANET each node operates independently and there is no centralized server or computer to manage this network. Instead this job is distributed among all operating nodes. Each node works with another node in cooperation to implement functions such as security and routing.
- MANETs have lower bandwidth capacity as compared with wired networks. MANETs can experience a problem of bit error rate and lower bandwidth capacity because end-to-end link paths are used by several nodes in the network. Also, the channel used for communication can be affected by other factors such as fading and interference.
- Another feature of MANET that can be used is energy in mobile devices. As all mobile devices will get their energy from batteries, which is a limited resource, whatever energy the mobile nodes have, it has to be used very efficiently.
- Security is the most important concern in MANETs because the nodes and the information in MANETs are not secured from threats, for example, denial of service attacks. Also, mobile devices imply higher security risks compared with fixed operating devices, because portable

devices may be stolen or their traffic may insecurely cross wireless links. Eavesdropping, spoofing, and denial of service attacks are the main threats for security.

■ In MANETs the network topology is always changing because nodes in the ad hoc network change their positions randomly as they are free to move anywhere. Therefore, devices in a MANET should support dynamic topology. Each time the mobility of node causes a change in the topology and hence the links between the nodes are always changing in a random manner. This mobility of nodes creates frequent disconnection; hence, to deal with this problem the MANET should adapt to the traffic and transmission conditions according to the mobility patterns of the mobile network nodes.

■ A MANET includes several advantages over wireless networks, including ease of deployment, speed of deployment, and decreased dependences on a fixed infrastructure. A MANET is attractive because it provides an instant network formation without the presence of fixed base stations and system administration.

1.8 Classification of Ad Hoc Networks

There is no generally recognized classification of ad hoc networks in the literature. However, there is a classification on the basis of the communication procedure (single hop/multihop), topology, node configuration, and network size (in terms of coverage area and the number of devices).

1.8.1 Classification According to the Communication

Depending on the configuration, communication in an ad hoc network can be either single hop or multihop.

1.8.1.1 Single-Hop Ad Hoc Network

Nodes are in their reachable area and can communicate directly, as shown in Figure 1.7. Single-hop ad hoc networks are the simplest type of ad hoc networks where all nodes are in their mutual

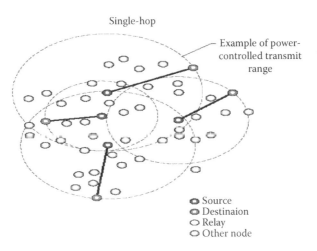

Figure 1.7   Single-hop ad hoc network.

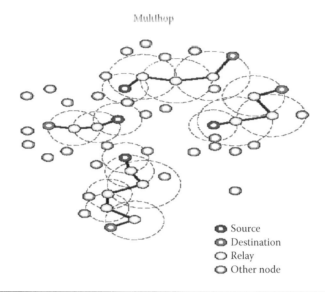

Figure 1.8 Multihop ad hoc networks.

range, which means that the individual nodes can communicate directly with each other, without any help of other intermediate nodes. The individual nodes do not have to be static; they must, however, remain within the range of all nodes, which means that the entire network could move as a group; this would not modify anything in the communication relations.

1.8.1.2 Multihop Ad Hoc Network

This class in the literature is the most examined type of ad hoc networks. It differs from the first class in that some nodes are far and cannot communicate directly. Therefore, the traffic of these communication endpoints has to be forwarded by other intermediate nodes. Figure 1.8 shows the communication path of far nodes as black lines. With this class also, one assumes that the nodes are mobile. The basic difficulty of the networks of this class is the node mobility, whereby the network topology is subjected to continuous modifications. The general problem in networks of this class is the assignment of a routing protocol. High-performance routing protocols must be adaptive to the fast topology modification.

1.8.2 Classification According to the Topology

Ad hoc networks can be classified according to the network topology. The individual nodes in an ad hoc network are divided into three different types with special functions: flat, hierarchical, and aggregate ad hoc networks.

1.8.2.1 Flat Ad Hoc Networks

In flat ad hoc networks, all nodes carry the same responsibility and there is no distinction between the individual nodes, as shown in Figure 1.9. All nodes are equivalent and can transfer all functions in the ad hoc network. Control messages have to be transmitted globally throughout the network, but they are appropriate for highly dynamic network topology. The scalability decreases when the number of nodes increases significantly.

1.8.2.2 Hierarchical Ad Hoc Networks

Hierarchical ad hoc networks consist of several clusters, each one represents a network and all are linked together, as indicated in Figure 1.10. The nodes in hierarchical ad hoc networks can be categorized into two types:

- Master nodes: Administer the cluster and are responsible for passing the data on to the other cluster.
- Normal nodes: Communicate within the cluster directly together and with nodes in other clusters with the help of the master node. Normal nodes are also called *slave nodes*.

One assumes that the majority of communication (control messages) takes place within the cluster and only a fraction between different clusters. During communication within a cluster, no forwarding of communication traffic is necessary. The master node is responsible for the switching of a connection between nodes in different clusters.

The no single point of failure is of great importance for a message to reach its destination. This means that if one node goes down, the rest of the network will still function properly. In the

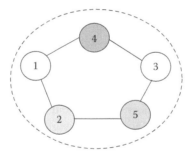

Figure 1.9 Flat ad hoc network

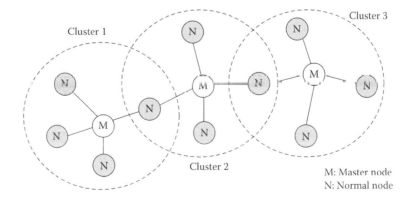

Figure 1.10 Hierarchical ad hoc networks.

hierarchical approach, this is altogether different. If one of the cluster heads goes down, that section of the network will not be able to send or receive messages from other sections for the duration of the downtime of the cluster head.

Hierarchical architectures are more suitable for low-mobility cases. The flat architectures are more flexible and simpler than hierarchical ones; hierarchical architectures provide a more scalable approach.

1.8.2.3 Aggregate Ad Hoc Networks

Aggregate ad hoc networks bring together a set of nodes into zones. Therefore, the network is partitioned into a set of zones as shown in Figure 1.11. Each node belongs to two levels of topology: low-level (node-level) topology and high-level (zone-level) topology. Also, each node may be characterized by two ID numbers: node ID number and zone ID number. Normally, aggregate architectures are related to the notion of zone. In aggregate architectures, we find both intrazone and interzone architectures, which in turn can support either flat or hierarchical architectures.

1.8.3 Classification According to the Node Configuration

A further classification of ad hoc networks can be performed on the basis of the hardware configuration of the nodes. There are two types of node configurations: homogeneous networks and heterogeneous networks. The configuration of the nodes in a MANET is important and can depend very strongly on the actual application.

1.8.3.1 Homogeneous Ad Hoc Networks

In homogeneous ad hoc networks, all nodes possess the same characteristics regarding the hardware configuration as processor, memory, display, and peripheral devices. Most well-known representatives of homogeneous ad hoc networks are wireless sensor networks. In homogeneous

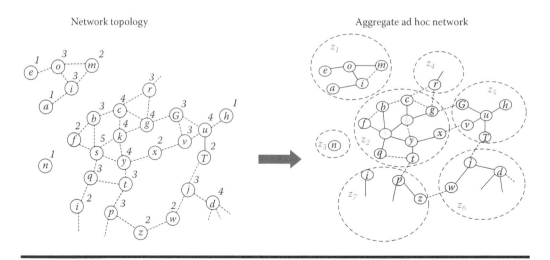

Figure 1.11   Aggregate network architecture.

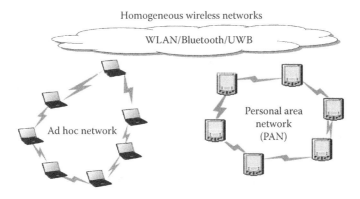

Figure 1.12 Homogeneous networks.

ad hoc networks, applications can proceed from certain prerequisites; for example, the localization is considerably facilitated by the presence of control components in each node, as shown in Figure 1.12.

1.8.3.2 Heterogeneous Ad Hoc Networks

In heterogeneous ad hoc networks, the nodes differ according to the hardware configuration. Each node has different characteristics, resources, and policies. In ad hoc networks of this class, all nodes cannot provide the same services, as shown in Figure 1.13.

1.8.4 Classification According to the Coverage Area

As shown in Figure 1.14, ad hoc networks can be categorized, depending on their coverage area, into several classes: depending on their coverage area, into several classes: body area network (BAN), personal area network (PAN), local area network (LAN), metropolitan area network (MAN), and wide area network (WAN) [13,14]. WAN and MAN are mobile multihop wireless networks presenting many challenges that are still being solved (e.g., addressing, routing, location management, and security), and their availability is not on immediate horizon.

A BAN is strongly correlated with wearable computers. The components of a wearable computer are distributed on the body (e.g., head-mounted displays, microphones, and earphones), and a BAN provides the connectivity among these devices. The communicating range of a BAN corresponds to the human body range, i.e., 1–2 m. As wiring around a body is generally cumbersome, wireless technologies constitute the best solution for interconnecting wearable devices. The PAN connects mobile devices carried by users to other mobile and stationary devices, while BAN is devoted to the interconnection of one-person wearable devices. A PAN has a typical communication range of up to 10 m. WPAN technologies in the 2.4–10.6-GHz band are the most promising technologies for the widespread PAN deployment. Spread spectrum is typically employed to reduce interference and utilize the bandwidth [15].

In the last few years, the application of wireless technologies in the LAN environment has become increasingly important, and WLAN can be found in different environments such as homes, offices, urban roads, and public places. WLAN, also called *wireless fidelity* (Wi-Fi), is

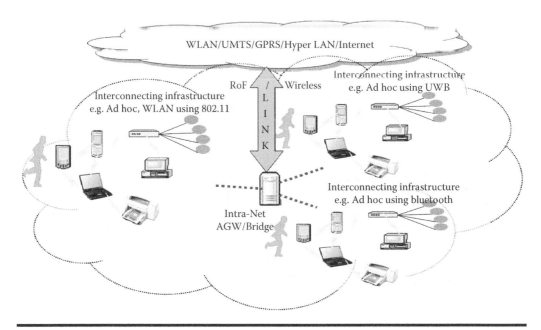

Figure 1.13 Heterogeneous networks.

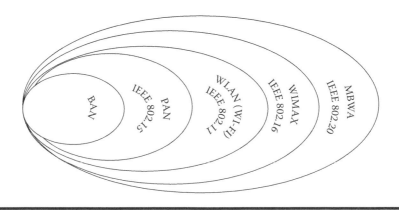

Figure 1.14 Ad hoc network taxonomy according to coverage area.

based on the 802.11 standard. It gives freedom to Internet users; also, they offer greater flexibility than the wired LANs. Most of the personal computers, laptops, phones, and PDAs are capable of connecting to the Internet via WLAN. Currently, there are five major specifications in the WLAN family 802.11 namely 802.11a, 802.11b, 802.11g and 802.11n. All use CSMA/CA (Carrier Sense Multiple Access with Collision Avoidance) for medium sharing which are standardized in 802.11c, 802.11d, 802.11e and 802.11f.

WIMAX is based on the 802.16 IEEE standard and defined as a wireless MAN technology that will provide a wireless alternative to wire and digital subscriber line (DSL) for last mile broadband access. WIMAX has a communication range of up to 50 km, which also allows the users to get broadband connections without directly connecting to the base station, and provides shared data rates of up to 70 Mbps, which is enough bandwidth to support more than 60 T1 link and hundreds of home and office DSL connections. Likewise, WIMAX fully supports the quality of service. Finally, the last but not the least wireless technology called mobile broadband wireless access (MBWA) is approved by the IEEE standard board and defined as 802.20. The MBWA is similar to the IEEE 802.16e in that its uses Orthogonal Frequency Division Multiple Access (OFDMA), provides very high mobility, and has a shared data rate of up to 100 Mbps. At present, no operator has committed to the MBWA technology.

Conclusion

The chapter has presented the overview of wireless networks and different aspects of MANET, such as, definition, application, classification, special features and various routing protocols of MANET. The applications of MANETs are described with examples and how those applications work with different environments. The MANET characteristic features are also pointed out such as distributed operation, lower bandwidth capacity, dynamic topology and security. This chapter also briefly covered the classification of MANETs in terms of communication procedure (single hop/multi hop), topology, (node configuration) and network size (coverage area and number of devices).

References

1. F. Dressler. Self-organization in ad hoc networks: overview and classification. Technical Report 02/06. Department of Computer Science 7, University of Erlangen.
2. C.E. Perkins. *Ad Hoc Networking*. Addison-Wesley, 2001.
3. S. Basagni, M. Conti, S. Giordano, and I. Stojmeovic. *Mobile Ad Hoc Networking*. Wiley-IEEE Press, 2004.
4. M. Gast. *802.11 Wireless Networks the Definitive Guide, Second Edition*. O'Reilly Media, 2005.
5. C. Siva Ram Murthy and B.S. Manoj. *Ad Hoc Wireless Networks Architectures and Protocols*. Prentice Hall, 2004.
6. M. Iiyas. *The Handbook of Ad Hoc Wireless Networks*. CRC Press, 2002.
7. M. Frodigh, P. Johansson, and P. Larsson. Wireless ad hoc networking: the art of networking without a network. *Ericsson Review*, no. 4, 2000, pp. 248–263.
8. B.-J. Kwak, N.-O. Song, and L.E. Miller. On the scalability of ad hoc networks. *IEEE Communication Letter*, vol. 8, no. 8, 2004, pp. 503–505.
9. J. Jubin and J.D. Tornow. The DARPA packet radio network protocols. *Proceedings of the IEEE*, vol. 75, no. 1, 1987, pp. 21–32.
10. R. Ramanathan and J. Redi. A brief overview of ad hoc networks: challenges and directions. *IEEE Communications Magazine*, vol. 40, no. 5, 2002, pp. 20–22.
11. http://www.ietf.org/html.charters/wg-dir.html (Accessed July 2011)
12. http://www.antd.nist.gov/ (Accessed July 2011)
13. Wireless local area network hits the public. Available at http://www.touchbriefings.com/pdf/744/wire041_vis.pdf (Accessed March 2011).
14. Chlamtac, M. Conti, and J.N. Jennifer. Mobile ad hoc networking: imperatives and challenges. In: *Elsevier Proceeding for Ad Hoc Networks*, vol. 1, 2003, pp. 13–64.
15. M. Abolhasan, T. Wysocki, and E. Dutkiewicz. *A Review of Routing Protocols for Mobile Ad Hoc Networks*, vol. 2, no. 1, 2004, pp. 1–22.

Chapter 2

Mobile Ad Hoc Routing Protocols

Jonathan Loo, Shafiullah Khan, and Ali Naser Al-Khwildi

Contents

The development of omnipresent mobile computing devices has fueled the need for dynamic reconfigurable networks. Mobile ad hoc network (MANET) routing protocols facilitate the creation of such networks, without a centralized infrastructure. One of the challenges in the study of MANET routing protocols is the evaluation and design of an effective routing protocol that works at low data rates and responds to dynamic changes in network topology due to node mobility. Several routing protocols have been standardized by the Internet Engineering Task Force to address ad hoc routing requirements. The classification of these protocols and some existing ad hoc routing protocols are discussed in this chapter [1–3].

2.1 Taxonomy of Ad Hoc Routing Protocols

Several ad hoc protocols have been designed for accurate, fast, reliable routing for a high volume of changeable network topology. Such protocols must deal with the typical limitations of changeable network topology, which include high power consumption, low bandwidth, and high error rates. As shown in Figure 2.1, these routing protocols may generally be categorized into three main types: proactive or table-driven, reactive or on-demand-driven, and hybrid. This classification differentiates the routing protocols according to their technique, their hop count, link state, and source routing in a route-discovery mechanism. In protocols based on a hop count technique, each node contains next-hop information in its routing table, linked to the destination. Link state routing protocols maintain a routing table for complete topology, which is built up by finding the shortest path of link costs. In the source routing technique, all data packets carry their routing information as their header. The originating node can obtain this routing information, for example, by means of a source routing protocol. The next section will present details for each routing category, including some of the existing routing protocols used for those categories [4–11].

2.2 On-Demand Ad Hoc Routing Protocols

Reactive protocols are also called *on-demand routing protocols*. These protocols create routes to a destination only when required. The route discovery procedure is triggered whenever a source wants to send data to find a destination node, and the route is maintained through the route maintenance procedure until the route is no longer required. In this manner, communication overhead is reduced and battery power is conserved as compared to proactive routing protocols. As shown in Figure 2.2, there is no topology table in each node. When there is a request in node A to transmit data to node D, the route discovery process starts by broadcasting to all nodes searching for node D. When node D receives this message, it responds to the request to build the route to node A. The process is complete once a route is found or all possible route permutations have been examined. Once a route has been established, it is maintained by a route maintenance procedure

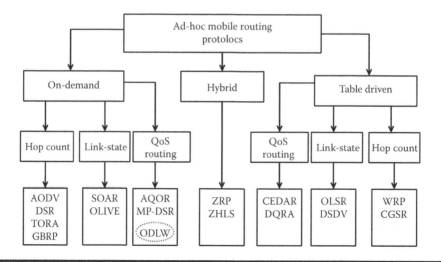

Figure 2.1   Categorization of ad hoc routing protocol.

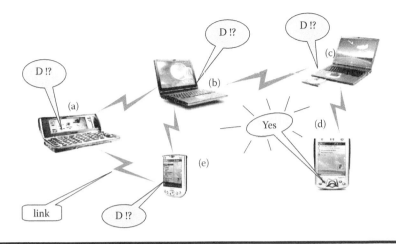

Figure 2.2 On-demand (reactive) ad hoc protocol.

until either the destination becomes inaccessible along every path from the source or the route is no longer desired.

A network using an on-demand protocol will not maintain correct routing information for all nodes at all times. As an alternative, such routing information is obtained on demand. If a node needs to transmit a message and does not have sufficient routing information to send the message to the destination, the necessary information has to be obtained. Typically, the node at least wants to identify the next hop (among its neighbors) for the packet. Although the node could just broadcast the packet to all neighbors, this leads to severe congestion in numerous instances. However, such broadcasts are used in a route discovery process, since there is no other next-hop information available yet.

The advantage of on-demand routing protocols lies in the fact that the wireless channel (a scarce resource) does not require to carry a large amount of routing overhead data for routes that are no longer used. This advantage may be reduced in certain scenarios where there is heavy traffic to a wide range of nodes. Thus, these scenarios have a strong impact on performance. In a scenario including large amounts of traffic to several nodes, the route setup traffic can rise higher than the constant background traffic to preserve the correct routing information at every node. Still, if sufficient capacity is available, the compact efficiency (increased overhead) may not influence other performance methods such as throughput or latency. Examples for on-demand protocols include the following: ad hoc on-demand distance vector (AODV), dynamic source routing (DSR), Temporally-Ordered Routing Algorithm (TORA) Associativity Based Routing (ABR), and Stability based Adaptive (SSA) [12–16].

2.3 Table-Driven Ad Hoc Routing Protocols

Proactive routing protocols enable each node to keep up-to-date routing information in a routing table. This routing table is exchanged periodically with all other nodes, as well as when network topology changes. Thus, when a node needs to send a packet, the route is readily available. However, most of the routing information that is exchanged is undesired. Proactive routing protocols are also called *table-driven routing protocols*.

Figure 2.3 illustrates the concept of proactive protocols. For example, if node A wanted to send some data to node D, all it would have to do is find node D on the previously prepared

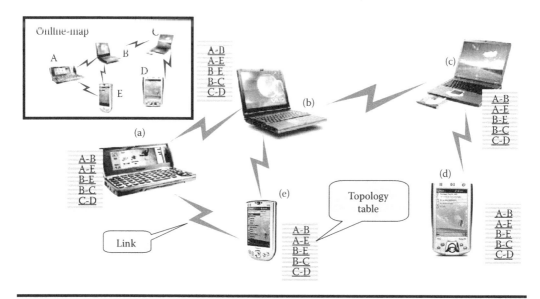

Figure 2.3 Proactive ad hoc protocol.

topology table, which is stored in node A. Table parsing is faster and requires less power than searching the entire network for a destination. If the network nodes do not have frequent mobility, then the topology table will not consume too much power.

In ad hoc networks based on proactive protocols, power and bandwidth consumption increase due to topology table exchange among nodes after each change in the nodes' location. This takes place even if the network is in stand-by mode (e.g., no data transmissions in the network). The best network context for proactive protocols is the low (or no) mobility networks. Some well-known proactive protocols include optimized link state routing (OLSR), destination sequenced distance vector Clusterhead Gateway Switch Routing (CGSR), and Wireless Routing Protocol (WRP) [11,17–19].

2.4 Hybrid Ad Hoc Routing Protocols

Based on the combination of both table- and demand-driven routing protocols, some hybrid routing protocols have been proposed to combine the advantages of both proactive and reactive protocols. The most typical hybrid protocol is a zone routing protocol [20]. With regard to the main division of routing protocols, Table 2.1 provides a comparison of table-driven, demand-driven, and hybrid routing protocols. Some hybrid routing protocols include zone routing protocol (ZRP), Zone-Based Hierarchical Link State (ZHLS), and core extraction distributed ad hoc routing(CEDAR) [21–23].

2.5 Description of Current Ad Hoc Routing Protocols

Many routing protocols have been proposed for ad hoc networks, which are different in the approach used for the routing discovery mechanism, maintaining the existing route when link failure occurs or the node moves away from the existing networks. In the next section, we will

Table 2.1   Characteristic Comparison of Proactive, Reactive, and Hybrid Routing Protocol

	Table-Driven	*Demand-Driven*	*Hybrid*
Network organization	Flat/hierarchical	Flat	Hierarchical
Topology dissemination	Periodical	On-demand	Both
Route latency	Always available	Available when needed	Both
Mobility handling	Periodic updates	Route maintenance	Both
Communication overhead	High	Low	Medium

present the operation and routing mechanism for well-known routing protocols, such as AODV, DSR, temporally ordered routing algorithm (TORA), OLSR, DSDV, ZRP, CEDAR, and ad hoc quality of service (QoS) on-demand routing (AQOR) [24].

The major differences between all the described protocols are shown in Table 2.2. The data were used for this investigation to enhance the overview of the interworking between the different protocols.

2.5.1 AODV

The AODV [12] routing protocol uses the on-demand approach for finding routes; that is, the route is established only when it is required by a source node for transmitting data packets. It employs a destination sequence number to identify the most recent path. In AODV, the source node and the intermediate nodes store the next-hop information corresponding to each flow for data packet transmission.

When a source requires a route to a destination, it floods the network with a *route request* (RREQ) packet. On its way through the network, the RREQ packet initiates the creation of temporary route table entries for the reverse path at every node it passes, and when it reaches the destination, a *route reply* (RREP) packet is unicast back along the same path on which the RREQ packet was transmitted. A mobile node can become aware of neighboring nodes by employing several techniques, one of which involves broadcasting *Hello* messages. Route entries for each node are maintained using a timer-based system. If the route entry is not used immediately, it is deleted from the routing table. AODV does not repair broken paths locally. When a path breaks between nodes, both nodes initiate *route error* (RERR) packets to inform their end nodes about the link break. The end nodes delete the corresponding entries from their table. The source node reinitiates the path-finding process with a new broadcast ID and the previous destination sequence number. The main advantage of this protocol is that the routes are established on demand and destination sequence numbers are used to find the latest route to the destination. The disadvantage of this protocol is that the intermediate nodes can lead to inconsistent routes if the source sequence number is very old and the intermediate nodes have a higher, but not the latest, destination sequence number, thereby hosting stale entries. This is illustrated in Figure 2.4.

Table 2.2 Characteristic Comparison of Proactive, Reactive, and Hybrid Routing Protocols

	AODV	DSR	TORA	OLSR	DSDV	ZRP	CEDAR	AQOR
Routing type	Reactive	Reactive	Reactive	Proactive	Proactive	Hybrid	Hybrid	QoS-Reactive
Alternative route	Not available	Not available	Available	Not available	Not available	Not available	Not available	Not available
Routing mechanism	Next hop	Source routing	Next hop	Next hop	Next hop	N/A	N/A	Next hop
Routing metrics	Shortest path	Shortest path	Shortest path	Shortest path	Shortest path	Shortest path	Shortest path	Shortest path
Update routing	Yes–Hello message	No	No	Yes	Yes	Half way Yes	Half way Yes	Yes–Hello message
Network size	Large	Small	Medium	Large	Large	Large	Large	Large
Routing adaption	No	No	No	No	No	No	No	Yes
QoS support	No	No	No	No	No	No	No	Yes
Security	No	No	No	No	No	No	No	No
Mobility	Good	Bad	Bad	Good	Good	Good	Good	Good

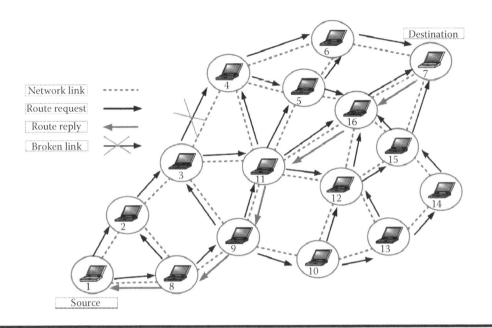

Figure 2.4   AODV routing mechanism.

2.5.2 DSR

DSR [13] is another on-demand protocol designed to restrict the bandwidth consumed by control packets in ad hoc networks by eliminating the periodic table-update message required in the table-driven approach. The key distinguishing feature of DSR is the use of source routing. The sender knows the complete hop-by-hop route to the destination, and those routes are stored in a route cache. The data packet carries the source route in the packet header. There are two major phases in this protocol. The first is route discovery, which is achieved by flooding the network with RREQ packets. The destination node, upon receiving an RREQ, responds by sending an RREP packet back to the source along the same route traversed by the incoming RREQ packet. Any node can update its cache when it receives or forwards a packet containing source route information. The route cache can be used to reduce the number of packets flooding the network. The second phase is route maintenance. If any link on a source route is broken, the source node is notified through an RERR packet. The source removes any route using this link from its cache. A new route discovery process must be initiated by the source if this route is still needed. The advantage of this protocol is that it reduces overhead on route maintenance. This is done by using route caching. The disadvantage is that the packet header size grows with the route length due to both source routing and RREQ flooding that may potentially reach all nodes in the network. The DSR routing mechanism is shown in Figure 2.5.

2.5.3 TORA

TORA [14] is a source-initiated on-demand routing protocol, which uses a link reversal algorithm and provides loop-free multipath routes to a destination node. In TORA, each node maintains its one-hop local topology information and also has the capability to detect partitions. TORA is proposed to operate in a highly dynamic mobile networking environment. The key design concept

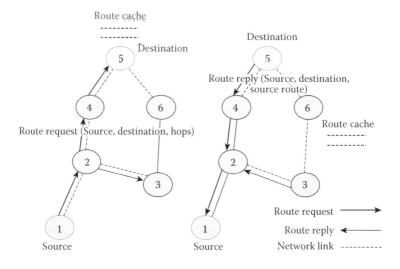

Figure 2.5 DSR routing mechanism.

of TORA is the location of control messages sent to a very small set of nodes near the occurrence of a topological change. The protocol performs three basic functions: (1) route creation, (2) route maintenance, and (3) route erasure.

During the route creation and maintenance phases, the nodes use a *height* metric, which establishes a direct acyclic graph (DAG) rooted at the destination. Therefore, links are assigned a direction (upstream or downstream) based on the relative height metric of neighboring nodes, as shown in Figure 2.6. The process for establishing a DAG is similar to the query/reply process in lightweight mobile routing. In times of node mobility, the DAG route is broken, and route maintenance is necessary to reestablish a DAG rooted at the same destination. Timing is an important factor for TORA because the height metric depends on the logical time of link failure. TORA assumes all nodes have synchronized clocks. In TORA, there is a potential for oscillations to occur, especially when multiple sets of coordinating nodes are concurrently detecting partitions, erasing routes, and building new routes based on each other. Because TORA uses internodal coordination, its instability problem is similar to the "count-to-infinity" problems.

2.5.4 OLSR

The OLSR protocol is a table-driven protocol [11]. In OLSR, nodes exchange messages with other nearby nodes of the network on a regular basis to update topology information on each node, as illustrated in Figure 2.7. Nodes determine their one-hop neighbors, i.e., nodes within their transmission radius, by transmitting Hello messages. Based on a selection criterion that will be elaborated upon in the subsequent sections, a set of nodes among the one-hop neighbors is chosen as multipoint relays (MPRs). Only these nodes forward topological information, providing every other node with partial information about the network. Furthermore, only these MPRs will generate link state information to be forwarded throughout the network. By these two optimizations, the amount of retransmission is minimized, thereby reducing overhead as compared to link state routing protocols. Each node will then use this topological information, along with the collected Hello messages, to compute optimal routes to all nodes in the network. In ad hoc radio networks,

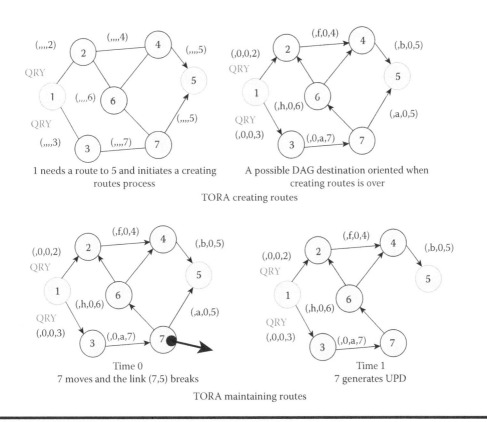

TORA creating routes

1 needs a route to 5 and initiates a creating routes process

A possible DAG destination oriented when creating routes is over

Time 0
7 moves and the link (7,5) breaks

Time 1
7 generates UPD

TORA maintaining routes

Figure 2.6 TORA routing mechanism.

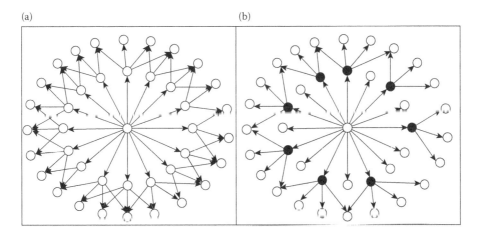

(a)

(b)

Figure 2.7 OLSR routing mechanism.

due to its limited bandwidth, a compromise has to be made between a small number of emissions and the reliability of delivery.

Every node periodically broadcasts Hello messages that contain one-hop neighbor information. The Time To Live (TTL) of a Hello message is 1, and so they are not forwarded by the neighbors. With the aid of Hello messages, every node obtains local topology information.

A node (also called a selector) chooses a subset of its neighbors to act as MPR nodes based on local topology information, which are later specified in the periodic Hello messages. MPR nodes have two roles:

a. When the selector sends or forwards a broadcast packet, only its MPR nodes among all its neighbors forward the packet.
b. The MPR nodes periodically broadcast the selector list throughout the MANET (again, by means of MPR flooding). Thus, every node in the network knows which MPR nodes could reach every other node.

Note that role (a) reduces the number of retransmissions of the topology information broadcast and role (b) reduces the size of the broadcast packet. As a result, much more bandwidth is saved compared with that saved by original link state routing protocols.

With global topology information stored and updated at every node, the shortest path from one node to every other node can be computed with Dijkstra's algorithm, which goes along a series of MPR nodes.

2.5.5 DSDV

DSDV [17] is a table-driven routing scheme for ad hoc mobile networks based on the Bellman–Ford algorithm. DSDV uses the shortest-path routing algorithm to select a single path to a destination. To avoid routing loops, destination sequence numbers have been introduced. In DSDV, full dumps and incremental updates are sent between nodes to ensure that routing information is distributed.

This protocol is the result of adapting an existing distance-vector routing algorithm to an ad hoc networking environment. DSDV is one of the first attempts to adapt an established routing mechanism to work with MANETs. Each routing table lists all destinations with their current hop count and a sequence number. Routing information is broadcast or multicast. Each node transmits its routing table to its neighbors. Routes with more recent sequence numbers render older routes obsolete. This mechanism provides loop freedom and prevents the use of stale routes. The routing information is transmitted every time a change in the topology has been detected (i.e., a change in the set of neighbors of a node). DSDV works only with bidirectional links. The drawback of this protocol is that it creates large amounts of overhead. Therefore, DSDV is not suitable for large networks, since it consumes more bandwidth than other protocols during the updating procedure.

2.5.6 ZRP

The MANET hybrid routing protocol is a combination of two ad hoc routing approaches: the reactive (on-demand) and the proactive (table-driven). The network in hybrid routing protocols such as ZRP [21] is divided into routing zones. The routing information within each routing zone is proactively distributed, while the global routing information is exchanged reactively. The ZRP approach has proved that it reduces the delay and the amount of routing overheads.

ZRP is a hybrid routing protocol suitable for a wide variety of MANETs, especially with a large network span and diverse mobility patterns. Around each node, ZRP defines a zone where the radius is measured in hops. Each node uses proactive routing within its zone and reactive routing outside its zone. Hence, a given node knows the identity of a route to all nodes within its zone.

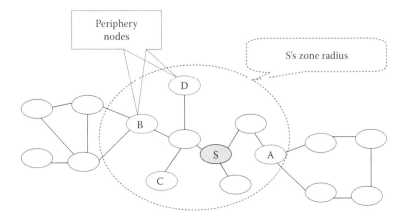

Figure 2.8   ZRP zone radius.

When a node has data for a particular destination node, it checks its routing table for a route. If the destination node is within the zone, a route will exist in the route table; if the destination node is not within the zone, the node will search for that specific destination to find the route.

The proactive maintenance of routing zones also helps to improve the quality of discovered routes by making them more sensitive to changes in the network topology. Zones in ZRP are configured by proper selection of only one parameter, the zone radius, which is measured in hops. The ZRP framework is designed to provide a balance between the contrasting proactive and reactive routing approaches. The proactive routing approach implemented in ZRP is the intrazone routing protocol (IARP). IARP is a link state protocol that maintains up-to-date information about all nodes within a zone. For example, S is a given node; the peripheral nodes of S are A, B, C, and D. The peripheral nodes with the shortest distance to S are defined, as shown in Figure 2.8. These nodes are important for reactive route discovery. The ZRP also utilizes the interzone routing protocol to discover routes to destination nodes outside the zone.

With regard to the route discovery process in ZRP, once the source node determines that the destination node is not within its zone, the source node broadcasts a query message to its peripheral nodes. This query message is relayed using trees constructed within the IARP topology. After receiving the message, the peripheral nodes check whether the destination is within their zone. If the destination is not located, the peripheral nodes broadcast the query message to their own peripheral nodes. This process continues until either the destination node is located or the entire network is searched. Once a node discovers the destination node, it unicasts a reply message to the source node, as shown in Figure 2.8.

The main concept for the ZRP is to integrate the features of both proactive and reactive routing protocols. With proactive (table-driven) protocols inside a limit zone, the connection establishment time can be reduced. On the contrary, reactive routing reduces the amount of control traffic by locating paths on demand for destinations outside the routing zone.

2.5.7 CEDAR

CEDAR is more of a routing framework scheme for QoS requirements than a MANET routing protocol. CEDAR dynamically establishes a core for the network and then incrementally propagates the link state of stable high-bandwidth links to the nodes of the core [23].

CEDAR has three key components. The first is the establishment and maintenance of a self-organizing routing infrastructure, called *the core*, for performing route computations. A subset of the nodes is selected to form a backbone within the network (the core). This structure is used for broadcast messages; hence, no flooding is needed. Each core node maintains the local topology of the nodes in its domain and also performs route computations on behalf of these nodes.

The second component of CEDAR is the propagation of the link state messages of high bandwidth and stable links in the core. The messages sent over the core network are *increase waves* (slow propagating) and *decrease waves* (fast), which notify the core of an increase or decrease in the available bandwidth. For unstable links that rise and fall frequently, the fast-moving decrease wave quickly overtakes and stops the slower-moving decrease wave from propagating, thus ensuring that the link state corresponding to dynamic links is kept local. Therefore, the propagation of these waves is dynamically limited, depending on the available bandwidth. As such, the relevant information for QoS is disseminated in an efficient way. Within the core network, any established ad hoc routing protocol may be used.

The last component of CEDAR is a QoS route computation algorithm that is executed at the core nodes using only locally available states. In order to establish QoS routes, the source node contacts its dominator (local core node) with an RREQ that contains the information on the source, destination, and required bandwidth. The source node then initiates a core broadcast to find the location of the receiver and simultaneously discover the core path. The dominator computes a QoS route, if this is feasible, and then continues to establish it. This includes the possible discovery of the dominator of the destination and a core path to it. Otherwise, if the dominator of the source node has already been cached and has a core path established to the dominator of the destination node, the source node's dominator proceeds with the QoS route establishment phase. If the dominator of the source node does not know the location of the destination node, it first locates the dominator of the destination node and simultaneously establishes a core path to it; this initiates the route computation phase.

A core path from source to destination results in a path in the core graph from the dominator of the source to the dominator of the destination. The dominator of the source then tries to find the shortest-widest-furthest admissible path along the core path. Based on its local information, the dominator of the source picks up the farthest reachable domain until it finds what it knows is an admissible path. It then computes the shortest-widest path to that domain, ending at some node, once again based on local information. Once this path is established, the dominator of the destination then uses its local state to find the shortest-widest-furthest admissible path to the domain along the core path, and so on. Eventually, either an admissible route to the destination is established or the algorithm reports a failure to find an admissible path.

2.5.8 AQOR

Ad hoc QoS on-demand routing (AQOR) provides end-to-end QoS support in terms of bandwidth and end-to-end delays in MANETs [24]. It is a resource reservation-based routing and signaling scheme that allows AQOR to make admission and resource reservation decisions. AQOR integrates on-demand route discovery between the source and the destination, signaling functions for resource reservation and maintenance and hop-by-hop routing. AQOR is also a source-initiated, on-demand routing protocol. It is built upon AODV routing, performing exploration of routes only when required.

The route discovery mechanism is in on-demand mode, broadcasting the RREQ and RREP packets between the source and destination nodes. The route discovery mechanism starts when the node broadcasts RREQ packets with QoS requirements to its neighboring nodes. The neighboring nodes that satisfy the requirement add a route entry to the source node's routing table and forward the RREQ until it reaches the destination. When the RREQ reaches the destination node, an RREP is sent back along the reverse route, reserving bandwidth at each node. Once the source node receives the RREP, it starts sending data out along the reserved route.

AQOR uses timers to detect route breaks and to trigger route recovery. If any node fails to receive a data packet before its reservation expires, a route recovery mechanism is triggered. The source node starts the route discovery process all over again by broadcasting an RREQ packet. Initiating a route discovery process each time a route break occurs can lead to high end-to-end delays. The QoS Multi-Mesh Tree (MMT) routing protocol, which is being designed for AQOR, allows each node to have multiple routes already discovered, with one primary route and other routes as secondary. In case a primary route fails, one of the secondary routes becomes the primary route and the node immediately starts sending data using the new primary route. This facility provided by QoS MMT leads to fast route failure recovery with immediate switch over and low end-to-end delays.

AQOR uses routing tables for keeping track of its routes. Every time a route failure occurs, AQOR must update its routing table entries, which may sometimes result in inconsistent entries due to the high dynamic nature of the network topology. On the contrary, QoS MMT will not use routing tables for facilitating its routing process. This leads to lower processing overheads and excellent consistencies.

None of the previously proposed routing protocols such as AODV and AQOR were able to satisfy all the different MANET application requirements. In this QoS MMT routing protocol being designed, the route discovery, bandwidth reservation, and forwarding is done at the data-link layer (Layer 2). This routing protocol includes all the features of other protocols, such as route discovery, link failure identification, bandwidth calculation, resource reservation, and resource release. It also takes the shared wireless medium and dynamic topology into consideration while providing QoS. To avoid possible loops during route exploration, AQOR uses a route sequence number to indicate the freshness of the control packets for each flow. The sequence number is maintained at each mobile node aware of the flow. The initial sequence number of any flow is 0. When sending out a route control packet for a flow (e.g., RREQ, RREP, or RERR), the initial node will increase its current sequence number by 1 and attach the value to the packet. When control packets are propagated through the network, only nodes with a lower sequence value of the flow will receive them. A node will forward only the first accepted control packet for a certain flow during one round of control packet propagation.

For discovered routes that meet QoS requirements, the admission control policy should guarantee for each flow the requested minimum flow bandwidth *Bmin* and the maximum end-to-end delay *Tmax*. A bandwidth admission control decision is made at every node in the exploration and registration phases, based on the detailed analyses of the traffic in shared channel wireless networks. In AQOR, the route with the shortest end-to-end delay, given it satisfies the bandwidth requirement, is selected.

2.6 Importance of Routing Protocols in MANET

The routing protocols for ad hoc wireless networks should be capable of handling multiple hosts with limited resources, such as bandwidth and energy. The main challenge for routing protocols is that they must also deal with host mobility. This means that hosts can appear and disappear in various locations. Thus, all hosts of the ad hoc network act as routers and must participate in the route discovery and maintenance of routes to the other hosts. For ad hoc routing protocols, it is essential to reduce routing messages overhead despite the increasing number of hosts and their mobility. Keeping the routing table small is another important issue, because the increase in the size of the routing table will affect the control packets sent in the network and this, in turn, will affect large link overheads [25].

The routing in MANETs has been of interest for quite some time in the research community. This section will present a short overview of the proposed work in MANET routing protocols. There are two main algorithms used and based on path selection for most of the existing MANET routing protocols: the Bellman–Ford algorithm [26] and Dijkstra's algorithm [27]. These algorithms are commonly used for the computation of a route in a link state routing protocol and distance vector routing protocols.

Many different routing approaches have been proposed so far to cope with different problems and meet different application requirements. For example, some routing protocols use proactive (table-driven) path discovery to reduce the route discovery delay (time) [11,17–19]. Other routing protocols use reactive (on-demand) path discovery to reduce and control overhead [12–16]. Some other approaches merge proactive and reactive path discoveries to reduce delay and control overhead. This type of protocol is called a *hybrid routing protocol* [21–23].

The hop count is used for path selection as an optimized metric in some routing protocols. Other cost metrics such as link quality and path quality have also been proposed [28,29]. The path filtering and path selection decisions can be made at different types of nodes, for example, at the source node, destination node, and intermediate node. Most routing protocols handle only single paths. However, some other protocols provide and maintain multiple paths [30,31]. The source-tree on-demand adaptive routing (SOAR) [32] is another routing approach that cannot be directly applied to the Bellman–Ford and Dijkstra's algorithms for path selection, because the standard for the choice of successor is determined both by the shortest path and by the set of neighbors that have advertised that route.

There are several approaches for QoS routing protocols based on the on-demand principle of route discovery. The first approach is based on a distributed on-demand path search, which uses a known link bandwidth between nodes [33]. Due to the distributed path calculation, this approach is scalable. Furthermore, by limiting the number of path search requests, flooding is prevented. The scalability and limited protocol overheads are clearly desirable in all ad hoc QoS routing techniques. There are some potential drawbacks to this approach. In particular, the path-finding procedure is not designed to take advantages of QoS information available at the Media Access Control (MAC) layer. The second approach of QoS implementation over ad hoc networks [34–37] focuses specifically on the MAC layer. It is based on the reservation of a node's MAC layer time. In this approach, single or multiple paths to the destination are discovered, and the path bandwidth to the destination node is calculated. However, acquiring the complete path information has several potential drawbacks, such as low scalability, poor tolerance to fast topology changes, and message flooding. The third approach is different from the above solution. It incorporates the QoS path-finding procedure, which is based on a bandwidth-scheduling mechanism. The routing protocol is made aware of the availability of bandwidth resources by coupling routing and MAC Time Division Multiple Access (TDMA) layers [38].

The other issue, which is considered deeply with regard to routing protocols, is the security of the routing protocol. Many proposed protocols are responsible for the creation of secure routing protocols (SRPs). An overview of secure routing in general can be found in the article by Xue and Ganz [39]. The first approach of securing the secure ad hoc on-demand distance vector protocol has been proposed by Gupte and Singhal [40]. In a second publication [41], the protocol was presented in greater detail. Further, related issues, such as key management, were presented briefly in the latter publication. Another secure routing protocol is ARIADNE [42], which is based on DSR. The security mechanism it uses is a broadcast encryption scheme called Timed Efficient Stream Loss-tolerant Authentication (TESLA). The other approach is called Authenticated Routing for Ad hoc Networks (ARAN), which is presented in [43]. ARAN is a reactive routing protocol based on AODV, using certificates. The SRP is another routing protocol with security, which is a reactive protocol relying on a shared secret exchanged *a priori* [44].

In our study, we have observed some shortcomings in existing MANET routing protocols:

- They have not covered all routing problems, such as reducing network load, data drop, and delay,, in some scenarios.
- They find the shortest path from the source to the destination, but for the worst-case scenario, when the shortest path is congested, a different path that might be longer but may be more efficient is used.
- Only the primary route is defined; however, if, for some reason, the primary route fails, then the protocol needs to rediscover the route, which will consume extra time and power.
- They exert extra load on the node in terms of memory size, processing power, and power consumption.
- They are not concerned with link reliability, such as the available data rate (bandwidth), delay, node battery life, and node selfishness, and thus, the path is not guaranteed to deliver the data from the source to the destination.
- Most existing MANET routing protocols find any path from source to destination, but it is not necessarily the optimum path. Such paths are not efficient for different applications.
- Most existing routing protocols send a Hello message or acknowledgment between the nodes, which increases the load and delay on the networks.

In view of the above shortcomings, we have drawn up a list of should-have features when designing a new routing protocol for a MANET. A new routing protocol should have the following features:

- Providing quick and high efficiency, for example, bandwidth, memory, and battery, in adapting to MANET topology change, especially in a high mobility environment
- Providing an alternative path in case the primary path fails; this will save time and power in an ever-changing MANET network topology
- Finding the optimum path instead of the shortest path when applications require QoS, for example, bandwidth, end to end delay, and packet losses, to deliver data from the source to the destination
- Providing quick establishment of paths, so that they can be used before an existing path becomes invalid
- Having a minimum control message overhead due to changes in the routing information when topology changes occur
- Consideration of QoS parameters, such as data rate, delay, and node battery life, when locating a path between the source and destination.

References

1. http://www.touchbriefings.com/pdf/744/wire041_vis.pdf
2. M. Abolhasan, T. Wysocki, and E. Dutkiewicz. *A Review of Routing Protocols for Mobile Ad hoc Networks,* vol. 2, no. 1, 2004, pp. 1–22.
3. E.M. Royer and C.K. Toh. A review of current routing protocols for ad hoc mobile wireless networks. *IEEE Personal Communications Magazine*, vol. 6, no. 2, 1999, pp. 46–55.
4. C.Z. Zhang, M.C. Zhou, and M. Yu. Ad hoc network routing and security: a review. *International Journal of Communication Systems*, vol. 20, no. 8, 2007, pp. 909–925.
5. H. Zhou. A survey on routing protocols in MANETs. Technical Report MSU-CSE-03-08, March 2003.
6. J. Broch, D.A. Maltz, D.B. Johnson, Y.-C. Hu, and J. Jetcheva. A performance comparison of multi-hop wireless ad hoc network routing protocols. In *Proceedings of the Fourth Annual ACM/IEEE International Conference on Mobile Computing and Networking*, October 1998, pp. 85–97.
7. L.M. Feeney. *A Taxonomy for Routing Protocols in Mobile Ad Hoc Networks*. SICS Technical Report. Swedish Institute of Computer Science, 1999.
8. M.Y. Lee, J.-L. Zheng, X.-H. Hu, H.-H. Juan, C.-H. Zhu, Y. Liu, J.S. Yoon, and T.N. Saadawi. A new taxonomy of routing algorithms for wireless mobile ad hoc networks: the component approach. *IEEE Communications Magazine*, vol. 44, no. 11, 2006, pp. 116–123.
9. S.R. Das, C.E. Perkins, and E.M. Royer. Performance comparison of two on-demand routing protocols for ad hoc networks. *IEEE Personal Communications*, vol. 8, no. 1, 2001, pp. 16–28.
10. S.R. Chaudhry, A.N. Al-Khwildi, Y.K. Casey, H. Aldelou, and H.S. Al-Raweshidy. A system performance criteria of on-demand routing in mobile ad hoc networks. In *Wireless and Mobile Computing, Networking and Communications (IEEE WiMob2005)*, August 2005, Montreal, Canada.
11. A.N. Al-Khwildi, K.K. Loo, and H.S. Al-Raweshidy. *Performance Evaluation of Proactive and Reactive Routing in Wireless Ad-Hoc Networks*. In *International Conference on Digital Telecommunications IEEE (ICDT 2006)*, Paris, 2006.
12. A.N. Al-Khwildi, T H. Sulaiman, K.K. Loo, and H.S. Al-Raweshidy. Performance comparison for proactive and reactive routing protocol for wireless ad hoc networks. In *IEEE Science of Electronic Technology of Information and Telecommunications (SETIT 2007)*, Tunisia, 2007.
13. C.E. Perkins and E.M. Royer. Ad-hoc on-demand distance vector routing. In *WMCSA Second IEEE Workshop*, 1999, pp. 90–100.
14. J.G. Jetcheva, Y. Hu, D. Johnson, and D. Maltz. The dynamic source routing protocol for mobile ad hoc networks (DSR). Internet Draft, IETF MANET Working Group, November 2001.
15. S. Corson and P. Vincent. Temporally ordered routing algorithm (TORA). Internet Draft, IETF MANET Working Group, August 2001.
16. C.-K. Toh. Associativity-based routing for ad hoc mobile networks. *Wireless Personal Communication*, vol. 4, no. 2, 1997, pp. 1–36.
17. P. Jacquet and T. Clausen. Optimized link state routing protocol (OLSR). Internet Draft, IETF MANET Working Group, October 2001.
18. C. Perkins and P. Bhagwat. Destination sequenced distance vector (DSDV). Internet Draft, IETF MANET Working Group, November 2001.
19. C.C. Chiang, H.K. Wu, W. Liu, and M. Gerla. Routing in clustered multi hop mobile wireless networks with fading channel. In *Proceedings of IEEE SICON*. Los Angeles: Computer Science Department, University of California, April 1997, pp. 197–211.
20. S. Murthy and J.J. Garcia-Luna-Aceves. An efficient routing protocol for wireless networks. *ACM Mobile Networks and Applications Journal*, vol. 1, no. 2 (Special Issue on Routing in Mobile Communication Networks), 1996, pp. 183–197.
21. L. Wang and S. Olariu. A two-zone hybrid routing protocol for mobile ad hoc networks. *IEEE Transactions on Parallel*, vol. 15, no. 12, 2004, pp. 1105–1116.
22. Z.J. Haas. The routing algorithm for the reconfigurable wireless networks. In *Proceedings of 6th International Conference on Universal Personal Communications (ICUPC)*, vol. 2, 1997, pp. 562–566.

23. M. Joa-Ng and I.T. Lu. A peer-to-peer zone-based two-level link state routing for mobile ad hoc networks. *IEEE Journal on Selected Areas in Communications*, vol. 17, no. 8, 1999, pp. 1415–1425.

24. R. Sivakumar, P. Sinha, and V. Bharghavan. CEDAR: a core-extraction distributed ad hoc routing algorithm. *IEEE Journal on Selected Areas in Communications*, vol. 17, no. 8, 1999, pp. 1454–1465.

25. Z.A. Ang Eu, K.-G. Seah, and K.-S. Tan. Experimental performance modeling of MANET interconnectivity. In *Proceedings of the 11th International Conference Parallel and Distributed Systems*, vol. 2, July 2005, pp. 155–161.

26. T.H. Ali Ahmed. Modeling and simulation of routing protocol for ad hoc networks combing queuing network analysis and ANT colony algorithms. PhD thesis, University of Duisburg Essen, April 2005.

27. R. Bellman. On a routing problem. *Quarterly of Applied Mathematics*, vol. 16, no. 1, 1958, pp. 87–90.

28. E.W. Dijkstra. A note on two problems in connexion with graphs. *Numerische Mathematik*, vol. 1, 1959, pp. 269–271.

29. B.S. Manoj, R. Ananthapadmanabha, and C. Siva Ram Murthy. Link life based routing protocol for ad hoc wireless networks. In *Proceedings of the 10th IEEE International Conference on Computer Communications and Networks 2001 (IC3N 2001)*, October 2001.

30. D.S.J. De Couto et al. A high-throughput path metric for multi-hop wireless routing. In *Proceedings of the 9th ACM International Conference on Mobile Computing and Networking (MobiCom'03)*, San Diego, CA, September 2003.

31. V.D. Park and M.S. Corson. A highly adaptive distributed routing algorithm for mobile wireless networks. *In Proceedings of InfoCom'97*, April 1997.

32. S. De, C. Qiao, and H. Wu. Meshed multipath routing with selective forwarding: an efficient strategy in wireless sensor networks. *Computer Networks*, vol. 43, 2003, pp. 481–97.

33. S. Roy and J.J. Garcia-Luna-Aceves. An efficient path selection algorithm for on-demand link-state hop-by-hop routing. In *Computer Communications and Networks, Eleventh International Conference*, October 2002, pp. 561–564.

34. S. Chen and K. Nahrstedt. Distributed quality-of-service routing in ad hoc networks. *IEEE Journal on Selected Areas in Communications*, vol. 17, no. 8, 1999, pp. 1488–1505.

35. T.-W. Chen, J.T. Tsai, and M. Gerla. QoS routing performance in multihop, multimedia, wireless networks. In *IEEE 6th International Conference Universal Personal Communications*, vol. 2, 1997, pp. 557–561.

36. C.H. Lin and J.S. Liu. QoS routing in ad hoc wireless networks. *IEEE Journal on Selected Areas in Communications*, vol. 17, no. 8, 1999, pp. 1426–1438.

37. C.R. Lin and C.C. Liu. An on-demand QoS routing protocol for mobile ad hoc networks. In *Proceedings of the IEEE Global Telecommunications Conference*, vol. 3, 2000, pp. 1783–1787.

38. Y.K. Ho and R.S. Liu. On-demand QoS-based routing protocol for ad hoc mobile wireless networks. In *Proceedings of the ISCC Fifth IEEE Symposium on Computers and Communications*, 2000, pp. 560–565.

39. Q. Xue and A. Ganz. Ad hoc QoS on-demand routing (AQOR) in mobile ad hoc networks. *Journal of Parallel and Distributed Computing*, vol. 63, 2003, pp. 54–165.

40. S. Gupte and M. Singhal. Secure routing in mobile wireless ad hoc networks. *Ad Hoc Networks*, vol. 1, no. 1, 2003, pp. 151–174.

41. M.G. Zapata. 2002. Secure ad hoc on-demand distance vector routing. *SIGMOBILE Mob. Comput. Commun*, Rev. 6, 3 (June 2002), http://doi.acm.org/10.1145/581291.581312, 2002, pp. 106–107.

42. M.G. Zapata and N. Asokan. Securing ad hoc routing protocols. In *Proceedings of the 2002 ACM Workshop on Wireless Security*, September 2002, pp. 1–10.

43. Y.-C. Hu, A. Perrig, and D.B. Johnson. Ariadne: A Secure On-Demand Routing Protocol for Ad Hoc Networks. In *Proceedings of the Eighth Annual International Conference on Mobile Computing and Networking (MobiCom 2002)*, ACM, Atlanta, September 2002, pp. 12–23.

44. K. Sanzgiri, B.N. Levine, C. Shields, B. Dahill, and E.M. Belding-Royer. A secure routing protocol for ad hoc networks. In *Proceedings of the 10th IEEE International Conference on Network Protocols*, November 2002.

Chapter 3

Modeling and Simulation Tools for Mobile Ad Hoc Networks

Kayhan Erciyes, Orhan Dagdeviren, Deniz
Cokuslu, Onur Yılmaz, and Hasan Gumus

Contents

A *model* is a simplified representation of a system that aids the understanding and investigation of the real system. *Simulation* is the manipulation of the model of a system that enables one to observe the behavior of the system in a setting similar to real life. By modeling and simulation of a mobile ad hoc network (MANET), it is possible to simplify many difficult real-life problems associated with them. Modeling and simulation of a MANET have limitations, and providing further flexibility in them such that a general MANET without much limitations can be modeled and simulated is an important research topic. In this chapter, we review network models, topology control models, mobility models, and simulators for MANETs by investigating their current limitations and future trends.

3.1 Introduction

Ad hoc networks are a key to the evolution of wireless networks. MANETs are nonfixed infrastructure networks that consist of dynamic collection of nodes with rapidly changing topologies of wireless links. Although military tactical communication is still considered the primary application for ad hoc networks, commercial interest in these types of networks continues to grow. Applications such as rescue missions in times of natural disasters, law enforcement operations, commercial and educational use of sensor networks, and personal area networking are just a few possible commercial examples. MANETs have the problems of bandwidth optimization, transmission quality, discovery, ad hoc addressing, self-routing, and power control. Power control is a very important issue in MANETs because nodes are powered by batteries only. Therefore, amount of communication should be minimized to avoid a premature dropout of a node from the network.

Links in a MANET change dynamically over time; thus, a functioning network must be able to deal with this dynamic nature. One key problem in MANETs is to model the mobile nodes and the communication edges of the network to provide a solution step for well-known problems such as medium access control (MAC) design, clustering, and backbone formation. These models should capture the behavior of the wireless transmission in different conditions since wireless transmissions in a MANET operating on a flat unobstructed environment may totally differ from the wireless transmissions in an ad hoc network of nodes each located on a building.

MANETs may be of large scale, consisting of even hundreds of nodes operating for the completion of an application. These nodes may be small and cheap devices as well as expensive military vehicles designed for operating in harsh conditions. Scientists aim to improve the operation quality and decrease the resource usage in MANETs by researching the various topics in communication layers. These studies may include theoretical analysis and extensive experiments to validate the superiority of the work. Since it is not feasible for researchers to afford the experiments on hundreds of moving nodes located on large areas, another key problem arises in MANETs: providing suitable simulation test beds.

3.1.1 Challenges

A MANET can be modeled as a graph $G(V, E)$, where V is the set of vertices and E is the set of edges. Two vertices (nodes) of a graph are connected only if there is a communication link between them. Once a MANET is represented as a graph, the next issue at hand is whether any graph property has any implications for the MANET. For example, a dominating set (DS) D of a graph is the set of vertices where a vertex $v \in V$ is either in D or adjacent to a vertex in D. If vertices of a DS are connected, the DS is called a *connected dominating set* (CDS), and forming a CDS in the graph model of the MANET provides a communication backbone for routing purposes in the actual mobile network. However, finding a minimum DS or a CDS is an NP-complete problem in graph theory, and hence approximation algorithms for such problems where suboptimal solutions using some heuristics are usually the only choice. However, designing an approximation algorithm with a favorable approximation ratio to the optimum solution to the problem is not sufficient, since one is dealing with a real network without any global information. Any algorithm employed must be distributed without any global knowledge. A distributed algorithm is run by all nodes of a MANET, provides exchange of information with its neighbor nodes by message passing only, and eventually results in reaching a determined state of the network [1]. Based on the above discussion, the challenge is in fact designing of distributed approximation algorithm with a favorable approximation ratio that can be implemented on the graph model of the MANET, which provides a solution to a graph problem that is usually extremal and has implications in the real MANET environment. Some other real-life considerations such as the battery lifetime of nodes in sensor networks or the mobility of the nodes in a MANET may have to be incorporated to the distributed approximation algorithm as the final adjustment.

Another example would be the vertex cover problem in a graph. A vertex cover of a graph is the set of vertices $S \in V$ such that any edge e is incident to at least one vertex in S. Finding a vertex cover of minimum size is NP-complete. For a distributed robot network such as a Special Weapons and Tactics (SWAT), finding a vertex cover is equivalent to placing robots at the corners of a maze such that every robot is in sight of at least another robot, which means all robots remain connected.

3.1.2 Scope

The scope of this chapter is to first specify basic models for MANETs. One such useful model is the graph representation, and once this is done, all of the graph theoretic results become available for the MANET. The key point then is the proper choice of some useful properties of graphs for the MANET as described earlier and designing of efficient and scalable distributed approximation algorithms. We show in Section 3.3 the external graph problems that have direct or indirect implementations for MANETs. We then provide detailed descriptions of the simulation platforms for MANETs. Section 3.2 outlines models for MANETs, whereas simulators for MANETs are discussed in Section 3.3. The conclusions are drawn in Section 3.4.

3.2 Modeling

In this section, we explain the network models, topology control models, and mobility models with their current limitations and future trends in modeling.

3.2.1 Network Models

3.2.1.1 Unit Disk Graph

A unit disk graph (UDG) is a special instance of a graph in which each node is identified with a disk of unit radius $r = 1$, and there is an edge between two nodes u and v if and only if the distance between u and v is at most 1 [2,3]. The model is depicted in Figure 3.1a. Each node's transmission range is drawn as a dotted circle. The edges, which connect nodes, are drawn as straight lines. The neighbors of node u are node v, node w, node y, and node z as shown in the simplified graph in Figure 3.1b.

This model is very simple yet captures the behavior of broadcast radio transmission; thus, it is good for modeling ad hoc and sensor networks [3]. It may be also suitable for modeling ad hoc networks located on unobstructed environments. Moreover, since this model is open for theoretical analysis due to its geometric properties, it is an important playground for the approximation algorithm designers. Efficient distributed approximation algorithms targeting to solve NP-complete network topology control problems such as finding minimum dominating set and maximum independent set, which will be described in the following sections, are studied by the researchers. Although UDG is a widely used networking model, it has drawbacks caused by its simplicity. In

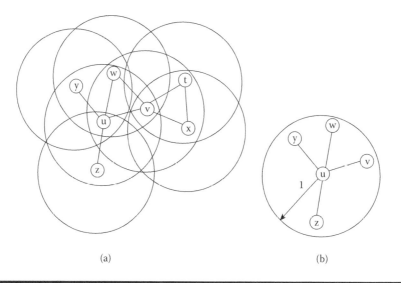

(a) (b)

Figure 3.1 (a) Unit disk graph model. (b) Node *u*'s neighbors.

real configurations, the wireless transmission may be disturbed by even small obstacles between communicating parties; therefore, UDG is not a realistic model for ad hoc networks located on areas consisting of heterogeneous objects. It does not model the signal quality between nodes, so it may result in poor topology control for multihop communication. Also it lacks modeling node weights consisting of node mobility, energy, etc., which makes UDG not suitable for the selection of routes with high weighted nodes.

3.2.1.2 Quasi Unit Disk Graph

In a quasi unit disk graph (QUDG), each node is identified with two disks: one with unit radius $r = 1$ and other with radius $q = (0, 1]$. It can be observed that a QUDG with $q = 1$ is a UDG [4]. The edges between nodes d away from each other are identified with respect to the below listed rules:

- There is an edge between two nodes if $d = (0, q)$.
- There is a possible edge connecting two nodes if $d = (q, 1]$.
- There is no edge between two nodes if $d = (1, \infty]$.

The model is depicted in Figure 3.2a. The inner circles are drawn with the dashed lines. The bold lines are communication edges, and the other lines are possible edges. In Figure 3.2b, the connections of node w are shown. Node y is the neighbor of node w; other nodes are the candidates for being the neighbors of node w.

QUDG is an extended model of UDG in which probabilistic links can be modeled. Also in QUDG model, the effect of the small obstacles located in the network area can be handled by adjusting the parameter q. Although the QUDG model has these advantages over the UDG model, the other disadvantages of the UDG model given in the previous section still exist in the QUDG model.

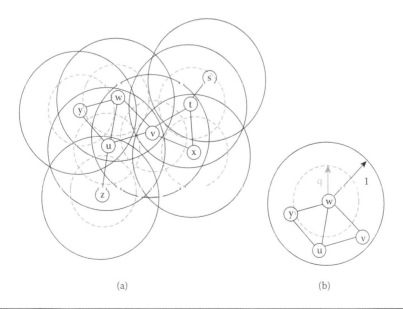

(a)   (b)

Figure 3.2   (a) Quasi unit disk graph model. (b) Node w's neighbors.

3.2.1.3 Undirected Graph

An undirected graph (UG) is described as $G = (V, E)$, where V is the set of vertices or nodes ($V = \{V_1, V_2, V_3, ..., V_N\}$) and E is the set of edges between these vertices ($E = \{E_{12}, E_{21}, ...\}$). E_{xy} is an edge starting from vertex x and ending at vertex y. If there is a communication link between V_1 and V_2, then $E \in E_{12}$ and E_{21}. Since the graph is undirected, links are assumed to be starting and ending on both sides. An example network with 10 nodes modeled with UG is depicted in Figure 3.3. In this model, the set of vertices is $V = \{V_1, V_2, V_3, ..., V_{10}\}$ and the set of edges is $E = \{E_{16}, E_{61}, E_{26}, E_{62}, ..., E_{910}, E_{109}\}$.

The UG model is simple and very common for various types of networks. There are many cases where modeling ad hoc networks with a UG is suitable. Also there is a significant amount of research on the UG model. In this model, the geometric properties of the wireless networks cannot be applied. Thus, this model results in more complicated approximation algorithm designs with probably higher resource requirements compared with the models with defined geometric property like a UDG. By not assuming a geometric wireless transmission pattern, this model may also be defined as pessimistic. One of the most important disadvantages of this model compared with the UDG and partially QUDG model is the undirected link assumption wherein real networks it may not be realistic. Also in a UG, node and edge weights cannot be modeled.

3.2.1.4 Directed Graph

A directed graph (DG) is described as UG: $G = (V, E)$, where E may contain one of E_{xy} and E_{yx}. A sample DG model is given in Figure 3.4. In this model, the set of vertices is $V = \{V_1, V_2, V_3, ..., V_{10}\}$ and the set of edges is $E = \{E_{16}, E_{26}, E_{38}, ..., E_{107}\}$. DG is an extended model of UG that captures the behavior of the heterogeneous ad hoc networks of nodes with different transmission ranges. In Figure 3.4, the transmission ranges of the nodes are depicted with the dotted circles of different sizes. Like a UG, a DG cannot assume a geometric transmission property and does not model the edge and node weights.

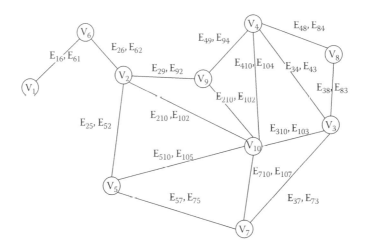

Figure 3.3 Undirected graph model.

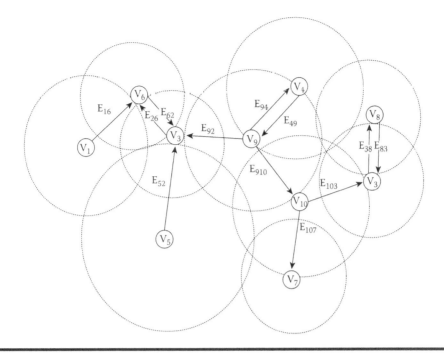

Figure 3.4   Directed graph model.

3.2.1.5 Weighted Graph

A weighted graph (WG) G_w can be *node weighted graph*: $G_{nw} = (V_w, E)$; *edge weighted graph*: $G_{ew} = (V, E_w)$; or the combination of both: $G_{new} = (V_w, E_w)$. Also a WG can be *undirected weighted graph* (UWG) or *directed weighted graph* (DWG). The weight of a node can be mobility, energy, nodal degree, etc., or a combination of all. The weight of an edge can be the signal strength, distance, etc. The weights are usually positive numbers, but negative numbers may also be used. Sometimes the cost term is used instead of weight.

An example directed node and edge weighted graph is depicted in Figure 3.5. The transmission ranges of the nodes are not identical; thus, connectivity between two nodes may be both unidirectional and bidirectional. The weight of a node in this figure is assumed to be 1/energy. An edge is represented with the signal strength. Like UG and DG, a DWG does not use geometric properties of the wireless transmission; thus, it is a pessimistic model.

3.2.2 Topology Control Models

3.2.2.1 Independent Set

An independent set (IS) is a set of nodes in which none of the nodes are adjacent. If this set cannot be extended by adding a new node, then IS is called the maximal IS. The IS with the greatest number of nodes is called the maximum IS. In Figure 3.6a, six gray-filled nodes are the elements of the maximal IS. However, this set cannot be extended by adding a new node; removing some nodes from this set and adding other nodes may increase the size. In Figure 3.6b, the maximum IS with eight nodes is shown. In the weighted version of this problem, a weight is assigned to each node,

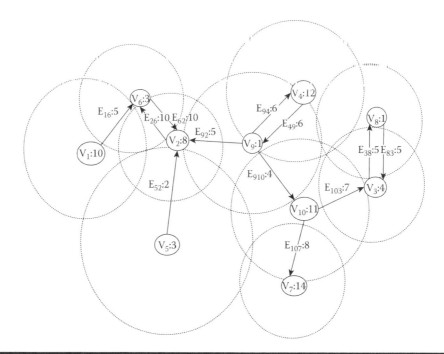

Figure 3.5   Directed weighted graph model.

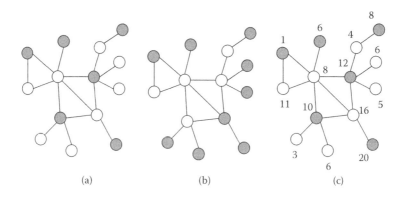

Figure 3.6   (a) Maximal independent set. (b) Maximum independent set. (c) Maximum weighted independent set.

and maximizing the total weight of this set is targeted. In Figure 3.6c, the maximum weighted IS having a total weight of 56 is depicted.

The IS structure is used in ad hoc networks for facility location and backbone formation. The facility location problem covers the optimal placement of facilities in order to minimize the related cost. The selected nodes for IS may communicate with each other by increasing their transmission range to deliver and route the data from the nodes not in the IS. This example also covers the backbone formation operation, which is the virtual path between the selected nodes. The selected nodes may be cluster heads if network is clustered.

The maximum IS problem and its weighted version is NP-hard; thus, maximal IS algorithms are studied in the literature. One practical distributed algorithm proposed by Chatterjee et al. [5] finds maximal IS and may find maximal weighted IS if node weights are used instead of node ids. At the beginning of the algorithm, all the nodes are in the *WHITE* state and all of them are the candidates for the IS. If a node's id is greater than its neighbors, it sends a *IamInTheSet* message and becomes an element of IS. When a node receives an *IamInTheSet* message, it sends a *NotInTheSet* message and does not become a candidate. When a candidate node receives a *NotInTheSet* message, it first deletes the source node from its neighborhood and then sends a *IamInTheSet* message and becomes an element of IS if a node's id is greater than its neighbors. Two sample outputs of this algorithm are shown in Figure 3.7. In Figure 3.7a, only node *u* constitutes IS, but when the nodes are replaced as seen in Figure 3.7b, node *u* and node *w* are the elements of IS.

3.2.2.2 Dominating Set

A dominating set (DS) is a subset *S* of a graph *G* such that every vertex in *G* is either in *S* or adjacent to a vertex in *S* [6]. Minimum DS problem is NP-complete. A maximal IS is a DS. DSs are widely used for topology control where elements in DSs are selected as cluster heads [7]. DSs can be classified into three main classes: independent dominating sets (IDSs), weakly connected dominating sets (WCDSs), and connected dominating sets (CDSs), as shown in Figure 3.8.

3.2.2.2.1 Independent Dominating Sets

An independent dominating set (IDS) is a DS *S* of a graph *G* in which there are no adjacent vertices. Figure 3.8a shows a sample IDS filled with gray. By using IDSs, one can guarantee that there are no adjacent cluster heads in the entire graph. This minimizes the number of dummy clusters in the network.

Figure 3.7   **(a) Node *u* is in independent set. (b) Node *u* and node *w* are in independent set.**

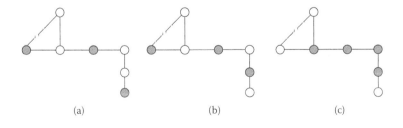

Figure 3.8   **(a) Independent dominating set. (b) Weakly connected dominating set. (c) Connected dominating set.**

3.2.2.2.2 Weakly Connected Dominating Sets

A weakly induced subgraph S_w is a subset S of a graph G that contains the vertices of S, their neighbors, and all edges of the original graph G with at least one endpoint in S. A subset S is a weakly connected dominating set (WCDS) if S is dominating and S_w is connected [8]. Gray nodes in Figure 3.8b show a WCDS example. Although IDSs are suitable for constructing optimum-sized DSs, they have some deficiencies such as lack of direct communication between cluster heads. In order to obtain the connectivity between cluster heads, WCDSs can be used to construct clusters.

3.2.2.2.3 Connected Dominating Sets

A connected dominating set (CDS) is a subset S of a graph G such that S forms a dominating set and is connected. Figure 3.8c shows a sample CDS. CDSs have many advantages in network applications such as ease of broadcasting and constructing virtual backbones [9]; however, when we try to obtain a CDS, we may have undesirable number of cluster heads. So, in constructing CDSs, our primary problem is the minimum CDS decision problem. Wu's distributed algorithm [10] is an important study in this field where researchers attempted to improve the performance of this work later [11,12]. The steps of Wu's algorithm are given below:

1. Each node u finds the set of neighbors $\Gamma(u)$.
2. Each node u transmits $\Gamma(u)$ and receives $\Gamma(v)$ from all its neighbors.
3. If node u has two neighbors v, w and w is not in $\Gamma(v)$, then u marks itself being in the set CDS. In Figure 3.9, a sample output of this algorithm is shown. A detailed survey about CDS can be found in [13].

3.2.2.3 Spanning Tree

A graph $G_S = (V_S, E_S)$ is a spanning subgraph of $G = (V, E)$ if $V_S = V$. A spanning tree of a graph is an undirected connected acyclic spanning subgraph. The spanning trees are very important structures for ad hoc networks, and they are widely used for data delivery from a source to sink or to multicast. Erciyes et al. [14] showed that constructing a spanning tree with clusters in ad hoc networks initiated from a root node is simple. The depth parameter is provided by the algorithm to adjust the diameter of the clusters. The root periodically sends a *PARENT(nhops)* message to its neighbors to reinitiate the operation. Each node sends the *PARENT((nhops + 1)mod depth)* message to its neighbors upon the first reception of the *PARENT(nhops)* message. The recipients of the message with *nhops = 0* are the *SUBROOTS*; *nhops < depth* are the *INTERMEDIATE* nodes; *nhops = depth* are *LEAF* nodes. A sample spanning tree with clusters is shown in Figure 3.10.

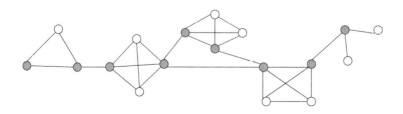

Figure 3.9   Sample output of Wu's connected dominating set algorithm.

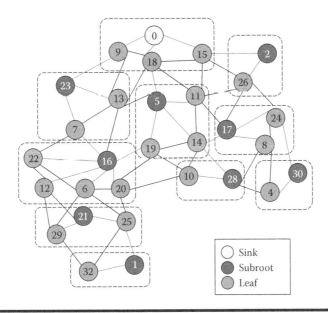

Figure 3.10   Sample output of Erciyes's spanning tree algorithm.

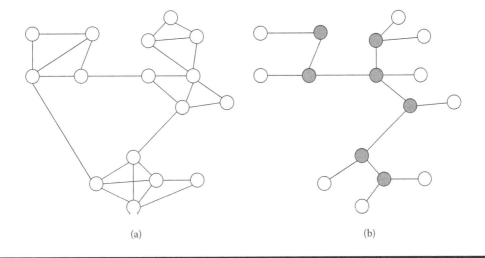

(a)   (b)

Figure 3.11   (a) An ad hoc network. (b) Its minimum spanning tree.

A minimum spanning tree (MST) for a graph is a subgraph that has the minimum edge weight for maintaining connectivity. A sample MANET and an MST constructed can be seen in Figure 3.11a and b, respectively. In Figure 3.11b, any nodes, other than the leaf nodes, that are shown by black color depict a connected set of nodes.

Gallager et al. [15] proposed a distributed algorithm that determines a minimum weight spanning tree for a UG that has distinct finite weights for every edge. The aim of this algorithm is to combine small fragments into larger fragments with outgoing edges. A fragment of an MST is a subtree of the MST. An outgoing edge is an edge of a fragment if there is a node connected to the edge in the fragment and one node connected that is not in the fragment. In this algorithm, each node starts with itself as a fragment and ends with the MST as the final fragment. The authors defined three possible node states: *Sleeping*, *Find*, and *Found*. Initially all nodes are in the *Sleeping* state and are either spontaneously awaken to initiate the overall algorithm or awakened by a message from another node. A node in the *Find* state searches for the minimum weight outgoing edge to combine with another fragment. The combination rules of fragments are related to levels. A fragment with a single node has the level $L = 0$. Assume two fragments F at level L and F' at level L':

■ If $L < L'$, then fragment F is immediately absorbed as part of fragment F. The expanded fragment is at level L'.
■ Else if $L = L'$ and fragments F and F' have the same minimum weight outgoing edge, then the fragments combine immediately into a new fragment at level $L + 1$.
■ Else fragment F waits until fragment F' reaches a high enough level for combination.

3.2.2.4 Graph Matching

A matching in a graph G is a set of nonloop edges with no shared endpoints. The vertices incident to the edges of a matching M are saturated by M. A maximal matching is a set of edges that cannot be extended by adding an extra edge. A perfect or maximum matching in a graph is a matching that saturates every vertex [6]. Maximum matching problem is in the P complexity set. An example matching, maximal matching, and perfect matching are shown with bold edges in Figure 3.12a, b, and c, respectively.

If the graph is weighted and we are looking for the maximum selected edge weight, then the problem is called *weighted matching*. Figure 3.13a shows an example weighted matching, and Figure 3.13b shows an example maximum weighted matching. Like maximum matching, maximum weighted matching is in the P complexity set. Although approximation algorithms are studied for performance considerations, Hoepman proposed an 1/2-approximation algorithm for maximum weighted matching. In Hoepman's algorithm, it is assumed that all of the nodes know their neighbors and the weights of the edges incident to them. Each node sends a *request* message to its candidate. The candidate is the neighbor node that is connected on the heaviest edge. If two of the nodes both receive the *request* message from each other, the edge connecting them is selected and they will be matched. If a node sends a *request* message to a matched node, the matched node will reply with a *drop* message. Each node uses two sets: the set of neighbors (N) and the set of requested neighbors (R). A node deletes the source of the *drop* message in its N.

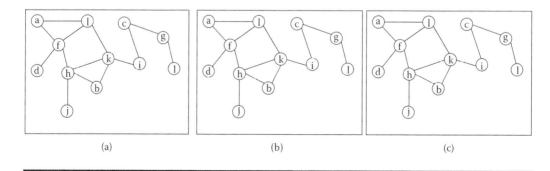

Figure 3.12 (a) Matching. (b) Maximal matching. (c) Perfect matching.

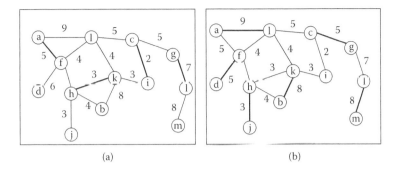

Figure 3.13 (a) Weighted matching. (b) Maximum weighted matching.

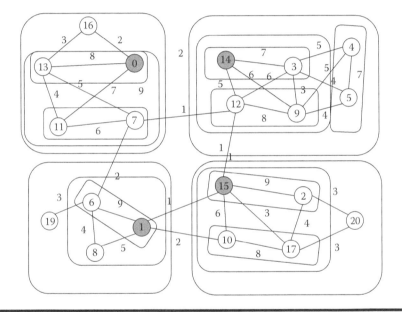

Figure 3.14 Clustering by iterative weighted matching.

To control the topology of an ad hoc network, weighted matching operation can be iteratively run to construct paths between nodes. Dagdeviren and Erciyes [16] proposed to construct the clusters and the backbone of an ad hoc network by using weighted matching. Their algorithm works in rounds and each round starts after the previous one and is completed by all cooperating nodes. At each round, each cluster head tries to merge with the adjacent cluster over the maximum weight edge connecting them. They used Hoepman's work as the basis algorithm and also provided asynchronous operation. An example clustered network in four rounds with this algorithm is shown in Figure 3.14. The selected edges are bold, and the cluster heads are filled with gray.

3.2.2.4.1 Interference Tree

Interference is one of the major challenges in wireless networks and thereby in MANETs. It alters or disrupts the message as it transmits along a channel between the source and the destination.

Since the message is disrupted when the interference occurs, it has to be detected and the interfered message has to be retransmitted. In particular, in multihop communications, the nodes dissipate energy and time due to interference. The interference in MANETs mostly occurs from concurrent message transmission. Since the nodes in MANETs generally use the omnidirectional antennas, the sent messages from a node are received by all nodes that are in transmission range of the sender node. When two messages are sent concurrently by two neighboring nodes, they affect each other and interference occurs.

Early studies on topology control algorithms mostly deal with the connectivity and sparseness of network. These studies consider the interference reduction implicitly and consider that if the resulting topology of topology control algorithms has low node degree, the interference is reduced intuitively. However, it is proved that this intuition is wrong. Recent topology control algorithms emphasize the interference reduction explicitly [17].

The first topology control algorithm for interference reduction tries to find an answer to the question "How many nodes are affected by communication over a certain link?" The proposed algorithms for this question are classified as sender-centric interference perspective. On the contrary, the resulting topology must have connectivity and spanner properties as well. The MANET may be modeled as a UDG. A node can adjust its transmission range from zero to the maximum level. A directed edge (s, r) may exist only if the maximum transmission radius of s is at least $|sr|$ Euclidean distance. In order to provide good service to the upper layer, undirected edges are used in the sender-centric perspective. The goal of topology control with explicit interference reduction is to generate a subgraph via eliminating the edges that have high coverage. The coverage of the undirected edge (s, r) is the number of nodes covered by disks of both s and r. In other words, the coverage of the undirected edge (s, r) is the number of nodes that are affected while nodes s and r communicate.

The Low Interference Forest Establisher (LIFE) algorithm is proposed in order to generate an interference optimal topology [18]. The LIFE activates the edges regarding the coverage in

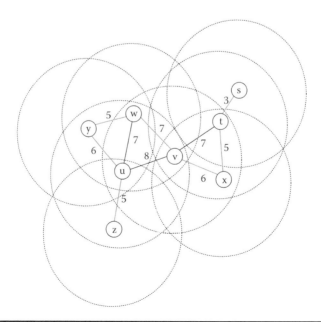

Figure 3.15 Sender-centric interference model and sample topology.

increasing order. Figure 3.15 displays a sample of the sender-centric interference model on a given UDG. The values of edges indicate the coverage of the link, and the bold edges indicate a sample of the resulted topology of the LIFE algorithm. Activating the edges only regarding the coverage can cause long-distance communications. In order to balance the interference and communication costs, the LISE algorithm is proposed, which is a *t-spanner* algorithm. The LISE activates the edges regarding the coverage and the distance of an edge with a constant *t-stretch* factor. Thus, an interference and communication cost-aware subgraph is constructed.

A new perspective for interference reduction is receiver-centric. In this approach, the topology control algorithms try to find an answer to the question "How many other nodes in a given network node can be disturbed?" By this question, topology control algorithms change their attention from the sender-centric to the receiver-centric perspective. It is proved that the receiver-centric perspective generates lower interference optimal topologies than do the sender-centric perspective. In the sender-centric perspective, the topology control algorithms compute the subgraph regarding the coverage of a certain communication link, but in the receiver-centric perspective, the goal is to minimize the interference at each possible receiver. The interference of node s is then defined as the number of other nodes that affect the message reception at node s. Simply, the *interference of a node* can be defined as the number of disks that include node s in a given UDG. Figure 3.16 displays a sample of the receiver-centric interference model on a given UDG. The values of vertices indicate the number of disks that include the corresponding nodes. The nearest component connector (NCC) algorithm [18] is proposed, which is based on the receiver-centric perspective, particularly for wireless sensor networks. The NCC algorithm generates a subgraph via connecting the components to their nearest neighbors. A component is a single or group node. The algorithm constructs a tree toward sink in several rounds.

A new trend in interference-aware topology control algorithms is to construct a topology regarding the signal-to-interference-noise ratio (SINR) model instead of graph theoretic models.

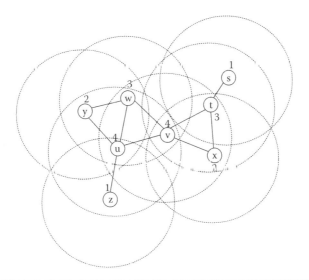

Figure 3.16 Receiver-centric interference model.

3.2.2.4.2 Vertex Cover

A vertex cover is a subset of vertices of V that includes at least one of the vertices (u, v) in set E. Vertex cover can be used for the facility location problem. Finding minimum vertex cover problem is NP-complete. The weighted version of the problem works on a vertex weighted graph model and aims to minimize the total cost of the set. Figure 3.17a shows a sample vertex cover, and Figure 3.17b shows a sample weighted vertex cover with gray-filled nodes.

One of the most important central methods for finding the weighted vertex cover is the pricing method that has an approximation ratio of 2. In this method, initially all vertices are relaxed. Then the edges are chosen randomly and they are given weights as much as possible, provided that the total weight of the edges incident to a vertex does not exceed the weight of the vertex. When there is no edge obeying this rule, the algorithm is finished. When the total weight of the edges incident to a relaxed vertex exceeds the weight of the vertex, it becomes tight. The set of tight nodes is the vertex cover. The WG matching is the other method for finding a weighted vertex cover. Selecting one endpoint of each edge of the maximal WG matching yields a weighted vertex cover. Figure 3.18a shows a sample output of the pricing method, and Figure 3.18b shows a sample output by the graph matching method.

3.2.2.4.3 Steiner Tree

A Steiner tree connects the nodes in T, which is a subset of V. A Steiner tree may be used to connect the nodes in a DS to construct CDS. Finding minimum Steiner tree is an NP-complete problem. An example Steiner tree is shown in Figure 3.19a, where $V = \{a, b, c, d, e, f\}$ is connected by vertices v and x.

In the Steiner tree problem, nodes and edges may be weighted. Klein and Ravi [19] proposed a central algorithm for a node weighted Steiner tree problem working on UWGs, which has an approximation ratio of $2 \ln(S)$, where S is the minimum weight. In this algorithm, the vertices in V that have a path only consisting of them are inserted in a tree. If a vertex has no neighbor in V,

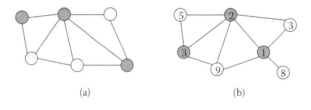

(a)   (b)

Figure 3.17   **(a) Vertex cover. (b) Weighted vertex cover.**

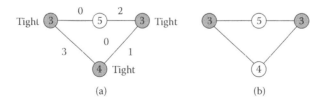

(a)   (b)

Figure 3.18   **(a) Pricing method. (b) Graph matching method.**

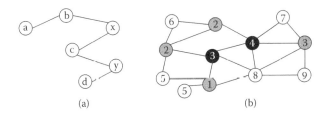

Figure 3.19 (a) Steiner tree. (b) Weighted Steiner tree.

then the vertex itself constitutes a tree. The weight of each vertex in V is assigned 0. The weight ratio of each vertex in S/V is calculated as the weight of the vertex divided by the number of total vertices to be connected. The vertex with the smallest weight ratio in S/V is chosen as the connector vertex; the neighboring trees are connected; and the connector is inserted to V. This operation continues until all the vertices in V are in the same tree. Figure 3.19b shows an example output of this algorithm, where black-filled nodes are the connectors and gray-filled vertices constitute the initial V.

3.2.3 Mobility Models

Mobility models are designed to represent the movements of mobile users, and how their location and acceleration change over time. They are used to evaluate the performance of ad hoc network protocols. Since the performance of a protocol depends on the mobility model, it is important to choose a suitable model for the evaluated protocol. Various mobility models have been proposed so far, but the most common ones are random waypoint, random direction, and Gauss–Markov mobility models [20].

3.2.3.1 Random Waypoint Model

The random waypoint model is currently a benchmark mobility model. It is a basic model that describes the movement pattern of nodes where mobile nodes randomly designate a destination in the simulation plane. Then, each mobile node goes to a designated destination with a constant velocity that each mobile node chooses randomly. Every node is independent in the plane. When the node arrives at the destination, it waits for a designated time, and if the pause time is equal to 0, then this means that the node has a continuous mobility. The two important parameters of the random waypoint model is velocity and pause time of each node. These parameters affect the performance of the evaluated protocol. If the simulation of velocity is small and pause time is long, a stable topology is formed. Otherwise, a dynamic topology can be formed. Various topologies can be obtained by varying these parameters [20].

3.2.3.2 Random Direction Model

The random direction model is similar to the random waypoint model, but in the random direction model, the node randomly and uniformly selects a direction in which it travels until it reaches the boundary, instead of selecting a random destination. Upon reaching the boundary, the node waits for a designated time. Then, it randomly and uniformly selects another direction to travel. Therefore, a uniformly distributed mobility model is provided for evaluations [20].

3.2.3.3 Gauss–Markov Model

The Gauss–Markov model is a different model from the random waypoint and direction models in terms of velocity management. In this model, the velocity of a mobile node is correlated over time and Gauss–Markov stochastic process. The Gauss–Markov stochastic process satisfies the requirements for both Gaussian processes and Markov processes. The velocity of a mobile node at time slot t is dependent on the velocity at time $t - 1$. Therefore, the Gauss–Markov model is a dependent mobility model, where dependency is determined by the parameter α, which affects the randomness of the Gauss–Markov process. By tuning this parameter, different mobility models are provided [20].

3.2.4 Current Status and Future Trends in Modeling

In this section, we describe the current status, limitations, and future trends in modeling. We discuss the network models, topology control models, and mobility models.

The UDG model is not realistic in some cases, does not model node and edge weights, does not provide probabilistic link modeling, and does not model heterogeneous ad hoc networks where nodes have different transmission ranges. Although a UDG has these limitations, it is popular for modeling ad hoc networks located on unobstructed environments since it models the simple wireless transmission. A QUDG has the same limitations as those of a UDG except it provides probabilistic link modeling in order to provide the wireless transmission behavior when the networking area has small obstacles; however, it is not a popular model like UDG. A UG model does not use geometric properties of the wireless transmission, does not model node and edge weights, does not provide probabilistic link modeling, and does not model heterogeneous ad hoc networks. Although a UG has these limitations, it is a very popular model that fits various types of networks in many cases. We think that this model will preserve its popularity and researchers will continue studying this model. The DG model has the same limitations as the UG model except it models heterogeneous ad hoc networks. A DG is not a widely used model, one of the main reasons being the hardness of the algorithm design on this model. Since heterogeneous ad hoc networks are increasingly becoming more popular, this model will probably receive more attention in the future. The weighted directed graph model extends the DG by modeling node and edge weights. Although it has so many advantages over the other models, algorithm design on this model is very hard compared with that on the other models. However, it will be an important model for advanced protocol design in heterogeneous ad hoc networks. Table 3.1 provides a summary of current limitations and future trends in network modeling.

The distributed algorithms proposed for IS and DS topology control models generally work on UDGs and UGs. With the increasing types of realistic network models, we think that the researchers will study these problems on other network models other than UDGs and UGs. Weighted versions of these problems are open research topics and, in general, are more difficult to model in a MANET. Fault-tolerant structures such as a k-connected m-dominating set will also be an important research topic. The proposed spanning algorithms generally work on UGs. Like IS and DS problems, this problem will be studied in other networking models and node and edge weighted tree construction may be one of the important research topics in future. Graph matching is a new topology control model in which there are only very few studies [16]. The approximation ratio and the resource usage of this study are open for improvements. Interference tree is another new model in which there is no algorithm for preventing interference, and currently, an SINR-based scheduling algorithm for preventing interference is an important research topic. There are

Table 3.1 Current Limitations and Future Trends in Network Modeling

Model	Current Status and Limitations	Future Trends
Unit disk graph	• Not realistic • Lacks modeling node and edge weights • Lacks providing probabilistic link modeling • Lacks modeling heterogeneous ad hoc networks where nodes have different transmission ranges	Will still be a popular model
Quasi unit disk graph	Same as unit disk graph except it provides probabilistic link modeling	This model will attract researchers
Undirected graph	• Lacks using geometric properties of the wireless transmission • Lacks modeling node and edge weights • Lacks providing probabilistic link modeling • Lacks modeling heterogeneous ad hoc networks where nodes have different transmission ranges	Will still be a popular model
Directed graph	Same as undirected graph model except it models heterogeneous ad hoc networks	With the increasing popularity of the heterogeneous ad hoc networks, this model will take more attention
Weighted directed graph	Same as directed graph model except it models node and edge weights	With the increasing popularity of the heterogeneous ad hoc networks, it will be an important model for advanced protocol design in heterogeneous ad hoc networks

few studies related to constructing vertex cover in MANETs. Since vertex cover is an appropriate solution for the facility location problem, we think that the researchers will study this model on different network models and also on MANETs. We also believe that the node and edge weighted version of the Steiner tree problem is immature. Since the Steiner tree is used as a step for the CDS problem, especially the weighted version as related to the CDS problem is an open research area that requires further investigation. Table 3.2 summarizes the current limitations and future trends in topology control modeling.

The random waypoint and random direction mobility models limited by the poor choice of velocity distribution and the mobility behavior of the nodes are independent and there is no geographic restrictions of movements. In spite of these disadvantages, these models are used as

Table 3.2 Current Limitations and Future Trends in Topology Control

Model	Current Status and Limitations	Future Trends
Independent set and dominating set	Distributed algorithms proposed in this area generally work on unit disk graph and undirected graph	• Network models other than unit disk graph and undirected graph need further research • Weighted versions of the problems are open research topics • Fault-tolerant structures such as k-connected m-dominating set will be important research topics
Spanning tree	Distributed algorithms proposed in this area generally work on undirected graph	• These problems on network models other than undirected graph need further research • Weighted tree formation will be an important research topic
Graph matching	There is only one proposed study	The approximation ratio and the resource usage of the previous work is open for improvements
Interference tree	There is no algorithm for preventing interference	Signal-to-interference-noise ratio-based scheduling algorithm for preventing interference is an important research topic
Vertex cover	There are few studies for constructing vertex cover in mobile ad hoc networks	Research on different network models is needed
Steiner tree	The node and edge weighted version of the problem is immature	The Steiner tree as related to the connected dominating set problem should be studied

benchmark models since they are simple and easy to implement. Due to these facts we think that the researches will continue to use these models. The Gauss–Markov model is limited by the geographic restrictions of the movements. Since this model has a temporal mobility dependency, we think that this model will gain attention of the researchers. Table 3.3 summarizes the current limitations and future trends in mobility modeling.

3.2.5 Distributed Algorithm Models

A distributed system consists of communicating devices over a network to achieve a common goal. In contrast to central systems where the data may be stored in a single device, in a distributed system, generally a device has a limited and incomplete view of the network such as knowledge of its immediate neighbors. A distributed system is prone to faults in the individual devices that should be tolerated for robust operation. MANET, grid, and wireless sensor networks are examples of the distributed systems.

Table 3.3   Current Limitations and Future Trends in Mobility Models

Model	Current Status and Limitations	Future Trends
Random waypoint and random direction	• It provides poor choice of velocity distribution • The mobility behavior of the nodes are independent • There is no geographic restrictions of movements	The researches will continue to use these models
Gauss–Markov	There is no geographic restrictions of the movements	This model will take attention

A distributed algorithm is designed to run on a distributed system in which each device starts in an initial state $S_i \in S$, can be in one of the states in S at any execution instance, changes its state from S_i to S_j, and may output an $O_n \in O$ after receiving an input $I_k \in I$ according to defined state transition procedures. In this manner, a distributed algorithm can be modeled as a Mealy finite state machine. An input can be in one of three forms. Firstly, it can be an internal event such as a timer interrupt. Secondly, it can be an external event such as detection of a failure in one of the neighbors. Lastly, it can be a message received from a neighbor or from another process running in the same device. In asynchronous communication, the devices are not coordinated; however, the operation is divided into rounds in synchronous communication, where each operation step is executed in a round.

A distributed algorithm may be initiated by a single device or by multiple devices and synchronously or asynchronously. For example, in a synchronous distributed spanning tree algorithm initiated by a single device, a *START* message may be sent from a root device in the first round, which is then forwarded to all devices in the later rounds. Each device may set its parent upon receiving the first *START* message. The algorithm is terminated after all the rounds are executed. Such an algorithm is called *synchronous single initiator distributed algorithm*. In an asynchronous distributed spanning tree algorithm initiated by a single device, which is an example of the *asynchronous single initiator distributed algorithm*, the root device initiates a flooding of a *START* message, where the algorithm may have no bound on the time to complete this operation. Each device sends an *ACCEPT* message upon receiving the first *START* message and otherwise sends a *REJECT* message. In an asynchronous distributed algorithm initiated by multiple devices, called *asynchronous concurrent initiator distributed algorithm*, multiple root devices may flood a *START* message, in which ordinary devices may receive *START* messages from different originators that may cause fragmentation of the network at an instance of the execution. These fragments may be merged by applying additional rules on the asynchronous spanning tree algorithm with a single initiator; for example, the device with the greatest id can be selected as the root. These three operation modes form the basis for distributed algorithm models in a MANET.

3.3 Simulation

Wireless communication technology has become an essential framework for today's communication environments. This interest leads the researchers to improve the current wireless technology. Most of today's wireless devices are, in some sense, limited by their need for infrastructure. Today's most popular approach for the nonfixed infrastructure is the ad hoc networking that can be considered as a solution to the infrastructure limitations of the wireless technology. Generally in a MANET, the nodes are dynamically moving without any boundary limitations. With the growth of the scale of today's MANETs, evaluations, experiments, and comparative studies become harder to implement in real environments. Even if the underlying environment permits distributed applications to be tested, it is not practical to realize everyday experimentation of distributed applications on the real environments, especially if the application is still under construction. Moreover, implementing and testing some applications in the real environment might not be possible at all due to practical considerations. To cope up with such problems, simulators are widely used in today's research community. In this chapter, we describe different simulation environments. In Section 3.3.1, an overview of the simulation environments is given. In later sections, POSIX thread-based simulators, Network Simulator 2 (ns2), TOSSIM, OPNET, OMNET, and other simulators are examined. Finally, current limitations and future trends are discussed in Section 3.3.8.

3.3.1 Overview of Simulation Environments

The main purpose of simulation environments is to provide a virtual platform to the distributed applications with the required components. For MANETs, the detailed level of simulation environments, in most cases, determines the literalness of the experiments. However, it does not mean that the simulation environments are evaluated regarding to their detailed levels. Mostly, simulation is a resource-consuming job. The challenge is to provide equilibrium between detail and performance. The existing simulators provide users the different levels of detail by providing different realistic layers in the open systems interconnection model (OSI model). The decision is left to users since each application needs different components of simulated environment. Some researchers prefer to use their own simulation environments such as threads or shared memory architectures, whereas some others use other simulators, which provide more details in the OSI layers.

3.3.2 POSIX Thread-Based Simulator

In today's research community, especially in distributed computing domain, simulators are highly utilized to test and evaluate studies. Many researchers create their own simulation environment for many reasons. For that purpose, threads, which are provided by operating systems, are very powerful tools to create application-specific simulation environments. A *thread* is the smallest unit of processing that can be scheduled by the operating systems. Threads are created at their parent process context and can be scheduled as different processes. They can share the same resources provided by their parent process and are very suitable for simulating message-passing environments using the shared memory space. POSIX thread is an application programming interface (API) for creating and manipulating threads. Implementations of the API are available on many Unix-like POSIX systems, such as FreeBSD, NetBSD, GNU/Linux, and Mac OS X. Some variations of POSIX threads also exist for Microsoft Windows and Solaris-based operating systems.

POSIX thread-based simulators have many advantages and disadvantages. They can be preferable in many cases in which existing simulators remain insufficient for the user's protocol. They can also be favorable because of their simplistic implementation. However, since they do not provide any standardized environment, comparative studies may not be trustworthy compared with those on known existing simulators.

In POSIX thread-based simulators, each node is represented by a thread. Protocols are coded into the thread's start routine, and communication between threads is handled by shared memory spaces such as globally defined arrays. The issue that must be taken into consideration in thread-based simulators is the mutual exclusion problem. For this problem, an interprocess communication module is recommended to handle mutual exclusion between threads as in [21].

Topology and scenario generation in such simulators can be developed explicitly from the simulator. Generally, the topology is stored in two-dimensional arrays, indicating the neighboring status of nodes. However, mobility scenarios can be realized by reasonably altering the neighbor matrix.

3.3.3 ns2 Simulator

ns2 (version 2.34) [22] is a discrete event simulator that is developed at ISI, California. ns2 provides substantial support for the simulation of transmission control protocol (TCP), routing, and multicast protocols over wired and wireless (local and satellite) networks. The development of ns2 began in 1989 as a variant of the REAL network simulator [23]. In years it has evolved substantially, and in 1995, the development of ns2 was supported by Defense Advanced Research Projects Agency (DARPA) through the Virtual InterNetwork Testbed (VINT) project at Lawrence Berkeley National Laboratory (LBNL), Xerox PARC, University of California, Berkeley (UCB), and University of Southern California, Information Sciences Institute (USC/ISI). Currently, ns2 development is supported by DARPA with Simulation Augmented by Measurement and Analysis for Networks (SAMAN) and by NSF with Collaborative Simulation for Education and Research (CONSER), both in collaboration with other researchers, including AT&T Center for Internet Research at ICSI (ACIRI). ns2 has included substantial contributions from other researchers, including wireless code from the UCB Daedelus and CMU Monarch projects and Sun Microsystems.

3.3.3.1 ns2 Overview

The ns2 is a *de facto* standard simulator in experimenting wired and wireless network applications, since it is an open source and powerful tool in simulating networks. Although its early versions are aimed at wired networks, the wireless network support is added later by various extensions. It closely follows the OSI model. It is an object-oriented simulator written in C++, with an OTcl interpreter as a front end. ns2 uses two languages: C++ and OTcl. Users implement their applications in C++, and simulation scenarios and setups are written in OTcl.

Presently, the simulator is a single-threaded event-driven simulator, and only one event can be executed at a time. If more than one event is scheduled to execute at the same time, the built-in scheduler executes them in a first-come first-serve fashion and the next earliest event in the queue is executed to completion.

Nodes in ns2 are OTcl classes, which contain an address or id, a list of neighbors, a list of agents, a node type, and a routing module. The id of nodes is a monotonically increasing integer starting from 0. In later versions of ns2, the node class is extended to cover mobility support. The extended class is named mobilenode. The network stack for a mobilenode consists of a link layer, an address resolution protocol module connected to a link layer, an interface priority queue, a MAC layer, and a network interface, all connected to the channel. These network components are created and brought together in OTcl.

Routing of messages is handled by five alternative ad hoc routing protocols: destination sequence distance vector, dynamic source routing, temporally ordered routing algorithm, ad hoc on-demand distance vector, and protocol for unified multicasting through announcements. By default, nodes in ns2 are constructed for unicast simulations. In order to enable multicast simulation, the simulation should be created with an option "multicast on."

3.3.3.2 Protocol Implementations

ns2 allows users to implement different protocols. Besides newly created protocols, all the internal protocols can be modified according to users' needs. Simulation in ns2 is realized in three steps. In the first step, the user implements the protocol by writing C++ and OTcl codes to the ns2 source base. In the second step, simulation is described by using an OTcl script. In the last step, user runs its protocol in ns2 and collects results by using either trace files, which are generated by ns2, or protocol's outputs, which are generated by output commands coded into the protocol's implementation. The protocol should be written in C++ under any folder structure as long as the path is added to the makefile of the ns2. The protocol is tied to mobile nodes individually by using the OTcl script prepared for the simulation. The OTcl script contains nodes' initializations such as channel type, radio propagation model, network interface type, MAC type, interface queue type and size, link layer type, antenna model, and routing protocol. It also contains nodes' initial locations and movement scenarios, dimensions of the area in which the nodes are moving, and start–end times of the simulation. After the creation of the nodes in the OTcl file, each node can be individually tied to different (or the same) protocols using loops. Upon execution of the simulation, the OTcl script file is traced sequentially, and nodes are created and configured as indicated in the script. Then the simulator executes indicated protocols in tied nodes as indicated in the script. From this point, the nodes execute their related protocol and can communicate to each other within their transmission range. The execution of the implemented protocols can be monitored in a number of ways. Generally, trace data are either displayed directly during execution of the simulation or stored in a file to be analyzed later on. There are two primary types of monitoring capabilities currently supported by the ns2. The first, called traces, records each individual packet as it arrives, departs, or is dropped at a link or queue. The second, called monitors, records various interesting quantities such as packet and byte arrival and departure numbers. Independently from these two trace methods, users may implement their own debugging methods into their protocols to output user-specific events.

3.3.3.3 Scenario Generations

The mobile node is designed to move in a three-dimensional topology in ns2. However, the third dimension (Z) is not used. That is, the mobile node is assumed to always move on a flat terrain with Z always equal to 0. The node movement scenarios are handled in two different alternatives: in the first alternative, nodes' speeds and starting and ending positions are selected randomly. At a predetermined time, the node would start moving from its initial position toward its destination at the defined speed. The movement scenario is generally stored in a separate file. In the second alternative, nodes' starting positions are generated initially; the destination and speed values are dynamically updated in a random manner during simulation. ns2 also provides network traffic patterns. The traffic generator can be used in order to generate both constant bit rate and TCP connection scenarios. Besides the internal scenario generator, third-party applications are available to generate wireless mobile scenarios for ns2 in the Internet.

3.3.3.4 Mobility Support

ns2 has mobile networking support by extending the standard ns2 release using CMU's Monarch group's mobility model. In the extended version, the node class is improved by adding more supporting features such as ability to move within a given topology and ability to send and receive packets over a wireless channel. The core difference between a node and a mobile node is that a mobile node is not connected to other nodes by links. The connectivity of nodes are determined by their positions and transmission ranges. Routing mechanisms and network components are also extended to include all mobile networking properties such as wireless channel, network interface, radio propagation model, MAC protocols, interface queue, link layer, and address resolution protocol model. The original mobility model of CMU's Monarch group was designed to provide pure wireless support. The extensions in ns2 allow users to combine wired and wireless networks together to create more realistic simulation environments.

3.3.4 TOSSIM Simulator

3.3.4.1 TinyOS and TOSSIM Overview

TinyOS [24] is a sensor network operating system that runs on motes, and TOSSIM is a discrete event simulator [25] for TinyOS for wireless sensor networks that were developed at University of California, Berkeley. Motes are tiny sensor nodes with limited resources. TinyOS has a component-based programming model and uses a language called nesC, which has a C-like syntax; and it supports the TinyOS concurrency model as well as mechanisms to provide robust embedded programming [24].

TOSSIM is designed considering four requirements, which are essential for efficient TinyOS simulation environment: (1) scalability: the simulator must be able to handle large-scale sensor networks; (2) completeness: the simulator must cover as many system interactions as possible; (3) fidelity: the simulator must capture the network behavior accurately; (4) bridging: the simulator must bridge the test implementation and the real implementation. TOSSIM has visual components to improve its usefulness. The TOSSIM architecture is composed of five parts: a compilation support for simulation, a discrete event queue, a small number of hardware abstraction components, mechanisms for extensible radio and analog–digital converter (ADC) models, and communication services. Figure 3.20 shows the graphical overview of the TOSSIM architecture [26].

A TinyOS program is composed of components that are independent computational entities. Components have three computational concepts: commands, events, and tasks. Commands and events are used for intercomponent communication, while tasks provide intracomponent concurrency. The component model, which is designed in TinyOS, allows the target platform to be easily changed from mote hardware to simulation mode. TOSSIM models the characteristics of the underlying hardware in software. To model hardware interrupts, a simulator event queue is located in TOSSIM that delivers the interrupts. This event queue is one of the most important components of TOSSIM. TinyOS abstracts each hardware resource as a component, and TOSSIM emulates the behavior of the underlying raw hardware, such as ADC, clock, EEPROM. The TOSSIM architecture includes two different models: ADC models for the ADC and radio models for all kinds of transmissions. In TOSSIM, a network signal can be either 0 or 1. TOSSIM provides two built-in radio models: simple radio model, which simulates error-free transmission, and lossy radio model, which simulates packet loss in probabilistic manner. The radio model can be easily changed for specific simulations. The default TOSSIM radio model is a signal-strength-based lossy model. In the default model, the propagation strengths, noise floor, and receiver sensitivity are provided to the simulator. TOSSIM models ADC hardware in two different ways: the first one is random, which generates a 10-bit random value upon sampling the ADC channel. The other is the generic way that works like a random one, and also it can be actuated by external

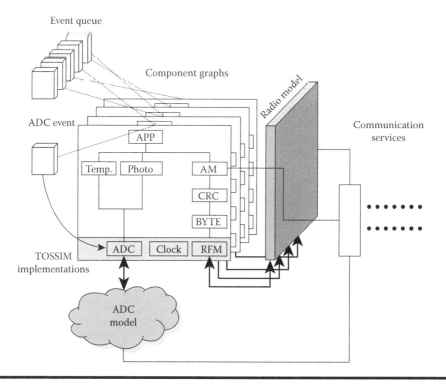

Figure 3.20 TOSSIM architecture: frames, events, models, components, and services. (From Levis, P., et al., TOSSIM: accurate and scalable simulation of entire TinyOS applications, paper presented at the Proceedings of the 1st International Conference on Embedded Networked Sensor Systems, Los Angeles, CA, 2003.)

applications. TOSSIM provides the communication of the applications running on a personal computer with simulation over TCP/IP [26].

3.3.4.2 Protocol Implementations

TOSSIM simulates TinyOS's networking stack, which is the most complex system of TinyOS. It is composed of 12 components. A networking stack uses Carrier Sense Multiple Access (CSMA) protocol and single-error-correction/double-error-detection data encoding with a full-packet Cyclic Redundancy Check (CRC). One of the most interesting features of TOSSIM is the bit-level simulation of the network stack. Understanding network stack simulation of TOSSIM requires knowledge of TinyOS network stack implementation. A TinyOS network stack handles MAC and single-hop packet transmission. It uses an active message (AM) as packet abstraction, and AM packets provide an unreliable data-link protocol. These packets provide precise time stamps and synchronous data-link acknowledgments. The TOSSIM's bit level simulation of the TinyOS networking stack results in a realistic simulation environment. A hidden terminal problem is accurately simulated, and errors at all phases of the packet reception are included in the environment. Signal corruption is observed by a listening node upon two sender nodes' concurrent transmissions. A listener node receives union of the two sender's bits as a result of interference. Moreover, a delay is occurred when motes repeatedly enter CSMA wait, because they continue to hear a signal on the channel. A single-bit error during the data phase can be handled with the data encoding, but a single-bit error during start symbol detection will prevent reception and a single-bit error during acknowledgment transmission will cause failure. This granularity changes the methodology with which one normally approaches network simulation. For example, instead of modeling latency, by modeling the network itself, TOSSIM simulates contention and backoff, which are causes of latency [26].

3.3.4.3 Scenario Generations

Scenario generations in TOSSIM consist of four stages. In the first stage, the predefined topology file is loaded to the nodes. In the second stage, noise traces are assigned to the nodes. In the third stage, nodes are booted at particular times, and finally in the fourth stage, packets are injected to simulation if necessary. Network topology has to be specified before in order to simulate the network behavior. Topology definition can be in different formats and can be stored in text files. TOSSIM loads the topology file using Python scripts. One example topology format defines each link in a line with three values: the source, the destination, and the gain; for instance, {1 2 −54.0} indicates that node 2 receives transmissions from node 1 at −54.0 dBm. A sample topology file is given in Table 3.4.

The topology file shown in Table 3.4 can be generated using "LinkLayerModel," a general link-layer model proposed by the Autonomous Networks Research Group at University of Southern California. It is valid for static and low-dynamic environments. The configuration file contains various channel, radio, and topology parameters that can be modified. Using topology parameters, different deployments can be specified as grid, uniform, or random. Number of nodes, terrain dimensions, and so on are also specified as topology parameters in the topology file [27].

In addition to the radio propagation model, TOSSIM also simulates the radio frequency noise and interference a node hears, from both other nodes and outside sources. It uses the closest pattern matching (CPM) algorithm. CPM takes a noise trace as the input and generates a statistical model from it. This model can capture bursts of interference and other correlated phenomena, such that it greatly improves the quality of the radio frequency simulation. CPM requires a noise trace to be configured. TOSSIM provides some sample noise traces, which are a series of noise readings. For example, in Table 3.5, first 10 lines of a sample noise trace file is given, which is a noise trace taken from Meyer Library at Stanford University [28].

TOSSIM is capable of injecting packets into the network dynamically. Packets can be scheduled to arrive at any time. A node can receive an injected packet even if it is in the middle of receiving a packet from another node over radio [28].

Table 3.4 Sample Topology File for TOSSIM

1 2 −54.0
2 1 −55.0
1 3 −60.0
3 1 −60.0
2 3 −64.0
3 2 −64.0

Table 3.5 Noise
Trace Sample from
Meyer Heavy Model

39
98
98
98
99
98
94
98
98
98

3.3.4.4 Mobility Support

In earlier versions of TOSSIM, before version 2.1.0, mobility was supported. However, mobility support is not included in the last version of TOSSIM (version 2.1.0) due to radio model changes in implementation. Considering lack of mobility in TOSSIM 2.1.0, Stevens et al. designed and implemented a mobility extension for the last release of TOSSIM. Evaluating the results of TOSSIM 2.x with mobility extension, Stevens et al. concluded that their implementation correctly demonstrates the expected behavior of mobile wireless nodes [29].

3.3.5 OPNET Simulator

OPNET is a high-level event-based network simulator [30]. The development of OPNET was started by MIL3 Inc., but nowadays, OPNET Technologies Inc. is conducting the development. It is a commercial simulator, but there is free license for educational purposes. Originally, the simulator was developed for military operations.

OPNET accelerates the R&D process for analyzing and designing communication networks and protocols. It is a very large and powerful network simulator with a variety of possibilities such as simulating entire heterogeneous networks with various protocols. It consists of user-friendly user interface, which is constructed from C and C++. The graphical user interface-based debugging and analysis simplify the simulation process. OPNET provides various tools for simulation including network model editor, node model editor, and process model editor. It has the fastest discrete event simulation engine among leading industry solutions. One of the significant features of OPNET is the supports grid computing for distributed simulations.

3.3.6 OMNeT++ Simulator

OMNeT++ is a modular object-oriented discrete event network simulator [31]. It has been developed by András Varga at the Department of Telecommunications, Technical University of Budapest. OMNeT++ is the extended version of OMNET, which was developed by Dr. György Pongor. OMNeT++ can perform various network simulations such as traffic modeling of telecommunication networks, protocol modeling, and modeling of queuing network, MANET, etc.

An OMNeT++ model consists of hierarchically nested modules that are managed with a high-level script language (NED). These modules communicate through message passing. The modules can have their own parameters and they can be used to customize a module. The modules are programmed in C++ using the simulation library. OMNeT++ simulations can feature varying user interfaces for different purposes, including debugging, demonstration, and batch execution. The Eclipse-based Integrated Development Environment (IDE) provides an easy control over simulation execution. In addition, the IDE allows the user development/debugging phase for the simulation project. User interfaces also facilitate demonstration of how a model works. The simulator as well as user interfaces and tools are portable: they are known to work on Windows and on several Unix flavors, using various C++ compilers. OMNeT++ also supports parallel and distributed simulation. OMNeT++ runs on Linux, Mac OS X, other Unix-like systems, and Windows, but the IDE runs only on Linux32/64, Mac OS X 10.5, or Windows XP. OMNEST is the commercially supported version of OMNeT++. OMNeT++ is free only for academic and nonprofit use, but for commercial purposes, OMNEST licenses have to be obtained from Omnest Global, Inc.

3.3.7 Other Simulators

Other than the well-known network simulators including ns-2 and OMNeT++, there are other network simulators that have wide use, such as GloMoSim [32], Sinalgo [33], and GTNetS [34]. These network simulators are increasingly being used by academic research groups.

GloMoSim is a scalable parallel discrete-event simulator for wired and wireless networks. The protocols are coded by Parsec, which is a C-based simulation language, developed in University of California, Los Angeles. GloMoSim has a layered approach like the OSI network architecture. Currently, GloMoSim supports protocols for a purely wireless network, but the wired and hybrid networks have not yet been accomplished. Sinalgo is a network simulator for testing and validating network protocols and algorithms. It was developed by the Distributed Computing Group at ETH Zurich and published under BSD license. It focuses on the verification of network algorithms. It tends to simulate the wireless networks and offers itself as a first test environment, prior to the deployment of the algorithm to the hardware. It provides a close environment to the real hardware devices. It has important features including quick prototyping of the network algorithms, many plug-ins, working over 100,000 nodes, 2D and 3D support, asynchronous and synchronous simulation, and customizable visualization of the network graph. Sinalgo contains the most frequently used modes such as UDG and QUDG. The GTNetS is a simulation environment that provides a protocol stack like the OSI reference model. It has a variety of features that offer researchers the ability to experiment in large-scale networks.

3.3.8 Current Status and Future Trends

Each simulator tool has its own advantages and disadvantages in different aspects. In these simulators, many protocol implementations can be realized and simulated. Even these simulators provide powerful environments; nearly all of them lack important details related to real-life experiments.

In ns2, the most important missing detail may be considered as the signal propagation and environmental aspects. The simulated environment is assumed to be a flat empty area in which mobile nodes move without interference in their signals. In reality, environment consists of buildings, people, moving vehicles, plants, etc., that affects the signal propagation of wireless mobile devices crucially. Moreover, the distance between wireless devices affects the signal level and transmission speeds drastically, which is not considered in the current version of ns2. Instead, the transmission capacity drops from full to zero immediately as the nodes move apart from their coverage areas. This phenomenon brings the impossibility to simulate proximity and environmental noise-related simulations.

In addition to these limitations, ns2 cannot easily handle large-scale simulation scenarios. ns2 simulations can be very resource consuming when the number of simulated nodes increase. It may consume a very large amount of memory once the number of nodes exceeds several hundreds. Although there is a workaround on parallelizing ns2 [35], users prefer to keep using ns2 and restrict their evaluations to smaller networks. Besides the memory leak problem, in general, depending on the simulation setup, ns2 is limited to simulate hundreds of nodes because of stack overflow problems. A comparative study and scalability limitations of ns2 can be found in [36]. This scalability issue can be a decisive factor in some situations in which large-scale simulations are required.

Another limitation with ns2 is its complex implementation. Developing new networking protocols and creating simulation scripts are complex tasks, which require understanding of the ns2 class hierarchy, C++, and Tcl programming. ns2 has a very complex hierarchy that is hard to learn at a glance. Learning ns2 in its simplest form may take weeks, depending on the background of the implementer. Moreover, debugging and output tracing is not straightforward and requires additional efforts. Although there are visualization tools for ns2 such as Nsnam [22], in general, tracing the results requires parsing an output file generated by the protocol itself.

TOSSIM can simulate large-scale sensor networks up to thousands of nodes. Its bit-level radio modeling allows simulating signal interference. Moreover, bridging capability of TOSSIM allows developers to test and verify the code that will run on hardware motes. As a drawback, TOSSIM is currently implemented only for mica platforms. In addition, TOSSIM lacks mobility support in its latest version TOSSIM 2.x. One more notable drawback of TOSSIM is a simulator-specific and simplified implementation of the MAC layer.

Similar to ns2 and TOSSIM, OMNeT++, GloMoSim, Sinalgo, and GTNetS lack in terms of the environmental effect to signal propagation. Between these network simulators, OMNeT++ comes forward since it provides testing environments to all kinds of networks. In addition, the Eclipse-based IDE makes it easy to debug, develop, and test the network algorithms and protocols. The limitation of OMNeT++ is that it supports up to nearly 2,000 nodes in simulations. OMNeT++ has found acceptance from the network research community, and it will be widely used by researchers. GloMoSim is a developing network simulator, and it currently cannot simulate wired networks. Sinalgo offers a good simulation environment specifically for MANETs. It supports large numbers of nodes in the simulations and contains well-known models. We believe that Sinalgo will take the attention from researchers. One of the disadvantages of Sinalgo is that it is not being widely used. Thus, the performance and usage of Sinalgo are not known enough. Currently, GTNetS has also not found enough acceptance from the research community. Lastly, OPNET lacks to support the recent wireless networks, and it is a commercial simulator. Table 3.6 shows a summary of current limitations and future trends in simulation.

Table 3.6   Current Limitations and Future Trends in Simulation

Model	Current Status and Limitations	Future Trends
ns2	• The simulated environment is assumed to be a flat empty area in which mobile nodes move without interference in their signals • The distance between wireless devices affects the signal level and transmission speeds drastically • It is hard to simulate proximity and environmental noise-related simulations • Limited nearly 500 nodes • It is hard to develop new protocol • Debugging and output tracing is not straightforward	ns2 is the most popular network simulator in the network research community. New version, Ns3, has been just released. We believe that this model will preserve its popularity and researchers will continue to study ns2 or the new version Ns3
TOSSIM	• Currently implemented only for mica platforms • Lack of mobility support in its latest version TOSSIM 2.x • Simulator-specific and simplified implementation of the MAC layer	TOSSIM will continue to support the mica platforms and the mobility support will be added. It is the primary simulator testing the software before embedding to the sensor motes
OPNET	• It is a commercial simulator	Although it is a commercial simulator, it provides various tools. We think that OPNET is one of the best network simulators. It can be used in commercial projects
OMNeT++	• Lack of the environmental effect to signal propagation • Limited nearly 2,000 nodes	It takes place in academic network research projects. We believe that OMNeT++ will be the most popular network simulator
GloMoSim	• Lack of the environmental effect to signal propagation • No support for wired networks • Currently in progress • Not being widely used	It is a developing network simulator. Currently, it is not being widely used and its success depends on the future improvements

(Continued)

Table 3.6 (Continued)   Current Limitations and Future Trends in Simulation

Model	Current Status and Limitations	Future Trends
Sinalgo	• Lack of the environmental effect to signal propagation • Only provides simulations for mobile ad hoc networks (MANETs) • Not being widely used	Currently it is not being widely used, but since it proposes various tools for MANETs, we think that it will be widely used in researches on MANET
GTNetS	• Lack of the environmental effect to signal propagation • Not being widely used	It is not well known and has not find enough acceptance from network research community

3.4 Conclusions

In this chapter, we described contemporary modeling and simulation methods in MANETs. We classified the network models as UDG, QUDG, UG, DG, and WG. We showed that each of these models have various advantages and disadvantages. Although the UDG model is simple, effective, and popular for ad hoc networks on unobstructed environments, it may be a poor model in realistic environments. A QUDG may model probabilistic links but has the same limitations as the UDG model. A weighted directed graph may be a good choice for heterogeneous ad hoc networks since it extends DG and WG models, but it is a pessimistic model since it does not use the geometric properties of the wireless transmission. We think that, in the future, the popular models such as UDG and UG will still be used widely by the researchers due to their simplicity, and besides that other models will attract researchers to model heterogeneous ad hoc networks with probabilistic links, for example, wireless sensor and actuator networks. We also provided detailed descriptions of IS, DS, spanning tree, matching, and interference tree topology control models. Generally, the common, most important limitation of these topology control models is that they were studied on simple network models. The mobility models in this study are classified as random waypoint, random direction, and Gauss–Markov models, and although mobility modeling has been improved by the researchers, all of these still lack modeling the behavior of real mobile nodes. We studied ns2, TOSSIM, OPNET, and OMNeT++ simulators in detail and mentioned other simulators such as GloMoSim, Sinalgo, and GTNetS. We listed the features of each simulator with regard to their protocol implementations, scenario generations, and mobility support. We think that the limitations in the network and mobility models reflect to the simulation environments; thus, we believe that the improvements in modeling will spark the enhancements in simulation. It is also highly probable that graph models of MANETs will continue to dominate modeling of MANETs, and distributed approximation algorithms that yield approximation ratios better than the existing ones for extremal graph problems that have direct implications for MANETs will attract researchers in the foreseeable future.

References

1. D. Peleg. *Distributed Computing: A Locality-Sensitive Approach.* Society for Industrial and Applied Mathematics, Philadelphia, PA, 2000.
2. B.N. Clark, C.J. Colbourn, and D.S. Johnson. Unit disk graphs. *Discrete Mathematics*, vol. 86, 1990, pp. 165–177.

3. F. Kuhn, T. Moscibroda, and R.Wattenhofer. Unit disk graph approximation.Paper presented at the Proceedings of the 2004 Joint Workshop on Foundations of Mobile Computing, Philadelphia, PA, 2004.

4. S. Schmid and R. Wattenhofer. Algorithmic models for sensor networks. In *20th IEEE International Parallel and Distributed Processing Symposium (IPDPS 2006)* 2006, p. 11.

5. M. Chatterjee et al. WCA: a weighted clustering algorithm for mobile ad hoc networks. *Cluster Computing*, vol. 5, 2002, pp. 193–204.

6. D. West. *Introduction to Graph Theory*, 2nd ed. Upper Saddle River, NJ: Prentice-Hall, 2001.

7. Y.P. Chen and A.L. Liestman. Approximating minimum size weakly-connected dominating sets for clustering mobile ad hoc networks. Paper presented at the Proceedings of the 3rd ACM International Symposium on Mobile Ad Hoc Networking and Computing, Lausanne, Switzerland, 2002.

8. Y. Chen et al. *Clustering Algorithms for Ad Hoc Wireless Networks*. Nova Science Publisher, Hauppauge, NY, 2004.

9. I. Stojmenovic, M. Seddigh, and J. Zunic. Dominating sets and neighbor elimination-based broadcasting algorithms in wireless networks. *IEEE Transactions on Parallel and Distributed Systems*, vol. 13, 2002, pp. 14–25.

10. J. Wu and H. Li. A dominating-set-based routing scheme in ad hoc wireless networks. *Telecommunication Systems Journal*, vol. 3, 1999, pp. 63–84.

11. D. Cokuslu, K. Erciyes, and O. Dagdeviren. A dominating set based clustering algorithm for mobile ad hoc networks. In *International Conference on Computational Science 2006 (ICCS 2006)*. Springer Lecture Notes in Computer Series Series. Reading Springer, 2006.

12. D. Cokuslu and K. Erciyes. A hierarchical connected dominating set based clustering algorithm for mobile ad hoc networks. In *15th International Symposium on Modeling, Analysis, and Simulation of Computer and Telecommunication Systems (MASCOTS'07)*. Istanbul: IEEE, 2007.

13. K. Erciyes, O. Dagdeviren, D. Cokuslu, and D. Ozsoyeller. Graph theoretic clustering algorithms in mobile ad hoc networks and wireless sensor networks. *Applied and Computational Mathematics*, vol. 6, no. 2, 2007, pp. 162–180.

14. K. Erciyes, D. Ozsoyeller, and O. Dagdeviren. Distributed algorithms to form cluster based spanning trees in wireless sensor networks. Paper presented at the Proceedings of the 8th International Conference on Computational Science, Part I, Krakow, Poland, 2008.

15. R.G. Gallager, P.A. Humblet, and P.M. Spira. A distributed algorithm for minimum-weight spanning trees. *ACM Transactions on Programming Languages and Systems*, vol. 5, 1983, pp. 66–77.

16. O. Dagdeviren and K. Erciyes. Graph matching based distributed clustering and backbone formation algorithms for sensor networks. *the Computer Journal*, vol. 53, no.11 2010, pp. 1553–1575.

17. T. Locher, P.V. Rickenbach, and R. Wattenhofer. Sensor networks continue to puzzle: selected open problems. Paper presented at the Proceedings of the 9th International Conference on Distributed Computing and Networking, Kolkata, India, 2008.

18. P.V. Rickenbach, R. Wattenhofer, and A. Zollinger. Algorithmic models of interference in wireless ad hoc and sensor networks. *IEEE/ACM Transactions on Networking*, vol. 17, 2009, pp. 172–185.

19. P. Klein and R. Ravi. A nearly best-possible approximation algorithm for node-weighted Steiner trees. *Journal of Algorithms*, vol. 19, 1995, pp. 104–115.

20. F. Bai and A. Helmy. A survey of mobility modeling and analysis in wireless adhoc networks. In *Wireless Ad Hoc and Sensor Networks*. Kluwer Academic Publishers, 2004.

21. K. Erciyes. A formal and practical method to develop distributed and critical software. Paper presented at the National Software Engineering Symposium, Bilkent University, Ankara, Turkey, 2007.

22. K. Fall and K. Varadhan. The ns manual (formerly ns Notes and Documentation). Unpublished manuscript, 2010.

23. Real Network Simulator, Overview, 2010. Available at http://www.cs.cornell.edu/skeshav/real/overview.html Accessed August 2011.

24. Lesson 1: Getting Started with TinyOS and nesC, 2010. Available at http://www.tinyos.net/tinyos-1.x/doc/tutorial/lesson1.html Accessed August 3, 2010.

25. Simulating TinyOS Networks, 2010. Available at http://www.cs.berkeley.edu/~pal/research/tossim.html Accessed August 3, 2010.

26. P. Levis, N. Lee, M. Welsh, and D. Culler. TOSSIM: accurate and scalable simulation of entire TinyOS applications. Paper presented at the Proceedings of the 1st International Conference on Embedded Networked Sensor Systems, Los Angeles, CA, 2003.

27. Building a Network Topology for TOSSIM, 2010. Available at http://www.tinyos.net/tinyos-2.x/doc/html/tutorial/usc-topologies.html Accessed August 4, 2010.

28. TOSSIM, 2010. Available at http://www.tinyos.net/tinyos-2.x/doc/html/tutorial/usc-topologies.html Accessed July 24, 2010.

29. C. Stevens et al. Simulating mobility in WSNs: bridging the gap between ns-2 and TOSSIM 2.x. Paper presented at the Proceedings of the 2009 13th IEEE/ACM International Symposium on Distributed Simulation and Real Time Applications, 2009.

30. OPNET, 2010. Available at http://www.opnet.com/ Accessed August 1, 2010.

31. OMNET++, 2010. Available: http://omnetpp.org/ Accessed August 1, 2010.

32. X. Zeng et al. GloMoSim: a library for parallel simulation of large-scale wireless networks. *SIGSIM Simulation Digest*, vol. 28, 1998, pp. 154–161.

33. Sinalgo—Simulator for Network Algorithms, 2010. Available at http://disco.ethz.ch/projects/sinalgo/ Accessed August 1, 2010.

34. G.F. Riley. The Georgia Tech Network Simulator. Paper presented at the Proceedings of the ACM SIGCOMM Workshop on Models, Methods and Tools for Reproducible Network Research, Karlsruhe, Germany, 2003.

35. G.F. Riley, R.M. Fujimoto, and M.H. Ammar. A generic framework for parallelization of network simulations. Paper presented at the Proceedings of the 7th International Symposium on Modeling, Analysis and Simulation of Computer and Telecommunication Systems, 1999.

36. F. Kargl and E. Schoch. Simulation of MANETs: a qualitative comparison between JiST/SWANS and ns-2. Paper presented at the Proceedings of the 1st International Workshop on System Evaluation for Mobile Platforms, San Juan, Puerto Rico, 2007.

Chapter 4

Study and Performance of Mobile Ad Hoc Routing Protocols

Raquel Lacuesta, Miguel García,
Jaime Lloret, and Guillermo Palacios

Contents

Routing is a fundamental component in the operation of ad hoc networks. Nodes must cooperatively create and manage the network without the support of any centralized infrastructure, taking into account network changes due to node mobility, nonfixed topology, and restrictions related to node resources. In addition, security must be established in routing processes to avoid attacks of malicious nodes while protecting the communication of nonallowed accesses. In this chapter, we introduce the main routing protocols for ad hoc networks: on the one hand, the protocols that do not introduce security measures, and on the other hand those that introduce security measures. We also deal with security criteria as well as the main ways to prevent security attacks. Finally, we show a performance comparison test of the most well-known routing protocols in order to know which performs better with respect to some parameters. These parameters are the received and sent routing messages with fixed and mobile nodes with failures, the load at the Media Access Control (MAC) level, the throughput of the mobile nodes with failures, and the average traffic received and sent and the average delay. Moreover, the instantaneous routing traffic received, the average traffic received, and the average load carried out by the node for each routing protocol are also used.

4.1 Introduction

A *network* can be defined as the group of people or systems or organizations who tend to share their information collectively for personal or professional purpose. Nature of data communication, node mobility, and new unregulated wireless data communication technologies such as Bluetooth [1,2] or IEEE 802.11 [3] are changing the network deployment. Fast and reliable information exchange is increasingly required. Nowadays, we use general-purpose devices with wireless communication capabilities, ranging from deskside computers and laptops to personal digital assistants and cell phones. There are also special-purpose devices such as wireless cameras and microsensors. Mobility is the major service for these devices. Users can communicate within fixed networks or similar devices without any infrastructure network. In this case, devices have been enabled to talk to each other directly via "ad hoc" networks, which must offer the same services as those offered by an infrastructure network. Ad hoc wireless networks will enhance communication capabilities by providing connectivity from anywhere at any time. On these networks, routing will allow connecting a call from an origin node to a destination node and will also play an important role in architecture, design, and networks operation. Devices will need to be able to identify each other to communicate, and so they need an address as well as an appropriate level of security to avoid attacks and protect both data and communications. This is the main header.

A mobile ad hoc network is a group of mobile wireless nodes that cooperatively form a network without the support of any centralized infrastructure. These networks are set up when needed. They are made up of similar nodes with no hierarchy. In these networks, there is no stationary infrastructure or base station for communication. If the networks cover a small area, routing would not be necessary. If our network covers a large area, connectivity must be obtained using an ad hoc routing. Each node has to act as a router sending and forwarding packets from other nodes. The configuration services required vary significantly, depending on the size of the network, the nature of the participants, and the applications that it supports. Each host must be able to produce, to route, and to consume data. Preferable assistance of any user or human administrator should not be required.

Routing is the act of moving information from a source to a destination in an internetwork. This concept has been used since early 1970s. Nowadays, it is gaining more and more popularity due to the advancement in networks and telecommunication technologies, new heterogeneous environments, and the advent of the Internet. Routing will be in charge of determining optimal routing paths and also transferring the packets among the nodes. Route information is stored in routing tables. Routing algorithms allow nodes to select the best paths for routing the packets to its destination.

In MANETs, infrastructure support or centralized administration could not exit, for example, in the case of wireless networks. Moreover, a destination node could be out of range of a source node. In these cases, routing protocols will always be needed to send the packets from a source to a destination; each node must be able to forward data for other nodes, mainly because of changing ad hoc network topology or node failure.

The characteristics that limit these types of networks, such as dynamic topology, restricted bandwidth, different capacity connections and high rate of errors, limited physical security, and limited capacity of batteries [4], make routing protocols fundamental to their operation. A dynamic routing protocol will be needed for these networks to function properly. The routing protocols used should perform the main objective of using routes among the different nodes that conform to the network correctly and efficiently. Ideally, messages should be sent following a protocol that permits the smallest possible consumption of bandwidth and energy. In addition, we will have to study other problems [5] such as asymmetric links, links quality (that could be different when sending and receiving), or routing overhead (that could be generated because of the node mobility). In this case, obsolete routes could be stored in routing tables. Interferences and dynamic topology should also be considered.

Routing is a challenging task because of the constant change in network topologies. Nodes will change their positions usually due to the node mobility. Organization, operation, and management of the network depend on the network's characteristics. There is no fixed topology; the devices are very mobile and may move around freely in and out of each other's range. As each node is a router, it has a limited communication range. The energy is limited; mobile devices generally operate on battery power, which is exhaustible, and the amount of energy available for each device may vary. To conserve energy, many nodes must not be available all the time. Other resources such as CPU are all limited. The transmission medium is accessible to anyone in range of the appropriate equipment, and so they have to share physical medium transmission. Identity has to be related with address; the address is given dynamically as it is difficult to associate a fixed identity. The devices have more physical vulnerability due to the small size of mobile wireless devices, and so they can be easily stolen and possibly modified. Hence the node owner relation is not stable; the node may be subverted. The devices to connect can come from anywhere, with no assumptions of central administration. As s result, these properties have two important constraints: Nodes should not be trusted without proper authentication of the node, the user, or both. Centralized services may not be available since they may be out of reach or powered down.

Classification of routing protocols can be done in many ways. The existing MANET (mobile ad hoc network) routing protocols can be classified into three categories—on-demand, proactive, and proactive–on-demand hybrid protocols—depending on how and when they find out the routes and store them (strategy and structure). According to strategy, we can classify protocols as table-driven (proactive) and source-initiated (reactive or on-demand). We will follow this classification to introduce algorithms.

The challenge of providing security in ad hoc wireless networks derives from the properties mentioned previously. Services such as confidentiality, integrity, and availability must be provided under these conditions. Confidentiality mechanisms, i.e., encryption and access control with authentication should be given without central administration and with energy restrictions.

They require key generation, management, and distribution schemes that can be run on small CPUs. Integrity solutions have similar problems. Availability could be more difficult given that new attacks are possible, such as energy starvation attacks. Some topics or challenges that are being studied on wireless ad hoc networks relate with security routing are interference, hidden terminals and exposed terminals, mobility, node failures, self-forming, self-configuration, topology maintenance, routing and self-healing, node localization and time synchronization, and end-to-end reliability and congestion control.

Two fundamental areas must be addressed when searching for wireless network security, which is comparable with that of traditional networks. The first one is trust establishment, key management, and membership control; the second deals with network availability and routing security.

The remainder of this chapter is structured as follows. Section 4.2 explains the need of routing protocols and gives their main features. Section 4.3 introduces secure routing and provides its main issues. Section 4.4 details the most well-known routing protocols for networks with infrastructure, nonsecure routing protocols for ad hoc networks, and secure routing protocols for ad hoc networks. Section 4.5 shows some routing protocols performance in terms of routing traffic received and sent by all nodes in the network with fixed nodes and mobile nodes with failures, the load at the MAC level, the throughput of the mobile nodes with failures, the average traffic received and sent and the average delay, the instantaneous routing traffic received, the average traffic received, and the average load carried out by the node for each routing protocol. Finally, Section 4.6 draws the conclusions and discusses about our future research.

4.2 Routing Protocols

Classical routing protocols are not well suited for ad hoc networks (e.g., IP), because their routing tables are unstable under frequent connectivity changes due to node mobility. For this reason, in this chapter we are going to analyze the currently proposed routing protocols for ad hoc networks, taking into account both characteristics of ad hoc networks and the required security in the communication.

Among the tasks that must be carried out, we find those of routing, address administration, and routing itself along a suitable route. To accomplish this, the nodes will control the throughput of the network, node mobility, and battery consumption, which leads to a protocol that bears the smallest consumption of both bandwidth and energy.

Routing protocols can be classified into three big groups: proactive, reactive, and hybrids. In proactive or table-driven routing (e.g., optimized link state routing, topology dissemination based on reverse-path forwarding, and Landmark Routing for Large Scale Wireless Ad Hoc Networks with Group Mobility (LANMAR)), all routes to all destinations are calculated *a priori* and are updated by periodic update messages; this category can be subdivided into two new categories: distance vector and link state. Most classic routing protocols are distance vector protocols, such as connection state. Proactive protocols maintain fresh lists of destinations and their routes by periodically distributing routing tables throughout the network. The main disadvantages of such algorithms are the respective amount of data for maintenance and the slow reaction on restructuring and failures. In reactive or on-demand routing (e.g., dynamic source routing, ad hoc on-demand distance vector, Temporally Ordered Routing Algorithm (TORA), and Relaxed Lee Gerla (RLG)), routes for a certain destination are calculated only when necessary. These protocols try to reduce the overload introduced by proactive protocols by minimizing the number of periodic upgrade packages sent in the network and by calculating the routes only when necessary. The main limitation of on-demand protocols is the initial delay that they introduce, which is a limitation of interactive applications that must ensure service quality (e.g., audio and interactive video). Finally, hybrid protocols combine the properties of the first two protocols in order to obtain flexible routing protocols with parameters, which adapt to a wide range of applications and environments. Hybrid protocols look for a balance between reactive and proactive protocols.

The above classification is not closed, and there are many protocols based on one of the above protocols that use additional parameters in the routing process to improve results.

The protocols presented do not introduce security measures. Usually, these mature ad hoc routing algorithms only check if the receiver's network interface is accepting packets, focusing on covering the changes in the dynamic topology in a good way [6,7]; otherwise, they assume that routing nodes do not misbehave. They assume that the involved nodes will cooperate for the network to operate and emerge. However, these networks are characterized by their spontaneous formation, formed in some cases by unknown nodes that may behave selfishly to obtain advantages, thus saving battery power and reserving more bandwidth for their own traffic. If a large number of nodes start behaving noncooperatively, the network would break down. For this reason, when the routing is carried out, one should keep in mind the necessary requirements of security to resist the attacks of malicious nodes and to protect the communication of non-allowed accesses. The attacks of malicious nodes may be frequent in the establishment of the routes where messages are exchanged among the routers or among the nodes that negotiate that routing.

In many of the protocols proposed in the related literature, the necessary security requirements are satisfied neither to resist the attacks of these malicious nodes nor to protect the communication of nonallowed accesses. In this situation, a malicious node could try to obtain this information in order to cause errors in the communication or even to be able to drop the network. The importance of the detection of the erroneous information is fundamental. A malicious node could also carry out an external attack, sending false routing information, repeatedly sending information that is no longer correct, or distorting the routing information. It would also be able to split the network or even to introduce an overload in the network, causing retransmissions and inefficient routing. An internal attack could cause more severe attacks: What would happen in case the nodes belonging to the network do not want to use their resources to forward packets of other nodes? Or, if these nodes introduce false routing information? These attacks could be carried out either by noncooperative users or by compromised nodes. In case a noncooperative node is detected, actions should be taken to remove the node from the network or to keep in mind when the routing process is being carried out. In the case of a compromised node, the detection of incorrect information would be difficult to be discovered in case the node has not already been detected. The information of the required routing could be signed by it and would seem to be correct for the rest of the nodes in the network.

Among the main attacks related with routing protocols, we may quote the following: nonforwarding, traffic deviations and route modifications, lack of error messages, and frequent route updates. Nonforwarding attack is the one carried out by a node when it does not forward packets to other nodes, generating in this way delays that could overload the network due to the unnecessary generation of error messages. A way of fighting against it could be by giving a reward to the collaborative nodes so that they can send their own messages and punishing those that do not cooperate by not allowing them to send or receive their own packets. A malicious node could send a packet to a nonappropriate node, introducing an unnecessary delay in the network and causing nodes to use energy unnecessarily. We will name it as "deviation attack." A way to control the deviation is to include the whole route in the head of the packet and check whether it has been received in the destination the appropriate route; in case it has not been done correctly it can be discovered which node has not acted correctly. If the updates are carried out in an inadequate way, a completely unnecessary overload would be introduced in the node absorbing their energy and processing resources since the node is busy upgrading/updating the routes (e.g., it will not be able to forward a packet). A node will think that its packet has been received correctly when after a certain time no error message has been received; so, if an error message should have been received but has not, it will be considered that there has been no error. That is, there will be an absence of error messages, which will cause loss of information that has not been received in the node.

Security requirements in a wireless network will be the same as the ones required in traditional networks: confidentiality, integrity, authentication, nonrepudiation, and availability. We must protect both data and routing information. In ad hoc networks, the attainment of these requirements will be much more complex due to the following characteristics of these types of networks: dynamic topology, restricted bandwidth, links of different capacities and high error rates, energy and processing capacity limitations, absence of a central server, and, in many cases, no prior knowledge of the nodes that will form the network. These limitations will have to be covered by administration mechanisms and by cooperation among the nodes that allow maintaining the quality of service, the security, and the discovery and access to service mechanisms in an almost automatic way. Transmission security will also be an important part of communication security. It will prevent data in transmission from being disclosed to unauthorized recipients. In addition, we will have to avoid compromised nodes that can be used to attack a network from inside. The situations that imply noncollaboration by a node will be solved in the first instance by repealing the trust in these nodes and then by expelling these nodes from the network. Thus, we will have to introduce security measures in ad hoc networks to guarantee good routing protocols performance.

4.3 Secure Routing

If the process of a routing algorithm is manipulated, the normal operation of MANET can be seriously affected. Prevention measures such as data encryption and user authentication can establish a defense against some of these attacks. We must also take into account internal attacks, which come from committed nodes belonging to the network. This is a more serious threat because such attacks are normally more difficult to detect and counterattack.

There are three main ways to prevent security attacks: the first is the "node identification." The problem is that nodes can be spoofed by other nodes and therefore correct node identification is very important. A useful solution deals with the use of pseudonyms to protect user privacy. These identities are commonly coded. Another method is to use a network of trust, such as Pretty Good Privacy (PGP) [8]. Each participant itself creates a couple of public–private keys. When the node is sure of the identity of the other, it signs the corresponding public key, certifying its identity. Thus, if A certifies the identity of B, the new node's identity can be verified. Other solutions are based on the use of cryptography mechanisms. In [9], the use of threshold cryptography through the creation of an authority of distributed certification, where n participants know a secret and k of them are able to reassemble it, has been proposed. Variations of this outline allow the public key pair, and the process of signatures, to be created distributed.

The second is the "preventing proud behavior." In [10], several methods have been proposed. One of them is to create a virtual currency called "nugget." We obtain nuggets by redirecting a packet from one node to another and spend them when we try to send our own data. We cannot send our packets if we do not have nuggets to use as payment. The drawback is that trusted hardware is required to secure the currency. Another method is to detect and expel a "proud" node using a "guard dog," which checks that data are transmitted via the correct route. Several systems have also been proposed. The first is the IDS (intrusion detection system) [11], which consists in local components being responsible for collecting data, detecting and responding global components cooperatively. Another system is CORE (collaborative reputation mechanism) [12], which is similar to IDS and consists in local observations that are combined and distributed to calculate the value of each node's reputation. The last system proposed here is CONFIDANT [13], a protocol that causes bad behavior to be considered as not attractive, since nodes that do not cooperate are excluded from the network. Another suggestion is to use MobIDS (mobile intrusion detection system), which focuses on integration with other mechanisms and sensors to detect proud nodes.

The last way to prevent security attacks is to establish "security for routing protocols against manipulation." Some works have focused on this aspect as in [10], where secure dynamic source routing (SDSR) is proposed. This protocol is simply a part of SAM (security architecture for mobile ad hoc networks) [14] and provides security by using MANET-IDs to identify the nodes, which are just a couple of Rivest, Shamir y Adleman (RSA) sign keys that prevent nodes from forging new identities. Coded SDSR [10] is used for routing protocols and MobIDS for detecting proud nodes.

4.4 Current Routing Protocols

The routing protocols generally use routing algorithms based either on the distance vector or on the state of the connection (link state). Both need announcements to be broadcasted periodically to all the nodes. In the routing based on the distance vector, each router (a router is not necessarily needed, because each node can act as a router) broadcasts to its neighboring router its view from the distance to all the nodes. Each router calculates the shortest path based on this information. In the state routing, each router sends to the other routers the state of the links to its adjacent network and each router works out the shortest distance according to the links.

All protocols presented have in common for the routing process broadcasting a message of route request to some of its neighboring nodes, and those nodes, in turn, will be in duty to guide the packet until the destination node is reached. This message will be replied with a reply message that will include the requested information or an error message if the node is not found. Among the discussed protocols, we will first present those that do not take into account the possibility of the existence of malicious nodes and/or external attacks. These protocols are focused only on establishing appropriate routes for each communication. In this category, we will make a distinction among those that consider the existence of clients of low resources, those that are more appropriate for ad hoc networks, and those that have been defined without considering this property and adapted fundamentally for networks with infrastructure. Later on, we will analyze routing protocols that include security in the communication, considering the possible existence of malicious nodes or another type of attackers.

4.4.1 Routing Protocols for Networks with Infrastructure

As far as these types of routing protocols are concerned, IP is the main protocol since it is used as the Internet protocol [15]. Based mainly on global routing, it is not adapted for its use in ad hoc networks since in these networks the nodes are in reduced areas and router nodes do not exist properly, which is the reason why all the nodes carry out the routing task by themselves.

Other protocols based on the existence of server nodes are FIDRAN [16], the Handoff protocol [17–19], the protocol of interconnection of wireless cells with a distributed system [20], and the protocol for mixed networks; these protocols are in charge of organizing the routing with a hierarchical structure, and in the existence of central servers, they are not appropriate for ad hoc networks. This is the level 2 header.

4.4.2 Nonsecure Routing Protocols for Ad Hoc Networks

In the protocols designed for ad hoc networks, all the nodes work as routers and they are involved both in the discovery and in the maintenance of routes. Examples of proactive protocols are DSDV (destination-sequenced distance vector), CGSR (clusterhead gateway switch routing), WRP (wireless routing protocol), and OLSR (optimized link state routing). Among the more common reactive protocols, we can find DSR (dynamic source routing), LMR (lightweight mobile routing), or AODV (ad hoc on-demand distance vector). We introduce some of them here and refer others to be consulted elsewhere. Since our study is focused on ad hoc networks in which neither the number of nodes nor the topology of the network is known, proactive protocols are of no interest in this study.

One example of a proactive protocol is DSDV, a proactive protocol that is a modification of the conventional Bellman–Ford routing algorithm. In DSDV a new attribute—the sequence number—is added. It will be introduced to each routing table entry at each node. In DSDV [21], each node of the network maintains a routing table that contains all the possible destinations and the number of hops that a given packet would need to get to the specified destination; however, this protocol can produce an overload of the nodes and messages in the network in order to maintain the up-to-date routes. For each route, the data broadcast by a node will contain its new sequence number, the destination address, the number of hops required to reach the destination, and the new sequence, originally stamped by the destination. The advantages of DSDV are as follows: it allows us to guarantee loop-free paths; it reduces count to infinity problem; it avoids extra traffic with incremental updates and maintains only the best path, reducing the amount of space in routing tables. However, some limitations arise: DSDV has unnecessary advertising of routing information; it does not support multipath routing; it is difficult to maintain the routing table's advertisements for larger networks. Other examples of proactive algorithms are AWDS (ad hoc wireless distribution service) [22], CGSR (clusterhead gateway switch routing) [23], DFR (direction forward routing) [24], DBF (distributed Bellman–Ford) [25], DSDV (destination-sequenced distance-vector) [26], Guesswork [27], HSR (hierarchical state routing) [28], IARP (intrazone routing protocol/proactive part of the Zone Routing Protocol (ZRP)) [29], LCA (linked cluster architecture) [30], MMRP (mobile mesh routing protocol) [31], OLSR (optimized link state routing) [32], TBRPF (topology dissemination based on reverse-path forwarding) [33], WAR (witness aided routing) [34], and WRP (wireless routing protocol) [35].

AODV [36,37] is an evolution of the DSDV protocol but has a reactive behavior. It is a very simple, efficient, and effective routing protocol for MANETs. AODV introduces the concept of "under request routing"; that is, it keeps information of only those nodes that are involved in data transmission. The algorithm makes sure that the nodes that are not in the active path do not maintain information about this route. AODV also maintains loop-free routes even when links change on active routers. Route discovery is achieved with source-initiated broadcast message (RREQ: route request). When this packet arrives at either the destination or an intermediate node that has a valid route to the destination, a new message (RREP: route reply) is unicast back to the source. Intermediate nodes also update their routing tables while traveling back to the source. In this way the time of the process and the use of memory are decreased as well as controls traffic. This protocol works quite well when the mobility is not very high. Some disadvantages of AODV are as follows: the algorithm requires that the nodes in the broadcast medium can detect each others' broadcasts. In the same way, an overhead on the bandwidth could occur when an RREQ travel from node to node in the process of discovering the route information. Other disadvantages could be consulted in [38]. DSR [39–41] is a simple algorithm based on the concept of source routing: source nodes determine routes dynamically and only as needed. It allows sender to determine the packets' travel path toward a destination. Nodes maintain cache memories whose entrances include the destination and the list of the nodes to get these destinations; these entrances are updated as they learn new routes. It is based on two main mechanisms: route discovery and route maintenance. If there is a valid entry for the destination, the node sends the packet using that route. Otherwise, the source node initiates the route discovery process. DSR does not use periodic messages, and this is the reason why it reduces the high throughput of the bandwidth and preserves the battery of hosts. These messages are used only when they are necessary. DSR is also able to adapt quickly to changes. However, when storing a single route, the network can be overloaded when the mobility is high; in Super Restrictive (SR) mode (based on DSR but with some optimizations), a list is added to store several routes to the same node; the problem that could arise here would be that the storage list is too large that the time of process in each node, besides its energy consumption, would be increased. Another protocol based on DSR [36,39] is Lee Gerla (LG). In this protocol the grade of disjoint paths is analyzed while they are found in the process of discovering routes; this protocol allows that two packets that have arrived to an intermediate node are transmitted until the end of the route; in this way we are able to find disjointed routes for a certain destination. Nevertheless, it has been probed that LG works worse than DSR or SR. RLG [36] is a modification to LG; in RLG, what is done is to interrupt the propagation of a route when it is detected that the last received packet came from the same source than the previous one. According to [11], RLG introduces too much overload without providing huge benefits. The ADSR (abbreviated dynamic source routing) protocol [42] uses the temporal information of nodes to calculate the fitness of the candidate paths. Routes are selected based on the relative stability of the intermediate nodes, which is based on time-averaged nodal connectivity and nodal mobility, residual battery life, signal stability, buffer occupancy rate, storage capacity, processing power, etc. It attempts to solve the problem of the TCP protocol for MANETs. This protocol does not work well in wireless networks due to the high bit error rate. If the rediscovery of the route uses more time than the "retransfer time over" (RTO), the retransfer of this packet is developed and the RTO will drop exponentially, which can cause a saturation of the networks. To avoid this, helping nodes are used; they take charge of upgrading the routes by means of the use of active packets. This protocol improves the congestion for TCP traffic; however, it needs a helping node to be used. ADSR offers both a good packet delivery ratio and a reduced normalized routing traffic overhead [43]. TORA is an adaptive routing protocol; this type of protocol combines the advantages of proactive routing and reactive routing. Initially, the routing is initialized with proactive routes and then servers demand for additional routes from nodes through reactive flooding. In TORA [21], the aim is to minimize the load of the networks. It is based on the impossibility of always maintaining the shortest route; for each possible route toward a node, it keeps the route being based on a measure called weight. Nevertheless, this protocol has high-energy consumption in contrast with other protocols of this type. The advantage of this protocol depends on the amount of nodes activated; reaction to traffic demand will depend on the traffic volume.

Other reactive protocols are ESAODV (extra secure ad hoc on-demand vector) [44], SENCAST [45], multiroute ad hoc on-demand distance vector routing protocol [46], reliable ad hoc on-demand distance vector routing protocol [47], minimum exposed path to the attack (MEPA) [48], anr-based routing algorithm for MANETs [49], ACOR (admission control enabled on-demand routing) [50], associativity-based routing [51], CHAMP (caching and multipath routing) [52], Ad hoc On-demand Multipath Distance Vector [53], dynamic nix-vector routing [54], dynamic MANET on-demand routing [55], mobile ad hoc on-demand data delivery protocol [56], on-demand routing in MANET [57], and secure routing protocol (SRP) for ad hoc networks based on trust [58]. Other protocols without definite security requirements can be consulted in [59–62].

4.4.3 Protocols for Secure Routing for Ad Hoc Networks

In the protocols that add security to the routing SAODV [10,63] can be found based on AODV; it adds means to obtain authentication, integrity, and nonrepudiation of the routing control packets by using asymmetric cryptography and hash chains. Nevertheless, it is vulnerable to attacks of malicious nodes that would want to forward the packets through themselves. To do that, these malicious nodes would make the others think that the route through them is the shortest route. ARIADNE [63–65] is a secure protocol based on DSR; it is also based on the use of TESLA [66,67] for the authentication. Basic symmetrical cryptography and hash chains are also used to add security in the route and in the error messages. A time counter is used to avoid loops. However, when retransferring the packet, it also introduces additional information, which causes an overload of bytes without proper information. SEAD (secure efficient distance vector) [63,68] is a modification of DSDV. It uses hash chains to provide security in the route as well as counters to avoid loops. However, the existence of a secret is supposed to be shared for the authentication of the source of each message. SRP [10] uses symmetrical cryptography; in this protocol, it is not necessary that each intermediate node carries out any cryptographic operation so that it does not overload the intermediate nodes. SRP can be easily added to a large number of protocols; it includes redundancy in the routes, which allows us to assure the communication between the destination and the source in case any of the routes fail. Nevertheless, if this redundancy is high, the traffic in the network could excessively be increased. The ARAN (authenticated routing for ad hoc networks) [10,63] protocol establishes authentication, message integrity, and nonrepudiation. It is based on the existence of a public key from a trust server of keys per node. Furthermore, each intermediate node will validate and sign the messages it receives, which causes a high consumption of the batteries of the devices in the processing of this cryptographic operations as well as generation of some considerable delays. MAODDP (mobile ad hoc on-demand data delivery protocol) [69] is based on the establishment of routes and simultaneous data delivery. For the establishment of the security, the use of a certification server of trust is required whose public key is known by all the nodes. Each node has a certificate with definitive lifetime. In each routing, the intermediate nodes will extract and verify the certificate of the previous node and will attach their own certificate. It can cause delays in the communication and a great consumption of the batteries of the devices. SDSR [10] is based on DSR; it is a part of SAM. Security is provided for the identification of the nodes with MANET-IDs, for the routing protocols such as SDRS, and for the detection of nodes in MobIDS. It uses hash functions and symmetrical and asymmetrical cryptography, and, of course, the nodes should be able to support only a limited number of asymmetric operations. In this protocol, keys are exchanged in the delivery of all the packets, losing time and battery in these operations. With some improvements, SDSR could be one of the best options for the fulfillment of secure routing in ad hoc networks. Some of these improvements deal with the use of piggyback, obtaining that the best route to the source arrives to the destination, again using the security information for the same communication and introducing mechanisms that allow excluding from the communication those nodes that do not want to take part in the routing process.

When we want to use one of these protocols, we must take into account whether we prefer a reliable communication or a more efficient communication. We believe that the best option will be that which maintains a balance between both security and efficiency; in this situation, the nodes do not have to perform many asymmetric cryptographic operations when they are routing the nodes. If asymmetric operations are necessary, only the sender and receiver nodes should make these operations.

4.5 Routing Protocols Performance

In this section, we show the simulations performed and discuss the measurements taken from the performance comparison test of AODV, DSR, and OLSR in order to know which performs better with respect to some parameters. The parameters taken into account for this discussion are the received and sent routing messages with fixed and mobile nodes with failures, the load at the MAC level, the throughput of the mobile nodes with failures, the average traffic received and sent and the average delay, the instantaneous routing traffic received, the average traffic received, and the average load carried out by the node for each routing protocol.

4.5.1 Test Bench

This subsection presents the test bench used for all the evaluated protocols. We have varied the number of nodes and the coverage area of the network. Each protocol has been simulated in two scenarios: (1) network with fixed nodes and (2) network with mobile nodes with failures.

Figures 4.1 through 4.3 present four mobile topologies. From each node (determined by a circle in the figures) emerges a line that suggests the mobility direction for the node. In simulations of 50, 100, and 250 nodes, all nodes move into an area delimited by the dotted points shown in each figure.

The topology of 50 nodes has a coverage area of 500 m² (Figure 4.1), 100 nodes cover 750 m² (Figure 4.2), and finally 250 nodes cover 1 km² (Figure 4.3). 250 nodes with a coverage radius of 50 m cover more than an area of 1 km², but in such networks, not only it is necessary to cover a given area, but also we need to communicate with other nodes, i.e., those nodes must be under our coverage area.

Each scenario has been simulated to observe the system scalability. Instead of a standard structure, we have chosen a random topology. It has been obtained by using OPNET simulator [19]. The nodes have a random mobility model. The physical topology does not follow any known pattern. The obtained data do not depend on the initial topology of the nodes or on their movement pattern, because all of it has been fortuitous.

The nodes have a 40-MHz processor, a 512-KB memory card, a radio channel with less than 1 Mbps, and their working frequency is 2.4 GHz. Their maximum coverage radius is 50 m.

We have forced node failures, with the consequent recovering processes, to take measurements from the mobile node simulation and to observe the network behavior against changes in the physical topology.

The traffic generated by OPNET has been used as the traffic load of the simulation. We injected this traffic 100 s after the simulation starts. We have configured the traffic arrival with a Poisson distribution (with a mean time between arrivals of 30 s). The packet size follows an exponential distribution with a mean value of 1024 bits. The destination address of the injected traffic is random to obtain a simulation independent of the traffic address. We have simulated both

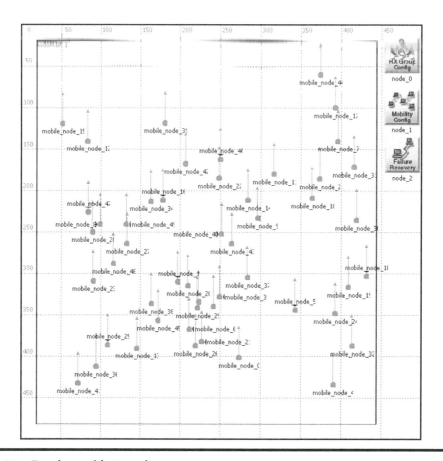

Figure 4.1   Topology with 50 nodes.

scenarios for AODV, DSR, and OLSR protocols. The results obtained are shown in the following subsections.

4.5.2 Routing Traffic Received and Sent by All Nodes in the Network with Fixed Nodes and Mobile Nodes with Failures

Figure 4.4 shows the routing traffic received in the fixed topology of 50 nodes for each protocol. We see that the OLSR protocol is the most stable, but the DSR protocol has less routing traffic.

Figure 4.5 shows the measurements when the traffic is sent. We note that the average traffic sent of the OLSR and DSR protocols is almost the same, although DSR is much less stable.

For the fixed topology of 100 nodes, we have obtained Figures 4.6 and 4.7. They show that the traffic is very similar for the routing protocols AODV and DSR. In this case, DSR is the most unstable. In contrast, the OLSR protocol is the best one for this topology because the traffic is very stable and also significantly lower than in the rest of the exposed routing protocols.

Finally, we observe the behavior of the OLSR, DSR, and AODV protocols when the number of nodes is equal to 250 (see Figures 4.8 and 4.9). In this case, the same behavior is observed

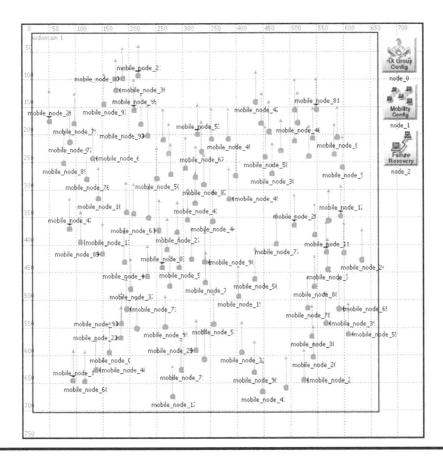

Figure 4.2   Topology with 100 nodes.

than with the fixed topology of 100 nodes. The OLSR protocol is the best one. In contrast, as the number of nodes increases, the routing traffic of DSR is higher as compared with that of the other routing protocols. In the case of AODV, when the number of nodes is low, it has the worst performance, but when the amount of nodes increases, it is better than DSR.

But what happens when we model a mobile stage with node failures and recoveries? This behavior can be observed in Figures 4.10 and 4.11 for the topology of 50 nodes. We have observed that DSR is the routing protocol that introduces less routing traffic in the network.

As stated earlier, we can see in Figure 4.11 that the OLSR protocol is best suited when the topology changes. It is the most stable.

When the topology has 100 nodes (see Figures 4.12 and 4.13), the difference of the routing traffic received of the three protocols is minimal. Once the network converges, they range between 175 and 210 Kbps (see Figure 4.12).

However, when we look at the traffic sent (see Figure 4.13), we do not obtain the same behavior. In this case, the OLSR protocol is the best one, while AODV and DSR have approximately the same routing traffic sent.

Finally, we are going to see what happens in the topology of 250 nodes. In Figure 4.14, we show the routing traffic received by all nodes. When the network has several failures, the highest mean value is given by the AODV protocol. The DSR protocol also introduces a high amount of traffic, but it has higher variations in its behavior. Finally, the OLSR protocol introduces less

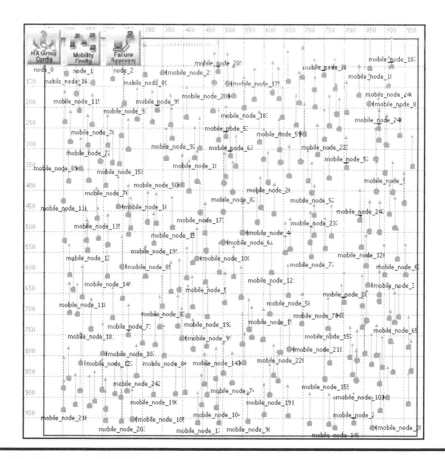

Figure 4.3   Topology with 250 nodes.

traffic, while its behavior is very stable when there are failures and recoveries in the network. So, it is the best for this case.

In Figure 4.15, we observe the same behavior described above but in this case for the routing traffic sent.

4.5.3 Load at the MAC Level

Taking into account the load at the MAC level, we analyze the topologies of 100 and 250 nodes because they are the most restrictive. In Figure 4.16, we see that when we have a topology of 100 nodes, the AODV and DSR protocols have lower mean load (80 Kbps) than that of the OLSR protocol (110 Kbps). This gap is getting smaller when the number of nodes increases in the network.

In Figure 4.17, we show that the AODV and DSR protocols have an average load of 180 Kbps and the OLSR protocol has 200 Kbps.

4.5.4 Throughput

In the throughput occurs the same behavior as in the load at the MAC level. In Figure 4.18, we show the values obtained for AODV, DSR, and OLSR. The best ones are obtained for DSR and AODV. There is a difference of 150 Kbps between the values of OLSR and them.

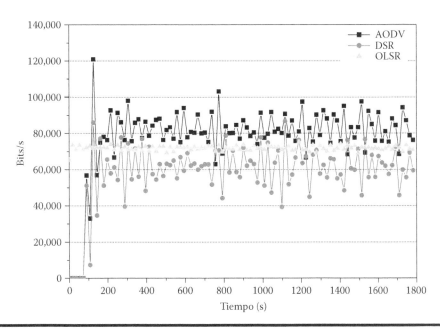

Figure 4.4 Routing traffic received by all nodes in the fixed network topology of 50 nodes.

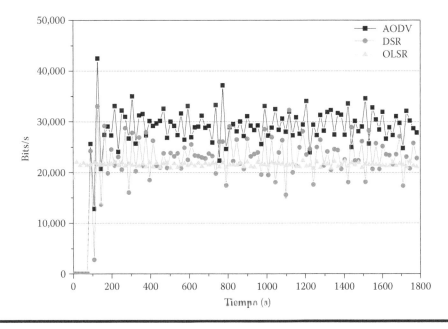

Figure 4.5 Routing traffic sent by all nodes in the fixed network topology of 50 nodes.

In contrast, in the topology of 250 nodes this difference decreases to 100 Kbps (see Figure 4.19). This leads us to believe that in DSR and AODV protocols the throughput increases more quickly with the number of nodes, unlike in the OLSR protocol. That is, when we increase the number of nodes in the network, there is a value where OLSR would have a throughput lower than that of DSR and AODV.

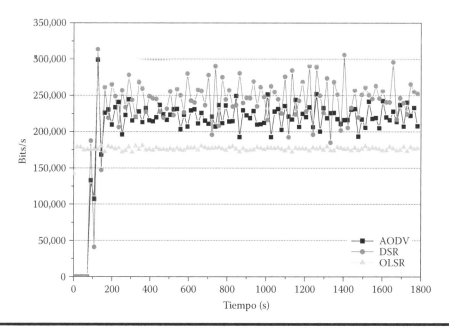

Figure 4.6 **Routing traffic received by all nodes in the fixed network topology of 100 nodes.**

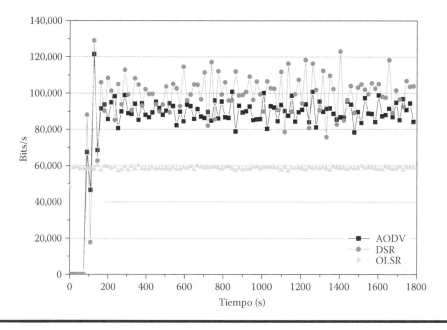

Figure 4.7 **Routing traffic sent by all nodes in the fixed network topology of 100 nodes.**

4.5.5 Average Traffic Received and Sent

In this subsection we study how the behavior of the network is for different routing protocols when the traffic is injected into the network.

In Figure 4.20, we observe the average traffic received. The AODV protocol is the one that receives less traffic (700 bps). However, the DSR and OLSR protocols have increased traffic up to

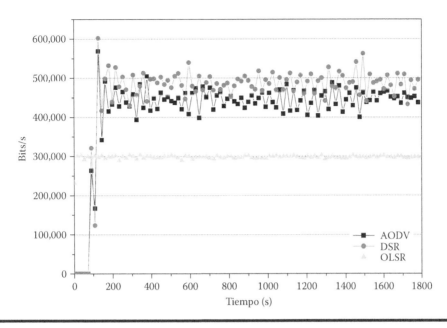

Figure 4.8   Routing traffic received by all nodes in the fixed network topology of 250 nodes.

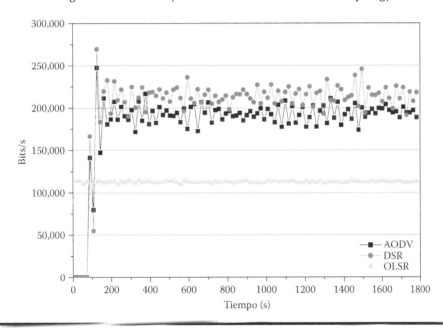

Figure 4.9   Routing traffic sent by all nodes in the fixed network topology of 250 nodes.

950 bps and 1 Kbps, respectively. This indicates that the AODV protocol is the one that adds less data traffic to the network. It is interesting to know this if we are planning to implement these routing protocols in ad hoc networks with power restrictions.

When we take into account the average traffic sent (see Figure 4.21), we do not appreciate differences between routing protocols, because all analyzed protocols have the same behavior. This happens because we introduce the same type of traffic in all scenarios.

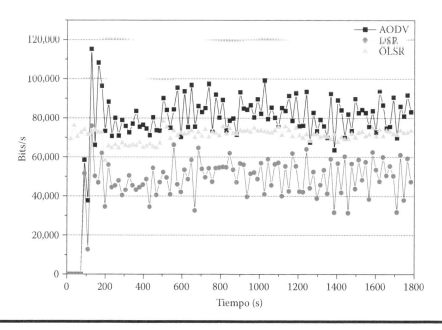

Figure 4.10 Routing traffic received by all nodes in the network topology of 50 mobile nodes with failures.

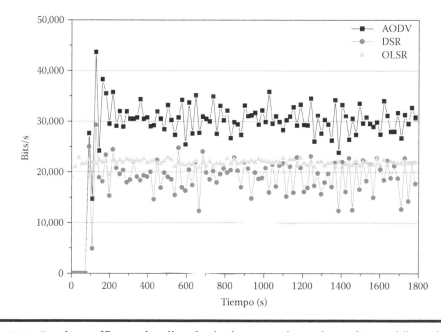

Figure 4.11 Routing traffic sent by all nodes in the network topology of 50 mobile nodes with failures.

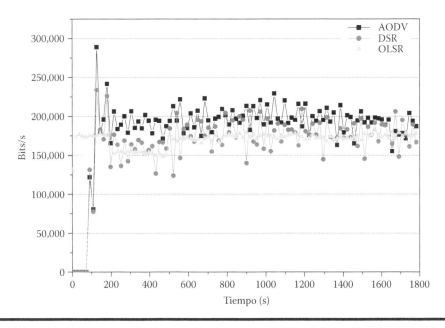

Figure 4.12 Routing traffic received by all nodes in the network topology of 100 mobile nodes with failures.

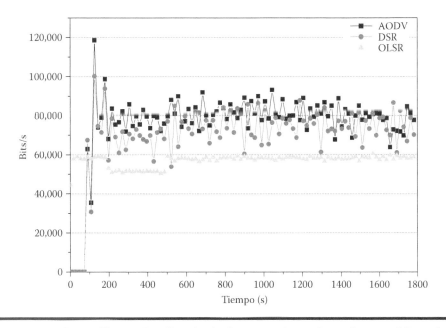

Figure 4.13 Routing traffic sent by all nodes in the network topology of 100 mobile nodes with failures.

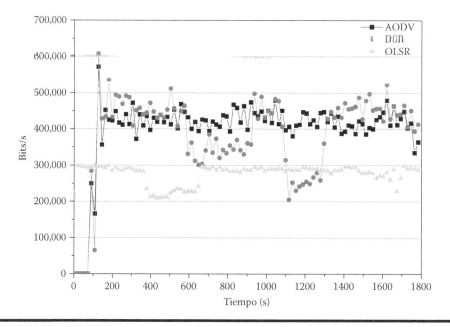

Figure 4.14 **Routing traffic received by all nodes in the network topology of 250 mobile nodes with failures.**

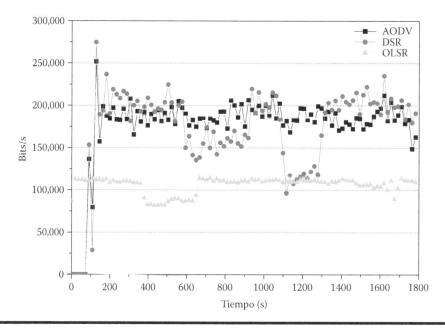

Figure 4.15 **Routing traffic sent by all nodes in the network topology of 250 mobile nodes with failures.**

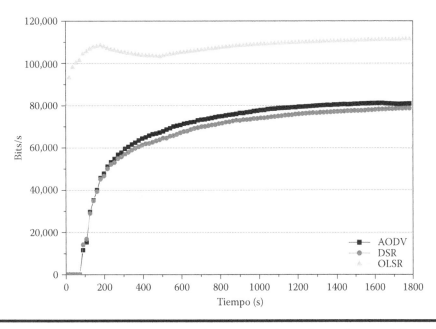

Figure 4.16   Average load at the MAC level in the network topology of 100 mobile nodes with failures.

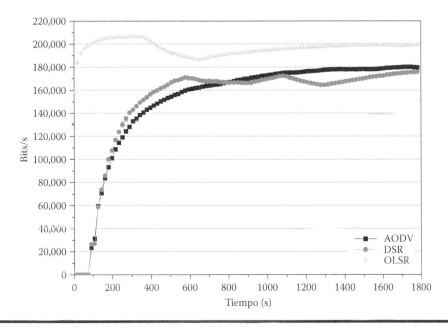

Figure 4.17   Average load at the MAC level in the network topology of 250 mobile nodes with failures.

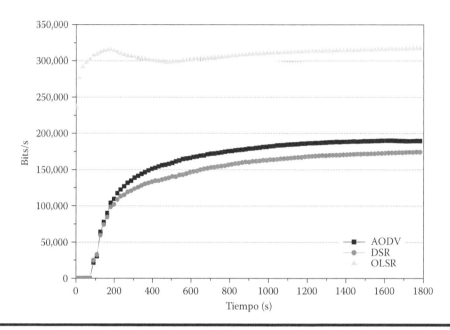

Figure 4.18 Average throughput in the network topology of 100 mobile nodes with failures.

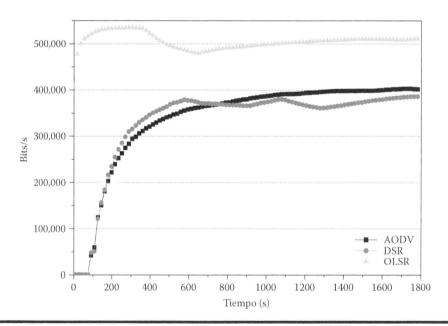

Figure 4.19 Average throughput in the network topology of 250 mobile nodes with failures.

In the topology of 250 mobile nodes with failures (see Figures 4.22 and 4.23), we obtain the same behavior as the one obtained by 100 mobile nodes with failures (see Figures 4.20 and 4.21). There are several differences between them: Figure 4.22 has less traffic than Figure 4.20, and in Figure 4.23 the traffic sent increases when the number of nodes increases as compared with Figure 4.21 (there are no variations when we inject the traffic into the network).

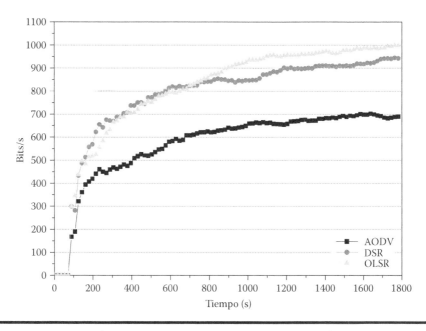

Figure 4.20 Average traffic received in the network topology of 100 mobile nodes with failures.

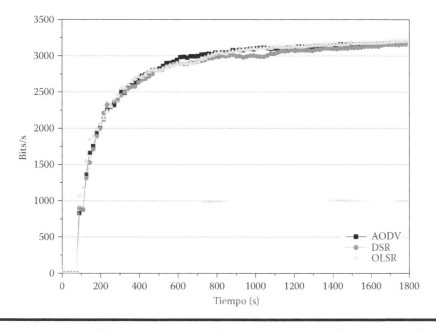

Figure 4.21 Average traffic sent in the network topology of 100 mobile nodes with failures.

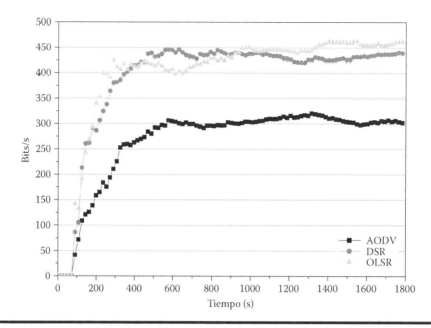

Figure 4.22   Average traffic received in the network topology of 250 mobile nodes with failures.

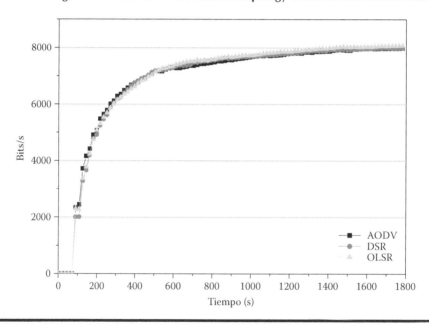

Figure 4.23   Average traffic sent in the network topology of 250 mobile nodes with failures.

4.5.6 Average Delay

Looking at the delay experienced by the injected traffic (see Figure 4.24), we conclude that the DSR and OLSR protocols introduce a delay close to zero. In contrast to the AODV protocol, there is a typical delay in the initiation phase, and once the network has converged, the delay is maintained at approximately 1 s.

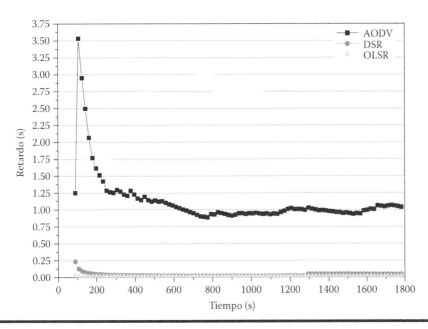

Figure 4.24 Average delay in the network topology of 250 mobile nodes with failures.

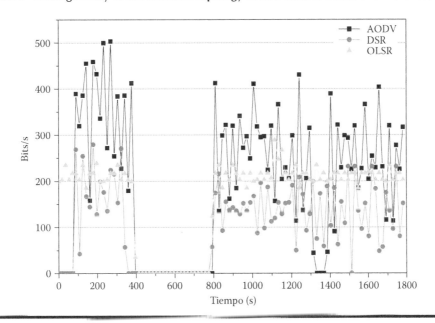

Figure 4.25 Instantaneous routing traffic received by the node.

4.5.7 Measurements of a Network Node

In this subsection, we see the results of a single node when it uses each protocol. In Figure 4.25, we see the routing traffic received by the node. A more stable traffic is sent by the OLSR protocol. We can observe that there are no traffic peaks in the initial phase or after the node failure. The DSR protocol provides a lesser traffic but is very unstable.

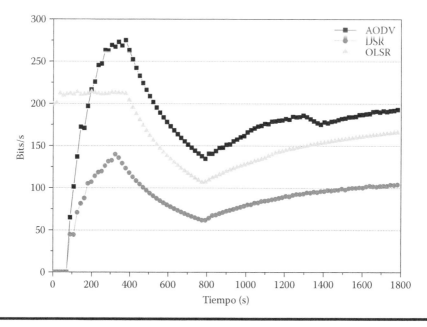

Figure 4.26   Average routing traffic received by a node.

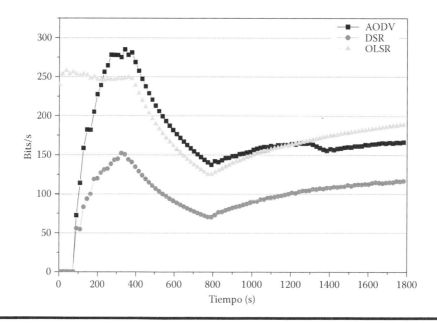

Figure 4.27   Average load supported by a node.

We analyze the average routing traffic in Figure 4.26. We draw the same conclusions as before, but in this case it is shown in a better way. So, as regard the amount of traffic received by the node, the best protocol is DSR.

In Figure 4.27, we show the average load supported by a node at the MAC level. The protocol that provides a lower load is DSR (125 bps). OLSR has 190 bps of load. The maximum difference between the three protocols is when the network has converged, and it is around 75 bps. The differences in the initial phase are given because of the characteristics of each routing protocol (reactive vs. proactive protocols).

4.6 Conclusions

Routing is needed for a suitable operation of ad hoc networks. A network designer must choose the appropriate routing protocol for the network; that is, he or she must select the routing protocol that best fits the purpose of the network. In order to achieve this goal administrators should take into account several issues such as Long-Hop and Short-Hop Routing [70] and Localization [71]:

The following developments produced the greatest impacts on the modern society:

■ Robust routing and mobility administration algorithms to increase network confidence and availability
■ Protocols and adaptive algorithms prepared to frequent radio propagation changes, to frequent network changes, and to work under different traffic conditions
■ Protocols and algorithms with low load to preserve the communication resources
■ A robust network architecture to establish security in order to avoid its vulnerability when having network attacks, network failures, node congestions, as well as a punishment due to inefficient routings

The study of the routing protocols carried out in this chapter provides a general vision of them. This analysis is the starting point for future investigations to outline a unique protocol that integrates the advantages of all of them, looking for a balance between security and the absence of an overload, not only in nodes but also in the network.

The MANET standard protocols introduce very few traffic, and so they can be used in other types of networks such as wireless sensor ad hoc networks.

In this chapter, we have simulated the AODV, DSR, and OLSR routing protocols performance by using the OPNET simulator in terms of routing traffic received and sent by all nodes in the network with fixed nodes and mobile nodes with failures, load at the MAC level, throughput, the average traffic received and sent and the average delay, the instantaneous routing traffic received, the average traffic received, and the average load carried out by the node for each routing protocol. We have seen that the routing protocol that fits best and has better results for almost all the parameters is the OLSR protocol. The traffic is stable and lower than in the other two simulated protocols even when there is a high quantity of nodes with mobility and failures/recoveries. But when we simulate the consumed throughput rate and the average load at the MAC level, the worst protocol is the OLSR because it is a proactive protocol. Moreover, it does not depend on the number of nodes in the network. AODV shows the best behavior (in terms of average traffic received and sent) when traffic is injected into the network.

All the routing protocols have an average delay close to zero for the simulated scenarios (the worst case varies between 1 and 3 ms).

In our future research we will add more routing protocols to this performance comparison test in order to have a larger overview of the routing protocols behavior with respect to the parameters selected for this test. Moreover, we will add secure routing protocols in order to know how the security affects the performance in comparison with nonsecure routing protocols.

References

1. Bluetooth SE. Bluetooth Specification, 2010. Available at Iittp://www.bluetooth.com (Accessed 01/07/2011).
2. D. Agrawal and Q. A. Zeng. *Introduction to Wireless and Mobile Systems*, Pacific Grove, CA: Brook/Cole Publishing, 2002, 438 pp.
3. B.P. Crow, I. Widjaja, L.G. Kim, and P.T. Sakai. IEEE 802.11 wireless local area networks. *IEEE Communications Magazine*, vol. 35, no. 9, 1997, pp. 116–126.
4. S. Corson and J. Macker. Mobile ad hoc networking (MANET): routing protocol performance issues and evaluation considerations. Technical Report, RFC2501, 1999.
5. K. Gorantala. Routing protocols in mobile ad-hoc networks. Master's Thesis in Computing Science, Department of Computing Science, Umea University, Sweden, 2006.
6. P. Jacquet, P. Muhlethaler, and A. Qayyum. Optimized link state routing protocol. IETF MANET, Internet Draft, November 1998. Available at http://tools.ietf.org/search/draft-ietf-manet-olsr-00 (Accessed 01/07/2011).
7. S. Jarecki. Proactive secret sharing and public key cryptosystems. Master's thesis, Department of Electrical Engineering and Computer Science, Massachusetts Institute of Technology, September 1994.
8. R. Lacuesta and L. Peñalver. Spontaneous networks: trust in a world of equals. In *Proceedings of the International Conference on Networking and Services (ICNS 2006),* Silicon Valley, California: IEEE Computer Society Press, 2006.
9. F. Stajano and R. Anderson. The resurrecting duckling security issues for ad-hoc wireless networks. In B. Christianson, B. Crispo, and M. Roe (eds) *7th International Workshop on the Security Protocols*. Lecture Notes in Computer Science 1796. Berlin: Springer-Verlag, 1999.
10. F. Kargl, A. Geiss, S. Schlott, and M. Weber. *Secure Dynamic Source Routing*. Ulm, Germany: University of Ulm, 2005.
11. S. Marti, T.J. Giuli, K. Lai, and M. Baker. Mitigating routing misbehavior in mobile ad hoc networks. In *International Conference on Mobile Computing and Networking*, 2000.
12. P. Michiardi and R. Molva. Prevention of denial of service attacks and selfishness in mobile ad hoc networks. Institut Eurecom Research Report, RR-02-063, January 2002.
13. S. Buchegger and J.Y. Le Boudec. *Performance Analysis of the CONFIDANT Protocol Cooperation of Nodes—Fairness in Dynamic Ad Hoc Networks*. Rüschlikon, Lausanne, Switzerland: IBM Zurich Laboratory, 2002.
14. F. Kargl, S. Schlott, and M. Weber. *Identification in Ad hoc Networks*. Ulm, Germany: Media Informatics Department, Ulm University, 2006.
15. A. Engst and G. Fleishman. *Introducción a las redes inalámbricas*, Primera edición. Anaya Multimedia, 2003.
16. A. Hess, M. Jung, and G. Schäfer. *Combining Multiple Intrusion Detection and Response Technologies in an Active Networking Based Architecture*. Berlin: Telecommunication Networks Group, Technische Universität Berlin, 2003.
17. X. Fu, T. Chen, A. Festag, H. Karl, G. Schäfer, and C. Fan. Secure, QoS-enabled mobility support for OP-based networks. In *Proceedings of the International Conference on IP Based Cellular Network Conference (IPCN 2003)*. Munich: Siemens AG, 2003.
18. M. Boghe and W. Trappe. *An Authentication Framework for Hierarchical Ad Hoc Sensor Networks*. Wireless Information Network Laboratory, Rutgers, The State University of New Jersey, 2003.
19. J.A. García-Macías and F. Rousseau. *Servicios Activos para Abordar Problemas en Ambientes Móviles*. Saint Martin d'Hères, France: LSR-IMAG, 1999.
20. J. Weinmiller. *Grouping Wireless Picocells with a Distribution System*. Berlin: Department of Electrical Engineering, Telecommunication Networks Group, Technische Universität Berlin, 2007.
21. C. Gahlin. Secure ad hoc networking Master thesis in Computing Science, Umeå University, Sweden, 2004.
22. AWDS (Ad Hoc Wireless Distribution Service) Web site: http://awds.berlios.de

23. C.-C. Chiang, H.-K. Wu, W. Liu, and M. Gerla. CGSR (clusterhead gateway switch routing protocol)—routing in clustered multihop, mobile wireless networks with fading channel. In *IEEE Singapore International Conference on Networks*, SICON'97, April 1997, pp. 197–211.

24. Y.-Z. Lee, M. Gerla, J. Chen, J. Chen, B. Zhou, and A. Caruso. DFR ("direction" forward routing). *Ad Hoc & Sensor Wireless Networks*, vol. 2, no. 2, 2006, pp. 151–168.

25. D.P. Bertsekas and R.G. Gallager, DBF (distributed Bellman–Ford routing protocol): distributed asynchronous Bellman–Ford algorithm. In *Data Networks*. Englewood Cliffs, NJ: Prentice-Hall, 1987, pp. 325–333.

26. C.E. Perkins and P. Bhagwat. Highly dynamic destination-sequenced distance vector (DSDV) for mobile computers. In *Proceedings of the SIGCOMM 1994 Conference on Communications Architectures, Protocols and Applications*, August 1994, pp. 234–244.

27. T. Parker and K. Langendoen. Guesswork: robust routing in an uncertain world. In *2nd IEEE International Conference on Mobile Ad Hoc and Sensor Systems (MASS 2005)*, November 2005.

28. G. Pei, M. Gerla, X. Hong, and C.-C. Chiang. A wireless hierarchical routing protocol with group mobility. In *IEEE WCNC'99*, New Orleans, September 1999. Available at http://wiki.uni.lu/secan-lab/Hieracical+State+Routing.html

29. Z.J. Haas, M.R. Pearlman, and P. Samar. The intrazone routing protocol (IARP) for ad hoc networks. Internet Draft, July 2002. Available at http://tools.ietf.org/html/draft-ietf-manet-zone-iarp

30. M. Gerla and J.T. Tsai. Multicluster, mobile, multimedia radio network. *ACM Wireless Networks*, vol. 1, no. 3, 1995, pp. 255–265.

31. K. Grace. Mobile mesh routing protocol (MMRP). Available at http://www.mitre.org/work/tech_transfer/mobilemesh/

32. T. Clausen and P. Jacquet. Optimized Link State Routing Protocol (OLSR), RFC 3626, October 2003. Available at http://www.ietf.org/rfc/rfc3626.txt

33. B. Bellur, R.G. Ogier, and F.L. Templin. Topology dissemination based on reverse-path forwarding (TBRPF), RFC 3684, February 2004.

34. I.D. Aron and S. Gupta. A witness-aided routing protocol for mobile ad hoc networks with unidirectional links. In *Proceedings of the First International Conference on Mobile Data Access*, 1999, pp. 24–33.

35. S. Murthy and J.J. Garcia-Luna-Aveces. A routing protocol for packet radio networks. In *Proceedings of the ACM International Conference on Mobile Computing and Networking*, November 1995, pp. 86–95.

36. Calafate, Malumbres, Manzoni. A flexible and tunable discovery mechanism for on-demand protocols. Department of Computer Engineering, Universidad Politécnica de Valencia.

37. Calafate, Malumbres, Manzoni. Performance issues of H.264 compressed video streams over IEEE 802.11b based MANETs. Department of Computer Engineering, Universidad Politécnica de Valencia.

38. K. Gorantala. Routing protocols in mobile ad-hoc networks. Master's Thesis in Computing Science, Department of Computing Science, Umea University, Sweden, June 2006.

39. D.B. Johnson, Y.-C. Hu, and D.A. Maltz. The Dynamic Source Routing Protocol (DSR) for Mobile Ad Hoc Networks for IPv4, February 2007. Available at http://tools.ietf.org/html/rfc4728 (Accessed 01/07/2011).

40. L. Zhou and Z.J. Haas. *Securing Ad-Hoc Networks*. IEEE Network, Special Issue on Network Security. Ithaca, NY: Cornell University, 1999.

41. Y. He, C.S. Raghavendra, S. Berson, and B. Braden. TCP performance with active dynamic source routing for ad hoc networks. *In 2nd IECE International Workshop on Active Network Technologies and Applications (IECE ANTA)*, Osaka, Japan, May 2003.

42. S.U. Rehman, W.-C. Song, and G.-L. Park. Associativity-based on-demand multi-path routing in mobile ad hoc networks. *KSII Transactions on Internet and Information Systems*, vol. 3, no. 5, 2009, pp. 475–491.

43. S. Mandala, M.A. Ngadi, A.H. Abdullah, and A.S. Ismail. A variant of Merkle signature scheme to protect AODV routing protocol, recent trends in wireless and mobile networks. *Communications in Computer and Information Science*, vol. 84, part 1, 2010, pp. 87–98.

44. P. Appavoo and K. Khedo. SENCAST: a scalable protocol for unicasting and multicasting in a large ad hoc emergency network. *International Journal of Computer Science and Network Security*, vol. 8, no. 2, 2008, pp. 87–98.

45. R. Guimaraes and Ll. Cerda. Improving reactive routing on wireless multirate ad hoc networks. In *Proceedings of 13th European Wireless*, 2007. Available at https://upcommons.upc.edu/e-prints/bitstream/2117/1173/1/mr-aodv.pdf (Accessed 01/07/2011).

46. S. Khurana, N. Gupta, and N. Aneja. Reliable ad hoc on-demand distance vector routing protocol. In *International Conference on Networking, International Conference on Systems and International Conference on Mobile Communications and Learning Technologies (ICNICONSMCL'06)*, Morne, Mauritius, April 23–29, 2006.

47. S. Khurana, N. Gupta, and N. Aneja. Minimum exposed path to the attack (MEPA) in mobile ad hoc network (MANET). In *Sixth International Conference on Networking ICN '07 Sainte-Luce*, Martinique, France, April 22–28, 2007.

48. M. Günes, U. Sorges, and I. Bouazizi. ARA - the ant-colony based routing algorithm for MANETs. In *International Workshop on Ad Hoc Networking (IWAHN 2002)*, Vancouver, British Columbia, Canada, August 18–21, 2002, pp. 79–85.

49. N. Kettaf, A. Abouaissa, T. Vuduong, and P. Lorenz. Admission control enabled on demand routing (ACOR). Internet Draft, July 2006. Available at http://tools.ietf.org/html/draft-kettaf-manet-acor (Accessed 01/07/2011).

50. C.-K. Toh. A novel distributed routing protocol to support ad hoc mobile computing. In *Proceedings of the IEEE 15th Annual International Phoenix Conference on Computers and Communications (IEEE IPCCC 1996), Phoenix, AZ*, pp. 480–486.

51. A.C. Valera, W.K.G. Seah, and S.V. Rao. Cooperative packet caching and shortest multipath routing in mobile ad hoc networks. In *22nd Annual Joint Conference of the IEEE Computer and Communications Societies (INFOCOM 2003)*, San Francisco, California, March 30–April 3, 2003.

52. M. Marina and S. Das. On-demand multipath distance vector routing in ad hoc networks. In *9th International Conference on Network Protocols (ICNP)*, Mission Inn, Riverside, California, November 11–14, 2001, pp. 14–23.

53. Y.J. Lee and G.F. Riley. Dynamic NIx-vector routing for mobile ad hoc networks. In *Proceedings of the IEEE Wireless Communications and Networking Conference (WCNC 2005)*, New Orleans, 2005.

54. I. Chakeres and C. Perkins. Dynamic MANET on-demand routing protocol (DYMO), RFC 4728. Internet Draft, June 2008. Available at http://tools.ietf.org/html/draft-ietf-manet-dymo.

55. H. Bakht. Mobile ad hoc on-demand data delivery protocol (MAODDP). Available at http://tools.ietf.org/html/draft-bakht-maoddp-00.html (Accessed 01/07/2011).

56. H. Bakht. On-demand routing in mobile ad-hoc network. *Georgian Electronic scientific Journal: Computer Science and Telecommunications*, vol. 22, no. 5, 2009.

57. R.L. Gilaberte and L.P. Herrero. A secure routing protocol for ad hoc networks based on trust. In *Proceedings of the Third International Conference on Networking and Services ICNS '07*, Athens, Greece, June 19–25, 2007.

58. S. Murthy and J.J. Garcia-Lunca-Aceves. An efficient routing protocol for wireless networks. *ACM Mobile Networks and Applications Journal*, vol. 1, no. 2 (Special Issue on Routing in Mobile Communication Networks), 1996, pp. 183–197.

59. V.D. Park and M.S. Corson. A highly adaptive distributed routing algorithm for mobile wireless networks. In *Proceedings of INFOCOMM'97*, April 1997.

60. C.E. Perkins and E.M. Royer. Ad hoc on-demand distance vector routing. In *2nd IEEE Workshop on Mobile Computing Systems and Applications*, February 1999, pp. 90–100.

61. E.M. Royer and C.K. Toh. A review of current routing protocols for ad-hoc mobile wireless networks. *IEEE Personal Communications Magazine*, April 1999, pp. 46–55.

62. P. Papadimitratos and Z.J. Haas. Secure routing for mobile ad hoc networks. In *Proceedings of the SCS Communication Networks and Distributed Systems Modeling and Simulation Conference (CNDS 2002)*, San Antonio, TX, January 27–31, 2002.

63. Y.-C. Hu, A. Perrig, and D. Johnson. Ariadne: a secure on-demand routing protocol for ad hoc networks. *Wireless Networks Journal*, vol. 11, no. 1–2, January 2005, pp. 21–38.

64. I. Vidal, C. García, I. Soto, and J.I. Moreno. *Servicios de Valor Añadido en Redes Móviles Ad-Hoc.* Departamento de Ingeniería Telemática, Universidad Carlos III de Madrid, Jornadas Telecom I+D 2003, Actas de las XIII Jornadas Telecom I+D 2003, Madrid, Spain, November 2003.

65. C.T. Calafate, M.P. Malumbres, P. Manzoni. Improving H.264 real-time streaming in MANETs through adaptive multipath routing techniques. In *IEEE Workshop on Adaptive Wireless Networks, Global Telecommunications Conference (GlobeCom)*, Dallas, Texas, November 29–December 3, 2004, pp. 433–441.

66. M. Bohge and W. Trappe. TESLA certificates: an authentication tool for networks of computer-constrained devices. In *Proceedings of 2003 ACM workshop on Wireless Security (WISE '03)*, San Diego, CA, August 2003.

67. Y.-C. Hu, D.B. Johnson, and A. Perrig. SEAD: secure efficient distance vector routing for mobile wireless ad hoc networks. *Ad Hoc Networks*, 1, 2003, pp. 175–192.

68. H. Bakht, M. Merabti, and R. Askwith. A study of routing protocols for mobile ad hoc networks. In *1st International Computer Engineering Conference*, Cairo, Egypt, December 2004.

69. S.R. Afzal, S. Biswas, J.-b. Koh, T. Raza, G. Lee, and D.-k. Kim. RSRP: a robust secure routing protocol for mobile ad hoc networks. In *Proceedings of Wireless Communications and Networking Conference (WCNC 2008)*, 2008, pp. 2313–2318. Available at http://ieeexplore.ieee.org/iel5/4489030/4489031/04489439.pdf

70. Mohammed Tarique, Anwar Hossain, Rumana Islam, C. Akram Hossian. Issues of Long-Hop and Short-Hop Routing in Mobile Ad Hov Networks: a Comprehensive Study, Network Protocols and Algorithms, Vol 2, No. 2, 2010.

71. Amitangshu Pal, Localization Algorthims in Wireless Sensor Networks: Current Approaches and Future Challenges, Network Protocols and Algorithms, Vol 2, No 1, 2010.

Chapter 5

Ad Hoc Routing Modeling and Mathematical Analysis

Jose Costa-Requena

Contents

Ad hoc networking is a technology that is still under development, and there are several proposals for defining the most suitable routing protocol. No single routing protocol proposed thus far performs optimally under the kind of dynamic conditions possible in ad hoc networks. We analyze the performance of existing ad hoc routing protocols using simulations and a test bed. Based on the results, the goal of this chapter is to analyze the use of a hybrid routing approach for ad hoc networks. However, rather than proposing another protocol, this study considers the use of the well-known routing protocol ad hoc on-demand distance vector (AODV), with a new broadcast algorithm to accommodate the new routing design. The contribution of the nodes to the routing functionality is critical for establishing ad hoc networks. Therefore, in the last part of this work, we analyze the incentives to participate in the routing functions by using game theory. We conclude that a novel architecture that integrates a rewarding mechanism for the participating nodes with the routing protocol is required to deploy ad hoc networks. This solution facilitates the cooperation of the nodes in the routing functionality of ad hocnetworks.

5.1 Introduction

Ad hoc networks are envisioned as a key technology for ubiquitous networking. It is a suitable technology for embedded network devices in multiple environments, such as vehicles, mobile telephones, and personal appliances. As an infrastructure-less technology, it will allow users to create their own personal area networks. The benefit of ad hoc networks is that users can create the network automatically when needed and tear it down when it is not required anymore. The network can be created at any point in time for any communication purpose including leisure, military, or disaster situations. Ad hoc networks have an undefined lifetime since they can be up and running momentarily or permanently as long as there is a group of users that are willing to be part of the network. In ad hoc networks, the link state information changes whenever users move and create interferences to each other. Ad hoc networks are self-established without previous knowledge of the environment. Ad hoc nodes require a set of mechanisms to allow the devices to be autonomously integrated and configured as part of the ad hoc network.

Network scalability is the ability to expand or reduce the number of nodes and the size of the network while maintaining similar performance for each user. Ad hoc nodes have to perform the routing functionality and maintain the network topology information while keeping track of the connection with other nodes. They must also be able to react fast to network changes and dynamically adapt to the new topology. Therefore, the overall ad hoc network performance is affected by the size of the network, the number of nodes, the node mobility, and resources.

Ad hoc nodes cannot rely on a fixed server that provides information about the services available in the ad hoc network. Therefore, each node needs its own mechanism to discover the network capabilities and configure itself to the services available in the ad hoc network. Besides these issues, ad hoc networks have to interconnect with other IP-based technologies such as fixed wireless local area networks and 3G networks. For that reason, ad hoc nodes have to act as routers and constantly search for services available in the networks. The nodes that become part of ad hoc networks contribute to the overall network performance while spending their own resources. This leads to a high energy consumption that exhausts the batteries of the nodes.

5.2 Ad Hoc Routing Protocols Analysis

5.2.1 Ad Hoc Routing Review

Ad hoc nodes act as routers that cannot rely on any fixed infrastructure devices such as gateways, Dynamic Host Configuration Protocol (DHCP), or Domain Name Service (DNS) for addressing assistance. Therefore, ad hoc nodes have to include all necessary routing and addressing functionalities themselves. This means that they must store all routing information and need a mechanism to discover the routes to other nodes that are outside the local subnetwork.

Scalable ad hoc networks require a hierarchical addressing structure, where the network is partitioned into subnetworks or clusters. Figure 5.1 represents a cluster-based network with four clusters.

A cluster-based network is a network divided into several clusters. Each cluster consists of a single cluster head and multiple cluster nodes. The cluster head is a node that performs the routing functionality assigned to gateways in fixed networks. When a cluster node needs to find a route to a destination node not located in the same cluster, it will contact the cluster head that acts as a gateway. The cluster head communicates with other cluster heads in different clusters to find the route to the destination node.

The communication between nodes in the same cluster is known as intracluster communication. Cluster heads establish the intercluster communication with nodes outside their own cluster. Cluster heads require additional resources to perform the gateway functionality. The cluster-based routing decreases the network reliability because the cluster head may become the bottleneck. Moreover, the algorithm for selecting the optimal cluster head from among the existing cluster nodes is cumbersome. Nevertheless, from a preliminary analysis of the evolution of the public

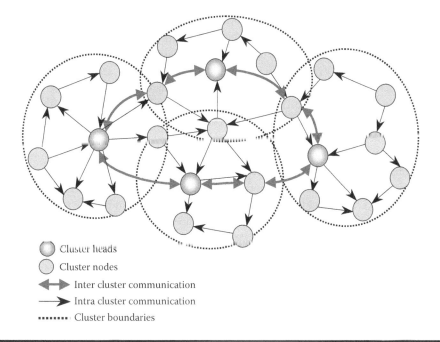

    ◯ Cluster heads
    ◯ Cluster nodes
    ◀▶ Inter cluster communication
    ⟶ Intra cluster communication
    ▪▪▪▪▪▪▪ Cluster boundaries

Figure 5.1   Cluster-based network routing.

Internet, a hypothesis can be formulated; a cluster-based routing protocol, where the changes in IP addresses and route updates are localized and do not span the entire network, is required to guarantee scalability in ad hoc networks.

The evolution path taken in the fixed Internet to solve the scalability problem might not be valid for ad hoc networks, and there is no mathematical analysis to prove that a cluster-based routing protocol is the only solution to make ad hoc routing scalable. Therefore, in order to verify this claim, the next section describes the state of the art of some of the existing ad hoc routing protocols and their performances. Ad hoc routing protocols can be classified into three categories: reactive, proactive, and hybrid.

5.2.2 Reactive Ad Hoc Routing

Reactive ad hoc routing protocols determine a path on demand only, meaning that they search for a single path when a message needs to be delivered. In this section, we briefly describe the ad hoc on-demand distance vector (AODV) [1], dynamic source routing (DSR) [2], and the temporally ordered routing algorithm (TORA) [3] as the most widely used reactive ad hoc routing protocols.

In AODV, the originating node initiates a *route request* (RREQ) message that is flooded through the network to the destination. The intermediate nodes in the route record the RREQ message. An RREP unicast message is sent back to the originating node as the acknowledgment following the reverse routes established by the received RREQ message. The intermediate nodes in the route also record the RREP message in their routing table for future use. Each node keeps the most recently used route information in its cache. Therefore, AODV is a simple protocol and does not require excessive resources on the nodes. However, the routing information available in the nodes is limited, and the route discovery process may take too much time. The initial RREQ is sent with time to live (TTL) = 1, and if no RREP is received within a certain time, the TTL is incremented and a new RREQ is sent. Thus, if the destination node is not close enough, the network is flooded several times during the RREQ process before a route is found or an error is notified.

DSR is similar to AODV, where RREQ and RREP messages are also used for discovering the route to the destination. The main difference is that in this case, these messages also include the entire path information (i.e., addresses of the intermediate nodes). The drawback is that the route information generates an overhead that can be excessive when the number of hops or node mobility increases.

TORA is a reactive routing protocol with some proactive enhancements where a link between nodes is established, creating a directed acyclic graph of the route from the source to the destination. The routing messages are distributed to a set of nodes following the graph around the changed topology. TORA provides multiple routes to a destination quickly with minimum overhead. In TORA, the optimal routes are of secondary importance versus the delay and overhead of discovering new routes.

5.2.3 Proactive Ad Hoc Routing

The proactive protocols are the traditional routing protocols used in fixed IP networks. These protocols maintain a table with the routing information and perform periodic updates to keep it consistent. In this section, we will introduce the destination sequenced distance vector routing (DSDV) [4] and the optimized link state routing (OLSR) [5] as the most representative proactive ad hoc routing protocols.

DSDV looks for the optimal path using the Bellman-Ford algorithm [6]. It uses a full dump or incremental packets to reduce the traffic generated by the routing updates in the network topology. However, it creates an excessive overhead because it constantly tries to find the optimal path.

OLSR defines multipoint relay (MPR) nodes for exchanging the routing information periodically. The nodes select the local MPR node that will announce the routing information to other MPR nodes in the network. The MPR nodes calculate the routing information for reaching other nodes in the network.

5.2.4 Hybrid Ad Hoc Routing

This section introduces a hybrid model that combines reactive and proactive routing protocols as well as a location-assisted routing protocol.

The zone routing protocol (ZRP) [7] is a hybrid routing protocol that divides the network into zones. The intrazone routing protocol implements the routing within the zone, while the interzone routing protocol implements the routing between zones. The ZRP provides a hierarchical architecture where each node has to maintain additional topological information, which requires extra memory.

The location aided routing (LAR) [8] is a location-assisted routing protocol that uses location information for the routing functionality. LAR works similarly to DSR, but it uses location information to limit the area where the RREQ is flooded. The originating node knows the neighbor's location and, based on that, selects the closest nodes to the destination as the next hop in the RREQ.

5.2.5 Ad Hoc Routing Evaluation

We have described different routing protocols, and based on the basic characteristics of reactive and proactive routing protocols, we can formulate a set of propositions. The propositions will consider the impact of system variables, such as used routing protocol type, node mobility, and number of nodes (i.e., node density), on performance measures such as routing overhead, percentage of packet loss, end-to-end packet delay, and percentage of optimal routes. At this stage, we are not able to indicate whether there is a linear or polynomial relationship between the system variables and the performance measures.

AODV, DSR, OLSR, and Topology Broadcast Reverse Forwarding (TBRF) are the experimental protocols standardized in the Internet Engineering Task Force (IETF) as reactive and proactive routing protocols. The routing protocols under consideration in this evaluation are AODV and OLSR as the most representative of the reactive and proactive categories. In our propositions, we assume that the following conditions do not change: bit rate, number of flows, and size of the ad hoc network. Let us now formulate the set of propositions using the notations introduced in Tables 5.1 and 5.2.

Proposition 5.1: Routing overhead increases with node mobility in both proactive and reactive routing protocols.

For
$$M_1 > M_2, \Omega_p(M_1) > \Omega_p(M_2) \tag{5.1.1}$$

For
$$M_1 > M_2, \Omega_R(M_1) > \Omega_R(M_2) \tag{5.1.2}$$

For
$$M > M_{\text{threshold}}, \Omega_p(M) > \Omega_R(M) \geq 0 \tag{5.1.3}$$

Table 5.1 System Variables

Proactive Routing Protocol	Proactive Routing Protocol and UDP (User Datagram Protocol) Flows	Proactive Routing Protocol and TCP (Transmission Control Protocol) Flows	Reactive Routing Protocol	Reactive Routing Protocol and UDP Flows	Reactive Routing Protocol and TCP Flows	Number of Nodes in the Network or Node Density	Node Mobility
P	Pu	Pt	R	Ru	Rt	N	M

Table 5.2 Performance Metrics

Routing Overhead	End-to-End Packet Delay	Percentage of Packet Loss	Percentage of Optimal Rroutes
W	D	L	Π

M_1 and M_2 represent different values for mobility. The derivatives $\Omega_P'(M) \geq 0$ and $\Omega_R'(M) \geq 0$ are used to demonstrate that overhead function increases with mobility, and these derivatives will be applied in the mathematical analysis in the rest of the section.

The routing overhead increases with node mobility due to the extra route discovery transactions generated in reactive protocols and the route updates required in proactive routing protocols. We expect that the routing overhead of proactive routing protocols increases more than the routing overhead of reactive protocols, because the route updates need to span all nodes when links break due to mobility. We assume that the routing overhead of reactive routing protocols is lower than the routing overhead of proactive protocols because only the existing routes need to be reestablished during a link break.

Proposition 5.2: End-to-end packet delay increases with node mobility in both proactive and reactive routing protocols.

For
$$M_1 > M_2, D_P(M_1) > D_P(M_2) \tag{5.2.1}$$

For
$$M_1 > M_2, D_R(M_1) > D_R(M_2) \tag{5.2.2}$$

For
$$M > M_{\text{threshold}}, D_P(M) > D_R(M) \geq 0 \tag{5.2.3}$$

M_1 and M_2 represent different values for mobility. The derivatives $D_P'(M) \geq 0$ and $D_R'(M) \geq 0$ are used to demonstrate that delay function increases with mobility, and these derivatives will be applied in the mathematical analysis in the rest of the section.

In proactive routing protocols, the end-to-end packet delay increases when there is network congestion because of the increment in the number of transactions required to exchange topology information with all the nodes. The end-to-end packet delay increases with node mobility in reactive routing protocols because of the increment of route discovery transactions. We expect that the packet delay in reactive routing protocols is lower than in proactive protocols because the route information is fresh since it is acquired right before starting the flow. We assume that the packet delay in proactive routing protocols is higher than in reactive protocols because the routing information may be stale when starting the packet flow, and the link breaks due to mobility create additional traffic, increasing the congestion in all nodes.

Proposition 5.3: Percentage of packet loss increases with node mobility in both proactive and reactive protocols.

For
$$M_1 > M_2, L_P(M_1) > L_P(M_2) \tag{5.3.1}$$

For
$$M_1 > M_2, L_R(M_1) > L_R(M_2) \tag{5.3.2}$$

For
$$M > M_{\text{threshold}}, L_P(M) > L_R(M) > 0 \tag{5.3.3}$$

M_1 and M_2 represent different values for mobility. The derivatives $L_P'(M) \geq 0$ and $L_R'(M) \geq 0$ are used to demonstrate that packet loss function increases with mobility, and these derivatives will be applied in the mathematical analysis in the rest of the section.

When mobility increases, links are more frequently broken and the percentage of packet loss increases. We expect that mobility will increase the link breaks, which, in proactive protocols, will result in additional traffic and congestion in all nodes. The reactive protocols have fresher routing information when starting the packet flow, which will result in lower packet loss than in proactive protocols.

Proposition 5.4: Percentage of optimal routes decreases in both proactive and reactive routing protocols when node mobility increases.

For
$$M_1 > M_2, \Pi_P(M_1) < \Pi_P(M_2) \tag{5.4.1}$$

For
$$M_1 > M_2, \Pi_R(M_1) < \Pi_R(M_2) \tag{5.4.2}$$

M_1 and M_2 represent different values for mobility. The derivatives $\Pi_P'(M) \leq 0$ and $\Pi_R'(M) \leq 0$ are used to demonstrate that the optimal route function decreases with mobility, and these derivatives will be applied in the mathematical analysis in the rest of the section.

When the nodes move, new, shorter routes may appear, and it takes time for a routing protocol to discover those optimal routes. This problem occurs more often when node mobility increases.

Proposition 5.5: Percentage of optimal routes obtained with proactive routing protocols is higher than with reactive protocols.

$$\Pi_P(M) > \Pi_R(M) \tag{5.5.1}$$

The routing protocols obtain the network topology based on periodic routing updates (i.e., proactive) or on-demand route discovery (i.e., reactive). The proactive routing protocols apply an additional algorithm over the discovered routes to select the most optimal route (e.g., lower number of hops). As a consequence, proactive routing protocols obtain a higher percentage of optimal routes compared to the routes obtained with reactive routing protocols. When mobility increases, the routes obtained become stale due to frequent link breaks.

Proposition 5.6: Routing overhead increases with the number of nodes in both proactive and reactive routing protocols.

For
$$N_1 > N_2, \Omega_P(N_1) > \Omega_P(N_2) \tag{5.6.1}$$

For
$$N_1 > N_2, \Omega_R(N_1) > \Omega_R(N_2) \tag{5.6.2}$$

N_1 and N_2 represent different values for the number of nodes. The derivatives $\Omega_P'(N) \geq 0$ and $\Omega_R'(N) \geq 0$ are used to demonstrate that the routing overhead function increases with the number of nodes, and these derivatives will be applied in the mathematical analysis in the rest of the section.

The proactive routing protocols have to share the routing information with all the other nodes in the network, which increases the routing information per node as a function of the total number of nodes in the network. The reactive routing protocols have to increase the TTL in the RREQ to reach all the nodes in the network. Therefore, when the node density increases, the RREQs are sent by a higher number of nodes, but few of the messages reach new nodes, thus decreasing the route discovery efficiency.

Proposition 5.7: For the same number of nodes and mobility conditions, the routing overhead is higher in proactive than in reactive protocols.

$$\Omega_P(M, N) \geq \Omega_R(M, N) \tag{5.7.1}$$

The routing overhead increases with the number of nodes due to the additional topology information required in proactive protocols and the additional RREQs forwarded by each of the intermediate nodes in reactive protocols.

Proposition 5.8: End-to-end packet delay increases with the number of nodes in both proactive and reactive routing protocols.

For
$$N_1 > N_2, D_P(N_1) > D_P(N_2) \tag{5.8.1}$$

For
$$N_1 > N_2, D_R(N_1) > D_R(N_2) \tag{5.8.2}$$

N_1 and N_2 represent different values for the number of nodes. The derivatives $D_P'(N) \geq 0$ and $D_R'(N) \geq 0$ are used to demonstrate that the delay function increases with the number of nodes, and these derivatives will be applied in the mathematical analysis in the rest of the section.

In this proposition, N denotes both the density and the number of nodes on the end-to-end path.

Proposition 5.9: Percentage of packet loss increases with the number of nodes in both proactive and reactive routing protocols.

For
$$N_1 > N_2, L_P(N_1) > L_P(N_2) \tag{5.9.1}$$

For
$$N_1 > N_2, L_R(N_1) > L_R(N_2) \tag{5.9.2}$$

N_1 and N_2 represent different values for the number of nodes. The derivatives $L_P'(N) \geq 0$ and $L_R'(N) \geq 0$ are used to demonstrate that the packet loss function increases with the number of nodes, and these derivatives will be applied in the mathematical analysis in the rest of the section.

When the number of nodes increases, the network gets congested because of the additional signalling, causing an increment of the packet delay and a percentage of packet loss. According to Proposition 5.1, the routing overhead increases with mobility, and therefore the throughput will decrease, reducing the available bandwidth and increasing the percentage of packet loss.

Proposition 5.10: Percentage of optimal routes obtained with proactive and reactive routing protocols decreases with the number of nodes.

For
$$N_1 > N_2, \Pi_P(N_1) < \Pi_P(N_2) \tag{5.10.1}$$

For
$$N_1 > N_2, \Pi_R(N_1) < \Pi_R(N_2) \tag{5.10.2}$$

N_1 and N_2 represent different values for the number of nodes. The derivatives $\Pi_P'(N) \leq 0$ and $\Pi_R'(N) \leq 0$ are used to demonstrate that the optimal route function decreases with the number of nodes, and these derivatives will be applied in the mathematical analysis in the rest of the section.

When calculating the optimal routes, increasing the number of nodes will decrease the efficiency of the protocols because of the additional topology information collected from all the nodes that has to be processed.

5.2.6 Proactive versus Reactive Simulation Comparison

In the previous section, we formulated a number of propositions based on our qualitative understanding of the behavior of ad hoc routing protocols. In this section, we include the results from a large set of simulations, and in Section 5.2.8, we provide the measurements obtained from our test bed to seek confirmation of the accuracy of our propositions. In order to make the transformation from quantitative numeric results obtained from simulations to qualitative statements, we fit the simulation results into parametric equations that minimize the approximation error.

The purpose of the parametric equations is not to reflect the behaviors of all ad hoc networks under certain conditions. However, the goal is to qualitatively explore the behavior of ad hoc networks under different routing protocols in order to have a good understanding of the design trade-offs of routing protocols. Therefore, we use both simulations and measurements to study the behavior. Based on our own experience, we believe that too many simulation results have been published that fit poorly to the measured behavior gained from a test bed or a real network. The limitation of measurements, on the other hand, is that generalizing the results is difficult. Therefore, we do not believe that it would be possible to propose a grand theory and verify it with the means in our disposal. However, our aim is to improve on routing protocol design and justify design choices without having such a theory by using both measurements and simulations and by explaining the differences between the two and thus verifying our work on a qualitative level.

In this section, simulation results justifying the advantages and disadvantages of the reactive and proactive ad hoc routing protocols will be presented [9]. The routing protocols comparison has been done using the ns-2 simulator [10], version 2.27 with standard IEEE 802.11 MAC protocol, which is used in the simulations and test bed included in this thesis. We also verify some of the propositions introduced in Section 5.2.5.

The results are obtained from the average of three simulation rounds performed continuously in order to reduce any possible effect due to the initialization process of the simulator. In simulations, we consider the following parameters:

- Simulation area: 1500 m × 300 m
- Simulation time: 900 s
- Traffic flows:
  - Constant bit rate with UDP transport: 20 IP unidirectional flows
  - Traffic with TCP transport: 20 IP unidirectional flows
- Connection rate: 8 packets/s
- Packet size: 65 bytes
- Number of nodes: 50 nodes using random waypoint mobility pattern
- Pause time between node movements: 0, 30, 60, 120, 300, 600, and 900 s

In simulations, we consider mobility as the average speed of the node during the simulation.

$$M = \frac{M_{max} t_{moving} + 0 t_{pause}}{t_{simulation}} = \frac{M_{max} t_{moving}}{t_{simulation}} \text{ where } M = M_{max}\big|_{t_{moving}=t_{simulation}} \text{ and } M = 0\big|_{t_{moving}=0}$$

We run simulations with the same parameters but using either UDP or TCP as the transport protocol for the traffic flows to compare the effect of congestion and reliable traffic control mechanisms.

The literature shows that different mobility patterns affect the performance results of ad hoc networks [11]. Ad hoc networks will be deployed under different mobility patterns, and the routing protocols will have to perform in different environments. Therefore, in simulations, the nodes follow a different mobility pattern after each waiting time, as characterized in the random waypoint model.*

The simulations are made considering that the network is handling the traffic generated by 20 active connections transmitting 8 packets/s. The simulations reflect the performance of ad hoc networks with real-time applications under different mobility conditions and using different routing and transport protocols. The simulations last for 900 s; thus, a pause time of 900 s is equivalent to static nodes that do not move during the simulation.

Both reactive (i.e., AODV, TORA, and DSR) and proactive (i.e., DSDV and OLSR) routing protocols are covered in simulations. The simulation results presented in this section are inaccurate due to the random behavior of the nodes. Therefore, a deeper analysis will be made, extracting from each simulation the associated equation for the most representative reactive (i.e., AODV) and proactive (i.e., OLSR) routing protocols and specific transport protocol (i.e., TCP or UDP).

The simulation results can be associated with an equation that can be linear $f(x) = cx + b$, polynomial $f(x) = b + c_1 x + c_2 x^2 + \ldots + c_n x^n$, logarithmic $f(x) = c \ln x + b$, or exponential $f(x) = c\, e^{bx}$. The constants c and b of these equations are adjusted by using the r^2 value

---

* It has been demonstrated that the random waypoint model is not the most accurate mobility pattern, but we will use it for simplicity assuming that it is good enough.

$$r^2 = 1 - \frac{\sum (Y_i - \hat{Y}_i)^2}{\sum Y_i^2 - \left[\left(\sum Y_i\right)^2 / n\right]}$$

where Y_i represents the value obtained in the simulation and $\hat{Y}_i$ represents the estimated value from the associated equation. The r^2 value represents the approximation error; thus, it tends to 1 when the values from the simulation and the associated equation match. In the following sections, each simulation is associated with the equation that provides the lowest approximation error r^2.

5.2.6.1 Simulation Results on Mobility

Figure 5.2 shows the routing overhead generated by reactive and proactive routing protocols during the simulation time versus node mobility with UDP traffic flows.

Proactive protocols have a higher routing overhead than do reactive protocols, which can be caused by the additional topology information they exchange. In particular, AODV generates less routing overhead compared to OLSR in similar conditions.

From the different equations that can be associated with the results of the AODV routing overhead with UDP traffic flows, the one with the lowest approximation error $r^2 = 0.976$ is

$$\Omega_{\mathrm{Ru}}(M) = 120.9\,e^{0.025M} \quad \text{(Kbytes)} \tag{5.1}$$

The first derivative is

$$\Omega_{\mathrm{Ru}}{}'(M) = \frac{\mathrm{d}(\Omega_{\mathrm{Ru}})}{\mathrm{d}M} = 3.02\,e^{0.025M} = \left. \begin{array}{c} 3.02 \big|_{M \to 0} \\[4pt] +\infty \big|_{M \to \infty} \end{array} \right\} \geq 0$$

proving (5.1.2).

The associated equation to the OLSR routing overhead simulation results with UDP traffic flows and the lowest approximation error $r^2 = 0.835$ is

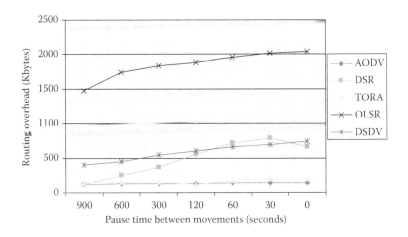

Figure 5.2   Routing overhead versus node mobility.

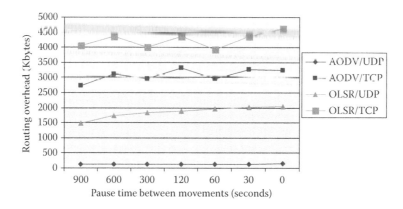

Figure 5.3   Routing overhead versus node mobility and transport protocol.

$$\Omega_{Pu}(M) = 1521 e^{0.047M} \text{ (Kbytes)} \qquad (5.2)$$

The first derivative is

$$\Omega_{Pu}'(M) = \frac{d(\Omega_{Pu})}{dM} = 71.4\, e^{0.047M} = \left. \frac{71.4}{+\infty} \right|_{\substack{M \to 0 \\ M \to \infty}} \geq 0$$

proving (5.1.1).

Figure 5.3 shows the routing overhead in AODV and OLSR using a transport protocol that includes reliability and congestion mechanisms such as TCP. The routing overhead increases in both AODV and OLSR compared with UDP traffic flows.

From the different equations that can be associated with the results of the AODV routing overhead with TCP traffic flows, the one with the lowest approximation error $r^2 = 0.456$ is

$$\Omega_{Rt}(M) = 2813.1 e^{0.022M} \text{ (Kbytes)} \qquad (5.3)$$

The first derivative is

$$\Omega_{Rt}'(M) = \frac{d(\Omega_{Rt})}{dM} = 61.88 e^{0.022M} = \left. \frac{61.88}{+\infty} \right|_{\substack{M \to 0 \\ M \to \infty}} \geq 0$$

proving (5.1.2).

The associated equation to the OLSR routing overhead simulation results with TCP traffic flows and the lowest approximation error $r^2 = 0.244$ is

$$\Omega_{Pt}(M) = 4014.7 e^{0.013M} \text{ (Kbytes)} \qquad (5.4)$$

The first derivative is

$$\Omega_{Pt}'(M) = \frac{d(\Omega_{Pt})}{dM} = 52.19 e^{0.013M} = \left. \frac{52.19}{+\infty} \right|_{\substack{M \to 0 \\ M \to \infty}} \geq 0$$

proving (5.1.1) and (5.1.3).

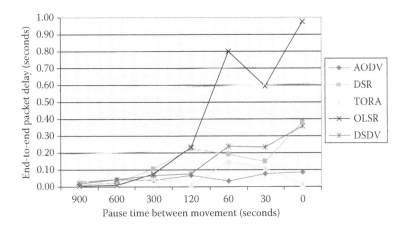

Figure 5.4 **End-to-end packet delay versus node mobility.**

The associated equations to AODV and OLSR using UDP are more accurate than the same equations when using TCP (i.e., higher r^2 value), and they show that proactive protocols have higher routing overhead than do reactive protocols under similar conditions, as stated in (5.1.3).

Figure 5.4 shows the end-to-end packet delay generated by reactive and proactive routing protocols during the simulation time versus node mobility with UDP traffic flows. In high mobility conditions, proactive routing protocols such as OLSR present higher delays than reactive routing protocols, as stated in (5.2.3). In the case of low mobility, the performance of reactive and proactive routing protocols is similar.

Node mobility affects the end-to-end packet delay because of different reasons such as network congestion and loss of connectivity. Network congestion increases with mobility due to the link breaks that generate new topology updates in proactive protocols and the additional RREQs initiated in reactive protocols. The connectivity is immediately reestablished after the link break by reactive protocols, but this is also performed after a periodic route update in proactive protocols.

The associated equation to the AODV end-to-end packet delay simulation results with UDP traffic flows and the lowest approximation error $r^2 = 0.625$ is

$$D_{\mathrm{Ru}}(M) = 0.008M + 0.021 \text{ (s)} \tag{5.5}$$

The first derivative is

$$D_{\mathrm{Ru}}'(M) = \frac{\mathrm{d}(D_{\mathrm{Ru}})}{\mathrm{d}M} = 0.008 \geq 0$$

proving (5.2.2).

The associated equation to the OLSR end-to-end packet delay simulation results with UDP traffic flows and the lowest approximation error $r^2 = 0.851$ is

$$D_{\mathrm{Pu}}(M) = 0.172M - 0.302 \text{ (s)} \tag{5.6}$$

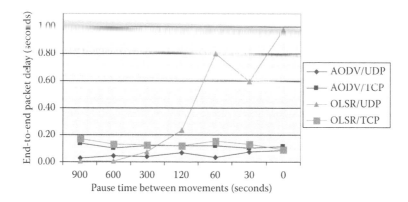

Figure 5.5   End-to-end packet delay versus node mobility and transport protocol.

The first derivative is

$$D_{P_u}'(M) = \frac{d(D_{P_u})}{dM} = 0.172 \geq 0$$

proving (5.2.1).

In (5.6), when $M = 0$, we obtain a negative value for the end-to-end packet delay $D_{pu}(0) = -0.302$, representing an approximation error.

Figure 5.5 shows that the end-to-end packet delay is reduced using TCP as a transport protocol. This can be due to the fact that with TCP, both ends maintain a connection state; thus, they will notice a link break immediately and either trigger a route update earlier than the normal periodic update or recalculate an alternative route in the routing table. The difference in reactive protocols when using either UDP or TCP is minor, because reactive protocols do not maintain routing tables. They do not have alternative routes available to reroute the traffic, and they just issue a RREQ when needed. The reactive protocols have similar behavior with UDP and TCP, because they detect the link break immediately and initiate the route discovery to provide an alternative path.

The associated equation to the AODV end-to-end packet delay simulation results with TCP traffic flows and the lowest approximation error $r^2 = 0.26$ is

$$D_{R_t}(M) = 0.0025M + 0.127 \text{ (s)} \tag{5.7}$$

The first derivative is

$$D_{R_t}'(M) = \frac{d(D_{R_t})}{dM} = 0.0003 \geq 0$$

proving (5.2.2).

The associated equation to the OLSR end-to-end packet delay simulation results with TCP traffic flows and the lowest approximation error $r^2 = 0.44$ is

$$D_{P_t}(M) = 0.0076M + 0.1619 \text{ (s)} \tag{5.8}$$

The first derivative is

$$D_{P_t}'(M) = \frac{d(D_{P_t})}{dM} = 0.0012 \geq 0$$

proving (5.2.1).

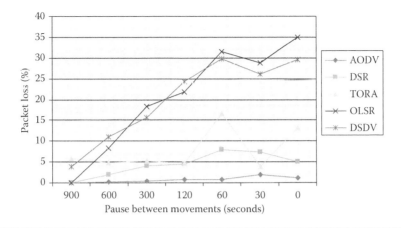

Figure 5.6 Percentage of packet loss versus node mobility.

In proactive protocols, the connection control in the traffic flow decreases the delay compared with nonreliable connections when using UDP as the transport protocol. The accuracy of the associated equations for UDP traffic flows is higher than the equations for TCP flows, but they still show that the end-to-end packet delay is higher in proactive routing protocols than in reactive routing protocols, as stated in (5.2.3).

Figure 5.6 shows the percentage of packet loss generated when reactive or proactive routing protocols are used during the simulation time versus node mobility with UDP traffic flows.

We measured the packet loss as the percentage of packets that did not reach the destination from the total number of packets sent. The percentage of packet loss is higher in the case of proactive routing protocols than in the case of reactive routing protocols and increases with mobility, as stated in Proposition 5.3.

The associated equation to the AODV percentage of packet loss simulation results with UDP traffic flows and the lowest approximation error $r^2 = 0.881$ is

$$L_{\mathrm{Ru}}(M) = 0.083\,e^{0.455M}\ (\%)\tag{5.9}$$

The first derivative is

$$L_{\mathrm{Ru}}'(M) = \frac{\mathrm{d}(L_{\mathrm{Ru}})}{\mathrm{d}M} = 0.038\,e^{0.455M} = \left.\begin{array}{l}0.038\big|_{M\to 0}\\[4pt]+\infty\big|_{M\to\infty}\end{array}\right\} \geq 0$$

proving (5.3.2).

The associated equation to the OLSR percentage of packet loss simulation results with UDP traffic flows and the lowest approximation error $r^2 = 0.56$ is

$$L_{\mathrm{Pu}}(M) = 0.225\,e^{0.89M}\ (\%)\tag{5.10}$$

The first derivative is

$$L_{\mathrm{Pu}}'(M) = \frac{\mathrm{d}(L_{\mathrm{Pu}})}{\mathrm{d}M} = 0.2\,e^{0.89M} = \left.\begin{array}{l}0.2\big|_{M\to 0}\\[4pt]+\infty\big|_{M\to\infty}\end{array}\right\} \geq 0$$

proving (5.3.1).

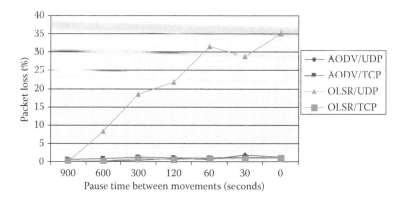

Figure 5.7   Percentage of packet loss versus node mobility and transport protocol.

Figure 5.7 shows that packet loss is reduced by using a transport protocol with connection control in the traffic flows (i.e., TCP).

The associated equation to the AODV end-to-end packet delay simulation results with TCP traffic flows and the lowest approximation error $r^2 = 0.488$ is

$$L_{Rt}(M) = 0.773\,e^{0.062M} \ (\%) \tag{5.11}$$

The first derivative is

$$L_{Rt}'(M) = \frac{d(L_{Rt})}{dM} = 0.048\,e^{0.062M} = \begin{matrix} 0.048\big|_{M\to 0} \\ +\infty\big|_{M\to\infty} \end{matrix} \ge 0$$

proving (3.2).

The associated equation to the OLSR end-to-end packet delay simulation results with TCP traffic flows and the lowest approximation error $r^2 = 0.779$ is

$$L_{Pt}(M) = 0.2418\,e^{0.221M} \ (\%) \tag{5.12}$$

The first derivative is

$$L_{Pt}'(M) = \frac{d(L_{Pt})}{dM} = 0.053\,e^{0.221M} = \begin{matrix} 0.053\big|_{M\to 0} \\ +\infty\big|_{M\to\infty} \end{matrix} \ge 0$$

proving (5.3.1).

TCP includes a connection control mechanism that reduces the end-to-end packet delay, as we can see when comparing (5.6) with (5.8), and it reduces packet loss as we can deduce from (5.10) and (5.12). Lower slopes in (5.11) than in (5.12) demonstrate that reactive protocols present shorter end-to-end packet delays than proactive routing protocols, proving (5.3.3).

Figure 5.8 shows the percentage of optimal routes obtained by reactive and proactive routing protocols during the simulation time versus node mobility. Proactive routing protocols perform better than reactive routing protocols when obtaining the optimal routes. Proactive routing

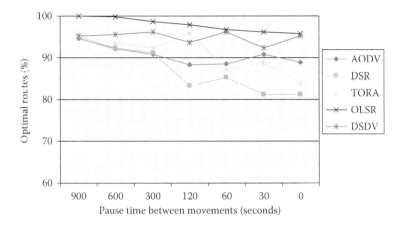

Figure 5.8 Percentage of optimal routes versus node mobility.

protocols maintain the routing information up to date and apply appropriate routing algorithms (e.g., shortest-path algorithm [6]). The percentage of optimal routes decreases in both reactive and proactive protocols with node mobility, as stated in Proposition 5.4.

The associated equation to the AODV percentage of optimal route simulation results with UDP traffic flows and the lowest approximation error $r^2 = 0.729$ is

$$\Pi_{Ru}(M) = 94.028 - 2.864\ln(M) \ (\%) \tag{5.13}$$

The first derivative is

$$\Pi_{Ru}{}'(M) = \frac{d(\Pi_{Ru})}{dM} = -\frac{2.864}{M} = \left.\begin{matrix} -\infty \big|_{M \to 0} \\ -0 \big|_{M \to \infty} \end{matrix}\right. \leq 0$$

proving (5.4.2).

The associated equation to the OLSR percentage of optimal route simulation results with UDP traffic flows and the lowest approximation error $r^2 = 0.902$ is

$$\Pi_{Pu}(M) = 100 - 2.381\ln(M) \ (\%) \tag{5.14}$$

The first derivative is

$$\Pi_{Pu}{}'(M) = \frac{d(\Pi_{Pu})}{dM} = -\frac{2.381}{M} = \left.\begin{matrix} -\infty \big|_{M \to 0} \\ -0 \big|_{M \to \infty} \end{matrix}\right. \leq 0$$

proving (5.4.1).

Figure 5.9 shows that the percentage of optimal routes has increased in reactive and proactive routing protocols when using a transport protocol with connection control in the traffic flows such as TCP.

The associated equation to the AODV percentage of optimal route simulation results with TCP traffic flows and lowest approximation error $r^2 = 0.504$ is

$$\Pi_{Rt}(M) = 96.85 - 2.708\ln(M) \ (\%) \tag{5.15}$$

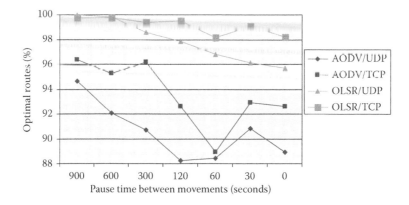

Figure 5.9 Percentage of optimal routes versus node mobility and transport protocol.

The first derivative is

$$\Pi_{Rt}'(M) = \frac{d(\Pi_{Rt})}{dM} = -\frac{2.708}{M} = \begin{matrix} -\infty|_{M\to 0} \\ -0|_{M\to\infty} \end{matrix} \le 0$$

proving (5.4.2).

The associated equation to the OLSR percentage of optimal route simulation results with TCP traffic flows and the lowest approximation error $r^2 = 591$ is

$$\Pi_{Pt}(M) = 100 - 0.7653\ln(M) \; (\%) \tag{5.16}$$

The first derivative is

$$\Pi_{Pt}'(M) = \frac{d(\Pi_{Pt})}{dM} = -\frac{0.7653}{M} = \begin{matrix} -\infty|_{M\to 0} \\ -0|_{M\to\infty} \end{matrix} \le 0$$

proving (5.4.1).

The associated equations show that 100% of the routes obtained with the proactive protocol can be optimal in the case of zero node mobility as compared with the case of reactive protocols, where with similar conditions, only 94% of the routes obtained are optimal, which proves Proposition 5.5. We can see that using a connection control transport protocol increases the percentage of optimal routes in reactive [(5.13) and (5.15)] and proactive [(5.14) and (5.16)] protocols. When the connection control detects a link break, it triggers either a route recalculation in proactive protocols or a route discovery in reactive protocols. However, proactive protocols obtain a higher percentage of optimal routes than do reactive protocols, as stated in (5.5.1).

5.2.6.2 Simulation Results on Scalability

We have verified some of the propositions based on the results from the simulations, but the scalability effect on the routing protocols when increasing the number or density of nodes remains to be demonstrated. The simulator has some limitations in terms of the number of nodes (i.e., maximum number of nodes is 100). Therefore, in order to study the impact on the performance

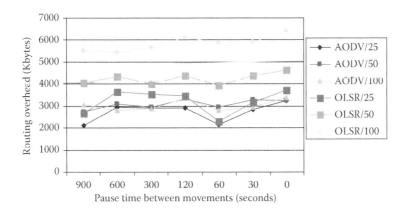

Figure 5.10   Routing overhead in reactive and proactive routing with 25, 50, and 100 nodes.

results when increasing the number of nodes, new simulations were performed with 25, 50, and 100 nodes, keeping the same value for the rest of the parameters. We selected TCP as the transport protocol for these simulations, because it provides similar results for proactive and reactive protocols regarding end-to-end packet delay and packet loss. However, we have to consider that the connection control mechanism in TCP creates additional overhead.

Simulation results presented in Figure 5.10 show that the routing overhead increases with the number of nodes in both proactive and reactive routing protocols, as stated in Proposition 5.6.

The associated equations with the AODV routing overhead simulation results for the different number of nodes with TCP traffic flows and the lowest approximation error $r^2 = 0.45$ are

$$\Omega_{Rt}(M, N = 25) = 2378.2\,e^{0.032M} \text{ (Kbytes)} \tag{5.17}$$

$$\Omega_{Rt}(M, N = 50) = 2813.1e^{0.022M} \text{ (Kbytes)} \tag{5.18}$$

$$\Omega_{Rt}(M, N = 100) = 2880.2\,e^{0.013M} \text{ (Kbytes)} \tag{5.19}$$

In reactive routing protocols, the routing overhead increases with the number of nodes, as stated in Proposition 5.6. The simulation results could be associated with linear equations, but it has a higher approximation error than the exponential equation. A major increase in the routing overhead takes place when incrementing from 25 (i.e., 2378.2 Kbytes) to 50 (i.e., 2813.1 Kbytes) nodes, while the values for 50 and 100 nodes are similar.

Next, we define a generic equation that includes both mobility and the number of nodes as variables. We take the equations obtained from the simulations for 25, 50, and 100 nodes, with mobility as the only variable, and we associate them with an equation that can be linear, polynomial, logarithmic, or exponential, depending on the associated error. The generic equation associated with the results of the AODV routing overhead with TCP traffic flows is drawn up using Equations 5.17 through 5.19 and obtaining the associated equation for the bases (i.e., 2378.2, 2,813.1, and 2,880.2) and the slope factors (i.e., 0.032, 0.022, and 0.013) with the lowest approximation error, which results in

$$\Omega_{Rt}(M, N) = (2188 + 251N)\,e^{(0.04+0.009N)M} \text{ (Kbytes)} \tag{5.20}$$

When comparing (5.18) and $\Omega_{Ru}(M) = 120.9e^{0.025M}$ (Kbytes) obtained to model the routing overhead for 50 nodes using TCP and UDP, respectively, we see that the results are different. This is due to the additional overhead in TCP compared with UDP. To model the routing overhead using UDP and considering as variables the mobility and the number of nodes, we take (5.20) and (5.18) as a reference to estimate the generic equation associated with the AODV routing overhead with UDP. The base of the equation with TCP changes from 2,188 in (5.18) to 2,813.1 in (5.20), which means an increment of 28.57%; thus, we can estimate that for UDP it will be $\Omega_{Ru}(M, N) = 155.4e^{0.025M}$ (Kbytes). The slope of the equation changes from 0.022 in (5.18) to 0.04 in (5.20), which means an increment of 81.82%; thus, we estimate that for UDP it will be $\Omega_{Ru}(M, N) = 155.4e^{0.045M}$ (Kbytes). The slope we obtain with UDP is similar to the one in (5.20), so we could extend the factor associated with N for UDP with the same value for TCP as in (5.20). We estimate that for UDP the final slope is $\Omega_{Ru}(M, N) = 155.4e^{(0.045+0.009)M}$ (Kbytes). The base of (5.18) for TCP is 2,813.1, which is 23.27 times bigger than the base of (5.1) for UDP. Therefore, we used the factor associated with N for TCP in (5.20) as a reference (i.e., $120N$) to estimate a similar value for UDP. Thus, we modeled the routing overhead for UDP using (5.18), (5.20), and (5.1) as a reference, resulting in (5.21), which represents the AODV routing overhead generic equation with UDP traffic:

$$\Omega_{Ru}(M, N) = (155.4 + 5.1N)e^{(0.045+0.009N)M} \text{ (Kbytes)} \tag{5.21}$$

The associated equations with the OLSR routing overhead simulation results for the different number of nodes with TCP traffic flows and the lowest approximation error $r^2 = 0.24$ are

$$\Omega_{Pt}(M, N = 25) = 3027.7\, e^{0.012M} \text{ (Kbytes)} \tag{5.22}$$

$$\Omega_{Pt}(M, N = 50) = 4014.7\, e^{0.013M} \text{ (Kbytes)} \tag{5.23}$$

$$\Omega_{Pt}(M, N = 100) = 5297.4\, e^{0.024M} \text{ (Kbytes)} \tag{5.24}$$

In proactive routing protocols, the routing overhead significantly increases with the number of nodes, as stated in Proposition 5.6. From the associated equations, the routing overhead value roughly increases by 1,000 Kbytes when doubling the number of nodes. The slope factor doubles when the number of nodes increases from 25 to 100.

The generic equation associated with the OLSR routing overhead with TCP traffic flows is drawn up using Equations 5.22 through 5.24 and obtaining the associated equation with the lowest approximation error, which results in

$$\Omega_{Pt}(M, N) = (1843 + 1134N)e^{(0.0037+0.0065N)M} \approx (1850 + 1130N)e^{(0.004+0.0065N)M} \text{ (Kbytes)} \tag{5.25}$$

When comparing (5.23) and $\Omega_{Pu}(M) = 1521e^{0.047M}$ (Kbytes) obtained to model the routing overhead for 50 nodes using TCP and UDP, respectively, the results are different. Both the base and slope factors are three times lower in UDP than in TCP. Thus, we modeled the routing overhead with UDP using (5.23), (5.25), and (5.2), resulting in (5.26), which represents the OLSR routing overhead generic equation with UDP traffic:

$$\Omega_{Pu}(M, N) = (615 + 375N)e^{(0.001+0.002N)M} \text{ (Kbytes)} \tag{5.26}$$

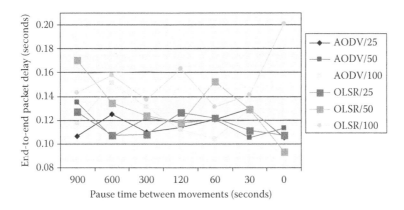

Figure 5.11   End-to-end packet delay in reactive and proactive routing with 25, 50, and 100 nodes.

Therefore, the routing overhead increases with the number of nodes, as stated in Proposition 5.6, and the proactive routing protocols present higher overhead than reactive protocols, as stated in (5.7.1). Increasing the number of nodes affects more of the proactive protocols routing overhead, while increasing the node mobility affects more of the reactive protocols routing overhead. For this reason, proactive routing protocols are not scalable in large ad hoc networks.

Figure 5.11 shows that the end-to-end packet delay is similar in the case of reactive and proactive routing protocols when the increase in the number of nodes is small (i.e., ±0.02 s end-to-end packet delay variation when $25 \leq N \leq 50$). When the number of nodes is increased (i.e., $N = 100$), the end-to-end packet delay is higher in proactive than in reactive routing protocols.

The associated equations with the AODV end-to-end packet delay simulation results for the different number of nodes with TCP traffic flows and the lowest approximation error $r^2 = 0.41$ are

$$D_{Rt}(M,\ N = 25) = 0.001M + 0.114\ \text{(s)} \tag{5.27}$$

$$D_{Rt}(M,\ N = 50) = 0.0025M + 0.127\ \text{(s)} \tag{5.28}$$

$$D_{Rt}(M,\ N = 100) = 0.0037M + 0.136\ \text{(s)} \tag{5.29}$$

The end-to-end packet delay is almost constant (i.e., between 114 and 136 ms for $M = 0$) for reactive routing despite an increase in the number of nodes when mobility is zero. However, the end-to-end packet delay increases with the number of nodes, as stated in Proposition 5.8.

The generic equation associated with the AODV end-to-end packet delay with TCP traffic flows is drawn up using Equations 5.27 through 5.29 and obtaining the associated equation with the lowest approximation error, which results in

$$D_{Rt}(M,\ N) = (0.0014N)M + 0.1 + 0.011N\ \text{(s)} \tag{5.30}$$

When comparing (5.28) and $D_{Pt}(M) = 0.008M + 0.21 (s)$ obtained to model the end-to-end packet delay for 50 nodes using TCP and UDP, respectively, the results are different. The values obtained with UDP in (5.5) are optimistic compared with (5.28), giving an end-to-end packet delay value of 21 ms when mobility is zero. The latest simulations using TCP provide more realistic values despite the higher approximation error. Thus, we modeled the end-to-end packet delay using the same (5.30), which represents the AODV end-to-end packet delay generic equation with UDP and TCP traffic:

$$D_{\text{Ru}}(M, N) = D_{\text{Rt}}(M, N) = D_{\text{R}}(M, N) = (0.0014N)M + 0.1 + 0.011N \text{ (s)}$$

The associated equations with the OLSR end-to-end packet delay simulation results for the different number of nodes with TCP traffic flows and the lowest approximation error $r^2 = 0.43$ are

$$D_{\text{Pt}}(M, N = 25) = 0.001M + 0.121E \text{ (s)} \tag{5.31}$$

$$D_{\text{Pt}}(M, N = 50) = 0.0076M + 0.161 \text{ (s)} \tag{5.32}$$

$$D_{\text{Pt}}(M, N = 100) = 0.0048M + 0.134 \text{ (s)} \tag{5.33}$$

From Equations 5.27 through 5.33, we observe that proactive and reactive protocols have similar end-to-end packet delays (i.e., between 114 and 136 ms delay for mobility zero), which contradicts (5.2.3). However, when the number of nodes is high ($N = 100$), the end-to-end packet delay in proactive routing protocols show more dependency with the mobility (i.e., a mobility incremental factor of 0.003) than in reactive routing protocols (i.e., a mobility incremental factor of 0.001).

The generic equation associated with the OLSR end-to-end packet delay with TCP traffic flows is drawn up using Equations 5.31 through 5.33 and obtaining the associated equation with the lowest approximation error $r^2 = 0.43$, which results in

$$D_{\text{Pt}}(M, N) = (0.0025N)M + 0.113 + 0.07N \text{ (s)} \tag{5.34}$$

When comparing (5.32) and $D_{\text{Pt}}(M) = 0.172M - 0.302$(s) obtained to model the end-to-end packet delay for 50 nodes using TCP and UDP, respectively, the results are considerably different, because UDP does not provide connection failure detection and so the routing protocol does not trigger a route update early enough. The latest simulations provide more realistic values despite the higher approximation error. Thus, we modeled the end-to-end packet delay using the same (5.34), which represents the OLSR end-to-end packet delay generic equation with UDP and TCP traffic.

$$D_{\text{Pu}}(M, N) = D_{\text{Pt}}(M, N) = D_{\text{p}}(M, N) = (0.0025N)M + 0.113 + 0.07N \text{ (s)}$$

Reactive and proactive routing protocols are not highly affected by the number of nodes from the end-to-end packet delay point of view. Proactive protocols present scalability issues when the number of nodes is high due to network congestion because of the additional routing overhead, as stated in Proposition 5.7.

Figure 5.12 shows that the percentage of packet loss increases with the mobility and the number of nodes in both reactive and proactive routing protocols. The left corner of Figure 5.12 shows that the percentage of packet loss in static conditions (i.e., the maximum mobility is represented in Figure 5.12 with zero pause time between movements) and for a small number or density of nodes (i.e., $N = 25$) is the same for reactive and proactive routing protocols. Moreover, when the

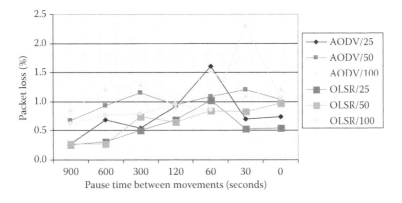

Figure 5.12 Percentage of packet loss in reactive and proactive routing with 25, 50, and 100 nodes.

number of nodes increases (i.e., $50 \leq N \leq 100$), the percentage of packet loss is higher for reactive routing protocols than for proactive routing protocols. This contradicts (5.3.3), which only stands in punctual cases with high mobility and number or density of nodes (i.e., OLSR with $N = 100$ and 30 pause time). This means that with regard to the percentage of packet loss, reactive routing protocols are less scalable than proactive routing protocols.

The associated equations with the AODV percentage of packet loss simulation results for the different number of nodes with TCP flows and the lowest approximation error $r^2 = 0.48$ are

$$L_{Rt}(M, N = 25) = 0.38 e^{0.146M} \ (\%) \tag{5.35}$$

$$L_{Rt}(M, N = 50) = 0.77 e^{0.062M} \ (\%) \tag{5.36}$$

$$L_{Rt}(M, N = 100) = 0.98 e^{0.036M} \ (\%) \tag{5.37}$$

Equations 5.35 through 5.37 show that the percentage of packet loss is low in reactive protocols, but it increases with the number of nodes, as stated in Proposition 5.9.

The generic equation associated with the AODV percentage of packet loss with TCP traffic flows is drawn up using Equations 5.35 through 5.37 and obtaining the associated equation for the bases (i.e., 0.38, 0.77, and 0.98) and the slope factors (i.e., 0.146, 0.062, and 0.036) with the lowest approximation error $r^2 = 0.48$, which results in

$$L_{Rt}(M, N) = (0.11 + 0.301N) e^{(0.192 - 0.05N)M} \ (\%) \tag{5.38}$$

When comparing (5.36) and $L_{Ru}(M) = 0.083 e^{0.455M}$ (%) obtained to model the packet loss for 50 nodes using TCP and UDP, respectively, the results are roughly 10 times lower with UDP than with TCP traffic. However, the dependency with the mobility is higher in UDP than in TCP as represented by the slope factor 0.445 in UDP versus 0.062 in TCP, which is 7 times lower. Thus, we modeled the AODV packet loss with UDP traffic using (5.38) as reference, resulting in (5.39), which represents the AODV packet loss generic equation with UDP traffic.

$$L_{Ru}(M, N) \approx (0.01 + 0.03N) e^{(1.34 + 0.35N)M} \ (\%) \tag{5.39}$$

The associated equations with the OLSR percentage of packet loss simulation results for the different number of nodes with TCP traffic flows and the lowest approximation error $r^2 = 0.77$ are

$$L_{P_t}(M, N = 25) = 0.283\,e^{0.143M} \ (\%) \tag{5.40}$$

$$L_{P_t}(M, N = 50) = 0.241e^{0.221M} \ (\%) \tag{5.41}$$

$$L_{P_t}(M, N = 100) = 0.551e^{0.137M} \ (\%) \tag{5.42}$$

The generic equation associated with the OLSR percentage of packet loss with TCP traffic flows is drawn up using Equations 5.40 through 5.42 and obtaining the associated equation with the lowest approximation error, which results in

$$L_{P_t}(M, N) \approx (0.091 + 0.134N)\,e^{(0.174 - 0.003N)M} \ (\%) \tag{5.43}$$

When comparing (5.41) and $L_{P_u}(M) = 0.0225e^{0.89M}(\%)$ obtained to model the packet loss for 50 nodes using TCP and UDP, respectively, the results show a major difference in the slope factor. However, assuming the inaccuracy of the simulations and the associated approximation error $r^2 = 0.77$, we can still use those results as a reference. Thus, we modeled the OLSR packet loss with UDP traffic flows using (5.43) as a reference, resulting in (5.44), which represents the generic equation associated with the OLSR packet loss:

$$L_{P_u}(M, N) = (0.09 + 0.13N)\,e^{(0.69 + 0.012N)M} \ (\%) \tag{5.44}$$

Figure 5.13 shows that the percentage of optimal routes obtained with reactive and proactive routing protocols with TCP traffic decreases with the number of nodes, as stated in Proposition 5.10.

Proactive routing protocols exchange topology information periodically and can implement different algorithms to optimize the routes. The reactive routing protocols implement route optimization during the RREQ based on the number of hops and sequence numbers to avoid loops.

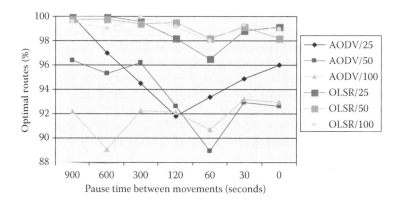

Figure 5.13   Percentage of optimal routes in proactive and reactive routing with 25, 50, and 100 nodes.

The associated equations with the AODV percentage of optimal route simulation results for the different number of nodes with TCP traffic flows and the lowest approximation error $r^2 = 0.504$ are

$$\Pi_{R_t}(M, N = 25) = 98.6 - 2.69\ln(M) \ (\%) \tag{5.45}$$

$$\Pi_{R_t}(M, N = 50) = 96.8 - 2.7\ln(M) \ (\%) \tag{5.46}$$

$$\Pi_{R_t}(M, N = 100) = 90.8 - 0.3\ln(M) \ (\%) \tag{5.47}$$

The generic equation associated with the AODV percentage of optimal routes with TCP traffic flows is drawn up using Equations 5.45 through 5.47 and obtaining the associated equation with the lowest approximation error, which results in

$$\Pi_{R_t}(M, N) = (103 - 3.9N) - (4.2 - 1.2N)\ln(M) \ (\%) \tag{5.48}$$

When comparing (5.46) and $\Pi_{R_u}(M) = 94.028 - 2.864\ln(M)$ (%) obtained to model the percentage of optimal routes for 50 nodes using TCP and UDP, respectively, the results are similar because the transport protocols used for the traffic flows do not affect obtaining optimal routes. Moreover, in both cases, the approximation error is similar: $r^2 = 0.729$ and $r^2 = 0.504$. Thus, we modeled the AODV percentage of optimal routes with UDP traffic flows using (5.48), which represents the generic equation associated with the AODV percentage of optimal routes:

$$\Pi_{R_u}(M, N) = \Pi_{R_t}(M, N) = \Pi_R(M, N) = (103 - 3.9N) - (4.2 - 1.2N)\ln(M) \ (\%)$$

The associated equations with the OLSR percentage of optimal route simulation results for the different number of nodes with TCP traffic flows and the lowest approximation error $r^2 = 0.61$ are

$$\Pi_{P_t}(M, N = 25) = 100 - 1.04\ln(M) \ (\%) \tag{5.49}$$

$$\Pi_{P_t}(M, N - 50) = 100 - 0.76\ln(M) \ (\%) \tag{5.50}$$

$$\Pi_{P_t}(M, N = 100) = 99.6 - 0.36\ln(M) \ (\%) \tag{5.51}$$

The generic equation associated with the OLSR percentage of optimal routes with TCP traffic flows is drawn up using Equations 5.49 through 5.51 and obtaining the associated equation with the lowest approximation error, which results in

$$\Pi_{P_t}(M, N) = (98.6 - 0.13N)\ln(M) \ (\%) \tag{5.52}$$

When comparing (5.50) and $\Pi_{P_u}(M) = 100 - 2.381\ln(M)$ (%) obtained to model the percentage of optimal routes for 50 nodes using TCP and UDP, respectively, the results show that the logarithmic factors are three times lower in TCP than UDP. However, we modeled the OLSR percentage of optimal routes using the more optimistic equation with the lower logarithmic factor and (5.52) to represent the generic equation associated with the OLSR percentage of optimal routes.

$$\Pi_{P_u}(M, N) = \Pi_{P_L}(M, N) = \Pi_{P}(M, N) = (98.6 - 0.13N)\ln(M) \ (\%)$$

In reactive protocols, the percentage of optimal routes decreases with the number of nodes, while in proactive protocols, the impact of the number of nodes is low. Therefore, when obtaining optimal routes, the reactive routing protocols are not scalable.

5.2.6.3 Complexity in Reactive and Proactive Routing Protocols

Table 5.3 compares reactive and proactive protocols in terms of complexity. The *storage complexity* indicates the size of the routing table required by each protocol. The *communication complexity* indicates the processing resources required to find routes or to perform a route update operation. N denotes the number or density of nodes in the ad hoc network, and complexity is represented with the O notation.

5.2.7 Ad Hoc Routing Protocol Simulation Conclusions

The reactive routing protocols under analysis have clear drawbacks such as the excessive flooding traffic in the route discovery and the route acquisition delay. When the network is congested, the routing information is lost and a consecutive set of control packets are issued to reestablish the links, increasing the routing latency (i.e., the time the routing protocol requires to obtain the route to the destination node) and the percentage of packet loss. If the Hello messages are not received, then error requests are issued and new RREQs are sent to reestablish the link. Thus, the reactive protocols do not scale when the load and node density increase. Moreover, the reactive routing protocols do not have knowledge of the Quality of Service (QoS) in the path before the route is established, and the routes are not optimized.

The reactive routing protocols suffer from high routing latency and percentage of packet loss, which increases with mobility and large networks. The percentage of optimal routes calculated with reactive protocols is lower than in proactive protocols, and it decreases in large networks. An advantage of reactive protocols such as AODV is that they maintain only the active routes in the routing table, which minimizes the memory required in the node. Moreover, the protocol itself is simple and so the computational requirements are minimal, extending the lifetime of the node in

Table 5.3 Comparison of Reactive and Proactive Routing Complexities

	Reactive Routing		Proactive Routing		
	AODV	DSR	OLSR	TORA	DSDV
Storage complexity	O(e)[a]	O(e)	O(N)[b]	O(N)	O(N)
Communication complexity	O(2N)[c]	O(2N)	O(N)[d]	O(N)	O(N)

[a]Requires maintaining in the cache only the most recently used routes.
[b]Requires maintaining tables with entries for all the nodes in the network.
[c]Requires additional route discovery and maintenance that increases with high mobility.
[d]Routing information is periodically maintained up to date in all the nodes.

the ad hoc network. The routing overhead is equivalent to additional packet processing; thus, reactive protocols will have lower power consumption than do proactive protocols. In simulations with a small number of nodes, AODV has a lower percentage of packet loss than does OLSR. Therefore, in networks with light traffic and low mobility, reactive protocols are scalable because of the small bandwidth and storage requirements.

The proactive routing protocols under analysis maintain topology information up to date with periodic update messages. The proactive routing protocols minimize the route discovery delay, which minimizes the percentage of packet loss since the routes are known in advance and no additional routing overhead and processing are required. However, under high mobility conditions, more and more routes established, based on the previous periodic update, become stale, leading to an increased percentage of packet loss.

The proactive routing protocols have low routing latency since all the routes are available immediately, even in large networks. The proactive routing protocols calculate the most optimal routes since they apply hop count-based routing algorithms. Proactive routing protocols have a higher percentage of packet loss than do reactive routing protocols in networks with a reduced number of nodes and high mobility, as depicted in Figure 5.6. However, if the transport protocol includes a connection control mechanism (i.e., TCP) that detects link breaks and triggers route updates or route recalculation, then proactive protocols present a lower percentage of packet loss than do reactive protocols, as depicted in Figure 5.12.

A drawback of proactive routing protocols is that they require a constant bandwidth and cause a processing overhead to maintain the routing information up to date. This overhead increases with the number of nodes and mobility since the updates have to be more frequent to maintain accurate routing information. The proactive routing protocols have lower routing latency, but they do not react quickly enough to topology changes. The proactive routing protocols have been enhanced toward hybrid and hierarchical solutions to deal with this scalability problem in ad hoc networks. OLSR reduces the control and processing overhead by selecting some nodes (i.e., MPR nodes) within the network to maintain the routing information. The link information updates are propagated between MPR nodes only, relieving the rest of the nodes from participating in the topology maintenance. Other optimizations consist of exchanging only the differential updates, implementing hybrid solutions such as ZRP [7] that combine reactive and proactive routing protocols or routing protocols that use the nodes' location data, such as LAR [8].

In order to analyze the performance of the hybrid protocols versus reactive and proactive protocols, we ran additional simulations in the ns-2 simulator with similar parameters.

- Simulation area: 1500 m × 300 m
- Transmitter range: 250 m and 2 Mbit bandwidth
- Simulation time: 900 s
- Constant bit rate traffic with UDP transport: 15 IP unidirectional connections
- Connection rate: 5 packets/s
- Packet size: 65 bytes
- Number of nodes: 50 nodes using random waypoint mobility pattern
- Pause time between node movements: 0, 30, 60, 120, 300, 600, and 900 s

Figures 5.14 and 5.15 show the results of the additional simulations, including hybrid routing.

Figures 5.14 and 5.15 show the throughput and routing overhead for AODV, DSR, DSDV, LAR, and ZRP, comparing two scenarios: zero node mobility and random pause time (i.e., static nodes and random mobility). Mobility similarly affects the throughput of the different routing

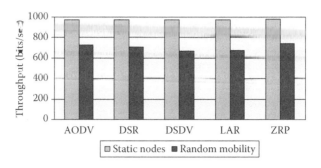

Figure 5.14   Throughput versus mobility in reactive, proactive, and hybrid routing.

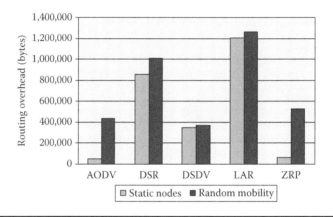

Figure 5.15   Routing overhead versus mobility in reactive, proactive, and hybrid routing.

protocols, while the routing overhead is different for both static and mobile nodes. The simulations have been executed for ZRP with a radius of 1 hop, and they show the same throughput results as for AODV. If we extend the ZRP radius to several hops, where proactive routing is used, then it will display behavior similar to DSDV, where the routing overhead is not affected by mobility. The routing overhead with static nodes is the same for AODV and ZRP, but it is 15% higher for ZRP with random mobility. LAR introduces the highest routing overhead for the same mobility conditions.

In addition to the hybrid routing protocols such as ZRP and LAR, other alternatives have been proposed to improve the reaction time to link breaks of the proactive routing protocols. One of them is a cross-layer architecture to receive information directly from the link layer in order to react quickly to topology changes when route breaks happen [12]. Despite this, when the network size increases, the bandwidth and processing overhead can still reach limits that cannot be afforded by ad hoc nodes. Another alternative consists of moving from a flat to a more scalable hierarchical routing as proposed in the fuzzy sighted link state (FSLS) routing [13]. FSLS defines a multilevel routing update hierarchy where each level has a different routing packet size and frequency of routing updates. FSLS minimizes the flooding traffic but increases the complexity when defining levels with different update frequencies. In this chapter, we will analyze a third alternative, which consists of a new hybrid routing approach based on AODV. AODV is extended with scalability optimizations in order to reduce the routing latency and the percentage of packet loss and to increase the routing efficiency when mobility, the number of nodes, or the network size increases.

The performance of ad hoc networks cannot be easily modeled due to the number of variables and the uncertainty of their values. In the literature, there are several attempts to provide a performance analysis of ad hoc networks based on an imprecise network state model. The existing models are not reliable due to the unpredictable behavior of the nodes. An accurate attempt to model the ad hoc networks' performance should consider the node mobility and the unpredictability of network conditions. However, our objective is to validate the hybrid routing protocol in terms of the impact on scalability rather than to define an accurate ad hoc network model. Therefore, we will define a generic ad hoc network model and apply the fully distributed virtual backbone (FDVB) concept on top of it.

Variables such as the location of the smart nodes within the network are relevant, but in order to simplify the model, we will consider an area of the network where the smart nodes appear randomly and remain stable for a certain period of time.

Our main objective is to prove that network scalability increases when we apply the FDVB concept and to determine the density of smart nodes required to build an optimal FDVB independent of the nodes' location. For this purpose, we defined the *smart nodes access control algorithm*.

In ad hoc networks, the nodes exhaust their resources because they perform packet forwarding and routing functions that, in fixed networks, are normally implemented in static servers or routers. In order to define a generic ad hoc network model, we will identify the metrics required to evaluate the performance.

We have identified the need to introduce smart nodes performing extra routing functionality in ad hoc networks. However, the preferred routing protocol to be implemented is the most critical part to improve scalability in ad hoc networks, and it remains to be selected.

Based on the simulation results and the test bed analysis, the combination of a reactive protocol that responds quickly to link breaks and a proactive protocol that provides optimal routes seems to be the optimal solution. Therefore, we consider that a hybrid approach is needed to overcome the drawbacks of existing routing protocols to scale up to large ad hoc networks. In the hybrid approach, the nodes are grouped into clusters and the cluster heads provide scalability by taking care of the heavy routing functionality between clusters. The drawbacks in cluster-based routing protocols are the additional complexity required in the nodes to implement the clustering algorithm. These protocols have additional overhead required for selecting the cluster head and having a single node acting as a bridge between clusters may become a bottleneck. The required hybrid solution should be dynamic to allow ordinary nodes to run reactive routing protocols while other smart nodes abstract the network and run a hybrid routing protocol (i.e., reactive together with proactive routing).

Each node interested and capable of becoming a cluster head (i.e., smart node) will create its own cluster and will try to become part of the distributed backbone. There should not be any cluster selection logic that forces the nodes to become cluster heads depending on their location (i.e., in the center of the cluster) or other metrics. The algorithm should allow the nodes to become cluster heads solely based on their available resources. A node can measure the environment (i.e., local traffic and channel utilization) and, based on its available resources, decide to become a cluster head or not. Therefore, there is no network-wide logic for selecting the cluster heads. Instead, any node can become a cluster head at any point in time. The nodes have the possibility of becoming cluster heads (i.e., smart) randomly, and they can fall back and act as cluster nodes (i.e., ordinary) after exhausting some of their resources. Thus, smart nodes have enough resources and willingness to maintain route and service information. Ordinary nodes are devices with limited resources, running an ad hoc MANET protocol with low complexity and computational requirements (i.e., a reactive protocol such as AODV).

Only the nodes that become cluster heads (i.e., smart nodes) will engage in additional control transactions for exchanging cluster information. The distributed backbone is composed of the smart nodes that exchange link state information between them in order to share the network topology information using a proactive protocol such as OLSR or DSDV or a reactive protocol such as AODV with new extension messages.

The cluster is set up by the TTL, and all the nodes that are close to the cluster head (i.e., nodes within TTL = 1 or 2) will be just ordinary nodes. The hybrid routing does not impose any additional requirements on the ordinary nodes, and they perform reactive routing and packet forwarding functionality as usual. In the same area, we can have several smart nodes, each of them controlling its own cluster; thus, the clusters can overlap, and the ordinary nodes can be part of multiple clusters. This leads to a fully distributed cluster creation that will benefit the ordinary nodes. A cluster head will receive an RREQ from a cluster node, and if the cluster head has the route information available, it will return an RREP to the cluster node. If the route information is not available in the cluster head, it will initiate a request to other cluster heads in the FDVB, which reaches to all clusters.

Figure 5.16 shows the concept of an FDVB, where several cluster heads are randomly distributed to form an FDVB.

5.2.8 Performance Metrics in Fixed Networks

Fixed networks are modeled as graphs $G(N, A)$, where N is the set of nodes and A is the set of arcs in the network [6]. The arcs are denoted as (i, j) representing the communication link between nodes i, j. A scalar value x_{ij} represents the flow between nodes i and j through the arc (i, j). In a

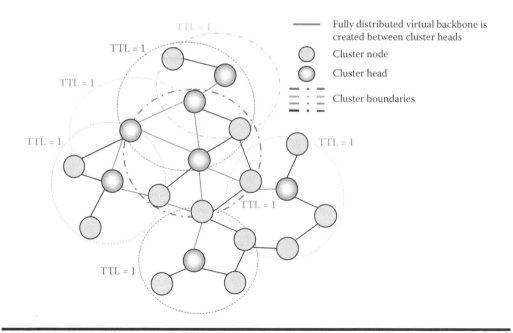

Figure 5.16 Fully distributed virtual backbone created with multiple cluster heads.

graph $G\ (N,\ A)$, the set of flows $x_{ij}|_{(i,\ j)\in A}$ is referred to as the flow vector. A path P in a graph is a sequence of arcs $P_{1,\ k}\equiv(1,\ 2,\ \dots,\ k)$, where $k\geq 2$. A graph is connected if, for each pair of nodes i and j, there is a path starting at i and ending at j. The routing algorithms calculate the optimal routes obtaining paths where the flow vectors x_{ij} are constrained between given lower and upper bounds (i.e., $b_{ij}\leq x_{ij}<c_{ij}$) in order to limit the available bandwidth for that flow. In fixed networks, stability is good enough and the end-to-end packet delay and throughput capacity are the only metrics to be optimized. These metrics are known in fixed networks, providing an NP-complete performance model that can be solved using some approximations. The routing algorithms in fixed networks aim to find a path that connects source and destination nodes through a set of arcs that minimize a linear cost function $\Sigma_{(i,\ j)\in A}\,a_{ij}x_{ij}$, where a_{ij} denotes, for example, the average packet delay to cross the arc $(i,\ j)$. The shortest path is the path with a minimum average delay that can be used for packet forwarding. Therefore, we can model performance in fixed networks using

$$f\left(a_{ij}\right)\equiv\min\sum_{(i,j)\in A}a_{ij}x_{ij} \tag{5.53}$$

In order to enhance network performance, the generic shortest path algorithms used in fixed networks try to maintain and adjust a vector $(d_1,\ d_2,\ \dots,\ d_N)$, where each d_i is the node label and can be either a scalar or ∞.

Let $\qquad\qquad d_1,\ d_2,\ \dots,\ d_N$ be scalars satisfying $d_j\leq d_i+a_{ij},\ \forall(i,\ j)\in A$ $\qquad\qquad$ (a)

and let P be a path starting at a node n_i and ending at a node n_j.

If $\qquad\qquad d_j=d_i+a_{ij}$ for all arcs $(i,\ j)$ of P, then P is the shortest path from i to j $\qquad\qquad$ (b)

where (a) and (b) are called the complementary slackness (CS) [6] conditions for the shortest-path problem. The routing algorithms use the CS conditions to calculate the shortest path. These algorithms successively select the arcs that violate the CS conditions, meaning $d_j>d_i+a_{ij}$. If an arc that violates CS is found, the routing algorithms will set $d_j:=d_i+a_{ij}$ and continue the processing through the available arcs until the CS condition $d_j\geq d_i+a_{ij}$ is satisfied for all the arcs $(i,\ j)$ in the path. The routing algorithms reiterate the calculation over an existing graph, and if they terminate then there is a node j with $d_j>\infty$. This means that d_j is the shortest distance with minimum delay (i.e., based on the cost a_{ij} assigned to each arc) from i to j. If the algorithm does not terminate, then a node j exists such that all sequences of paths that start at i and end at j will have lengths that diverge to $-\infty$. The algorithm terminates if and only if there is no path that starts at i and contains a cycle with negative length. In fixed networks, connectivity (i.e., the probability of having active links) seldom changes. However, in ad hoc networks, connectivity and many other metrics impact network performance. Connectivity in the path between source and destination is often lost because links are broken due to node mobility. Ad hoc networks cannot rely on fixed routes, and the frequent topology changes can make connectivity close to zero. Thus, network performance optimization cannot be solved within a limited processing time.

5.2.9 Performance Metrics in Ad Hoc Networks

In fixed networks, performance is modeled with one equation that will be minimized by the routing algorithm. However, in ad hoc networks, several metrics will affect the performance

Table 5.4   Ad Hoc Network Model Basic Variables

Number of Nodes	Node Mobility	Number of Hops
N	M	γ^a

ᵃ Represents the number of hops in the path, as identified in the test bed.

independently, and there is no single equation that considers all the metrics. A nontrivial problem like this can be resolved by approximation, heuristics, or probabilistic methods. Thus, we need to identify the metrics and variables with a major impact on the ad hoc networks' performance and define the relationship between them. To simplify the resolution, we will first find and compare the values of the variables that optimize the performance for each metric separately. After that, we will select those values that give the best performance in all metrics.

Table 5.4 represents the basic variables in the ad hoc network model.

Node mobility and the number of hops are variables that can be considered linear (e.g., node mobility can vary between 0 and 10 m/s and the number of hops, γ, depends on the selected path). The number of nodes is a critical variable for measuring the ad hoc network scalability, so we will analyze its impact. We consider that the probability of nodes joining the ad hoc network follows a Poisson arrival time distribution (5.54), where λ is the average number of node arrivals in a given time interval t and $f(k)$ is the probability of having k nodes in a given time.

$$f(k,\ \lambda t) = \frac{e^{-\lambda t}(\lambda t)^k}{k!} \tag{5.54}$$

In the FDVB concept, we defined two types of nodes: ordinary and smart. The initial assumption is that the nodes do not earn incentives to become smart and implement hybrid routing functionality. In this case, we assume that the nodes select randomly with equal probability to be either ordinary or smart $p(t)|_{node=ordinary}=p(t)|_{node=smart} = 0.5$. Thus, $p_{S_a}(k,\ \lambda t)$ in (5.55) represents the smart nodes' arrival time distribution, considering (5.54) and the probability to be smart, $p(t)|_{node=smart} = 0.5$.

$$p_{S_a}(k,\ \lambda t) = 0.5\frac{e^{-\lambda t}(\lambda t)^k}{k!} \tag{5.55}$$

The smart nodes may exhaust their battery after some time in the network and become ordinary or die. The battery consumed by a node is modeled using the Peukert Equation 5.56. The consumed battery capacity (C_b) increases with time (t, hours), depending on the discharge current (I, amperes) and the Peukert constant ($n = 1.1$ or 1.2, typically).

$$C_b = I^n t \ (\text{ampere} \times \text{hour}) \tag{5.56}$$

The residual battery capacity in a node is $C_r = C_t - C_b$, where C_t is the full capacity of the battery. Based on the residual battery, we can model the node death process with an exponential $p_d(t) = e^{-\partial t}$, where the slope d depends on the battery age and the processing consumption on each node among other variables. Nevertheless, we consider that all nodes have a similar battery age, but the processing consumption will be higher in smart nodes due to their participation in the hybrid routing functionality.

Figure 5.17 represents the battery consumed by ordinary ($n = 1.1$) and smart nodes ($n = 1.15$) besides their residual battery capacity. The equations associated with the residual battery capacity

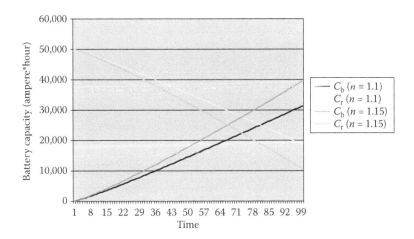

Figure 5.17 Consumed and residual battery capacities in smart and ordinary nodes.

for C_r ($n = 1.1$) and C_r ($n = 1.15$) with the lowest approximation error $r^2 = 0.95$ and $r^2 = 0.97$ result in exponentials with $\partial(1.15) = -0.0144$ and $\partial(1.1) = -0.0097$ slopes, respectively. Thus, we assume that the slope of smart nodes is approximately $\partial_S \approx 0.015$ while it is $\partial_O \approx 0.01$ for ordinary nodes.

We assume that the nodes' arrival and death processes are independent. The $p_S(t)$ in (5.57) represents the probability of having smart nodes in the network. In (5.57), we consider that initially the number of smart nodes that are part of the FDVB is high, but after a period of time, the nodes exhaust their resources, and the smart nodes' death is not compensated with the new smart node arrivals. We also consider as new node arrivals those smart nodes that exhaust their batteries and become temporarily ordinary since the node may become smart again after recharging the battery.

$$p_S(t) = p_{S_a}(t)\, p_{S_d}(t) = e^{-\partial_S t} \sum_{t=0}^{t} 0.5 \frac{e^{-\lambda t}(\lambda t)^k}{k!} \tag{5.57}$$

Figure 5.18 shows $p_{S_a}(t)$ as the smart node arrival cumulative probability (i.e., considering an average node arrival of $\lambda = 5$ nodes and equal probability to become smart or ordinary). $p_{S_d}(t)$ represents the smart node survival probability and $p_S(t)$ represents the probability of having smart nodes left in the network. Figure 5.18 shows that if we consider only the Poisson distribution of arrivals, then it will result in the probability of having a constant share of smart nodes in the network, as represented with $p_{S_a}(t)$. After adding the node survival probability due to battery consumption $p_{S_d}(t)$, the probability of having smart nodes in the network $p_S(t)$ after reaching an initial peak level decreases over time.

Table 5.5 represents the metrics under study to model the scalability of ad hoc networks.

We focus on real-time communications, which require an end-to-end packet delay below 200 ms and a percentage of packet loss lower than 5%.

The metrics can be grouped based on how they affect the ad hoc network performance. Performance can be modeled using multiplicative, $m(i, j)$, concave, $cm(i, j)$, and additive, $a(i, j)$, groups of metrics. Connectivity and packet loss can be considered multiplicative; bandwidth is concave; and end-to-end packet delay and jitter are additive. We will obtain an equation that

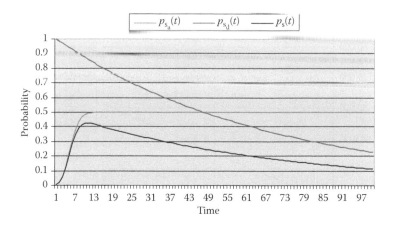

Figure 5.18 Probability of arrival, death, and smart nodes left in the ad hoc network.

Table 5.5 Ad Hoc Network Model Metrics

Connectivity	Bandwidth	End-to-End Packet Delay	Percentage of Packet Loss	Jitter
C	B	D	L	J

defines the relationship between each group of metrics and the ad hoc network model basic variables presented in Table 5.4.

We start the analysis by defining a theoretical function to model the performance based on the multiplicative metric of a path between the source and destination nodes. In this first step, we obtain a performance equation that depends on a single metric, which makes the routing analysis tractable. With this equation, we obtain a list of optimal routes, in a manner similar to the routing algorithm used in fixed networks. Later, a theoretical function is defined for the concave metric, and from the list of optimal routes obtained for the multiplicative metric, we select the ones that also provide the optimal values for the concave function. In the last step, a new equation that models the additive metric is defined, and the remaining routes are prioritized based on the values obtained from the additive function.

5.2.9.1 Multiplicative Metric of the Ad Hoc Networks Model

Connectivity can be modeled as a multiplicative metric, $m(i, j)$, since it is defined as the probability of having active links leading to a successful packet delivery through all the links on the path. It has a critical impact on the ad hoc network performance. If connectivity is null, the rest of the metrics are irrelevant.

Connectivity strongly depends on the mobility of all the nodes in the path. Node mobility can break a link in the path without time to update the network topology. Connectivity is inversely proportional to the percentage of packet loss L. Thus, the percentage of the packet loss will be measured based on the results from the connectivity metric.

The connectivity c_{kl} of a link (k, l) is the probability that the link is active in a communication network. Each link is active and independent of the other links. Thus, the connectivity of nodes i, j, $C(i, j)$, is the product of the connectivity of the links (i, l_1), ... , (k_n, j) on the path from i to j:

$$C(i, j) = m(i, j) = m(i, l_1) \times m(k_1, l_2) \times m(k_2, l_3) \times \cdots \times m(k_n, j) = c_{i,l_1} \times c_{k_1,l_2} \times c_{k_2,l_3} \times \cdots \times c_{k_n,j}$$

The routing algorithm has to find a path with the maximum value for

$$m(i, j) = \prod_{k=i}^{l=j} c_{kl} \quad \forall k, l \tag{5.58}$$

where n is the number of links on the path such that links (i, l_1), ... , (k_n, j) form a path from i to j and c_{kl} is the connectivity of the link (k, l), which depends on the mobility M_{kl}. If $M_{kl} \to 0$, then $\lim_{M_{kl} \to 0} m(i, j) = 1$; and if $M_{kl} \to M_{max}$, then $\lim_{M_{kl} \to M_{max}} m(i, j) = 0$.

Based on (5.58) and the limits, we can model the link connectivity as an exponential function (5.59) that depends on the nodes' relative mobility M_{kl}.

$$c_{kl} = c_O \, e^{-\alpha M_{kl}} \tag{5.59}$$

where c_O is the connectivity of the link (k, l) when the mobility is zero ($M_{kl} = 0$) and α is the slope factor representing the dependency from the mobility of the connectivity function.

The maximum link connectivity between two nodes k and l is obtained when both are completely static ($M_{kl} = 0$), which rarely happens.

$$\lim_{M_{kl} \to 0} c_{kl} = 1 \Rightarrow c_0 \, e^{-\alpha M_{kl}} \Big|_{M_{kl}=0} = c_O \Rightarrow c_O \approx 1$$

The minimum link connectivity is reached when the nodes k and l are moving ($M_{kl} = M_{max}$).

$$\lim_{M_{kl} \to M_{max}} c_{kl} = 0 \Rightarrow c_0 \, e^{-\alpha M_{kl}} \Big|_{M_{kl}=M_{max}} \approx 0$$

The connectivity will be null when the mobility is ∞ ($e^{-\alpha M_{max}} = 0 \Rightarrow \alpha M_{max} \to \infty$). This scenario is not feasible in practice, but we consider that the probability of connectivity is almost null in high mobility conditions

The aim of the FDVB architecture under study is to improve the connectivity by introducing nodes with enough resources and low mobility (i.e., smart nodes). These nodes will support the nodes with limited resources and higher mobility (i.e., ordinary nodes) in terms of routing functionality. The smart nodes will reduce the routing latency, find the optimal routes, and also provide more stability where they are part of the routes.

The link connectivity between two smart nodes is higher than between two ordinary nodes $\left(c_{kl} \big|_{k,l=\text{Smart}} > c_{kl} \big|_{k=\text{Ordinary},l=\text{Smart}} > c_{kl} \big|_{k,l=\text{Ordinary}} \right)$. Thus, connectivity will increase with the introduction of smart nodes on the path. The link connectivity between two smart nodes is represented by

$$c_{kl} \big|_{k,l=\text{Smart}} = c_{O_S} \, e^{-\alpha_S M_{kl}} \tag{5.60}$$

where c_{O_S} is the connectivity of a link (k, l) between smart nodes when mobility is zero ($M_{kl} = 0$) and α_S is the slope factor representing the dependency with mobility in the connectivity function of a link (k, l) between smart nodes.

Applying the FDVB concept on the top of the generic ad hoc network model, the multiplicative metric is represented by

$$m_F(i, j) = \prod_{k,l \in O} c_O \, e^{-\alpha M_{kl}} \prod_{k \in O, l \in S\_or\_k \in S, l \in O} c_O \, e^{-\alpha_{S_O} M_{kl}} \prod_{k,l \in S} c_{O_S} \, e^{-\alpha_S M_{ij}} \; \forall k, l \qquad (5.61)$$

where c_O is the connectivity of a link (k, l) when the node mobility is zero ($M_{kl} = 0$), c_{O_S} is the connectivity of a link (k, l) between smart nodes when the node mobility is zero ($M_{kl} = 0$), n is the number of links (k, l) on the path (i, j), α is the slope factor representing the dependency with mobility in the connectivity function of the link (k, l) between ordinary nodes, α_S is the slope factor representing the dependency with mobility in the connectivity function of the link (k, l) between smart nodes, α_{S_O} is the slope factor representing the dependency with mobility in the connectivity function of the link (k, l) between a smart and an ordinary node, and M_{kl} is the relative mobility of the nodes in the link (k, l).

Equation 5.57 shows that when the smart nodes' energy decreases, the probability of having smart nodes left in the network decreases. Therefore, the connectivity in (5.61) will decrease. Increasing the number of hops in the path decreases the connectivity regardless of the number of nodes in the network. Therefore, a small number of hops and smart nodes in the path will improve the connectivity in ad hoc networks, providing the highest value of the multiplicative metric.

5.2.9.2 Performance Simulation Based on the Multiplicative Metric

Once we have obtained the equations for modeling the connectivity as a multiplicative metric, we compare the results to evaluate the performance difference between the generic and the FDVB ad hoc network models.

We set $c_0 \approx 0.7$ as the value for the connectivity in ad hoc networks with ordinary nodes, assuming static conditions ($M_{kl} = 0$). The connectivity decreases with mobility, so taking as a reference the equation $L_{Ru}(M, N) \approx (0.01 + 0.03N) \, e^{(13.4 + 0.35)M} \ln(M)$ (%) that models the packet loss in reactive routing protocols, we set $\alpha \approx 1.34$ as the slope factor for the ordinary nodes.

We set $c_{O_S} \approx 0.9$ as the value for connectivity in ad hoc networks with smart nodes assuming static conditions ($M_{kl} = 0$). The connectivity between smart nodes decreases with mobility, so taking as a reference the equation $L_{Pu}(M, N) \approx (0.09 + 0.13N) \, e^{(0.69 + 0.012N)M}$ (%) that models the packet loss in proactive routing protocols, we set $\alpha_S \approx 0.69$ as the slope factor for the smart nodes. Figure 5.19 shows the results of the connectivity probability on paths with 2 hops in five scenarios. Each scenario considers a different percentage of smart nodes in the network (i.e., $p_S(t) = 1, 0.7, 0.5, 0.3, 0$). In all these scenarios, we vary the mobility from 0 to 4 m/s, with 0.5 m/s increments (each of them represented with a different curve). The curve on the top represents the highest connectivity obtained when the mobility is 0 m/s, while the curve on the bottom represents the lowest connectivity obtained when the mobility is 4 m/s. The results in Figure 5.19 show that the connectivity probability decreases when the percentage of smart nodes is low. However, when 50% of nodes are smart (S = 50%) and 50% are ordinary (O = 50%), the connectivity probability is similar to the scenario where all the nodes are ordinary (O = 100% and S = 0%). A low percentage of smart nodes (O = 70% and S = 30%) does not improve much the connectivity probability, because it is mostly provided by the ordinary nodes.

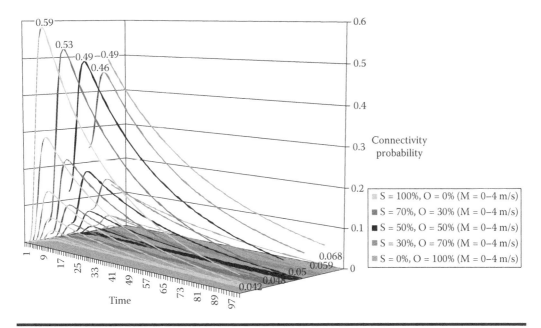

Figure 5.19 Connectivity probability on routes with two hops in five different scenarios.

Figure 5.19 shows that when the percentage of smart nodes is higher than the percentage of ordinary nodes, the connectivity probability is affected by mobility. Thus, in the scenario with 100% of smart nodes (S = 100%) and nonstatic conditions (i.e., second curve from the top represents mobility M = 0.5 m/s), the connectivity probability is lower than in the scenario with 100% of ordinary nodes (O = 100%) and static conditions. The time units are not shown in the figure because time represents the total network lifetime. As we can observe from the value of the connectivity probability for each scenario in the simulation, the network lifetime is much shorter in the scenario only with smart nodes than with ordinary nodes, regardless of mobility. At the end of the network lifetime (i.e., $t = 100$ hours), the scenario only with ordinary nodes (O = 100%) is the one with the highest connectivity probability value (i.e., C = 0.068%).

We conclude that in ad hoc networks with low mobility, a higher percentage of smart nodes than of ordinary nodes in the path increases the connectivity but reduces the network lifetime. On the contrary, in high mobility conditions, a higher percentage of ordinary nodes than of smart nodes increases the connectivity and extends the network lifetime. From the connectivity point of view, the benefit of a high percentage of smart nodes is considerable when the node mobility is low and they can guarantee stable routes. If the node mobility is high, the fact of being smart does not improve the connectivity, because proactive routing might provide routes that are stale because of the node mobility.

5.2.9.3 Concave Metric of the Ad Hoc Network Model

The path bandwidth is the minimum available bandwidth in any of the links $(i, l_1), \ldots, (k_n, j)$ on the path from node i to node j.

$$b(i,j) = \min\left\{b(i,l_1), b(k_1,l_2), b(k_2,l_3), \ldots, b(k_n,j)\right\}$$

The optimal bandwidth metric $B(i, j)$ is the maximum available bandwidth on the paths from node i to node j. It is modeled as a concave metric because its minimum value is the minimum available bandwidth in any of the links on the path.

$$B(i, j) = \max\{b(i, j)\} = \max\{\min\{b(i, l_1), \dots, b(k_n, j)\}\} \forall k, l$$

Throughput is directly proportional to the available bandwidth and inversely proportional to the routing overhead $W(i, j)$, which decreases the available bandwidth for data transmission. We consider an ad hoc network scalable if the performance metrics do not change when the number of nodes increases. Thus, the available bandwidth for data transmission is directly proportional to the network scalability and inversely proportional to the routing overhead generated to keep the same connectivity while increasing the number of nodes.

The bandwidth metric decreases with the number of active connections that nodes i and j maintain with their neighbors because they share the same channel. For simplicity, we assume that regardless of the number of active connections, the available bandwidth (B_T) on each node is $B_T \equiv B_N/n_e$, where n_e is the number of neighbors and B_N is the nominal bandwidth provided by the wireless technology. The available bandwidth in the ad hoc network depends on the selected wireless technology (e.g., 802.11b: 11 Mbs, 802.11a: 54 Mbs). Moreover, the available bandwidth on each link (k, l) in the path from node i to node j is modeled in (5.62), which is the available bandwidth on the node, $B_T \equiv B_N/n_e$, minus the routing overhead on each link $W(k, l)$:

$$b(k, l) = \frac{B_N}{n_e} - \Omega(k, l) \tag{5.62}$$

In order to maximize the available bandwidth on any of the links on the path, we have to find the percentage of smart nodes that minimizes the routing overhead. Equations 5.63 and 5.64 represent the concave metric for the generic ad hoc network model for reactive and proactive routing, respectively, where $\Omega_R(M, N)$ and $\Omega_P(M, N)$ represent the reactive and proactive routing overhead, respectively.

$$B_R(i, j) = \max\{b_R(i, j)\} = \max\left\{\min\left\{\frac{B_N}{n_e} - \Omega_R(M, N)_{kl}\right\}\right\} \forall k, l \in \text{Path} \tag{5.63}$$

$$B_P(i, j) = \max\{b_P(i, j)\} = \max\left\{\min\left\{\frac{B_N}{n_e} - \Omega_P(M, N)_{kl}\right\}\right\} \forall k, l \in \text{Path} \tag{5.64}$$

Based on the simulation results, we approximated the reactive routing overhead in (5.20) and (5.21) and the proactive routing overhead in (5.25) and (5.26). However, the test bed proved that these equations were not accurate and that the number of hops had an impact on the routing overhead. Therefore, we will define new equations to approximate the routing overhead using some of the results from the simulations but also from the test bed. We assume that reactive and proactive protocols increment the routing overhead exponentially with mobility because, when the links break, the route recovery control messages are triggered mainly in reactive routing protocols. When mobility is zero, the routing overhead is minimum as in fixed networks, which leads to

$$\Omega(M, N, \gamma) = f(N, \gamma)e^{\alpha M_{kl}}$$

where N is the number of nodes in the ad hoc network, M is the node mobility, γ is the number of hops on the path, and α is the slope factor representing the dependency from mobility.

We consider the following limits for the routing overhead:

If $\qquad M_{kl} \to 0$, then $\lim_{M \to 0} \Omega(k, l) = A$; and if $M_{kl} \to M_{\max}$, then $\lim_{M \to \infty} \Omega(k, l) = \infty$.

where A is a constant value equivalent to the routing overhead with zero mobility.

In the test bed, the mobility variable was zero. Thus, we assume that the slope factors representing the dependency with mobility are still valid, and we model the routing overhead with the values from (5.21) and (5.26)

(5.21): $\Omega_{Ru}(M, N) = (155.4 + 5.1N) \, e^{(0.045 + 0.009N)M}$ (Kbytes)—mobility is affected by a slope factor of 0.045

(5.26): $\Omega_{Pu}(M, N) = (615 + 375N) \, e^{(0.001 + 0.002N)M}$ (Kbytes)—mobility is affected by a slope factor of 0.001.

Next, we will identify the rest of the parameters in equations for $f_R(N, \gamma)$ and $f_P(N, \gamma)$ to represent more accurately the routing overhead according to the test bed results. Let us first consider the impact of the number of hops and nodes, $f_P(N, \gamma)$, that represents the proactive routing overhead, which is based on the number of nodes and hops. In proactive routing, the routing overhead is affected mainly by the number of nodes and not by the number of hops, $f_P(N, \gamma) \equiv f_P(N)$, since each node has to periodically exchange topology information with the neighbors. We define Q_P as the route updates per second that the nodes running proactive routing protocols have to send to their neighbors. The route update will contain the entire routing cache that includes the topology information from all the available nodes. We define a variable $W_P(N)$ that represents the bytes per route update. $W_P(N)$ is represented as $W_P(N) = K + 4N$ including the fixed protocol information (i.e., K) plus a minimum of 4 bytes of link information (i.e., IP address: 4 bytes, number of hops: 1byte, etc.) associated with each node N in the network. Thus, the routing overhead per node in proactive protocols can be modeled by

$$f_p(N) = (N-1)Q_p W_p(N) \tag{5.65}$$

In order to evaluate the accuracy of (5.65), Table 5.6 compares the values obtained from the test bed with the values obtained from the equation after replacing the variables with the values used in the test bed.

Table 5.6 shows that the values from the model equation are similar to the ones obtained from the test bed, so we can conclude that the model Equation 5.66 accurately represents the OLSR routing overhead in real ad hoc networks.

$$\Omega_p(M, N, \gamma) = f_p(N, \gamma)e^{\alpha M_{kl}} = (N-1)Q_p W_p(N)e^{0.001 M_{kl}} \tag{5.66}$$

Let us now consider $f_R(N, \gamma)$ as the routing overhead for reactive routing based on the number of nodes that receive the RREQ and on the number of hops. Each node in the network will send a route discovery broadcast when the route is not available in the routing cache. We define Q_R as the number of requests per second that each ordinary node issues to find new routes. The RREQ message includes only the required information, W_R bytes, to find the destination. If the node does

Table 5.6   Proactive Routing Overhead Comparison between the Test Bed and the Model Equation

Performance Metrics	OLSR/1 Hop	OLSR/2 Hops	OLSR/3 Hops
N (number of nodes)	2	3	4
Q_P (route updates/s)[a]	0.4	0.4	0.4
W_P (bytes/route update)[b]	$60 + 4N = 68$	$60 + 4N = 72$	$60 + 4N = 76$
Routing overhead (model equation)			
Routing overhead (bytes/s)	27.2	57.6	91.2
Routing overhead (test bed)			
% routing overhead (Routing packets/Real Time Transport Protocol (RTP) packets)	3.39% (109/3215)	3.86% (181/4688)	3.58% (286/7969)
Routing overhead (bytes/s)	32.3	54.2	88.4

[a] OLSR sends 0.2 Topology Control (TC) updates/s and 0.5 Hello messages/s.
[b] OLSR has 60 bytes of fixed protocol information in the TC updates/s and in the Hello messages/s.

not receive a response to the request within a certain time, it will increase the TTL and resend the same RREQ that will reach new nodes several hops away from the originating node.

In the case of AODV, the RREQ process starts with TTL = 1; and if no response is received, the source node will increment the TTL by 2 and will resend the RREQ with TTL = 3. If no response is received, a new RREQ will be sent, incrementing the TTL by 2 (i.e., TTL = 5). The RREQ process is repeated until the maximum of TTL = 7 is reached.

The number of RREQs to be sent increases with the number of hops *g*. If the nodes receiving each RREQ do not have the address of the destination node nor have they seen the RREQ before, they will issue a new RREQ, increasing the routing overhead in the network. Thus, the routing overhead depends on the number of hops between the source and the destination. In order to measure this effect, we define n_γ as the average number of neighbors in the network within each hop *g* from the originating node.

The originating node will launch several attempts to find the destination until either a node responds with the route to reach the destination node or no route is found and a node-not-reachable error occurs; thus, the overhead created is

$$\text{Round 1:} \quad f_R\left(n_\gamma, 1\right)\big|_{TTL=1} = Q_R W_R$$

$$\text{Round 2:} \quad f_R(n_\gamma, 3)\big|_{TTL=3} = f_R(n_\gamma, 1)\big|_{TTL=1} + Q_R W_R n_1 + Q_R W_R n_2$$

$$\text{Round 3:} \quad f_R(n_\gamma, 5)\big|_{TTL=5} = f_R(n_\gamma, 3)\big|_{TTL=3} + Q_R W_R n_4 + Q_R W_R n_5$$

$$\text{Round 4:} \quad f_R(n_\gamma, 7)\big|_{TTL=7} = f_R(n_\gamma, 5)\big|_{TTL=5} + Q_R W_R n_6 + Q_R W_R n_7$$

The total overhead generated from any node will depend on whether the destination node is found close to the originating node or whether an additional request with higher TTL is needed to reach the destination node. If the destination node is far away, the number of nodes that receive the RREQ on each hop, n_γ, will retransmit the RREQ, causing a flooding explosion in the network as modeled in (5.67). In (5.67), we ignore the extra routing from the RREP sent to the originating node by all the neighbors that have a route to the destination node. The routing overhead depends on the probability of having the destination node within a certain number of hops away from the originating node. The reactive routing overhead depends on g and the number of nodes depends on each hop from the originating node n_γ, so $f_R(N, \gamma) = f_R(N_\gamma, \gamma)$. If the destination node is found with an odd number of hops from the originating node, then TTL = γ; and if it is an even number of hops, then TTL = $\gamma + 1$. We consider the number of rounds r needed to reach a destination at the distance of γ hops; then the total overhead is modeled as

$$f_R(n_\gamma, \gamma) = Q_R W_R \left[\rho + (\rho - 1)(n_1 + n_2) + (\rho - 2)(n_3 + n_4) + (\rho - 3)(n_5 + n_6) \right] \quad (5.67)$$

where negative terms of the sum are capped to zero and n_γ is the number of nodes being exactly γ hops away from the originating node. With AODV, the farthest we can reach are nodes that are, at most, exactly 7 hops away from the originating node. Equation 5.67 is pessimistic in the sense that we ignore the possibility that some intermediate node on a path to the destination may have a valid route to the destination when an RREQ reaches it. Nevertheless, despite the fact that the intermediate node has a valid route and sends the RREP, the rest of nodes that are not aware of a valid route will receive the RREQ and will forward it until TTL = 0. In order to evaluate the accuracy of (5.67), Table 5.7 compares the values obtained from the test bed with the values obtained from the equation after replacing the variables with values equivalent to the ones used in the test bed.

Table 5.7 Reactive Routing Overhead Comparison between the Test Bed and the Model Equation

Performance Metrics	AODV/1 Hop	AODV/2 Hops	AODV/3 Hops
Q_R (route request/s)	0.7	0.7	0.7
W_R (bytes/route request)[a]	68	68	68
N_γ	1	1	1
Routing overhead (model equation)			
Routing overhead (bytes/s)	47.6	142.8	142.8
Routing overhead (test bed)			
% routing overhead	7.22 %	7.38%	18.17 %
(routing packets/RTP packets)	(170/2,353)	(506/6,858)	(666/3,665)
Routing overhead (bytes/s)	48.1	49.2	121.8

[a] 68-byte message size for RREQ messages in AODV; 153-byte message size for RTP messages; 15 messages/s.

Table 5.7 shows that the values from the model equation are similar to the ones obtained from the test bed for 1 and 3 hops. The difference for the case of 2 hops is due to the fact that the RREQ will have TTL – 3, and in the test case, there is a single node 2 hops away from the originating node that will provide the RREP, so the RREQ will not be forwarded any further and no additional overhead is generated. On the contrary, the model measures the overall overhead generated with the RREQ that has TTL = 3. Thus, despite the destination node being 2 hops away from the originating node, the RREQ will be forwarded by other nodes in the network that are not aware of the destination node and an overhead similar to the case with 3 hops will be generated. Thus, the results from the test bed for 2 hops would be similar to the results obtained from the model for 1 hop. Therefore, despite the inaccuracy in some specific conditions, we can conclude that the model equation that represents accurately enough the AODV routing overhead in real ad hoc networks is

$$\Omega_R(M, N, \gamma) = f_R(n_\gamma, \gamma) e^{\alpha M_{kl}}$$

$$= \left[Q_R W_R \left[\rho + (\rho-1)(n_1+n_2) + (\rho-2)(n_3+n_4) + (\rho-3)(n_5+n_6) \right] \right] e^{0.045 M_{kl}}.$$

Using (5.63), (5.67), and (5.21), (5.68) represents the concave metric (i.e., the bandwidth) in the generic ad hoc network model where all the nodes are ordinary:

$$B_R(i, j) = \max\{b_R(i, j)\} = \max\left\{ \min\left\{ \frac{B_N}{n_e} - \left(\left[Q_R W_R \left[\rho + (\rho-1)(n_\gamma+n_2) + (\rho-2)(n_3+n_4) \right. \right. \right. \right. \right.$$
$$\left. \left. \left. \left. + (\rho-3)(n_5+n_6) \right] \right] e^{0.045 M_{ij}} \right)_{kl} \right\} \right\} \forall k, \, l \in \text{Path} \tag{5.68}$$

Using (5.63) through (5.67), (5.26), and (5.57), (5.69) represents the concave metric in the FDVB ad hoc network model where there are ordinary and smart nodes in the network:

$$B_F(i, j) = \max\{b_F(i, j)\} =$$

$$\max\left\{ \min\left\{ \frac{B_N}{n_e} - \left(\left[Q_R W_R \left[\rho + (\rho-1)(n_{1\gamma}+n_{2\gamma}) + (\rho-2)(n_{3\gamma}+n_{4\gamma}) + (\rho-3)(n_{5\gamma}+n_{6\gamma}) \right] \right] \right. \right. \right.$$
$$\left. \left. \left. \left(\left((1-p_S(t)) e^{0.045 M_{ij}} \right)_{kl} \right) - \left((N'-1) Q_P W_P(\hat{N}) p_S(t) e^{0.001 M_{ij}} \right)_{kl} \right) \right\} \right\} \tag{5.69}$$

$$\forall k, \, l \in \text{Path}$$

where N' is the number of smart nodes that will exchange topology information. The smart nodes in the FDVB will not maintain the link information from all nodes in the network but only from the nodes from which they have received RREQs, $\hat{N}$ (i.e., $\hat{N} \square N$). Thus, the size of the route updates will be proportional to the $\hat{N}$ number of nodes (i.e., $W_P(\hat{N}) = \hat{N} \times \text{Size of Route Entry}$).

5.2.9.4 Performance Simulations Based on the Concave Metric

Once we have obtained the equations for modeling the concave metric, we compare the results to evaluate the performance difference between the generic and the FDVB ad hoc network models. In order to simplify the equation, we consider a uniform distribution of nodes in all directions,

Table 5.8   Concave Metric Simulation Values for the Generic Ad Hoc Network Model

B_N	n	$B_T = B_N/n$	Q_R	W_R
11 Mbs	20 nodes	11/20 = 0.55 Mbs	0.7 route request/s	68 bytes

Table 5.9   Concave Metric Simulation Values for the FDVB Ad Hoc Network Model

Q_P	$\hat{N}$	W_P	N'	W_R
0.4 route updates/s	16	$60 + 4\hat{N}$ (bytes)	12 nodes	68 bytes

where n is the average number of one hop neighbor of a node: $n_1 = n$, $n_2 = 2n$, $n_3 = 3n$, $n_4 = 4n$, $n_5 = 5n$, $n_6 = 6n$.

Equation 5.67 becomes $f_R(n_\gamma, \gamma) = Q_R W_R \left[\rho + (\rho - 1)3n + (\rho - 2)7n + (\rho - 3)11n \right]$. In order to evaluate the network performance in terms of the concave metric, Table 5.8 shows the values used for the variables in the equations.

We will vary the percentage of ordinary and smart nodes in the network and their mobility to see the effect on the ad hoc network performance.

Equation 5.68 of the concave metric in the generic ad hoc network model after replacing the proposed simulation values is the following:

$$B_R(i, j) = 0.55(\text{Mb}) - \left(\left[Q_R 68\rho + Q_R 68(\rho - 1)3n + Q_R 68(\rho - 2)7n + Q_R 68(\rho - 3)11n \right] e^{0.045 M_{kl}} \right)_{k,l}$$

$$\forall k, l \in \text{Path}$$

The bandwidth in ad hoc networks including the FDVB concept is modeled with (5.69). In order to evaluate the network performance in terms of the concave metric, Table 5.9 shows the values used for the variables in the model.

OLSR defines a period of 2 s (i.e., 0.5 route updates/s) between Hello messages and 5 s (i.e., 0.2 route updates/s) between *topology* messages. Considering that each node will have around 20 neighbors (n_e) and that the smart nodes will keep information only from those ordinary nodes from which they received an RREQ in the past, we assume that each smart node will maintain information from 80% of their neighbors ($\hat{N} = 16$).

The Hello messages in OLSR are similar to RREQ in AODV, but the size of the *topology* messages in OLSR depend on the number of neighbors for which the smart node maintains link information [$60 + 4\hat{N}$ (bytes)].

We vary the percentage of ordinary and smart nodes in a range from 0 to 100%. Thus, since the total number of nodes within each hop is 20, we will have $N' = 12$ for 2-hop routes and 30% of smart nodes.

Equation 5.69 of the concave metric in the FDVB ad hoc network model after replacing the proposed simulation values is

$$B_R(i, j) = 0.55(\text{Mb}) - \left(\left[Q_R 68\rho + Q_R 68(\rho - 1)3n + Q_R 68(\rho - 2)7n + Q_R 68(\rho - 3)11n \right] e^{0.045 M_{kl}} \right)_{k,l}$$

$$\forall k, l \in \text{Path}$$

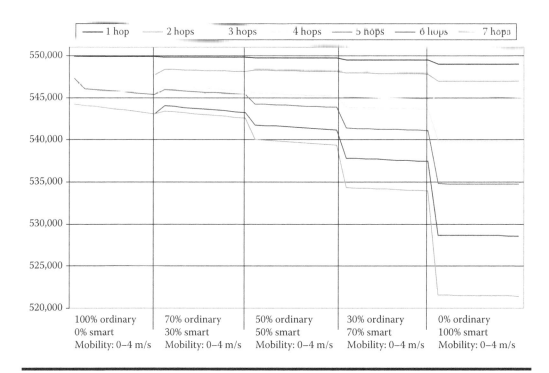

Figure 5.20 Available bandwidth in routes with 1, 2, and 3 hops.

$$B_F(i, j) = 0.55(\text{Mb}) - \left(\left[\rho + (\rho - 1)3n + (\rho - 2)7n + (\rho - 3)11n\right]Q_R 68(1 - p_S(t)) e^{0.045 M_{kl}}\right)_{k,l}$$

$$- \left((12 - 1)0.4(60 + 4 \times 16) p_S(t) e^{0.001 M_{kl}}\right)_{k,l} \quad \forall k, l \in \text{Path}$$

Figure 5.20 shows the available bandwidth in routes with 1, 2, and 3 hops with a different percentage of smart and ordinary nodes in the network.

Figure 5.20 shows that in routes with 1-hop size, the percentage of ordinary or smart nodes does not have much impact on the available bandwidth. We can see that with ordinary nodes only (O = 100% and S = 0%), the overhead is the same for 2–3, 4–5, and 6–7 hops, because the protocol uses the same TTL for the RREQ in those cases.

We can see that in all cases except in 1-hop networks, introducing a low percentage of smart nodes (S = 30–50%) increases the available bandwidth. However, when all the nodes in the route are smart (O = 0% and S = 100%), the bandwidth decreases. This effect has a higher impact in large networks, as we can see in Figure 5.20, where the bandwidth capacity is reduced 5% by the overhead when the destination is 7 hops from the originating node. We observed that introducing a low percentage of smart nodes (S = 30%) gives the highest value of the concave metric when the size of the network increases (i.e., route with 4–5 or 6–7 hops).

5.2.9.5 Additive Metric of the Ad Hoc Network Model

The next step in the analysis is to define the model equation for the additive metric. The end-to-end packet delay $D(i, j)$ is an additive metric because it is the sum of the packet delays on each link in the path from node i to node j. This metric depends on the number of hops in the path.

$$D(i, j) = a(i, j) = a(i, 1) + a(1, 2) + a(2, 3) + \cdots + a(k, j) = a_{i,1} + a_{1,2} + a_{2,3} + a_{k,j}$$

This model is similar to the one used in fixed networks $f_{ij} \equiv \min \Sigma_{(i,j) \in A} a_{ij} x_{ij}$, where a_{ij} is the average packet delay to cross the link (i, j).

However, in the ad hoc network model, we have to take mobility into account. The end-to-end packet delay in ad hoc networks is higher than in fixed networks, because an unstable network environment due to the node mobility and the topology information is constantly changing. For these reasons, having optimized routes from the end-to-end packet delay standpoint is difficult. Therefore, when considering the FDVB ad hoc network model, we have analyzed the impact of the types of nodes in the network (i.e., ordinary and smart). We concluded that the end-to-end delay is not affected by the type of nodes since all of them will have similar processing capabilities. However, the end-to-end delay is affected by the number of hops in the route and the node mobility despite having smart nodes in the path. Thus, having smart nodes in the network will decrease the end-to-end packet delay because their mobility is lower and they find optimal routes to reach the destination with a minimum number of hops. Therefore, we conclude that the routing optimization based on the minimum number of hops will provide the lowest end-to-end packet delay. However, an additional optimization based on the type of nodes in the path and their mobility should be considered. The routes with a higher number of nodes with low mobility might have lower end-to-end delay than routes with few nodes but high mobility and a higher number of hops.

5.2.9.6 *Ad Hoc Model Evaluation Conclusions*

The simulation results for the multiplicative and additive metrics represented in Figure 5.19 show that adding smart nodes will improve the network performance in terms of connectivity and end-to-end packet delay. However, excessively increasing the number of smart nodes will not be an optimal solution since smart nodes are severely affected by mobility, which decreases the probability of connectivity and the network lifetime, as shown in Figure 5.19. Moreover, the results for the concave metric depicted in Figure 5.20 show that a reasonable percentage of smart nodes (i.e., 30%) provides better performance than having only either ordinary or smart nodes in the network. In terms of the probability of connectivity, the optimal value results when all the nodes are smart with mobility zero, which is equivalent to the fixed network environment. However, after considering the rest of the metrics, we have seen that having a certain percentage of smart nodes joining the network will reduce the end-to-end packet delay and increase the available bandwidth, keeping connectivity at a certain level.

We also have to consider that the number of smart nodes joining the network decreases over time (i.e., Poisson arrival time); thus, a control mechanism is necessary to keep the percentage of smart nodes in the network at around 30%. From the mathematical ad hoc network models, we now conclude that we have obtained results that provide a first estimation of the optimal parameters to improve network scalability.

References

1. C. Perkins, E. Belding-Royer, and S. Das. Ad hoc on-demand distance vector (AODV) routing, RFC 3561, July 2003.
2. D. Johnson and D.A. Maltz. Dynamic source routing in ad hoc wireless networks. In T. Imielinski and H. Korth (eds.), *Mobile Computing*. 1996: Kluwer Academic Publishers, Chapter 5, pp. 153–181.

3. A. Iwata, C. Chiang, G. Pei, M. Gerla, and T.-W. Chen. Scalable routing strategies for ad hoc wireless networks. *IEEE Journal on Selected Areas in Communications*, vol. 17, no. 8 (Special Issue on Wireless Ad Hoc Networks), 1999, pp. 1369–1379.

4. C. Perkins and P. Bhagwat. Highly dynamic destination-sequenced distance-vector routing (DSDV) for mobile computers. *SIGCOMM*, August 1994.

5. T. Clausen and P. Jacquet. Optimized link state routing protocol (OLSR), RFC 3626, October 2003.

6. D.P. Bertsekas. *Network Optimization. Continuous and Discrete Models*. Nashua, NH: Athena Scientific, 1998, pp. 73–75.

7. M.R. Pearlman and Z.J. Haas. Determining the optimal configuration for the zone routing protocol. *IEEE Journal on Selected Areas in Communications*, vol. 17, no. 8 (Special Issue on Wireless Ad Hoc Networks), 1999, pp. 1395–1414.

8. Y.-B. Ko and N.H. Vaidya. Location-aided routing (LAR) in mobile ad hoc networks. In *MOBICOM'98*, October 1998, pp. 66–75.

9. J. Costa-Requena, T. Vadar, R. Kantola, and N. Beijar. AODV-OLSR scalable ad hoc routing proposal. *ISWPC*, January 2006.

10. The Network Simulator NS-2. Available at http://www.isi.edu/nsnam/ns/

11. X. Hong, T.J. Kwon, M. Gerla, D. Lihui Gu, and G. Pei. A mobility framework for ad hoc wireless networks. *MDM*, 2001, pp. 185–196.

12. M. Goto, S. Yoshida, K. Mase, and T. Clausen. A study of link buffering for OLSR. In OLSR Interoperability & Workshop, August 2004.

13. C. Santivanez, R. Ramanathan, and I. Stavrakakis. Making link-state routing scale for ad hoc networks. In *MobiHoc 2001*, October 2001.

COMMUNICATION PROTOCOLS OF MANET

Chapter 6

Extending Open Shortest Path First for Mobile Ad Hoc Network Routing

Katherine Isaacs, Julie Hsieh, and Melody Moh

Contents

Mobile ad hoc networks (MANETs) face many challenges due to the diverse nature of their nodes and the fluidity of their topology. Routing is central to the efficiency and scalability of MANETs.

Open Shortest Path First (OSPF) is the most widely used intradomain routing protocol on the Internet. Extending OSPF to handle the unique concerns of MANETs would be ideal as OSPF is well-known and well tested. Furthermore, this extension would more easily support the seamless traveling of nodes between MANETs and wired networks. To understand MANET routing in OSPF, we first review many of the techniques used in MANET routing and discuss widely researched protocols including ad hoc on-demand distance vector (AODV), dynamic source routing (DSR), optimized link state routing (OLSR), zone routing protocol (ZRP), and fuzzy-sighted link state (FSLS). Next, we review the standard OSPF protocol and the existing MANET extensions for it. Then, we compare the OSPF extensions for MANET, both in design and by summarizing individual simulation results. Finally, we discuss future directions by pointing out several promising approaches as well as emerging concerns.

6.1 Introduction

Routing—the determination of paths for data flow throughout the network—is especially challenging in a MANET. The wide variety of devices, topologies, and other network characteristics alters the issues the routing algorithm must address. With limited resources, the demands for efficiency and scalability are heightened.

Many examples of MANETs involve small wireless devices such as cell phones and laptops and exist in areas where users may not have reliable access to power. Thus, energy usage and battery life must be taken into consideration. With these small devices, processing power and storage capability are also issues. However, the computational ability and storage of current devices are better than those of early Internet nodes, so the main focus is now on energy [1].

Channel capacity adds another constraint. Due to attenuation, dispersion, and multipath fading, the wireless channel is bandwidth-constrained and error-prone; it is also seriously subjected to the influences of noise and other natural effects (mountains, buildings, air and road traffic, storms, etc.). As a result, as networks grow, congestion and quality of services issues are much worse than wired networks [2].

Furthermore, nodes are assumed to be in motion, constantly changing the topology of the network. A node may also enter or leave the network anywhere at any time. This unpredictability leads to a frequent breaking of links, loss of packets, and the need to reorganize the network. Still it is expected that a MANET should be self-configuring and self-healing, have low latency, and be reliable [2].

Finally, there is the issue of how a MANET or nodes in the MANET may eventually reconnect to the existing Internet. The most popular interior gateway routing protocol is OSPF [3,4]. Because of its extensive use, there have been several proposals for extensions to support the unique considerations of MANETs. Since most of the protocols have already been shown to work in other networks, MANET support may reach end users more quickly. Also, extending OSPF may provide a natural connection with the existing Internet and allow nodes to easily "plug in" to existing networks without a disruption of service [5].

The rest of the chapter is organized as follows. In Section 6.2, we discuss issues in general MANET routing and present a few major protocols: AODV [6], DSR [7], OLSR [8], ZRP [9], and FSLS [10]. Section 6.3 focuses on extending OSPF over MANET; the OSPF protocol is first reviewed and key challenges of the extensions are then described, followed by exiting proposals on OSPF extension for MANET. Finally, Section 6.4 discusses future directions and concludes the chapter.

6.2 MANET Routing

A large body of work exists in the field of MANET routing. We summarize some of the key techniques and features that routing protocols utilize to overcome some of the inherent difficulties of MANETs. We then introduce in further detail some well-known algorithms.

6.2.1 Techniques and Classification

There are many choices that may be made when designing a MANET routing protocol. Protocols may be distinguished by how much of the network they communicate with, how many routes they cache, or how they self-organize their topology.

The most common way to classify a protocol is based on whether they store all possible routes. If a routing scheme constantly keeps an up-to-date routing table, it is called *proactive* or *static*. If it only searches for routes on-demand, it is called reactive, dynamic, or adaptive.

Proactive routing generally incurs more overhead than does reactive routing, which is an issue for MANETs because many of the nodes are mobile devices with limited battery life. However, latency is much higher in reactive protocols because routes must be created after they are requested. Therefore, proactive routing is preferred for high-traffic networks where the benefit of keeping all routes is greater, while reactive routing is preferred for low- to mid-range traffic networks where many paths may never be used and therefore the overhead spent determining them would be wasted [11].

Many proactive MANET routing algorithms draw from well-established wired methods, such as distance vector and link state, and make optimizations for the characteristics of the mobile wireless environments. Reactive routing algorithms are also modified for MANETs. In DSR [2], instead of a table of recently used routes, a cache is kept that may contain multiple routes to a destination—allowing fast access to alternatives should one break.

Proactive–reactive hybrid algorithms, such as ZRP [2], also exist. In ZRP, a node will proactively maintain a routing table of its nearest neighbors and then use a reactive approach for all further destinations. This method also reduces the number of nodes outside the neighborhood (zone) used in routing; packets may be sent directly to the edge of that zone.

Another way to view routing protocols is whether they are global or localized. Localized algorithms may also be called decentralized, distributed, or online [12]. Localized algorithms are generally distance vector-based, while global schemes are generally link state based. The difference between distance vector and link state routing can be seen as this: a node in a distance vector network exchanges messages with its immediate neighbors about all the routes it knows, whereas a node in a link state network sends messages to the entire network about all of its neighbors. Link state routing generally results in a more stable network [13].

Routing algorithms may also be flat or hierarchical. To improve scalability, MANETs may make use of hierarchies and clustering. If leader nodes are determined for a cluster or hierarchy, they serve as a backbone between clusters. The drawback is that this may be unfairly draining on the leader node. Furthermore, care must be taken should a leader node suddenly drop from the network. Cluster-based addressing means less information must be sent and maintained for destinations. However, because routing based on clusters is more coarse-grained, absolute shortest hop paths may not always be found [14], and there is a trade-off between routing overhead and clustering overhead [11].

Another feature a routing protocol may have is storing multiple paths for handling broken routes. In multipath schemes, several routes are stored by the node. Algorithms may simply use the extra routes as alternatives should the main route not work or may use a dispersion technique. In dispersity routing, the multipaths must be independent, meaning they have no shared nodes or channels. Each packet is then split and repackaged with extra reconstructive information. The new packets are sent along all paths. This way if some number of paths fail, there is still enough information to retrieve the original packet from the dispersed ones that do arrive at the destination [11]. Overlapping multiple routes may also be used to create new routes that might be better, by using genetic algorithms. Fitness functions are based on link costs. Routes may exchange the remaining parts of their paths where they overlap [15].

Since many mobile devices are equipped with global positioning system capabilities, some routing protocols may incorporate geographical information to help prune their topology or elect cluster heads. They may also find routes statelessly by forwarding messages in the last-known direction of a node, falling back to flooding only when location information is unavailable. Velocity estimates of destinations may be used in forwarding determination as well. The use of geographical information may allow the adaptation of more schemes from cellular telephony [16]. Routing protocols using geographical information include greedy perimeter stateless routing [17] and location aided routing [18].

A routing protocol can be constructed with any assortment of these characteristics. There are many trade-offs to consider. Proactive schemes have lower latency but require more overhead. Hierarchical algorithms may decrease routing control overhead but increase hierarchy control overhead. Multipath algorithms increase robustness but require more storage.

We now review some of the well-known MANET routing protocols. Reactive protocols are represented by AODV and DSR. Proactive protocols are represented by OLSR and FSLS. Hybrid protocols are represented by ZRP.

6.2.2 AODV

AODV [6] is a reactive routing protocol. Instead of determining routes to all nodes, routing tables are updated with information gleaned only from routing requests or data traffic along preexisting routes. Destination sequence numbers are used to gauge the newness of information and to ensure loop-free routing.

Routing table entries include a destination sequence number, active and reparable statuses, the hop count to the destination, a list of precursors, and the lifetime of the route. The control messages used in AODV are route requests (RREQs), route replies (RREPs), route errors (RERRs), and route reply acknowledgments (RREP-ACKs). Any time a route is used successfully, its lifetime is refreshed.

Destination sequence numbers are incremented before originating an RREQ or RREP. The protocol uses the destination sequence numbers to discard incoming messages that are out of date and to avoid the *count to infinity* problem that plagues some distance vector algorithms. Because the control messages contain destination sequence numbers for both the originating node and the destination, receiving nodes may update their data as they pass on these messages. If two sequence numbers are found equal, the one associated with the shorter hop count will be used.

A node seeking a route broadcasts an RREQ in search of the destination. Upon receipt of an RREQ, an intermediate node with an active route to the destination with a newer destination sequence number than the RREQ will generate an RREP using the destination sequence number from its routing table and unicast it to the originating node. The intermediate node may also generate a "gratuitous" RREP to the destination node in order to set up the reverse route.

If the intermediate node does not have an active route, it will update its routing information from the previous hop and broadcast the updated RREQ. It will also create or update the reverse route to the originating node. If the RREQ should reach the destination node, it will generate the RREP.

To lessen the impact of flooding an RREQ, originating nodes may use an *expanding ring search*. This entails successively rebroadcasting the RREQ with updated sequence numbers and an increased maximum hop count until a route is discovered or the destination is determined unreachable. This allows initial flooding to be limited by radius. The RREQ rebroadcasts utilize a binary exponential backoff in waiting time for an RREP.

For RERR handling and repair, nodes keep track of the connectivity with their neighbors. When a broken link has been detected, existing route entries are invalidated and affected destinations are aggregated. The node detecting the break checks the precursor lists of the affected routes and sends an RERR to those affected neighbors. The RERR will list all broken routes for which there is at least one neighbor in the precursor list. RERRs are forwarded by intermediate nodes and sent as responses to requests for broken routes.

Local link repair may be attempted by an upstream node. While route repair is in progress, data sent along the route is buffered. The repairing node generates an RREQ. Since this operation is local and the repairing node does not send an RERR unless the repair fails or the hop count of the route is increased, the originating node need not be unaware of the process.

Like most reactive protocols, AODV has less overhead than do proactive solutions. Because AODV does not keep track of all intermediate hops like DSR, its control messages are small, meaning less overhead in terms of bytes [19]. Also, as a distance vector-based scheme, AODV finds routes closer to the optimal path than protocols based on other algorithms [20]. AODV has a flat topology and thus no extra overhead is required for a hierarchical one. Topological changes can be dealt with quickly, locally, and only as necessary [19].

As a trade-off, again due to its reactive nature, AODV has increased latency when new routes must be found [19]. In simulation, it has been found to offer inconsistent performance when compared with proactive algorithms in certain mobility models. Furthermore, it is not as robust as DSR because of its single route nature and tends toward high contention when flooding the network in search of routes [20].

AODV works well for small- and medium-sized networks, but is not scalable to large-sized networks. Because RREQs are broadcasted over the network, the number of control packets becomes untenable as the network grows [21].

6.2.3 DSR

DSR [7] is another reactive routing protocol. Unlike AODV, nodes use source routing for each packet, where each packet contains the full route record of node addresses in the path to the destination. This allows the source to consider load balancing when choosing a route and to recover quickly should the default route break. Optionally, DSR can use flow IDs instead of the full node list after a source route is established.

When a source node wants to send a packet to a destination not in its cache, it broadcasts an RREQ. The RREQ contains a sequence number and a route record. As the RREQ fans out from the source, intermediate nodes append their addresses to the route record and rebroadcast. When the destination node receives the RREQ, it sends an RREP with the entire accumulated route record.

The flooding is limited by the time-to-live (TTL) field in the Internet protocol (IP) header, and received RREQs with duplicate sequence numbers are discarded. In promiscuous mode, intermediate nodes cache routes overheard from other packets.

Multiple routes may be cached from the same RREQ. In the event that a network becomes partitioned, a sending node may reinitiate route discovery.

Each node is responsible for reporting its broken links. Broken links are removed from the route cache and an RERR message is generated. The RERR message is sent to all nodes that have attempted to use the affected link since its last acknowledgment.

Once a route has been discovered, a sender can establish a flow, whereby packets after the first will follow the same route without requiring full specification. This is done by setting a flow ID in addition to the source route in the first packet. Intermediate nodes store the information in their flow table. Subsequent packets then require only the flow ID.

6.2.4 OLSR

OLSR [8] is a proactive routing protocol. A second version was in an International Engineering Task Force (IETF) Internet draft at the time of this writing. The newer version simplifies the message structures and offers more options for efficient signaling.

In traditional wire link state routing, each node keeps track of the link state to its immediate neighbors and floods the network with its link state advertisement (LSA). Each LSA has a sequence number and a TTL. The sequence number identifies the LSA; it is incremented when the node generates a new LSA in response to a change in one or more of its links. This is necessary since multiple versions of LSAs from the same node may be circulating the network [13]. Using this information, each node is able to determine the topology of the network and routes to all destinations.

OLSR improves upon traditional link state routing primarily by introducing multipoint relays (MPRs). Instead of flooding the entire network with link state updates (LSUs), messages are sent to and from MPRs only, consolidating the information and decreasing the control overhead. The use of MPRs may be further extended as a form of topology reduction.

There are two types of control messages in OLSR: Hello and Topology Control (TC) messages. Hello messages contain the originating node's address, its willingness to perform as an MPR for other nodes, and a list of other nodes that have selected it as an MPR. Like other link state algorithms, TC messages will either contain the complete local topology information or only what has changed since the last complete message. The complete messages will contain all advertised neighbors, addresses, and gateways.

Nodes organize their information into information bases. This includes the latest TC messages and derived network topology, neighbor relationship information, addressing information, and outside network information. By storing control messages, the protocol guards against sending the same information twice.

Each node must select at least one MPR for each of its interfaces. The set of MPRs is chosen from the subset of symmetric neighbors that have broadcasted a positive willingness to be in the MPR set. The MPR set is required to be able to forward any messages to all 2-hop neighbors of the node. In other words, the MPR 1-hop neighbor set should cover the entire 2-hop neighbor set of the originating node.

The OLSR specification does not require any specific algorithm for MPR selection, although selecting the minimal number of MPRs is the most beneficial. The specification does suggest one algorithm, which is often referenced by other protocols. This algorithm is outlined below.

Other MPR selection algorithms may make use of geographical location information or other topological considerations.

A node can derive its MPR set in two phases. First, it waits for the periodic Hello messages from its set of neighbors, N, each telling the others with which it has direct links. Once the node has received Hello messages from all of its neighbors, it knows the set of its 2-hop neighbors, N_2.

Using the sensed neighborhood information, the node creates an MPR set. If there are nodes in N_2 that have only one neighbor in N, then those critical neighbors are added to the MPR set. Then, the node in N with the greatest willingness to be an MPR is added to the MPR set. If there are multiple nodes of equal willingness, the one that can reach the most still uncovered nodes in N_2 is selected. Further ties are broken with the greatest total degree and MPR selector status. This process is repeated until all nodes in N_2 are covered.

Once each node has calculated its MPR set, that information is disseminated through TC messages and nodes learn of their MPR selector sets. They will then broadcast their updated MPR selector sets in the next TC message to their neighbors.

As a proactive algorithm, OLSR enjoys low latency associated with route finding [19] and robustness of service in low-mobility scenarios [20]. Its reactivity may be further adjusted by changing the broadcast interval of the TC messages. The use of MPRs alleviates some of the control congestion problems of other link state algorithms. However, it may lengthen the shortest route in the network. In addition, it still requires a lot of overhead and is, therefore, not scalable to large networks. Furthermore, the memory required by the information bases is great [19].

6.2.5 FSLS

The FSLS [10] family of link state routing algorithms attempts to decrease overhead by relaying LSUs within a changing limited radius of the generating router. This is separate from OLSR's overhead reduction strategy of limiting the neighbor nodes that receive LSUs.

The protocol checks the amount of time elapsed since the last change in topology—the longer the elapsed time, the higher the hop limit is set on the LSU. Because routing is done on a hop-by-hop basis, topology changes in nodes far from the sender do not greatly affect the local next hop, so the number of updates those faraway nodes receive need not be as frequent.

The impetus for the FSLS family is finding a balance of proactive, reactive, and suboptimal routing overheads such that the total overhead is minimized. The *suboptimal routing overhead* is defined as the excess amount of bandwidth consumed by packets along the route traveled over the bandwidth that would be consumed if packets always traveled the shortest possible distance.

In FSLS, the TTL of an LSU sent at $(2^i - 1)t$ is set to s_i, where t is the LSU time interval. This means every odd-numbered LSU will have a TTL of s_1. The sequence s_i should be increasing. It is also suggested that the sequence provides an infinite TTL LSU to be sent regularly.

To determine the optimal overhead-reducing sequence s_i, several assumptions were made. Average node degree, network density, and traffic generated per node were assumed to be independent of network size. Each node was equally likely to be a destination for some sending node. The rate at which link statuses change was dependent solely on mobility. Finally, the node velocity was randomly distributed with decaying autocorrelation.

Based on the assumptions, the integer solution that most minimized total overhead is $s_i = 2^i$. The algorithm using this sequence is known as *hazy-sighted link state* (HSLS) routing. The value assumes a high-mobility scenario. To account for low-mobility scenarios, HSLS may fall back to a standard link state approach when link state changes are less common.

Although the algorithm was designed as a balance of the overhead incurred by proactive and reactive algorithms, members of the FSLS family are still classified as proactive because they calculate all routes before they are requested. The calculated routes are just more out-of-date for faraway destinations.

6.2.6 ZRP

The ZRP [9] is a hybrid routing algorithm designed to be proactive locally and reactive globally. The system works by combining three protocols: intrazone routing protocol (IARP) [22], interzone routing protocol (IERP) [23], and bordercast resolution protocol (BRP) [24]. The IARP can be modified from any proactive link state routing protocol. The IERP can be modified from any reactive routing protocol. This is modular since IARP and IERP can be selected independently of each other. The BRP makes use of a logical zone construct for more efficient control message dissemination.

Each node in the ZRP routed network maintains a zone consisting of all nodes that have a minimum distance within R hops. At first glance, ZRP may seem hierarchical, but it is not, because each zone corresponds to a single node. The value of R may be predetermined by the network or determined dynamically for each node. In sparse networks with slow mobility, a large value of R produces better results. In dense networks with fast mobility, a small zone radius is better. When R is set to zero, the network is fully reactive. Hence, R can be understood as a proactivity index.

Any existing global link state proactive routing protocol may be converted into an IARP. The scope of the routing protocol may be limited by setting the TTL of the control messages to the zone radius. Link state table transfers between neighboring nodes should not include the table entries of peripheral nodes (those at the edge of each zone), since these will likely be outside the zone or redundant. Update messages should also discard peripheral node information in this fashion.

Any existing global reactive routing protocol may be converted into an IERP. Local proactive elements of the protocol such as Hello beacons should be disabled since they will be handled by the IERP. The protocol must also be modified to be able to import or use information from the IARP table. The routing protocol should also be altered, making use of bordercasting for its RREQs. Any functionality in the routing protocol that can be handled by BRP, such as flood control or RREQ broadcast jitter, should also be disabled.

Instead of broadcasting route discovery messages, ZRP uses bordercasting. In this scheme, the route discovery messages are sent only in the direction of peripheral nodes that are yet to receive the message. The information required for this is determined by a *bordercast tree* sent along with each message.

When an originating node wishes to send data to a destination node, it first attempts to find the destination in its routing zone. If not found, it constructs a bordercast tree. The tree is rooted at the originating node and constructed to reach all peripheral nodes. (Thus, nodes that do not reach peripheral nodes are not included in the tree.) The routing message, with the bordercast tree, is sent to all 1-hop neighbors.

Each node receiving the route discovery message first checks to make sure it is in the bordercast tree of the message. If it is, it then searches its own zone for the destination node. If the node is found, it sends an RREP with the discovered path. If the destination node is not found, the intermediate node creates a bordercast tree rooted at itself and branching to all its peripheral nodes not already covered by the previous hop. Other nodes are considered covered if they are in the zone of a bordercasting node.

By acting proactively within local zones, ZRP is able to limit the effect of local changes in the network [9]. Broken links may be repaired quickly because another route within the node's zone can be determined quickly. If the topology changes such that paths get shorter, as in the case of nodes moving closer together, shortcuts will be determined by the proactive portion of the protocol [23].

The combination of local proactive information and bordercasting makes route discovery more efficient than many reactive protocols. Unlike a fully proactive protocol, ZRP does not use excessive control messages for routes that may never get used [22]. However, bordercasting does not fully alleviate the flooding the network problem, since it is necessary to perform a full search under IERP [16].

6.3 OSPF over MANET

Having reviewed MANET routing, we now turn our attention to OSPF. In order to take advantage of the widespread use of OSPF, techniques from MANET routing research can and have been used to extend the protocol. First we review the OSPF protocol, with reference to both its IPv4 and IPv6 incarnations. Then we discuss the difficulties with using OSPF directly for MANETs. The three MANET extensions are then explained and compared.

6.3.1 OSPF Protocol

OSPF [3,4] is a link state protocol. In OSPF, the basic link state algorithm runs over a unit known as an area. The protocol also directs routing between areas in an autonomous system (AS). OSPF router LSAs perform the duties of LSAs in other link state protocols, flooding updated information about a router's neighbors within a single area. In OSPFv3 (OSPF for IPv6), there are link LSAs that match IPv6 address prefixes with link-local addresses used in other LSAs. The link LSAs are flooded only on the local link. In OSPFv2, the full (IPv4) addresses are present in all LSAs. There are several other types of LSAs that provide routing information to destinations external to the ASs or routing between different OSPF areas. These other LSAs may be flooded AS-wide.

The use of areas provides a sense of hierarchy to the AS. The exact topology within the area need not be known by the other areas. This reduces the number and size of the LSAs that travel between areas. Routing between areas is handled by an elected designated router (DR) and any area border routers, which are routers spanning multiple areas. An area may also have a backup designated router (BDR). These special routers both process and originate the other types of LSAs mentioned. Changes in the identity of the DR will trigger the origination of router LSAs.

The shortest-path algorithm (Dijkstra's algorithm) is done in two stages in OSPF. The first stage considers only the transit links. OSPF saves all next hops for a route that results in an equal-cost path.

When two routers in OSPF synchronize their topological databases through a series of database description packets, they are said to be *adjacent*. On links without the need for DRs, such as point-to-point links and areas with a single router, all routers become adjacent. Otherwise the DRs determine which routers should become adjacent. Flooding then occurs across the adjacencies. As such, reducing the number of adjacencies, if possible, can lead to better performance. Changes in the adjacency state on a link will trigger the origination of router LSAs.

OSPF LSAs are uniquely identified by sequence number, advertising router, and age. The age is incremented upon reflooding and is periodically incremented within the link state database. When an LSA reaches the maximum age, it is reflooded with its maximum age, which indicates that it should be purged. Therefore, any max age LSA is considered the most recent. Otherwise, between two LSAs with the same sequence number, the smaller age is considered more recent

LSAs from a single router may be broken up into separate messages to avoid IP fragmentation. LSAs may also be coalesced into LSUs, though each LSA needs to be acknowledged separately. The acknowledgment may be implicitly implied by the reflooding of the LSA.

6.3.2 OSPF Challenges for MANET

Because each node in a MANET acts as a router, under the base OSPF protocol, each node will generate and flood LSAs and rebroadcast LSUs. This flooding results in an enormous amount of control overhead, redundant information, and wireless collisions. Furthermore, such a high density of routers leads to high database description packet overhead as adjacencies are formed.

Issues of control overhead become worse when mobility is taken into account. Links are constantly being broken and formed due to the movement of the nodes, resulting in a higher frequency of link state changes, which in turn results in more LSAs. Simulations have shown that the rate of link state changes may exceed the suggested minimum LSA interval [25].

In addition to the effect on LSA generation, links coming online incur the cost of adjacency formations. Database description messages are sent back and forth between each pair of nodes so that topology databases are synchronized. This may be especially wasteful when link lifetimes are short.

To address these issues, the OSPF extensions for MANET have focused on LSA flooding reduction, adjacency reduction, and topology reduction.

6.3.3 OSPF MPR Extension

The OSPF MPR [26] extension explicitly aims to reduce flooding, topology, and adjacency. Flooding is decreased by having only a subset of the router's relay packets. Topology is reduced by only advertising a subset of the links, which also decreases the size of the LSAs. Adjacencies are reduced by being required only between MPRs and MPR selectors.

To achieve its goals, the extension makes use of MPRs, similar to those previously described for OLSR. There are two types of MPRs in this protocol: flooding and path. Both types of MPRs are used in adjacency creation.

The flooding MPR is most similar to the MPRs in OLSR. The 2-hop neighborhood coverage requirement for flooding MPRs and the per-interface selection heuristic suggested by the Request for Comments (RFC) are as in OLSR and shown in Figure 6.1. Also, the flooding MPRs are tasked with retransmitting broadcast packets from their selectors.

The path-MPRs are used for the shortest-path calculation. They represent routers in the shortest path between a node and the members of its 2-hop neighborhood. To reduce the size and frequency of LSAs, only path-MPRs are advertised.

The suggested path-MPR selection heuristic first forms special 1-hop and 2-hop neighborhood sets over all MANET interfaces. These sets include only the links that describe minimum cost paths to nodes in the 2-hop neighborhood, as determined by metric information from the Hello messages and the SPF algorithm. The algorithm described for OLSR (and used for flooding MPRs) is then used again, only this time over the newly constructed minimum-cost graph. An example is shown in Figure 6.2.

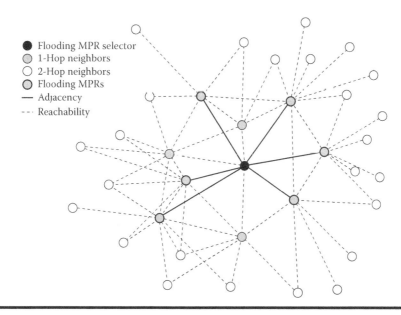

Figure 6.1   In OSPF MPR, a node (shown in black) selects its flooding MPRs from its 1-hop neighbors such that all 2-hop neighbors may be reached. Adjacencies are formed between the node and its selected flooding MPRs. Reachability is not shown between 2-hop neighbors in this figure.

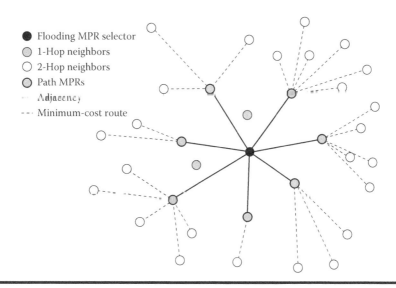

Figure 6.2   Link information is pruned to only the links required for minimum cost paths within the 2-hop neighborhood. The path-MPRS are then selected as those required for minimum-cost paths. Adjacencies are formed between the node and its selected path-MPRs.

Flooding MPR selections are recalculated whenever there is a neighborhood change on the interface. Similarly, path MPR selections are recalculated whenever there is a neighborhood or cost change over all the node's MANET interfaces. If the selected MPR set changes, the adjacency need not be torn down until the link is broken. MPR selectors notify their selected MPR using link local signaling in Hello messages.

If routers exist in the network with both MANET and non-MANET interfaces, they will select all their MANET neighbors as path-MPRs. If no such routers exist, at least one router must be selected as a synch router. The synch router becomes adjacent with all MANET neighbors. This serves to ensure connectivity in the network.

6.3.4 OSPF MDR Extension

OSPF MANET designated router (MDR) [27] is an extension to OSPF for MANET routers, proposed by Boeing. To support MANET interfaces, it generalizes the concept of DRs and BDRs to MDRs and backup MANET designated routers (BMDRs) without requiring the use of areas. The MDRs are selected such that they form a connected dominating set (CDS) over the MANET. Because of the focus on this technique, the extension is sometimes referred to as *OSPF CDS*.

The intent of the design is to reduce flooding by having only MDRs retransmit LSAs. To ensure that the information is properly flooded, the BMDRs listen for an acknowledgment. Should they not observe an acknowledgment, either explicit or implicit, within the specified interval, they will retransmit the LSA. Together with the MDRs, the BMDRS form a bi-CDS. To further reduce flooding, an MDR may choose not to flood if it determines all of its neighbors have received the message. This determination may be made by listening for acknowledgments and checking if the neighbor belongs to the bidirectional neighbor set of another MDR.

To select MDRs, each router uses the local information regarding their 2-hop neighborhood from the Hello messages. Routers are ordered lexicographically by a tuple consisting of their RouterPriority, MDR Level, and Router ID. The use of MDR Level biases the selection toward existing MDRs to avoid creating new adjacencies. If a router determines it has the highest tuple in its neighborhood, it will become an MDR. Otherwise, if there is a router with a higher tuple, the router will ensure that all of its neighbors can reach that router within k hops. If they cannot, the router will still become an MDR. If they can, but only via a single node disjoint path, the router will become a BMDR. If the router determines that it will not be an MDR or BMDR, it is known as an *MDR Other*.

The CDS formed is also used to reduce adjacencies. Each MDR determines which other MDRs it must become adjacent with in order to form a backbone. These MDRs are known as the *dependent neighbors*. Other routers select an MDR with which to become adjacent. Under certain protocol options, routers may select a BMDR and form a second adjacency for robustness. Figure 6.3 displays the adjacencies in an OSPF MDR network. Optionally, there may be no adjacency reduction—only full adjacencies between all links.

Routing calculations are performed using only hops between *routable* neighbors. A neighbor is routable if it is fully adjacent or if it is bidirectional and there exists a route calculated by the SPF. This second class, the nonadjacent bidirectional links, are used to shortcut packets directly when a route exists, similar to the way neighbors are handled in broadcast networks.

The OSPF MDR extension has five levels of advertised link state information. In the first level, only the minimum number of routable neighbors needed to form a backbone is advertised. In the fifth and last level, all routable neighbors are advertised. The second level, known as minimum-cost *LSAs*, advertises all routable neighbors necessary to provide the shortest paths. The third level

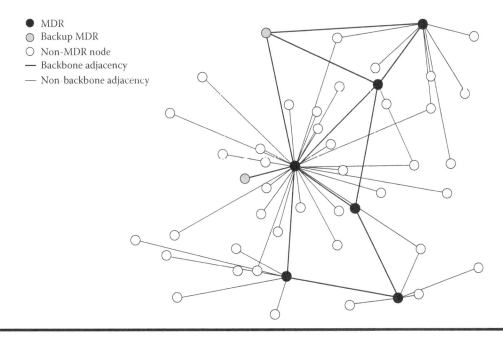

● MDR
◉ Backup MDR
○ Non-MDR node
— Backbone adjacency
— Non-backbone adjacency

Figure 6.3   In OSPF MDR, MDR and BMDR nodes form a backbone of adjacencies. MDR Other nodes form adjacencies with one or more MDR parents. In this small example, the MDR Others have few adjacencies while the MDRs have many. This diagram was adapted from a small OSPF MDR network in simulation.

is like the second but includes more neighbors for redundancy. In the fourth level, MDRs advertise all routable neighbors while all other routers send minimal LSAs. The fourth level does not guarantee the shortest paths but offers a reasonable approximation while decreasing overhead. All routers in the network do not need to synchronize this setting in order to receive some benefit.

Another feature of OSPF MDR is the use of *differential Hellos*. When using this feature, partial Hellos are normally sent, although a full Hello is sent once every set number of partial Hello messages. It is recommended that this be used when the number of neighbors is large. A partial Hello consists only of information on routers that have a different state than that of the last full Hello.

To support OSPF MDR features, the neighbor data structure is to be extended to store more information. This includes each neighbor's parent, bidirectional neighbor set, dependent neighbor set, selected advertised neighbor set, and link metric. Every non-MDR router selects and forms an adjacency with an MDR known as its parent. A selected advertised neighbor is a neighbor that the router includes in its LSA. The Hello messages have been updated to include this information by ordering the list of neighbors into five disjoint sets. The lengths of these lists, as well as other added Hello parameters and link metrics for each neighbor, are appended to the Hello using link local signaling.

6.3.5 OSPF OR/SP Extension

OSPF *overlapping relay* (OR)/*smart peering* (SP) [5,28] is an extension to OSPF for MANET routers, proposed by Cisco Systems. In this protocol, flooding is reduced by only having a subset of nodes, known as ORs, flood the LSAs. Adjacencies are reduced by only synchronizing with nodes to which no adjacent route already exists. This latter feature is referred to as SP

A node's ORs are 1 hop bidirectional neighbors whose own 1-hop neighborhood is not completely contained in the node's 1-hop neighborhood. In other words, these 1-hop neighbors have nodes in the original node's 2-hop neighborhood. Because multiple neighbors may reach the same 2-hop neighbor, they are considered overlapping in this respect, and hence the name "overlapping relay." The ORs relay routing information to the 2-hop neighborhood. However, the 2-hop neighborhood can be covered by a subset of the ORs; thus, the protocol focuses on determining which ORs should be active.

Each node selects a minimum set of active ORs that cover the 2-hop neighborhood. The extension suggests the selection algorithm described for MPR selection in OLSR. An example is constructed in Figure 6.4.

Relaying is done immediately by the active set of ORs. The nonactive set waits on a timer. If the timer expires and the nonactive OR has not overheard an active OR or the originating node, it will retransmit the LSA. It should be noted that all the nonactive ORs may choose to retransmit in this scheme.

SP reduces adjacencies by not requiring or maintaining the neighbor state. Adjacencies are brought up between neighbors only if a route to that neighbor does not already exist. Upon receiving a Hello message, a node will check its link state database to determine whether the neighbor is already present. If it is, no adjacency is formed. The neighbor may be considered an unsynchronized adjacency, because a decision may be made later to bring the state to fully adjacent later.

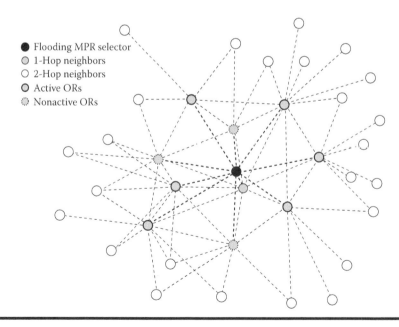

Figure 6.4   All 1-hop neighbors that can reach 2-hop neighbors are considered overlapping relays (ORs). A node selects the active set of ORs to cover the entire 2-hop neighborhood. Nonactive ORs act as backups. Adjacencies are determined by smart peering and not shown in this diagram.

Because SP drastically reduces the number of adjacencies, it may lead to longer calculated routes. To alleviate this problem, unsynchronized adjacencies may be used in route calculation. However, to avoid the circular dependency between route calculation and SP's use of the route calculation, this requires an addition SPF calculation on full adjacencies every time a link change is detected. Similarly, when determining the active OR set, unsynchronized adjacencies may need to be ignored.

Like OSPF MDR, OSPF OR/SP may send Hello messages with only state-change information. In this extension, they are referred to as *incremental Hellos*. This is in part because it is suggested that multiple copies of these Hellos be sent persistently to ensure delivery.

6.3.6 Qualitative Comparison of OSPF Extension

All three extensions are based off of the point-to-multipoint interface model. Each creates dominating sets of routers in order to reduce flooding. However, there are several differences between the extensions, which are summarized in Table 6.1.

Table 6.1 Comparison of OSPF MANET Extensions

	OSPF MPR [26,29]	*OSPF MDR* [27,29,30]	*OSPF OR/SP* [5,28,30]
Hello message size	None	Differential Hellos that may contain only change information	Incremental Hellos that may contain only change information
Control message reliability	Forwarding router retransmits if not acknowledged	Backup mobile ad hoc network designated routers retransmit if they do not overhear acknowledgment	Nonactive overlapping relays (ORs) retransmit
Topology reduction	Only the subset of nodes required for shortest paths are advertised	Five link state advertisement options including backbone only, min-cost path nodes only, and full topology	May advertise all two-way neighbors or just adjacencies
Adjacency reduction	Adjacencies between multipoint relay (MPR) selectors and MPRs are created	Mobile ad hoc network designated routers (MDRs) and BMDRs build an adjacency backbone. Others form an adjacency with their MDR parent	Two nodes form an adjacency if no route exists between them
Flooding reduction technique	Control packets are relayed by flooding MPRs as determined by the optimized link state routing (OLSR) algorithm	MDRs forward control packets. MDRs are determined by largest priority, status, and ID or are promoted when neighbors do not have an MDR in range	An active set of ORs relay packets first. These are determined by the OLSR algorithm

When determining the dominating set, OSPF MDR and OSPF OR/SP utilize the algorithm suggested by OLSR to create a *local* CDS of relaying nodes. OSPF MDR uses its own different method to create a *global* CDS, known as the *backbone.*

Each extension performs retransmission differently. OSPF MPR has the most straightforward approach in which the sending router retransmits if the LSA is not acknowledged. On the other end of the spectrum, in OSPF OR/SP, any nonactive (not selected) OR is eligible for retransmission. In OSPF MDR, specific BMDRs are selected to perform retransmission.

In terms of adjacency reduction, OSPF MPR creates adjacencies between MPRs and MPR selectors. Similarly, in OSPF MDR, adjacencies are created between MDR parents and MDR Other children. However, a backbone of adjacencies is also negotiated between the set of MDRs and BMDRs. In OSPF OR/SP, adjacencies are formed if no route between the nodes already exists.

Topology in OSPF MPR is advertised as the subset of links required to determine the shortest paths (links to path-MPRs). In OSPF OR/SP, adjacencies are advertised. Alternatively, all bidirectional neighbors may be advertised. OSPF MDR also has a range of topology advertising options from only the neighbors on the backbone to all routable neighbors.

Another difference is specialized Hello messages. OSPF MDR and OSPF OR/SP may send partial Hello information with differential Hellos and incremental Hellos, respectively. OSPF MPR does not implement a similar feature.

6.3.7 Performance Comparisons of OSPF Extensions

Spagnolo [25] compared OSPF MDR, OSPF OR/SP, and the nonextended OSPF protocol in simulation using the Georgia Tech Network Simulation (GTNetS) and the Quagga routing daemon. Each scheme was tested using full adjacency/full topology advertisement, minimal adjacency/full topology advertisement, and minimal adjacency/minimal topology advertisement. In addition, simulations were run using the other levels of adjacency formation and topology advertisement that OSPF MDR offers. Measurements included average control message bandwidth, delivery ratio, number of adjacencies, adjacency stability, route stability, route stretching, and LSA generation rate.

Both OSPF MDR and OSPF OR/SP performed similarly, although OSPF MDR was usually better. Both outperformed the base OSPF protocol. Spagnolo hypothesized that the integration of flooding, adjacency, and topology reduction in OSPF MDR, as compared with OSPF OR/SP's piecemeal approach, contributed to its success. Furthermore, OSPF MDR's backbone imposed a structure along the adjacencies that limited data plane route stretching, while OSPF OR/SP suffered from its more arbitrary construction and redundancy in route repair and reflooding. Spagnolo noted that the OSPF MDR backbone does not stretch a route more than a single hop while there is no limit on the OSPF OR/SP stretch factor. Also, OSPF OR/SP had a worse delivery ratio when using minimal topology advertisements due to the longer routes. The reflooding by multiple nonactive ORs was a source of extra control overhead traffic in OSPF OR/SP.

In nonextended OSPF, LSA generation was capped by the protocol's minimum-generation time parameter. Both simulated extensions did not hit that bound. In general, the reduced adjacency and topology options outperformed the full adjacency and topology options across both protocols. Spagnolo found that the extra MDR options that fall between the full and minimal options performed the best in terms of control overhead.

Jun and Sichitiu [31] compared nonextended OSPF, OSPF MPR, and OSPF MDR over networks from 40 to 80 nodes in high and moderate mobility conditions. He observed that while the packet delivery ratio for all three implementations was similar, nonextended OSPF resulted in a much larger number of relayed LSAs, especially as network size grew. Furthermore, OSPF MDR consistently relayed fewer LSAs than OSPF MPR.

Baccelli et al. along with Cordero and co-workers [32] applied some of the techniques from all three extensions to the base OSPF MPR selection and compared them with the simulated results, using GTNetS and Zebra, a predecessor to Quagga. Four protocol variants were created with MPR flooding optimization but using different mechanisms for flooding retransmission, adjacency selection, and topology reduction.

Two of the variants used the OR scheme of control message retransmission, where neighbors retransmit if they do not overhear an acknowledgment. Adjacencies were formed via the SP mechanism. The variants differed in their topology reduction scheme. One advertised both adjacencies and some of the bidirectional links, similar to some of the LSA Fullness options in OSPF MDR. The other listed solely the adjacencies.

The other two variants performed topology reduction in the same way, advertising links to path-MPR neighbors. The first one had sending routers perform LSA retransmission for adjacent neighbors and created adjacencies only between MPRs and their selectors. The second variant's flooding retransmission was also performed by the sending router but for both adjacent neighbors and MPR selectors. The second also had a different adjacency selection scheme—only creating adjacencies with neighbors not pruned during the creation of the relative neighbor graph. This method was referred to as *synchronized link overlay* (SLO-T) selection.

Baccelli et al. noted that both SP and SLO-T create a connected graph but do not guarantee the shortest paths, while creating adjacencies only between MPRs and their selectors guarantees the shortest paths but not a connected set. It was suggested that the synch router be used to ensure connectivity. Because SLO-T creates the smallest graph, it creates the least adjacencies per node, with SP creating the second least. However, SLO-T's adjacencies had a significantly shorter lifetime than that of SP or MPR selector adjacency selection.

Similar to Spagnolo's findings, Baccelli et al. also observed that the OR approach to retransmission yielded more retransmitted LSAs. It was also noted that this approach was more complex due to issues in synchronization and buffer management.

Examining route length, topology reduction by advertising path-MPR neighbors significantly outperformed the other two scenarios. In terms of data overhead, listing all the adjacencies in the LSA wasted more bandwidth when compared with the other two topology-reduction mechanisms.

In addition to the variants explored earlier, Baccelli et al. also examined the effect of Hello message reduction due to OSPF OR/SP's incremental Hellos and OSPF MDR's differential Hellos. It was found that these mechanisms reduce control traffic by less than 2% of the total, in part due to the small portion of the control traffic that normal Hellos contribute.

Based on the simulation findings, Baccelli et al. suggested that mechanisms that preserve the shortest paths are a necessity in protocol design. However, synchronization may be relaxed, especially because of the short lifetime of links in MANETs. As a synchronized protocol, the variant that retransmits along adjacencies, selects adjacencies based on MPR selection, and advertises links to path-MPRs was found to perform best. For unsynchronized protocols, Baccelli et al. suggested retransmitting both along adjacencies and MPR relationships, SP adjacency selection, and advertising links to all path-MPRs.

Ogier [29,30] compared OSPF MDR with the other two extensions separately by using updated GTNetS/Quagga code. When simulating OSPF MDR and OSPF MPR, he tested with OSPF MDR in both minimal LSA and min-cost LSA settings. For OSPF MDR and OSPF OR/

SP, he tested both protocols in minimal LSA and full topology LSA settings. A range of network sizes, from 40 to 200 nodes, were simulated.

Ogier observed that OSPF MDR resulted in the least number of adjacencies per node and adjacencies formed per second, especially when compared with OSPF MPR. As network sizes grew, OSPF MDR's adjacencies per node remained constant while OSPF MPR's grew linearly. OSPF OR/SP's adjacencies per node also grew with network size but more slowly than in OSPF MPR and starting at a fraction of the total number. The number of adjacencies per node affects the amount of overhead required for database description packets. As such, OSPF MPR had much greater overhead than that of OSPF MDR.

As Spagnolo found, with full topology LSAs and modifications to OSPF OR/SP's backup relay system, OSPF MDR and OSPF OR/SP performed similarly. When OSPF MDR's adjacency reduction was used, its route lengths were slightly longer due to more stringent conditions for nodes to recognize new neighbors.

With minimal LSAs, OSPF MDR exhibited significantly less overhead than OSPF OR/SP. This was in part because OSPF OR/SP's nonactive overlapping relays can result in redundant backup flooding from multiple nonactive ORs and because the greater number of adjacencies results in more database description packets being sent. Furthermore, it was observed that OSPF OR/SP generated twice as much overhead due to Hello packets.

Ogier found OSPF MPR's packet delivery ratio to be the lowest. He hypothesized this was due to OSPF MPR not having a direct backup relay system and instead relying on the originating MPR to retransmit and MPR selection to deal with link failure. This method requires an extra exchange of selector information via Hello messages. OSPF MDR and OSPF OR/SP's delivery ratio results were more similar under both LSA settings, with OSPF MDR performing slightly better in the minimal LSA simulations.

6.4 Conclusion and Future Directions

We have presented an overview of routing in MANETs, with a special focus on MANET extensions to the most popular Internet routing protocol, OSPF. We described common techniques and classifications of MANET routing algorithms and discussed well-known examples of proactive, reactive, and hybrid protocols. With that background in place, we described OSPF and its MANET extensions in detail. We then reviewed the differences between the OSPF extensions and summarized their performance in simulation.

There is still much to be explored in OSPF MANET extensions. In addition to the reductions in topology, adjacency, and flooding made by the existing extensions, there are other techniques that may be considered. There has been some success with applying scope reduction techniques similar to those used in FSLS [31]. Methods from geographical, hierarchical, and multipath protocols could also lead to improvements in the existing extensions.

Jun et al. [33] explored adapting OSPF areas, a hierarchical scheme from the base OSPF protocol, to MANET. In analyzing the balance between containing flooding within areas and increasing flooding due to nodes moving between areas, he estimated the optimal number of areas. In practice, he suggested that the network could consider using geography, activity, or topology to select areas dynamically.

Ogier [29,30] has explained that adjacency reduction is essential not only for the reduction in database description exchanges but also for reducing the control traffic associated with external networks. Simulations of how OSPF MANETs behave with other networks will be elucidating.

Security concerns should also be acknowledged by routing protocol designers. Possible techniques include having nodes obtain certificates through another network or negotiate their own certificates [34], having some nodes act as watchdogs [14], exploring network diagnostic path rating packets [14], establishing trust rankings [14,35], inserting packets to hide traffic patterns [2], using private–public keys or other digital signatures [2], and giving incentives for good behavior [11].

Because each node must maximize their level of service and determine methods of dealing with misbehaving nodes, game theoretical techniques may prove useful. The self-organization aspect of MANETs may also be improved by exploring learning algorithms for the nodes [11].

The existing OSPF MANET extensions provide a strong base upon which to build. By applying other promising techniques, they may be further improved, allowing for efficient and scalable MANETs.

Acknowledgment

This work was supported in part by the 2009–2010 collaborative research experiences for undergraduates award, sponsored by the Computing Research Association's Committee on the Status of Women in Computing Research and the Coalition to Diversify Computing and funded by the CRA and the National Science Foundation.

References

1. J.P. Macker and S.M. Corson. Mobile ad hoc networks (MANETs): routing technology for dynamic wireless networking. In S. Basagni, M. Conti, S. Giordano, and I. Stojmenovic (eds.), *Mobile Ad Hoc Networking*. New York: IEEE Press, 2004, pp. 273–292.
2. J.J.-N. Liu and I. Chlamtac. Mobile ad hoc networking with a view of 4G wireless: imperatives and challenges. In S. Basagni, M. Conti, S. Giordano, and I. Stojmenovic (eds.), *Mobile Ad Hoc Networking*. New York: IEEE Press, 2004, pp. 1–45.
3. J. Moy. OSPF version 2. IETF Request for Comments 2328, April 1998.
4. R. Colton, D. Ferguson, J. Moy, and A. Lindem. OSPF for IPv6. IETF Request for Comments 5340, July 2008.
5. A. Roy and M. Chandra. Extensions to OSPF to support mobile ad hoc networking. IETF Request for Comments 5820, March 2010.
6. C. Perkins, E. Belding-Royer, and S. Das. Ad hoc on-demand distance vector (AODV) routing. IETF Request for Comments 3561, July 2003.
7. D. Johnson, Y. Hu, and D. Maltz. The dynamic source routing protocol (DSR) for mobile ad hoc networks for IPv4. IETF Request for Comments 4728, February 2007.
8. T. Clausen, C. Dearlove, and P. Jacquet. The optimized link state routing protocol version 2. IETF Internet Drafts, September 2009.
9. Z.J. Haas, M.R. Pearlman, and P. Samar. The zone routing protocol (ZRP) for ad hoc networks. IETF Internet Drafts, July 2002.
10. A. Santivanez, R. Ramanathan, and I. Stravrakakis. Making link-state routing scale for ad hoc networks. In *MobiHoc'01: Proceedings of the 2nd ACM International Symposium of Mobile Ad hoc Networking & Computing*, 2001, pp. 22–33.
11. S.G. Glisic and B. Lorenzo. *Advanced Wireless Technologies: Cognitive, Cooperative & Opportunistic 4G Technology*, 2nd ed. New Jersey: John Wiley & Sons, 2009.

12. X.-Y. Li. Topology control in wireless ad hoc networks. In S. Basagni, M. Conti, S. Giordano, and I. Stojmenovic (eds.), *Mobile Ad Hoc Networking*. New York: IEEE Press, 2004, pp. 175–203.

13. L.L. Peterson and B.S. Davie. *Computer Networks*, 4th ed. New York: Morgan Kaufmann, 2007.

14. E.M. Belding-Royer. Routing approaches in mobile ad hoc networks. In S. Basagni, M. Conti, S. Giordano, and I. Stojmenovic (eds.), *Mobile Ad Hoc Networking*. New York: IEEE Press, 2004, pp. 293–318.

15. P. Kumar and S. Ramachandram. The performance evaluation of cached genetic zone routing protocol for MANETs. In *Networks, 2008. ICON 2008*. New York: IEEE, 2008, pp. 1–5.

16. H. Cheng, J. Cao, and X. Fan. GMZRP: geography-aided multicast zone routing protocol in mobile ad hoc networks. In *QShine'08: Proceedings of the 5th International ICST Conference on Heterogeneous Networking for Quality, Reliability, Security and Robustness*. Brussels: Institute for Computer Sciences, Social-Informatics and Telecommunications Engineering, 2008, pp. 1–7.

17. B. Karp and H.T. Kung. GPSR: greedy perimeter stateless routing for wireless networks. In *MobiCom'00: Proceedings of the 6th Annual International Conference on Mobile Computing and Networking*. New York: ACM, 2000, pp. 243–254.

18. Y.-B. Ko and N.H. Vaidya. Location-aided routing (LAR) in mobile ad hoc networks. *Wireless Networks*, vol. 6, no. 4, 2000, pp. 307–321.

19. A. Huhtonen. Comparing AODV and OLSR routing protocols, Helsinki University of Technology Telecommunications Software and Multimedia Laboratory, TML-C15, 2004.

20. A. Kumar B.R., L.C. Reddy, and P.S. Hiremath. Performance comparison of wireless mobile ad-hoc network routing protocols. *International Journal of Computer Science and Network Security*, vol. 8, 2008, pp. 337–343.

21. J. Xie, L. Quesada, and Y. Jiang. A threshold-based hybrid routing protocol for MANET. In *Wireless Communication Systems*, 2007 (ISWCS 2007). Norway: IEEE, 2007, pp. 622–626.

22. Z.J. Haas, M.R. Pearlman, and P. Samar. The intrazone routing protocol (IARP) for ad hoc networks. IETF Internet Drafts, July 2002.

23. Z.J. Haas, M.R. Pearlman, and P. Samar. The interzone routing protocol (IERP) for ad hoc networks. IETF Internet Drafts, July 2002.

24. Z.J. Haas, M.R. Pearlman, and P. Samar. The bordercast resolution protocol (BRP) for ad hoc networks. IETF Internet Drafts, July 2002.

25. P. Spagnolo. Comparison of proposed OSPF MANET extensions. In *Military Communications Conference*, 2006 (*MILCOM 2006*). Washington, DC: IEEE, 2006, pp. 1–7.

26. E. Baccelli, P. Jacquet, D. Nguyen, and T. Clausen multipoint relay (MPR) extension for ad hoc networks. IETF Request for Comments 5449, February 2009.

27. R. Ogier and P. Spagnolo. Mobile ad hoc network (MANET) extension of OSPF using connected dominating set (CDS) flooding. IETF Request for Comments 5614, August 2009.

28. A. Roy. Adjacency reduction in OSPF using SPT reachability. IETF Internet Drafts, November 2005.

29. R. Ogier. Comparison of OSPF-MDR and OSPF-MPR. IETF Internet Drafts, September 2010.

30. R. Ogier. Comparison of OSPF-MDR and OSPF-OR. IETF Internet Drafts, September 2010.

31. J. Jun and M.L. Sichitiu. Scalable OSPF updates for MANETs. In *Proceedings of the IEEE Globecom 06—WASNet*, November 2006.

32. E. Baccelli, J.A. Cordero, and P. Jacquet. Multi-point relaying techniques with OSPF on ad hoc networks. In *Systems and Networks Communications, 2009* (*INSNC'09*), September 2009, pp. 53–62.

33. J. Jun, M.L. Sichitiu, H.D. Flores, and S.J. Eidenbenz. The optimum number of OSPF areas for MANETs. In *Proceedings of the Fourth Annual IEEE Communications Society Conference on Sensor*, Mesh and Ad Hoc Communications and Networks (SECON 2007), June 2007.

34. R. Chang, S. Gundala, T.-S. Moh, and M. Moh. VESS: a versatile extensible security suite for MANET. In *Proceedings of the IEEE Pacific Rim Conference on Communications, Computers and Signal Processing (PACRIM'09), Victorian, Canada*, August 2009, pp. 944–950.

35. J. Li, T.-S. Moh, and M. Moh. Path-based reputation system for MANET routing. In *Proceedings of the 7th International Conference on Wired/Wireless Internet Communications, Enschede, The Netherlands*, May 2009, pp. 48–60.

Chapter 7

New Approaches to Mobile Ad Hoc Network Routing: Application of Intelligent Optimization Techniques to Multicriteria Routing

Bego Blanco Jauregui and Fidel Liberal Malaina

Contents

The growing interest in mobile ad hoc networks (MANETs) is reflected in the noticeable publication and standardization effort during the past decade. The new standards always include a certain routing protocol but accept the use of other possibilities as well; the reason is that there is no single routing protocol that perfectly adapts to any scenario. A feasible solution is the design of adaptive intelligent protocols capable of readjusting their routing strategy to the dynamic nature of MANETs according to the requirements of the end user. This next generation of routing protocols should combine features of both multicriteria optimization and artificial intelligence to provide the best global performance at the least computational cost.

7.1 Introduction

The growing interest in routing protocols in MANETs is reflected in the noticeable publication and standardization effort during the past decade. Since IETF MANET Working Group (MANET WG) was created in the late 1990s [1] to centralize the standardization task, dozens of routing solutions for ad hoc networks have been proposed. However, even though the advantages and potential applications of MANETs have been clear for a long time, their development has not been as successful as it seemed in the initial studies of the area.

The analysis of the literature leads to two main reasons for the slowing down in the definitive expansion of this technology. On the one hand, it is noticeable that a fraction of these proposed and published solutions are just proposals of routing techniques suitable to be included into an existing routing protocol. The difficulty of deploying real-world test beds to obtain reliable results has inclined researchers to simply suggest mathematically feasible algorithms that could eventually complement any existing routing protocol.

On the other hand, no suitable routing protocol has been designed flexible enough to provide required performance levels in every considered scenario. Although MANETs were initially conceived for a military/emergency use, their scope has been rather widened toward more commercial applications with quite different service requirements. As a result, the design of a single protocol with acceptable (at least) performance when running in any scenario has become a more complicated challenge.

This context causes the proliferation of numerous MANET routing protocol designs (see Table 7.1 for a compilation of the most relevant proposals). In such a scenario, the initial goal of the MANET WG (i.e., standardizing IP routing protocol functionality suitable for ad hoc mobile networks for both static and dynamic scenarios) became increasingly difficult. In view of the impossibility of finding a routing scheme that properly suited any scenario, this WG has decided to eventually develop two routing protocol specifications: one for a proactive routing protocol and another for a reactive routing protocol. At present, the selected protocols are Optimized Link State Routing protocol version 2 (OLSRv2) [2] as a proactive protocol and Dynamic Manet On-demand (DYMO) [3] as a reactive protocol. However, provided that a significant commonality between Reactive MANET Routing Protocol (RMRP) and Proactive MANET Routing Protocol (PMRP) protocol modules is observed, the WG may decide to go with a converged approach [1].

Table 7.1   Classification of MANET Routing Protocols

| Unicast | | | | | | | Multicast | |
| Proactive | | | Reactive | | Hybrid | | | |
Flat	Hierarchical	Position Based	Flat	Position Based	Flat	Hierarchical	Flat	Geocast
ARS	CBCCR	DFR	ABR	GRID	ADV	CBRP	ABAM	LBM
CCBR	CGSR	DREAM	ADQR	LAR	Ant-AODV	IZR	ADMR	GeoGRID
DBF	DST	GPSR	ADRA	LOTAR	AntHocNet	VBR	AMRIS	GeoTORA
DSDV	DDR	GRA	AEADMRA	VADD	CEDAR	ZRP	AMRoute	MRGR
IARP	FSR	NEAR	AMQR		DDR		CAMP	Mobicast
IQR	GSR	Octopus	AODV		HARP		CBM	
MMRP	HSR	QoS-GPSR	BEQR		HSLS		DCMP	
NSR	LANMAR	ZHLS	BSR		GLS		DDM	
OLSR	WHIRL		CAAODV		GPSAL		DSR-MB	
PERA			CACP				FGMP	
PLBQR			CHAMP				LAM	
STAR			DAR				MAODV	
TBRPF			DNVR				MCEDAR	
WRP			DSR				MZR	
			DYMO				ODRMP	
			EAAR				SOM	

(Continued)

Table 7.1 (Continued) Classification of MANET Routing Protocols

Unicast					Hybrid		Multicast	
Proactive			Reactive					
Flat	Hierarchical	Position Based	Flat	Position Based	Flat	Hierarchical	Flat	Geocast
			EADSR				SMF	
			FSDSR				SPBM	
			IERP				SRMP	
			ISAIAH					
			LUNAR					
			ODCR					
			OQR					
			OLMQR					
			PBAR					
			QoS-AODV					
			RDMAR					
			ROAM					
			SAMPLE					
			SBR					
			SSR					
			TBPTDR					
			TORA					
			WAR					

In addition to the aforementioned review of the state of the art of routing in MANETs, a deeper analysis of the temporal evolution of all this development is also noteworthy.

Figure 7.1 shows a first boom of the development of routing protocols for MANETs around the year 2000, after the first version of OLSR was published in 1999. It was now the time of PMRPs, which ended up with the release of RFC 3623 with the specification of OLSR in 2003. Around the year 2005, there was another peak that corresponds, to some extent, to the growing interest in reactive protocols. In fact, this peak coincides with the first publication of DYMO.

Nevertheless, the study of the evolution of the single-metric routing protocols for MANETs, that is, those routing protocols that implement an algorithm whose final aim is to find the best-effort route without considering any other requirement, indicates that the research on MANET-specific routing protocol experiments saw its maximum development evolution around the year 2000, with a subsequent loss of attention (Figure 7.2). The reason is the emergence of new research areas in the field of MANETs, produced by the radical shift in the requirements of the new applications of ad hoc networks. Ad hoc networks have gained growing interest in scenarios such as sensor or vehicular networks and the new portable game platforms. As a result, the prevailing traffic becomes multimedia-related and issues regarding quality of service (QoS) gather increasing relevance.

These new scenarios depict a situation that absolutely differs from that of wired networks and even infrastructured wireless networks, and it must be taken into account that in addition to the

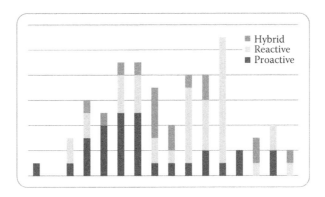

Figure 7.1 Evolution of MANET routing protocols.

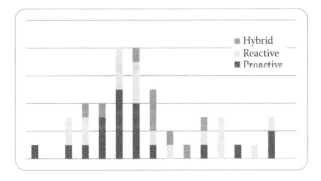

Figure 7.2 Evolution of single-metric routing protocols for MANETs.

habitual QoS issues of traditional networks, MANETs involve new challenges to guarantee a set of minimum quality levels [4–6].

All these new criteria, such as mobility or energy awareness, have been incorporated into different routing protocols, and therefore, from now on, the classical QoS parameters will not be the only ones taken into consideration. The QoS management systems must interoperate with those adopted routing schemes, and the resulting interactions must be carefully analyzed.

Table 7.2 represents a summary of the most relevant QoS routing protocols developed for MANETs [7,8].

The maximum interest in the development of protocols with QoS support for multihop MANETs was gained around 2003–2005, coinciding with the loss of attention in MANET routing protocols with no QoS awareness. However, in this case as well, the relevance of the simple inclusion of QoS criteria into routing protocols eventually decreases (Figure 7.3).

This new approach to MANET routing also does not succeed in providing a feasible solution to every scenario, because it does not take into account that the provision of QoS is not the same for all the applications or scenarios. Numerous metrics must be taken into account when assessing the QoS in a MANET.

Table 7.3 shows which routing type and routing strategy theoretically achieve the best performance in each scenario. The results obtained can be applied to the selection of the most suitable routing protocol for each particular case of application of a MANET.

Despite these theoretical values, generally speaking there are just two extreme scenarios where a proactive protocol performs better than a reactive protocol: applications with a highly delay-sensitive traffic and static or low-mobility scenarios. Therefore (and even from an historical point of view), research on MANET routing started with the proposal of proactive algorithms, and the research effort has since focused on the development of reactive MANET routing protocols.

However, not only real-world applications demand the provision of a single QoS parameter, but it is quite probable that the mentioned applications need the guarantee of several metrics simultaneously. Furthermore, very likely these metrics can be conflicting, which forces one to reach a trade-off solution. Here is where research on multicriteria optimization finds its place, widely applied in other fields, and whose contribution to QoS proves to be very valuable.

The next section studies this new approach to MANET routing in detail, describing available multicriteria optimization techniques and summarizing recent proposals of applying mathematical methods to MANETs.

7.2 Multicriteria Optimization Applied to MANET Routing

The need to satisfy multiple conflicting service requirements is a classic problem of resource assignment. This concept has been deeply studied in other areas, such as economics, under the terms *decision making* and *operational research*. In these areas, an optimization process is a constructive approach to a proper decision-making procedure, whose final goal is to allocate a limited number of resources in the most effective way.

This section analyzes the mathematical characteristics of a multicriteria optimization problem, as a first step to the analysis of the existing multicriteria optimization methods for decision making and its subsequent application to MANET routing.

Table 7.2 MANET Routing Protocols with QoS Support

Protocol	Provided QoS Guarantee
AAQR	Limited jitter and delay; bandwidth
ADQR	Bandwidth
AQOR	Bandwidth; limited delay
BFQR	Bandwidth
CAAODV	Bandwidth
CACP	Bandwidth
CBCCR	Bandwidth
CCBR	Bandwidth
CEDAR	Bandwidth
CLMCQR	Bandwidth; limited delay; limited packet police
DSARP	Limited jitter; limited delay
EBR	Route lifetime
GAMAN	Limited delay; limited packet police
HARP	Limited delay; limited congestion; route lifetime
IAR	Bandwidth
LSBR	Limited probability of route failure
MRPC	Route lifetime; reduced power consumption; limited packet police
NSR	Bandwidth
ODCR	Guaranteed maximum delay
PANDA	Route lifetime
QAODV	Bandwidth; limited delay
QGUM	Bandwidth; guaranteed maximum delay
QMRP	Guaranteed maximum delay
QOLSR	Improved bandwidth; limited delay
SIRCCR	Bandwidth; limited delay
TBR	Bandwidth; limited delay
TDR	Bandwidth

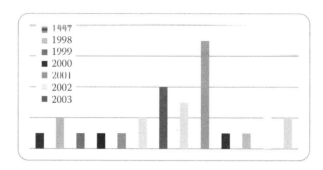

Figure 7.3 Temporal evolution of QoS routing in MANETs.

Table 7.3 Theoretical Comparison of the Different MANET Routing Types and Strategies

	Routing Type		Routing Strategy	
	Proactive	*Reactive*	*Flat*	*Hierarchical/Position Based*
Mobility				
High		X	X	
Low	X			X
Number of nodes				
High		X		X
Low	X		X	
Node density				
High		X	X	
Low	X		X	X
Number of connections				
High	X			X
Low		X	X	
Traffic density				
High		X		X
Low	X		X	

7.2.1 Characterization of an Optimization Problem

Optimization is the process of minimizing or maximizing an objective single- or multivariable function, which can be subject to a number of equality or inequality restrictions. Common components of any optimization problem are listed below [9]:

- Decision or design variable(s): These are the unknown parameters of the real-world problem that can be modified to obtain different results. Depending on the number of decision variables involved, the problem can be single variable or multivariable. In the case of routing protocols, these variables are related to the design parameters of the algorithms that those protocols implement.
- Objective functions: These functions are a representation of the decision maker that must be optimized. When there is only one objective function, the problem is named single-criterion or single-objective. On the contrary, a problem with several objective functions is known as a *multicriteria* or *multiobjective problem* (i.e., considering either a single-metric or multiple-metric routing protocol).
- Restrictions to decision or design space: The decision variables of the problem can be restricted or nonrestricted, depending on the existence of equality or inequality restrictions. In most routing scenarios, these constraints will be derived from bandwidth/delay/number of hops requirements.

For simplicity, from now on, the optimization problem will be analyzed first from a theoretical point of view and applied later (in Section 7.2.2) to the specific routing use case. Furthermore, for simplicity purposes it will be considered a minimization problem, since this is the most usual situation, keeping in mind that any conclusion could be extensible to the maximization problem.

7.2.1.1 Single-Variable and Single-Criterion Optimization Problem

The final aim of any optimization problem is to obtain a local or global minimum of the objective function $f(x)$ (Figure 7.4). A single-variable objective function $f(x)$ has a global minimum in $x = x^*$ when

$$f(x^*) \leq f(x^* + h) \quad \forall h \tag{7.1}$$

A single-variable objective function $f(x)$ has a local minimum in $x = x^*$ when

$$f(x^*) \leq f(x^* + h) \tag{7.2}$$

for values of h sufficiently close to 0. Therefore, the single-variable, single-criterion optimization problem provides a way of selecting the "best" possible value of the input parameter (design or decision variable) to optimize the single criterion or objective function $f(x)$.

The minimization problem can also be expressed as

$$\min\{f(x) = z\} \tag{7.3}$$

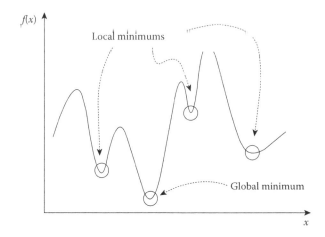

Figure 7.4   Local and global minimums.

7.2.1.2 Multivariable Single-Criterion Optimization Problem

In this case, the value of the single criterion analyzed depends on several factors or input parameters and, as a result, the multivariable single-criterion optimization problem provides a way of selecting the "best" possible combination of these input parameters to optimize the single criterion considered.

Assuming x is any combination of input parameters, the multivariable single-criterion optimization problem has the following definition:

$$\min\{f(x) = z\} \text{ with } x \in S \tag{7.4}$$

where S is the feasible region (often also called *feasible design space, feasible decision space,* or *restricted set*).

The concept of S leads to the classification of optimization problems according to another of their defining features: the format of the problem. Some problems do not need the solution to meet any prerequisite. In these cases, $S \equiv R^n$, and the optimization is generally called "restriction-less." However, usually x will be subject to a number of constraints and, hence, S will be a subset of $S \equiv R^n$ defined by a group of equality and/or inequality functions that members of S must satisfy. This kind of problem is known as "restricted" and is mathematically formulated as

$$\min\{f(x) = z\} \tag{7.5}$$

$$\text{with } g_i(x) \leq 0 \ \forall i = 1, 2, \ldots, m$$

$$\text{with } h_j(x) \leq 0 \ \forall j = 1, 2, \ldots, l$$

Therefore, the single-criterion optimization problem can be defined in short as "the search for x in S so that $f(x) = z$ is maximum."

7.2.1.3 Multicriteria Optimization Problem

Not only routing but many real-world problems are often based on the existence of several opposite objectives or criteria, which has led to the multicriteria or multiobjective problem approach. The basic aim of this approach is to minimize all the criteria simultaneously and meet the equality and inequality constraints of the feasible space S.

The multiobjective problem can thus be defined as an extension of the single-criterion problem:

$$\min\{f_1(x) = z\}$$
$$\min\{f_2(x) = z\}$$
$$\vdots \quad\quad\quad\quad (7.6)$$
$$\min\{f_k(x) = z\}$$
$$\text{with } x \in S$$

where f_i is the i-th objective function.

The simplest solution to this kind of problem lies in finding the input vector x^*, which satisfies

$$\exists x^* \in S \,\Big|\, \min\{f_i(x^*) = z_i^*\} \quad \forall i = 1, 2, \ldots, k \quad\quad (7.7)$$

Figure 7.5 depicts an optimization problem with two variables or parameters to optimize (x_1, x_2) and two criteria (f_1, f_2) for a space S constrained by two inequality functions (g_1, g_2). The feasible region S in the design space is mapped into Y in the criteria space, which is constrained by q_1, q_2, which correspond to g_1, g_2 in this space. The optimal solution $f^* = f(x^*)$ that minimizes all the criteria simultaneously is seldom achievable. In this case, it is known as a *utopian solution* [10]. The solution to the problem should then be as close as possible to such a utopian solution, resulting in the Pareto-optimum front ϕ.

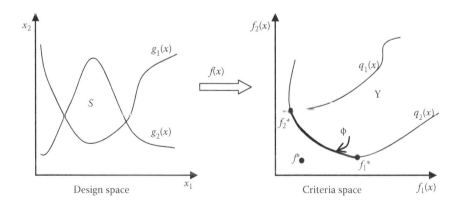

Figure 7.5 Graphic representation of a bivariable and bicriteria optimization problem in design and criteria spaces.

In most cases, there is no x^* that simultaneously minimizes all the criteria. Thus, it is necessary to redefine the optimization problem to attain, from within the whole feasible set of possible solutions, the solution that optimizes the global result, for example, that which is closer to the utopian solution. As shown in Figure 7.5, this solution often is not unique, and the Pareto front is composed of a set of possible solutions. The final aim of a high-quality multicriteria optimization problem is to find a sufficiently representative solution set, homogeneously distributed along that Pareto front.

7.2.1.4 Taxonomy of a Multicriteria Optimization Problem

Optimization problems can be classified according to a wide variety of perspectives, resulting in a general taxonomy of the multicriteria optimization problems composed from the works in [9,11–15]:

■ Derivative vs. nonderivative: This classification is according to whether the objective functions to be optimized can be derived using the resolution algorithm. This point settles the possibility for the optimization problem to explore the feasible region in a simple way by the use of a gradient.

■ Qualitative vs. quantitative criteria: This classification is according to the way of measuring the analyzed criteria. When the criteria that model the preferences of the decision maker are quantitative, they are represented by a numeric value, indicating the preference degree of one preference against the others. When the criteria are qualitative, preference lacks an exact numeric value and a descriptive value is assigned.

■ Preference articulation: The moment the preferences of the decision maker are given classifies optimization problems into another four categories:
  – *A priori* preference articulation: These preferences add additional restrictions to the problem (weighted sum and lexicographic methods).
  – *A posteriori* preference articulation: The final solution is selected from the set of optimal solutions that result from the optimization process (i.e., in evolutive and genetic methods).
  – Progressive preference articulation: The decision maker's preferences are gradually incorporated in an interactive way during the optimization process.
  – Without preference articulation: When the decision maker cannot (or does not want to) define the preference for the solution set (i.e., max–min formulation, global criterion method).

■ Continuous vs. discrete: This classification refers to the variable type to work with. Some optimization problems are restricted to the manipulation of discrete variables, such as integers, binary values, or other abstract objects. The goal of the discrete optimization is to attain the optimum point from a finite, but usually huge, set. Continuous optimization problems, on the contrary, operate with infinite variable values. Continuous problems are usually easier to solve due to their predictability, for the solution can be achieved with an approximative iterative process.

■ Constrained vs. nonconstrained: In some cases, the purpose of multiobjective optimization problems is not only to find a solution that optimizes a set of criteria, but also to must meet a series of requirements expressed through (in)equality equations. Nonconstrained methods are also used to solve constrained methods, substituting restrictions for penalizations on objective functions to prevent possible constraint violations.

- Global vs. local: Global optimization methods search for the solution all over the feasible space. But some applications just need to find a locally optimum solution near a certain point. A global solution is also locally optimal, but the contrary is not always true. However, many global optimization methods make use of local methods to locate a first approach to the solution.
- Stochastic vs. deterministic: In some cases, the model to be optimized is not completely described because some specifications remain unknown at the analysis of the problem. In these situations, and in contrast with deterministic methods, stochastic optimization problems use uncertainty quantifications to achieve an optimum solution for the expected model.

This classification does not result into disjoint categories, but a multicriteria optimization problem can fall into one or several of the categories listed above. This is the main difficulty in finding a definitive taxonomy of multicriteria optimization problems or methods in the existing literature.

7.2.1.5 Preference or Criteria Modeling

Before beginning with the process of decision-making analysis of a multicriteria optimization problem, there is a fundamental first step: the characterization of the criteria to be optimized, i.e., the preferences of the decision maker about the adequacy of the selected solution.

A key factor in the analysis for decision making is the fact that the functions that model the decision maker's preferences (criteria or objective functions) are not usually known *a priori*. In this sense, several methodologies have been proposed within the field of multicriteria mathematical programming in order to develop a successful interaction with the decision maker that makes it possible to provide relevant information about local solutions to the decision maker and to get useful information about the decision maker's preferences at the same time.

It must be kept in mind that the final goal is to establish an ordered ranking as a natural foundation to solve decision or selection problems.

7.2.1.6 Valuation Scale

A preference model establishes a formal representation of the comparison between alternatives that express both the structure of the described situation and the variety of manipulations that can be made on it [16]. Logical language proves to be appropriate for these descriptions. Classic logic, however, can turn out to be too rigid to sufficiently define expressive models. Hence, other formalisms must be considered to provide the model with the required flexibility.

The preference or criteria representation format plays an important role since it defines the nature and structure of the information on which the decision maker bases his inclinations toward the different alternatives. Experts from each field of knowledge use the representation format that better applies to their area of expertise. In some cases, preferences will be expressed through numerical values, and in other cases, more natural expressions, such as words or linguistic terms, are used.

The final objective is the comparison of potential actions in order to make a decision, so it becomes mandatory to determine a scale for every considered criterion. The elements of the scale are denoted degrees, levels, or ranks. When comparing the behavior of the different options, it is important to analyze the specific meaning in terms of preference associated with each criterion. This leads to the differentiation of two main types of scales [17]:

- Pure ordinal or qualitative scale: In these cases, the gap between two levels or degrees has no clear meaning regarding preference disparity. This has no relation to the levels expressed in a verbal or numerical way. For example, talking about temperatures, 20°C does not mean twice as hot as 10°C, and the increase from 5 to 8°C does not feel like the increase from 15 to 18°C, even when the numerical difference in both gaps remains the same. In this case, the aggregation of criteria using an arithmetical procedure makes no sense due to the lack of real numerical meaning of the graduation.

■ Pure cardinal or quantitative scale: It is a numerical scale with definite meaning regarding the relative difference of preference between two levels, regardless of the levels being considered.

Between these two opposite cases, there is a wide variety of possibilities, especially in the case of interval scales.

Once the scale is determined, the comparison between two alternatives boils down to one of the following scenarios [16]:

■ To determine whether option A is preferable to option B.
■ To check whether option A is close to option B in the sense of being considered indifferent from the decision maker's point of view.

7.2.1.7 Objectivity and Uncertainty

To build a model from reality is always an abstraction drill. Even in the case of a highly objective decision maker, this objectivity is limited, basically due to the following factors [18]:

■ The boundary between feasible and nonfeasible solutions is often fuzzy and varies as the analysis of the scenario gets deeper.
■ Even for a well-defined decision maker, the preferences are seldom perfectly characterized. Uncertainty zones, conflicts, or contradictions almost always exist and constitute a source of ambiguity or arbitrariness.
■ Available information is often imprecise, uncertain, or is poorly defined, leading to misinterpretation.
■ Optimization processes are objective by definition, avoiding subjective aspects associated with organizational, cultural, or social fairness considerations.

The goal is to compose a preference model based on incomplete or biased information and to know to what extent uncertainty or ambiguity is propagated along a model built in those circumstances. In this case, there are two possibilities:

■ When there is a well-known preference relation, but there is not enough available information to build the model, probability distributions are used to associate an uncertainty level to each statement.
■ When the concept of preference itself is poorly defined, independently from the available information, multivalued logic is used to associate each statement on preference with a value representing its "veracity intensity."

7.2.2 Multicriteria Optimization Methods

Once the optimization problem is formulated, an optimization method is used to find a solution, usually aided by a computer. An optimization algorithm usually implies an iterative process.

Starting from an initial estimation of the solution, the algorithm carries out consecutive approaches toward the solution until it reaches a termination point. Obtaining a solution is, however, not always guaranteed.

The search strategy is what differentiates each method, and there is no universal method applicable to any kind of problem, but a variety of methods are developed to fit each situation. The selection of the most appropriate method for a certain scenario is an important choice, since it affects the response time or even actually finding the solution. Any good optimization method should encompass the following features:

- Robustness: Assuring a good performance in a wide variety of similar problems for all reasonable initial values
- Efficiency: Without demanding too much computational time or storage space
- Accuracy: Being able to precisely identify a solution and without being too sensitive to data errors or arithmetical rounding in the computer

These features are often conflicting, due to, for example, the fact that a fast convergence may require a huge amount of storage space or that a robust method can be very slow at the same time. Both limitations are particularly significant in a highly dynamic environment such as a MANET. Therefore, the trade-off between opposite criteria must gain special consideration.

Keeping in mind the preceding objectives and due to the multicriteria optimization having been an intensively studied field during the last years, the development of multicriteria optimization approaches has been substantial from exact to metaheuristic methods, including algorithms of diverse nature.

There are plenty of partial classifications in [19,20–24], but since these families of multicriteria optimization methods do not constitute disjoint groups, their full categorization is a difficult task.

7.2.2.1 Weighted Sum Methods

The foundation of the weighted sum multicriteria optimization problems (also known as *objective function aggregation methods*) is the transformation of the vectorial (multiobjective) problem into a scalar (single-objective) problem as a result of the aggregation of all the criteria into a single expression by the assignment of weights [25].

These methods are easy to understand and use, and the concept of weight assigned to each objective function as a measure of its relative importance is a clear indicator of the decision maker's preferences. However, weighted sum methods have the following drawbacks [26]:

- A uniform distribution of the weights assigned to the objective functions does not guarantee a uniform distribution of the resulting Pareto points.
- These methods are not capable of finding solutions in nonconvex zones of the Pareto front.
- The need for the evaluation of a large number of designs (one for each weight distribution) implies a huge computation effort, even becoming impracticable in complex systems.

The complexity of this optimization method is the tuning of the weights in order to accurately model the situation to be optimized. Usually, weighted sum methods are implemented as a convex combination of objective functions. This means that no negative value is accepted for a weight, and the sum of all weights is a constant value.

This can be a disadvantage when, for example, the influence of metrics over the user's perception may vary or when it is necessary to introduce a negative metric, i.e., a metric that must be maximized, while the others must be minimized.

A solution to this problem may be an *adaptive weighted sum method* [25] that focuses on unexplored regions changing the weights in an adaptive way instead of using predefined values. Thus, this method "learns" the shape of the Pareto front, and the computational effort is focused on the regions where relevant information can be attained more efficiently. In addition, this method makes it possible to explore nonconvex regions, where the traditional weighted sum method fails.

Another option is the use of *physical programming*, which does not require the decision maker to specify optimization weights in the problem formulation phase. In its place, decision maker indicates the preference ranges over the design variables, graduating its inclination from unacceptability to high desirability [27]. The qualitative classification of preferences in physical programming avoids the need for playing with different weights to later reoptimize the system until the design variables match the decision maker's preferences. Works in [28–30] show various applications of physical programming in real-world scenarios.

Regarding its applicability to MANET routing, in [31] the authors introduce a proposal of a multiple objective dynamic routing (MODR) method, which enables the representation of several QoS-related metrics and requirements in a consistent manner. The MODR method is based on a weighted sum scheme with constraints and is prepared to use implied costs as one of the metrics. Alternative paths for each traffic flow are changes according to the periodic updates of QoS parameters estimated from real-time measurements on the network.

Similarly, Donoso [32] proposes a weighted sum method to aggregate several criteria into a single metric: the maximum link utilization, the hop count, the total bandwidth consumption, and the total end-to-end delay.

Integer programming methods are a particular case of weighted sum where all variables are integers. Reference [33] gives an example of integer programming as a multiobjective approach for energy consumption and link stability issues in mobile networks.

Usually, weighted sum methods are implemented as a convex combination of objective functions. This means that no negative value is accepted for a weight, and the sum of all the weights is a constant value. The disadvantage is that these methods often do not provide a solution set uniformly distributed along the Pareto front, so the solution set may not be representative. In addition, sometimes the methods are not able to find solutions in nonconvex regions of the feasible space.

7.2.2.2 Heuristic and Metaheuristic Methods

The search for exact solutions of a multicriteria optimization problem demands a computational effort that grows proportionally to the number of variables and alternatives to the solution of the problem. In fact, many exact methods fall into an NP-complete problem category, so there is no polynomial time solution.

In such circumstances we must do without the exact solution and search for a solution as close as possible to the ideal utopian one. This is the purpose of heuristic and metaheuristic methods. The difference is that each heuristic method is explicitly designed for its application to a certain situation and is not adequate for other problems, while metaheuristic methods are more general. These methods, also known as *stochastic* or *probabilistic methods*, make use of three major strategies [34,35]:

- Divide and conquer: The problem is divided into more manageable subproblems. Once the solutions to the subproblems are found, they are gathered together to select the optimum solution. This strategy is adequate for problems that are disjoint by nature.
- Iterative improvement: Starting from a single configuration or known solution, this method tries new configurations until it reaches a better one. The reconfiguration process goes on until no solution that improves the previous one is found.
- Distributed metaheuristics: A population of solutions is managed in parallel (evolutionary/genetic methods).

Regardless of the solution search strategy, heuristic and metaheuristic methods have some common features [35]:

- They do not need to use derivatives of objective functions to explore the feasible region.
- They are inspired by analogies with physics (simulated annealing), biology (evolutionary methods, tabu search), and others.
- They can be used to "guide" another exploration method, constituting a hybrid method that exploits the advantages of different approaches.
- They still have a high computational cost.

The greater difficulty lies in the adjustment of parameters in the iterative process.

7.2.2.2.1 Simulated Annealing

The simulated annealing optimization method is inspired by the metallurgical process of consecutive heating and cooling of materials to increase the size of the crystals and decrease the number of imperfections [34]. Heat forces atoms to disengage from their initial positions (local energy minimums) to go on to new random higher-energy states. The later, slower cooling makes it easier to reach new configurations with an energy state lower than the initial one.

The simulated annealing method imitates this process considering objective functions to be optimized as the energy of the material and using a controlled parameter of the algorithm as the pretended temperature [36].

Different application approaches of simulated annealing methods to multicriteria optimization are proposed in [37–43].

Simulated Annealing-Multi-Constrained Path (SA_MCP) [44] is a QoS routing algorithm that applies simulated annealing to Dijkstra's algorithm after converting multiple QoS metrics into a single one. Similarly, Simulated Annealing- Representation Analysis (SA_RA) [45] also uses an energy function to translate multiple QoS weights into a single metric and then searches the optimal path by simulated annealing.

7.2.2.2.2 Tabu Search

This method uses flexible memory cycles (tabu lists) to control the search process [10]. At each iteration, the best nontabu move is taken, even if it means an increase in the objective function values, to avoid local minimums. The previously visited solutions are classified as tabu and stored in memory, so that the algorithm does not return to a local optimum and the new search direction ensures the exploration of other regions in the feasible space.

Tabu search is an adaptive heuristic technique that has been applied to a wide range of problems, mostly of a combinatorial nature, and is able to manage both continuous and discrete parameters.

Different application approaches of tabu search methods in multicriteria optimization are proposed in [43,46–49]. Belfares et al. [49] propose an application of tabu search to multiobjective resource allocation.

Unfortunately, no relevant contribution of a tabu search applied to MANET routing has been proposed yet, probably due to its high computational cost.

7.2.2.2.3 Evolutionary Methods

Evolutionary methods are inspired by the natural evolution optimization process including all heuristic techniques derived from biological evolution, such as natural selection and genetic inheritance [50].

Evolutionary algorithms are characterized by introducing an initial candidate solution population that is recombined by means of a reproduction procedure to generate new solutions. Afterward, a natural selection process is used to separate the individuals or solutions that are a better fit (closer to the utopian ideal solution), getting rid of the rest [51]. The aim is to find the solution set that minimizes the distance to the Pareto front while maximizing the diversity of candidates to obtain a relevant set.

There are different techniques included in this family of optimization algorithms. *Evolution strategies* work with vectors of values and emulate the mutation process over the possible solution population to explore the feasible space and to avoid holding up in a local minimum. Therefore, it is an abstraction of the evolution at the individual behavior level. *Evolutionary programming* is a stochastic optimization technique that focuses on the behavioral link between ascendants and descendants instead of replicating genetic operators observed in nature.

Genetic algorithm (*GA*) reproduces the natural selection based on genetics. A multipath algorithm carries out several searches in parallel in order to avoiding getting stuck in a local minimum.

GAs encode each parameter or design criterion in a bit chain as if they were chromosomes carrying information about every individual. From this moment on, GA works with a parameter code in the place of the parameters themselves. This parameter code allows the genetic operator to evolve from the actual state to the following one with minimum computational effort. Finally, GA evaluates the suitability of each chain to decide whether it will lead the next iteration [52,53].

GA establishes a candidate solution set (also known as *individuals*) that constitutes a population. During the execution of the algorithm, a suitability function is applied to each individual, which returns a value [53]. In addition, new candidate solutions are generated, on the one hand accomplishing mutations on the solutions and on the other hand recombining or crossing the genetic characteristics of two parent solutions to engender a child solution to explore the whole feasible space. Each mutation alters from 3 to 5% of the population by randomly changing part of the genetic code. In the recombination, the result of the suitability function sets the percentage of information that each parent transmits to its children. Finally, a selection process chooses the better-fitted individuals that survive.

Finally, *genetic programming* (*GP*) is similar to GA [52]. The main difference against GA is that while GA tries to find the chain/chromosome that represents the solution to the problem, the objective of GP is to find a formula or computer program that provides that solution [54], making the computer capable of solving the problem without being programmed to find the solution [55].

Cui et al. [56] proposes GAMAN, a MANET QoS routing algorithm that employs GA to compute routes adaptively, flexibly, and intelligently. The primary concern of this approach is robustness rather than optimality. Thus, it is better to obtain an available route as soon as possible rather than the optimal one out of time. The authors emphasize that the algorithm supports soft QoS without hard guarantees, because in a mobile environment there may exist transient time periods when the required QoS is not ensured due to path breaking or network partition.

The work in [57] introduces a multiobjective model for QoS multicast routing based on GA. This approach finds a set of possible multicast paths from a source node to each destination by using the depth-first search algorithm. Then, each multicast tree is mapped to a chromosome string as a sequence of nodes along the path. This is the initial population of the genetic process. The suitability function checks whether each new individual exceeds the jitter bound, and if so, the solution is rejected.

7.2.2.3 Lexicographic Methods

Lexicographic methods become useful in cases in which there are several objectives or criteria in conflict but that, for whatever reason, cannot be scalarized (e.g., with a weighted sum) and must be treated hierarchically.

In lexicographic optimization [58], objective functions are optimized in a lexicographic order, for example, low-priority criteria are optimized only if they do not interfere with the optimization of higher-priority criteria. It is an intuitive technique similar to the alphabetic ordering of a name list, in which, initially, the ordering is performed according to the first letter of each name (highest priority). Then, for each group with the same first letter, the same process is repeated for the second letter (immediately next priority), and so on [35].

Reference [59] gives an application of lexicographic max-ordering to the optimization of network flows for load balancing. Reference [59] gives another example of lexicographic optimization applied to the routing of optic networks. Finally, Reference [60] introduces an application in Vehicular Ad-hoc NETworks (VANET).

7.2.2.4 Fuzzy Logic Methods

In contrast to the strict binary logic, which only accepts "true" or "false" values, the fuzzy logic introduced by Zadeh makes it possible to work with sets in which the concepts of uncertainty and inaccurate information, inherent to real-world situations, are present [61].

Alandjani and Johnson [62] apply fuzzy logic to differentiate resource allocation. Important messages can be routed redundantly over disjoint paths, while less important traffic may be suppressed at the source. Therefore, resource differentiation is performed, attending to traffic importance and the network state.

Zhang et al. [63] extend the DSR routing protocol, adopting fuzzy logic to select the appropriate QoS routing in multiple paths that are searched in parallel. This scheme considers the bandwidth, end-to-end delay, and the cost of the path to integrate the QoS requirements.

Gomathy and Shanmugavel [64] introduce a fuzzy-logic-based priority scheduler to set the priority of the packets according to multiple quantitative criteria, such as packet delivery ratio and average end-to-end delay.

7.2.2.5 Outranking Methods

Outranking methods are the European proposal for decision-making support in opposition to the utility theory, which is more common in American studies.

These methods build an ordered relation of the possible alternatives according to the preferences over a series of criteria to eventually supply a recommendation [65].

The preference relation is expressed using a utility function that allows stating a full and transitive binary relation to order the alternatives. Using this information, the classification built is exploited to produce a recommendation. However, obtaining the utility function that aggregates all the criteria to be evaluated is not a trivial process. Succeeding in all alternatives being comparable requires lots of information and a careful analysis of the interchange relations among attributes.

Outranking methods build the preference relation as a binary relation S over x set of alternatives, so that xSy if, according to the preferences of the decision maker, there are enough arguments to state that x is, at least, as good as y, while there are not enough reasons to refute such a statement [66].

Outranking methods are useful in cases where at least one of the following circumstances takes place [65]:

- At least one of the criteria is evaluated qualitatively. Therefore, it is difficult or artificial to establish an encoding that makes it possible to perform comparisons in terms of preference divergence.
- The heterogeneity of criteria makes the aggregation of them all into a common scale difficult.
- The decision maker does not always accept a compensation for the losses of one criterion with the gains of another.
- Small differences in the evaluation may not be relevant for some criteria in terms of relevance, but the accumulation of those differences may be. This enforces the use of indifference discrimination thresholds.

The disadvantage of this family of optimization methods is the need to use numerous parameters whose value must be settled by the decision maker. Although some of these parameters have a practical meaning, others such as concordance discrepancy, discrimination, or veto thresholds have only a technical connotation, and their influence on the final result is often misunderstood. Part of this family of optimization methods are ELECTRE [67] and PROMETHEE [68] methods.

Unfortunately, the complexity of these methods has discouraged researchers from using this technique in MANET routing and, therefore, no relevant contribution has been released.

7.2.3 Applicability of Multicriteria Optimization to MANET Routing

The study of the existing literature displays an eclectic variety of multicriteria optimization techniques, ranging from the simpler weighted sum algorithms to the more sophisticated evolutionary or genetic methods.

Generally speaking, the inclusion of advanced multicriteria optimization algorithms into the implementation of MANET routing protocols is unlikely to be feasible for mobile devices with relatively low battery supplies and low memory or computational resources.

However, in spite of its computational cost, the application of GAs to MANET routing has clearly gained particular relevance with a wide assortment of proposals.

The main problem with optimization processes is that as an iterative search, they take a large amount of time to reach an acceptable solution. This is because each point of the searching process involves a recompilation—a re-execution of the algorithm.

Artificial intelligence techniques may help in reducing the necessary effort required to solve optimization problems as a complement for the previously studied methods especially due to their distributed nature.

7.3 Artificial Intelligence

Most references define *artificial intelligence* as the analysis and design of rational or intelligent agents [69,70]. Actually, artificial intelligence is the science that allows machines to reproduce human behavior or, at least, some human capabilities.

In contrast with the more sophisticated mathematical multicriteria optimization algorithms, natural intelligence is based on very simple principles that, properly combined, become capable of solving complex situations. For example, biological neurons that are imitated in neural networks are composed of rather basic units with a highly limited computational power. However, when a large number of neurons work together, they are not only capable of solving a complex problem but even of learning how to solve it.

Some of the optimization techniques studied in the previous sections, such as evolutionary methods, genetic methods, simulated annealing, or tabu search, are also part of artificial intelligence since they imitate nature-based behavior.

The next sections analyze the application of artificial intelligence techniques directly to networking and especially to MANETs or indirectly as optimization techniques.

7.3.1 Machine Learning

One of the research areas of artificial intelligence is *machine learning* or *automatic learning* [71–73], whose final goal is to develop techniques that make it possible for machines to learn. The objective is as follows: starting from a given set of examples at a training stage previous to the operative stage, make the machines capable of generalizing their knowledge so as to obtain new knowledge.

Automatic learning is clearly an algorithmic science, traditionally classified into the six categories described in Table 7.4 [74]. Two strategies can be distinguished: *online learning* and *offline learning*. The offline learning occurs when all the examples or instances are introduced simultaneously. In contrast, in online learning, the samples are presented one by one. There are intermediate possibilities as well, such as gathering the instances in groups and introducing the groups one by one. Another distinction among the strategies of the learning algorithms is that they can be divided into two methods: *incremental* if instances are treated one by one and *nonincremental* if instances are handled simultaneously. It may appear that an online learning algorithm corresponds to an incremental one and an offline method to a nonincremental one, but it may not always be so. For example, it would be possible to adapt a nonincremental method to an online learning task, storing in memory all the previous cases and adding the new case to that existing set to later apply the algorithm again to the extended set.

For example, self-organized marketplace-based middleware for mobile ad hoc networks [75] is a middleware architecture for distributed applications over a MANET. It is based on the

Table 7.4   Classification of Machine Learning Optimization Methods

Supervised learning	The algorithm generates a function that maps inputs to desired outputs. The knowledge basis is built from previous examples (experience), and the output is contrasted or supervised, providing feedback to adjust the behavior.
Nonsupervised learning	It is performed by agents that only model a set of inputs without the availability of a series of examples that provide information about the categories of the inputs.
Semi-supervised learning	It combines labeled or preclassified examples with others without categorization to generate a classifying function.
Effort learning	The algorithm learns a behavior by observing the real world. Each action has an impact over the environment, and the agent gets a feedback from that environment to guide the learning process.
Transduction	It is similar to supervised learning but differs in that it does not return any explicit function, but it tries to predict the output from the inputs of the training, the outputs of those examples, and the acquired experience.
Multitask learning	This technique uses its own experience to solve cases, searching for similarities with other previously solved cases.

communication pattern in a market, in which negotiating agents gather spontaneously at a geographical point (market), where the probability of finding the required information grows exponentially. In this way, using geographical routing variations, the agents move around active nodes and "markets" to transmit data.

Similarly, Agakov et al. [76] introduce a methodology to accelerate the speed of the optimization using automatic learning to guide or focus the iterative process, pointing out to the search regions more inclined to give good results. This methodology is independent from the algorithm used, the search space, or the platform over which it is executed. Offline training is performed and evaluated iteratively, storing the results. These results are used to build a model that will later be correlated with the inputs to predict which regions of the optimization space show more leaning to contain the optimum solution. It also stores the characteristics of every program, and so when this methodology is used in other problems, it will look for the best model to perform the correlation.

SAMPLE [77] uses reinforced collaborative learning to solve in real time the optimization problems of dynamic and distributed networks. This protocol is able to adapt system routing behavior to the changing environment using positive and negative feedbacks.

7.3.2 Neural Networks

Neural networks belong to the field of automatic learning [78]. A neural network consists of a set of interconnected artificial neurons that imitate the features of biological neurons. These networks are used to solve artificial intelligence problems and have been successfully applied to tasks such as voice recognition, image analysis, or statistical estimation and optimization. The difference with other artificial intelligence techniques (such as GAs or fuzzy logic) is the lack of explicit rules.

A neural network comprises a network of simple processing elements (the artificial neurons) that, as a group, produce a complex global behavior thanks to a mathematical or computational model that makes it possible to process the information received by the system. Often, a neural network is an adaptive system that modifies its structure as the internal or external information flows through the network. The final result is that the neural network is able to infer a function from the observation and then use it to generate new results. In summary, it *learns* from the observation [79]; i.e., given a task that must be worked out and a class of functions F, the neural network employs observation to find the function $f \in F$ that optimally solves the task.

The work in [80] shows the 10 basic characteristics of neural networks that are summarized in Table 7.5, and it also shows a classification of the applications of neural networks, illustrated in Figure 7.6.

In the area of the application of a neural network to optimization problems, there have been plenty of proposals. Peterson and Soderberg [81] introduce a method to obtain solutions for complex optimization problems through neural networks. The distinctive element is the reduction of the feasible space in one dimension using differentiated neurons that avoid the usual redundancy of this kind of problem when direct neural network techniques are used.

Wang et al. [82] introduce an application of chaotic neural networks to combine the best features of two simulated annealing algorithms chaotic simulated annealing and stochastic simulated annealing.

Another example of the application of neural networks to the resolution of optimization problems is found in [83]. This work introduces a recurrent neural network for the resolution of convex nonlineal programming problems subject to inequality restrictions.

There are several contributions that apply neural networks to relieve the computational effort of multicriteria optimization. Some examples can be found in [84–86].

The work in [87] shows the advantages of using artificial neural networks (ANN) in their application in telecommunications. However this use of ANN is underexploited.

Yagan and Chen-Khong [88] introduce a scheme for the provision of QoS in MANETs that implements DiffServ with bandwidth reservation and buffer management. This scheme models the system as a semi-Markov decision system and uses a reinforced learning neural network that pursues the maximization of the utilization of the network while the violations of QoS regarding bandwidth, queue delay, and buffer losses are minimized.

Another work that applies neural networks to the routing in MANETs is presented in [89]. It presents a routing protocol whose routing tables are maintained in master devices while the radius of the routing zone for each table is dynamically adapted using evolutionary fuzzy neural networks.

7.3.3 Swarm Intelligence

Swarm intelligence emulates the behavior of a herd by gathering little pieces of intelligence in order to build an entity capable of solving problems in a way that collective requirements are considered above the individual ones. In this way, a more efficient distribution of resources is achieved from a global point of view, without the need for self-consciousness and thus avoiding a selfish behavior.

The most popular swarm intelligence technique employed in routing algorithms is ant colony optimization (ACO). This model using self-organization theories provides powerful tools to transfer knowledge about the social insects to the design of intelligent decentralized routing protocols. The basic principle of an ACO algorithm is the deposit of pheromones on the ground as the ants roam in their search for food. When the objective is reached, the ant makes its way back to the

Table 7.5 Fundamental Characteristics of Neural Networks

Informational activation patterns	Activation patterns are based on the vector that stores the required information or signal level to activate the synaptic union of all the neurons of the network.
Associative learning	The efficiency of the information transmission at the synaptic union that links a pair of neurons can be altered when both neurons are activated simultaneously. These neural changes are the basis of learning, and the knowledge is stored in the connections between neural units.
High-dimensional state spaces	Multiplying the number of neurons by the number of synaptic junctions for everyone, it is obvious that the activation pattern is a large multidimensional vector.
Local interactions	The activation frequency of a neuron at any given time is just a function of a few neurons of the system. This indicates that the learning process is based on local interactions.
Fast computing with slow computational units	Due to the simplicity of the neurons, the only way to achieve a fast and efficient computation is to work in parallel. Therefore, the state of many neurons should change simultaneously for each iteration of the classification process.
Distributed control and processing	Global behavior procedures imply distributed processing and control principles, so that a local "lesion" in the neural network does not prevent the recovery of correct behavior, at least partially. This fact forces neural network zones not to be devoted to just one specific task.
Quasilinear activation updating rule	Neurons are able to integrate the input activation information with previous information, affecting following activations of the neuron. This process is modeled quasilineally; so depending on the case, either the lineal or the nonlineal parts may stand out.
Homogeneity of processing	The mechanism used for learning in a neuron zone is basically the same as what is used in any other neuron zone. The most relevant processes of a neural network are spatially homogeneous. From a computational point of view, this homogeneity assumption makes it possible for many processing units to communicate efficiently in a parallel processing environment.
Tolerance to local inaccuracy	Neural systems are capable of performing precise computations from local and inaccurate information thanks to the spatial distribution of computation.
Presence of analog computing	As in biological neural networks, artificial neural networks are also analog machines that can occasionally work as digital machines.

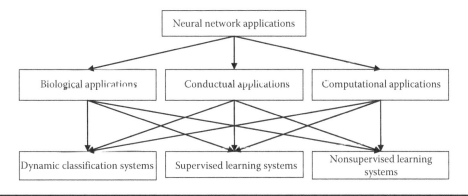

Figure 7.6 Applications of neural networks.

nest, while at the same time reinforcing the previously dropped pheromones. The pheromone trails allow the other ants to find the best way to the food source or the way back home by following those paths with a stronger pheromone concentration [90].

Similarly, when a source node searches for a route to a destination, at first there will be a flooding of packets that leave a pheromone trail in their way toward the destination. The multicriteria optimization is then implemented in the way pheromone value is set and reinforced or faded if no update is performed.

The ant colony system [91] is one of the most remarkable examples and is widely used in later research. A detailed study of ACO methods can be found in [92]. This kind of algorithm is also useful as a complement for optimization algorithms based on population, such as evolutionary and genetic methods, because they are capable of rapidly detecting the best individuals and focusing the search direction. Dorigo [93] introduce a collection of ACO optimization algorithms combined with static and dynamic algorithms.

Regarding the application of ACO methods to networking, AntNet [94] is one of the most popular ant colony-based routing protocols, existing with several variations.

AntHocNet [95] is a hybrid MANET routing protocol that combines reactive route setup with proactive route probing, employing ant agents in both the proactive and reactive phases.

AntNet-QoS [96] is another ACO routing mechanism based on AntNet that provides routes with QoS guarantees, both in a DiffServ and in a best-effort network. The DiffServ model will consider four classes of service: urgent retransmission for real-time traffic, two types of guaranteed retransmission for traffic with a certain bound in loss rate, and the traditional best effort for the traffic without QoS requirements.

Other relevant contributions of ant colony optimization to MANET routing are the works in [90,97–100].

ACO algorithms have definitely become a popular technique for the optimization of MANET routing thanks to the following key features: They work in a distributed way, are highly adaptive and robust, and provide automatic load balancing.

7.4 Conclusions

MANET routing is a key aspect in the development and eventual success of mobile ad hoc networking. Its original purpose was oriented just to provide the minimum service in emergency/military situations where any other infrastructure network (fixed or wireless) was not feasible. This technology must now adapt its operation to much more commercial applications whose users demand QoS levels that must be close to the ones of the infrastructure networks.

However, the dynamic nature of MANETs makes it extremely difficult to achieve the aforementioned expectations. Not only is it necessary to arrange valid routes from a source to a destination but the distribution of limited resources must also be as efficient as possible. That means conflicting requirements can be found when trying to satisfy the level of service the end users demand, and thus forcing one to find a trade-off solution.

This chapter has studied the possibility of using multicriteria optimization techniques to make an efficient distribution of the available resources. Nevertheless, these algorithms exhibit an unacceptable high computational cost for real-time MANET communications involving rapidly changing network topologies. Therefore, and generally speaking, there has been no successful application of this concept to MANET routing, with the exception of some proposals that include a genetic or weighted sum algorithm.

Further research has turned toward artificial intelligence as a plausible alternative to multicriteria optimization, since artificial intelligence offers more intuitive algorithms that reach solutions close enough to the optimal ones at a much lower computational cost compared to multicriteria optimization. Since deterministic multicriteria optimization algorithms with a polynomial time complexity are not suitable for the real-time computations of routes in MANETs, quite a few research works have been conducted to solve the routing problem using artificial intelligence techniques, for example, ANNs and particle swarm optimization.

In particular, ant colony algorithms have proved to be a pretty suitable option for optimal route learning, while at the same time they are highly adaptable to the intrinsically dynamic nature of MANETs. The principal reason is that the main objective of an ant colony perfectly models the circumstances of MANETs.

ACO methods constitute a promising technique, and many researchers have supported this initiative. However, the lack of conclusive results keeps this field open.

References

1. www.ietf.org. Accessed on June 2011.
2. T. Clausen, C. Dearlove, and P. Jacquet. The optimized link state routing protocol, Version 2. IETF Internet Draft, draft-ietf-manet-olsrv2-11, 2010.
3. I. Chakeres and C. Perkins. Dynamic MANET on-demand (DYMO) routing. IETF Internet Draft, draft-ietf-manet-dymo-21. 2010.
4. M. Mirhakkak, N. Schult, and D. Thomson. Dynamic quality-of-service for mobile ad hoc networks. In *Proceedings of the 1st ACM International Symposium on Mobile Ad Hoc Networking & Computing, Boston, Massachusetts*, 2000, pp. 137–138.
5. H. Israat Tanzeena, A. Chadi, and A.J. Atwood. Randomized energy aware routing algorithms in mobile ad hoc networks. In *Proceedings of the 8th ACM International Symposium on Modeling, Analysis and Simulation of Wireless and Mobile Systems, Montreal, Quebec, Canada*, 2005, pp. 71–78.
6. M. Mauve, A. Widmer, and H. Hartenstein. A survey on position-based routing in mobile ad hoc networks. *IEEE Network*, vol. 15, no. 6, 2001, pp. 30–39.

7. L. Hanzo and R. Tafazolli. A survey of QoS routing solutions for mobile ad hoc networks. *IEEE Communications Surveys & Tutorials*, vol. 9, no. 2, 2007, pp. 50–70.
8. C. Lei and W.B. Heinzelman. A survey of routing protocols that support QoS in mobile ad hoc networks. *IEEE Network*, vol. 21, no. 6, 2007, pp. 30–38.
9. J. Nocedal and S.J. Wright. *Numerical Optimization*, New York: Springer, 2006.
10. J. Andersson. *A Survey of Multiobjective Optimization in Engineering Design*. Linköping, Sweden: Department of Mechanical Engineering, Linköping University, 2000.
11. R.T. Marler and J.S. Arora. Survey of multi-objective optimization methods for engineering. *Structural and Multidisciplinary Optimization*, vol. 6, no. 26, 2004, pp. 369–395.
12. M. Ehrgott and X. Gandibleux. *Multiple Criteria Optimization: State of the Art Annotated Bibliographic Surveys*. Dordrecht: Kluwer Academic Publishers, 2002.
13. H.W. Hamacher, C.R. Pedersen, and S. Ruzika. Multiple objective minimum cost flow problems: a review. *European Journal of Operational Research*, vol. 176, no. 3, 2007, pp. 1404–1422.
14. M. Farina, K. Deb, and P. Amato. Dynamic multiobjective optimization problems: test cases, approximations, and applications. *IEEE Transactions on Evolutionary Computation*, vol. 8, no. 5, 2004, pp. 425–442.
15. R.T. Marler. *A Study of Multi-Objective Optimization Methods for Engineering Applications*. Iowa: Graduate College of the University of Iowa, 2005.
16. M. Oztürk, A. Tsoukias, and P. Vincke. Preference modelling. In J. Figueira, S. Greco, and M. Ehrgott (eds.), *Multiple Criteria Decision Analysis: State of Art Surveys*. Boston: Kluwer Academic Publishers, 2005, pp. 27–72.
17. B. Roy. *Multicriteria Methodology for Decision Aiding*. Dordrecht: Kluwer Academic Publishers, 1996.
18. B. Roy. Paradigms and challenges. *Multiple Criteria Decision Analysis: State of Art Surveys*. Boston: Kluwer Academic Publishers, 2005, pp. 3–24.
19. C.A. Coello Coello. Evolutionary multi-objective optimization: a historical view of the field. *IEEE Computational Intelligence Magazine*, vol. 1, no. 1, 2006, pp. 28–36.
20. M. Ehrgott and X. Gandibleux. A survey and annotated bibliography of multiobjective combinatorial optimization. *OR Spectrum*, vol. 22, no. 4, 2000, pp. 425–460.
21. M. Gen, A. Kumar, and J.R. Kim. Recent network design techniques using evolutionary algorithms. *International Journal of Production Economics*, vol. 98, no. 2, 2005, pp. 251–261.
22. C. Gil, R. Banos, M.G. Montoya, et al. Performance of simulated annealing, tabu search, and evolutionary algorithms for multi-objective network partitioning. *Algorithmic Operations Research*, vol. 1, 2006, pp. 55–64.
23. Z. Jia and P. Varaiya. Heuristic methods for delay constrained least cost routing using k-shortest-paths. *IEEE Transactions on Automatic Control*, vol. 51, no. 4, 2006, pp. 707–712.
24. I.H. Osman and G. Laporte. Metaheuristics: a bibliography. *Annals of Operations Research*, vol. 63, no. 5, 1996, pp. 511–623.
25. I.Y. Kim and O.L. de Weck. Adaptive weighted sum method for multiobjective optimization: a new method for Pareto front generation. *Structural and Multidisciplinary Optimization*, vol. 31, no. 2, 2006, pp. 105–116.
26. A. Messac and C. Mattson. Generating well-distributed sets of Pareto points for engineering design using physical programming. *Optimization and Engineering*, vol. 3, no. 4, 2002, pp. 431–450.
27. A. Messac. From the dubious art of constructing objective functions to the application of physical programming. *AIAA Journal*, vol. 38, no. 1, 2000, pp. 155–163.
28. A. Messac. Physical programming: effective optimization for computational design. *AIAA Journal*, vol. 34, no. 1, 1996, pp. 149–158.
29. A. Messac, M. Martinez, and T. Simpson. Effective product family design using physical programming. *Engineering Optimization*, vol. 34, no. 3, 2002, pp. 245–261.
30. C.D. McAllister, T.W. Simpson, K. Hacker, K. Lewis, and A. Messac. Integrating linear physical programming within collaborative optimization for multiobjective multidisciplinary design optimization. *Structural and Multidisciplinary Optimization*, vol. 29, no. 3, 2005, pp. 178–189.

31. J. Craveirinha, L. Martins, T. Gomes, C.H. Antunes, and J.N. Clímaco. A new multiple objective dynamic routing method using implied costs. *Journal of Telecommunications and Information Technology*, vol. 3, 2003, pp. 50–59.

32. Y. Donoso. *Multi-Objective Optimization Scheme for Static and Dynamic Multicast Flows*. Girona, Spain: Departament d'Electrònica, Informàtica i Automàtica, Universitat de Girona, 2005.

33. F. de Rango, F. Guerriero, S. Marano, and E. Bruno. A multiobjective approach for energy consumption and link stability issues in ad hoc networks. *Communications Letters, IEEE*, vol. 10, no. 1, 2006, pp. 28–30.

34. S. Kirkpatrick, C. Gelatt, and M. Vecchi. Optimization by simulated annealing. *Science*, vol. 220, no. 4598, 1983, pp. 671–680.

35. Y. Collette and P. Siarry. *Multiobjective Optimization: Principles and Case Studies*. Berlin: Springer, 2003.

36. G. Kuczera and E. Parent. Monte Carlo assessment of parameter uncertainty in conceptual catchment models: the Metropolis algorithm. *Journal of Hydrology*, vol. 211, no. 1–4, 1998, pp. 69–85.

37. P. Serafini. Simulated annealing for multiple objective optimization problems. In *International Conference on Multiple Criteria Decision Making*, 1994, pp. 283–292.

38. E. Ulungu, J. Teghem, P. Fortemps, and D. Tuyttens. MOSA method: a tool for solving multiobjective combinatorial optimization problems. *Journal of Multi-Criteria Decision Analysis*, vol. 8, no. 4, 1999, pp. 221–236.

39. P. Czyzak and A. Jaszkiewicz. Pareto simulated annealing—a metaheuristic technique for multiple-objective combinatorial optimization. *Journal of Multicriteria Decision Analysis*, vol. 7, 1998, pp. 34–47.

40. A. Suppapitnarm, K. Seffen, G. Parks, and J. Clarkson. A simulated annealing algorithm for multiobjective optimization. *Engineering Optimization*, vol. 33, no. 1, 2000, pp. 59–85.

41. B. Suman. Study of simulated annealing based algorithms for multiobjective optimization of a constrained problem. *Computers and Chemical Engineering*, vol. 28, no. 9, 2004, pp. 1849–1871.

42. B. Suman and P. Kumar. A survey of simulated annealing as a tool for single and multiobjective optimization. *Journal Operational Research Society*, vol. 57, no. 10, 2006, pp. 1140–1160.

43. R. Baños, C. Gil, B. Paechter, and J. Orteya. A hybrid meta-heuristic for multi-objective optimization: MOSATS. *Journal of Mathematical Modelling and Algorithms*, vol. 6, no. 2, 2007, pp. 213–230.

44. C. Yong, W. Jian-Ping, and X. Ke. A QoS routing algorithm by applying simulated annealing. *Journal of Software*, vol. 5, 2003, pp. 877–884.

45. L. Liu and G. Feng. Simulated annealing based multi-constrained QoS routing in mobile ad hoc networks. *Wireless Personal Communications*, vol. 41, no. 3, 2007, pp. 393–405.

46. F. Glover and M. Laguna. *Tabu Search, Modern Heuristic Techniques for Combinatorial Problems*. New York: John Wiley & Sons, 1993.

47. M.P. Hansen. Tabu search for multiobjective optimization:MOTS, In *13th International Conference on Multiple Criteria Decision Making*, University of Cape Town, 6–10 January, 1997.

48. J. Molina, M. Laguna, R. Marti, and R. Caballero. SSPMO: a scatter tabu search procedure for nonlinear multiobjective optimization. *Informs Journal on Computing*, vol. 19, no. 1, 2007, pp. 91–100.

49. L. Belfares, W. Klibi, N. Lo, and A. Guitouni. Multi-objectives tabu search based algorithm for progressive resource allocation. *European Journal of Operational Research*, vol. 177, no. 3, 2007, pp. 1779–1799.

50. S. Winkler, M. Affenzeller, and S. Wagner. New methods for the identification of nonlinear model structures based upon genetic programming techniques. *Systems Science – Wroclaw*, vol. 31, no. 1, 2005, pp. 386–393.

51. A. Abraham, L. Jain, and R. Goldberg. Evolutionary multiobjective optimization. In *Evolutionary Multiobjective Optimization. Theoretical Advances and Applications*. Berlin: Springer, 2005, pp. 1–6.

52. Z. Michalewicz. *Genetic Algorithms + Data Structures = Evolution Programs*. Berlin: Springer, 1996.

53. J. Holland. *Adaptation in Natural and Artificial Systems*. Cambridge, MA: MIT Press, 1992.

54. J. Koza. *Genetic Programming: On the Programming of Computers by Means of Natural Selection*. Cambridge, MA: MIT Press, 1992.

55. L. Barolli, A. Koyama, and N. Shiratori. A QoS routing method for ad-hoc networks based on genetic algorithm. In *Proceedings of the14th International Workshop on Database and Expert Systems Applications*, 2003, pp. 175–179.

56. X. Cui, C. Lin, and Y. Wei. A multiobjective model for QoS multicast routing based on genetic algorithm. In *International Conference on Computer Networks and Mobile Computing, 2003. ICCNMC 2003, Shanghai, China*, 2003, pp. 49–53.

57. H. Isermann. Linear lexicographic optimization. *OR Spectrum*, vol. 4, no. 4, 1982, pp. 223–228.

58. L. Georgiadis, P. Georgatsos, K. Floros, and S. Sartzetakis. Lexicographically optimal balanced networks. In *IEEE INFOCOM*, 2001, pp. 689–698.

59. W. Lin and R. Wolff. A lexicographically optimized routing algorithm for all-optical networks. In *Proceedings of the OCSN*, 2005, pp. 138–142.

60. S. Saliba. Heuristics for the lexicographic max-ordering vehicle routing problem. *Central European Journal of Operations Research*, vol. 14, no. 3, 2006, pp. 313–336.

61. L.A. Zadeh. Fuzzy logic. *IEEE Computer*, 1988, pp. 83–93.

62. G. Alandjani and E. Johnson. Fuzzy routing in ad hoc networks. *Proceedings of the IEEE International Conference Performance, Computing, and Communications*, 2003, pp. 525–530.

63. X. Zhang, S. Cheng, M. Feng, and W. Ding. Fuzzy Logic QoS Dynamic Source Routing for Mobile Ad Hoc Networks. *The Fourth International Conference on Computer and Information Technology (CIT'04)*, 2004, pp. 652–657.

64. C. Gomathy and S. Shanmugavel. Supporting QoS in MANET by a fuzzy priority scheduler and performance analysis with multicast routing protocols. *EURASIP Journal on Wireless Communications and Networking*, vol. 2005, no. 3, 2005, pp. 426–436.

65. J. Figueira, V. Mousseau, and B. Roy. ELECTRE methods. In J. Figueira, S. Greco, M. Ehrgott (eds) *Multiplecriteria decision analysis: state of the art surveys*. New York: Springer Science+Business Media, Inc, 2005.

66. D. Bouyssou. Outranking methods. *Encyclopedia of Optimization*. Dordrecht: Kluwer Academic Publishers, 2001.

67. B. Roy. The outranking approach and the foundations of electre methods. *Theory and Decision*, vol. 31, no. 1, 1991, pp. 49–73.

68. J.P. Brans. L'ingénièrie de la décision; Elaboration d'instruments d'aide à la décision. La méthode PROMETHEE. In R. Nadeau and M. Landry (eds) *L'aide à la décision: Nature, Instruments et Perspectives d'Avenir*. Presses de l'Université Laval: Québec, Canada, 1982, pp. 183–213.

69. M. Negnevitsky. *Artificial Intelligence: A Guide to Intelligent Systems*. New York: Addison-Wesley, 2005.

70. M. Hutter. *Universal Artificial Intelligence: Sequential Decisions Based on Algorithmic Probability*. Berlin: Springer, 2005.

71. E. Alpaydin. *Introduction To Machine Learning*. Cambridge, MA: MIT Press, 2004.

72. M. Bowling, J. Fürnkranz, T. Graepel, and R. Musick. Machine learning and games. *Machine Learning*, vol. 63, no. 3, 2006, pp. 211–215.

73. T.M. Mitchell. *Machine Learning McGraw-Hill Series in Computer Science*. New York: McGraw-Hill, 1997.

74. P. Langley. *Elements of Machine Learning*. San Francisco: Morgan Kaufmann, 1996.

75. D. Görgen, H. Frey, J.K. Lehnert, and P. Sturm. SELMA: A Middleware Platform for Self-Organizing Distributed Applications in Mobile Multihop Ad-hoc Networks. In *Western Simulation MultiConference WMC '04*, 2004.

76. F. Agakov, E. Bonilla, J. Cavazos, B. Franke, G. Fursin, Michael F.P. O'boyle, J. Thomson, M. Toussiant, and Christopher K.I. Williams. Using machine learning to focus iterative optimization. In *Proceedings of the International Symposium on Code Generation and Optimization (CGO'06)*, 2006, pp. 295–305.

77. J. Dowling, E. Curran, R. Cunningham, and V. Cahill. Using feedback in collaborative reinforcement learning to adaptively optimize MANET routing. *IEEE Transactions on Systems, Man, and Cybernetics—Part A: Systems and Humans*, vol. 35, no. 3, 2005, pp. 360–372.

78. J.A. Anderson. *An Introduction to Neural Networks*. Cambridge, MA: MIT Press, 1995.

79. J.S. Judd. *Neural Network Design and the Complexity of Learning*. Cambridge, MA: MIT Press, 1990.

80. R.M. Golden. *Mathematical Methods for Neural Network Analysis and Design*. Cambridge, MA: MIT Press, 1996.

81. C. Peterson and B. Soderberg. A new method for mapping optimization problems onto neural networks. *International Journal of Neural Systems*, vol. 1, no. 1, 1989, pp. 3–22.

82. J. Wang, S. Li, F. Tian, and X. Fu. A noisy chaotic neural network for solving combinatorial optimization problems: stochastic chaotic simulated annealing. *IEEE Transactions on Systems, Man, and Cybernetics, Part B: Cybernetics*, vol. 34, no. 5, 2004, pp. 2119–2125.

83. Y. Xia and J. Wang. A recurrent neural network for nonlinear convex optimization subject to nonlinear inequality constraints. *IEEE Transactions on Circuits and Systems I: Regular Papers*, vol. 51, no. 7, 2004, pp. 1385–1394.

84. A.L. DeCegama and J.E. Smith. Neural networks and genetic algorithms for combinatorial optimization of sensor data fusion. *Proceedings of SPIE*, vol. 1699, 2004, pp. 108–115.

85. T. Kwok and K.A. Smith. A noisy self-organizing neural network with bifurcation dynamics for combinatorial optimization. *IEEE Transactions on Neural Networks*, vol. 15, no. 1, 2004, pp. 84–98.

86. M. Forti, P. Nistri, and M. Quincampoix. Generalized neural network for nonsmooth nonlinear programming problems. *IEEE Transactions on Circuits and Systems I: Regular Papers,* vol. 51, no. 9, 2004, pp. 1741–1754.

87. T. Clarkson. Applications of neural networks in telecommunications. In *Proceedings of the ERUDIT Workshop on Application of Computational Intelligence Techniques in Telecommunications*, 1999.

88. D. Yagan and C.K. Tham. Adaptive QoS Provisioning in Wireless Ad Hoc Networks: A Semi-MDP Approach. In *IEEE Wireless Communications and Networking Conference*, New Orleans, Louisiana, March 2005.

89. H. Chenn-Jung, C. Liang-Chun, L. Yao-Chuan, C. Yi-Ta, L. Wei Kuang, and H. Sheng-Yu. A zone routing protocol for Bluetooth MANET with online adaptive zone radius. *Fifth International Conference on Information, Communications and Signal Processing*, Bangkok, 2005, pp. 579–583.

90. L. Rosati and B. Gianluca. On ant routing algorithms in ad hoc networks with critical connectivity. *Ad Hoc Networks*, vol. 6, no. 6, 2008, pp. 827–859.

91. M. Dorigo and L.M. Gambardella. Ant colony system: a cooperative learning approach to the traveling salesman problem. *IEEE Transactions on Evolutionary Computation*, vol. 1, no. 1, 1997, pp. 53–66.

92. M. Dorigo and C. Blum. Ant colony optimization theory: a survey. *Theoretical Computer Science*, vol. 344, no. 2–3, 2005, pp. 243–278.

93. M. Dorigo. Ant algorithms for discrete optimization. *Artificial Life*, vol. 5, no. 2, 1999, pp. 137–172.

94. G. Di Caro and M. Dorigo. AntNet: distributed stigmergetic control for communications networks. *Journal of Artificial Intelligence Research*, vol. 9, no. 2, 1998, pp. 317–365.

95. G. Di Caro, F. Ducatelle, and L. Gambardella. AntHocNet: an ant-based hybrid routing algorithm for mobile ad hoc networks. In *Parallel Problem Solving from Nature*, 2004, pp. 461–470.

96. L. Carrillo, J.L. Marzo, L. Fábrega, P. Vilá, and C. Guadall. Ant colony behavior as routing mechanism to provide quality of service. In *Proceedings of ANTS, volume 3172 of Lecture Notes in Computer Science*, Berlin: Springer, 2004, pp. 418–419.

97. M. Heissenbüttel and T. Braun. Ants-based routing in large scale mobile ad-hoc networks. In Kommunikation in verteilten Systemen (KiVS03), 2003.

98. O. Hossein and T. Saadawi. Ant routing algorithm for mobile ad hoc networks (ARAMA). In *Proceedings of the 22nd IEEE International Performance, Computing and Communications Conference*, Phoenix, Arizona, April 2003, pp. 281–290.

99. L. Liu and G. Feng. A novel ant colony based QoS-aware routing algorithm for MANETs. In L. Wang, K. Chen, and Y.S. Ong (eds) *Advances in Natural Computation*. Lecture Notes in Computer Science. Berlin: Springer, 2005, pp. 457–466.

100. Z. Wu and H. Song. Ant-based energy-aware disjoint multipath routing algorithm for MANETs. *The Computer Journal*, vol. 53, no. 2, 2008, pp. 166–176.

Chapter 8

Energy-Efficient Unicast and Multicast Communication for Wireless Ad Hoc Networks Using Multiple Criteria

Christos Papageorgiou, Panagiotis Kokkinos, and Emmanouel Varvarigos

Contents

Routing is a primordial task in wireless ad hoc networks due to the multihop nature of packet relaying for both unicast (one-to-one) and multicast (one to many) communication. Furthermore, the lack of a fixed infrastructure makes energy efficiency a top priority in all operational aspects of these networks. This chapter focuses on energy-aware solutions for unicast and multicast routing, presenting a novel class of routing algorithms based on multiple criteria. According to this approach, each link—and by extension, path—is assigned a cost vector consisting of several cost parameters. After enumerating the candidate nondominated paths, the optimal is selected based on an optimization function. The algorithms are evaluated against well-known unicast and multicast strategies and are shown to result in improved and more balanced energy consumption.

8.1 Introduction

Energy plays a central role in the operation of wireless ad hoc networks. Given the lack of any kind of fixed infrastructure in these networks, energy is a scarce resource that directly limits their performance. Therefore, energy conservation and efficiency is a top priority for all protocols designed for wireless ad hoc networks, regardless of the layer of their operation. Furthermore, the multihop nature of the communication, along with the cooperative applications that these networks are often applied to, makes routing a cornerstone operation. In this context, energy-efficient routing algorithms are critical for wireless ad hoc networks and have deservedly attracted a great deal of research effort in the last few years.

Generally, we can distinguish between two routing approaches: the *single-cost* and the *multicost*. Most routing protocols proposed to date are based on the single-cost idea, where a single metric is used to represent the cost of using a link. This link metric can be a function of several network parameters (including load-, energy-, and interference-related parameters), but it is still a scalar. Routing algorithms of this kind calculate the path that has the minimum cost for each source–destination pair. Single-cost routing algorithms cannot optimize performance with respect to general cost functions, and they do not easily support quality of service (QoS) differentiation. Also, they usually yield only one path per source–destination pair, leading to nonuniform traffic distribution and possible instability problems [1].

In the multicost routing approach, each link is assigned a cost vector consisting of multiple cost criteria. A cost vector can then be defined for a path by combining component-wise the cost vectors of its links according to some associative operator. Multicost routing consists of two phases: first a set of candidate nondominated paths is computed and then the path that minimizes a desired optimization function is selected. We will say that a path p_1 dominates another path p_2 that corresponds to the same source–destination pair if p_1 is better than p_2 with respect to all cost parameters. Also, the function to be optimized is chosen based on the interests of the network, but it may also depend on the user QoS requirements or on the amount of data that have to be transferred. We should note that applying similar optimization functions using the single-cost approach either would be impossible (for nonlinear metrics) or would require (for linear metrics) the rerunning of a minimum cost algorithm, using each time a corresponding link metric.

Multicost routing can be considered a generalization of *multiconstrained* routing, studied in earlier works mainly in the context of wired networks. In the multiconstrained routing problem, a constraint is specified for each of the cost parameters of a path and the target is to find paths that satisfy all the constraints. In the multicost routing problem, there are no hard constraints and the target is to find paths that are better than all other paths for *all* or *some* of the cost parameters and to select the one that is optimal with respect to the optimization function used. Multicost routing is, therefore, a generalization of multiconstrained routing, in the sense that the latter case can be obtained from the former case by choosing the optimization function so as to have infinite cost at the constraint points.

In this work, we apply the multicost approach in unicast and multicast routing following a unifying methodology. In our formulation, the cost parameters of a link $l = (i, j)$ include the hop count, the energy expended by the transmitting node i, and its residual energy. Other parameters of interest, such as the residual energy of the receiving node j, a measure of the interference caused to other nodes and links, and the available link capacity, can also be included in a straightforward manner. The proposed algorithms are based on the multicost routing approach that operates, as mentioned, in two phases. In the first phase, the set of candidate paths for each source–destination pair or the set of candidate nondominated sequences of node transmissions for each source node and destination group is enumerated for the case of unicast and multicast communication, respectively. A sequence of node transmissions is said to be nondominated if it is not worse than any other such sequence with respect to *all* parameters (precise definitions will be given later for the unicast and broadcast problems). In the second phase, the selection of the optimal path (unicast) or sequence of node transmissions (multicast) is then made based on the optimization function in use. In the unicast domain, we propose several multicost routing algorithms, each corresponding to a different choice for the optimization function used. Different cost parameters, domination relations, and optimization functions represent distinct sets of routing decisions (routing algorithms). In the multicast domain, we present an energy-efficient multicasting algorithm that optimizes any desired function of the total power consumed by the multicasting task and the minimum of the current residual energies of the nodes, provided that this optimization function is monotonic in each of these parameters. However, the proposed algorithm has nonpolynomial complexity. To address this drawback, we also present a relaxation of the optimal multicasting algorithm that produces a near-optimal solution in polynomial time. When the set of desired destinations in the multicast group contains every node in the network, the optimal and the near-optimal algorithms implement broadcasting in an optimal and near-optimal manner, respectively, regarding energy efficiency.

The remainder of the chapter is organized as follows. Section 8.2 discusses the current status and future trends of the fields of energy-efficient unicasting and multicasting for ad hoc networks. In Sections 8.3 and 8.4, the multicost approach for unicast and multicast routing is presented, respectively. In Section 8.5, the performance evaluation of the presented algorithms is described, while Section 8.6 concludes the chapter.

8.2 Current Status and Future Trends

The traffic generated in wireless networks may be of the unicast, multicast, or broadcast type. Energy efficiency in all types of communication tasks has previously been considered from the perspective of either minimizing the total energy consumption or maximizing the network lifetime [2]. Regardless of the methodology used, most energy-efficient protocols search for a path that minimizes an energy-related cost metric.

8.2.1 Unicast

Toward the direction of maximizing network lifetime, Chang and Tassiulas [3] propose a protocol where the link costs are defined based on the initial and the current energy at the transmitting nodes, while Toh [4] presents an algorithm that excludes the energy-starving nodes from the route selection. MacEdo et al. [5] employ transmission power and rate control to improve the network performance and energy consumption. The authors in [6] take into account energy harvesting to produce an energy efficient routing strategy, while in [7,8] two protocols are presented that incorporate current energy levels and energy consumption in making routing decisions. In [9], a cost metric is used for routing, which is a function of the remaining battery level and the number of neighbors of a node. Other works have focused on the discovery of energy-efficient routes under the constraint of a fixed end-to-end bit error rate (BER) [10] or by considering the expected number of retransmissions for reliable packet delivery [11]. In [12], the Local Energy-Aware Routing (LEAR) protocol is presented, where a node decides whether to forward traffic based on its residual energy.

A protocol that minimizes the network energy consumption is presented in [13], where the link costs are defined based on the energy expenditure for unit flow transmission. Ramanathan and Rosales-Hain [14] propose two routing algorithms that adjust the node transmission power in order to reduce the energy expenditure. In another work [15], a distributed algorithm is presented that incorporates power control in the routing of packets and tries to increase energy consumption at nodes with plenty of energy while reducing consumption at nodes with small energy reserves. Span [16] is a distributed randomized algorithm where nodes make local decisions on whether to sleep, thus reducing energy consumption, or to join a backbone infrastructure.

The routing protocols mentioned earlier follow the single-cost approach, in the sense that they base their decisions on a single scalar metric (which may be a function of several metrics). Multiconstrained routing algorithms have also been investigated, especially for wired networks [17–20]. Finding paths subject to two or more cost parameters/constraints is, in most cases, an NP-complete problem [21,22]. As a result, most algorithms proposed in this area concentrate on solving the multiconstrained path (MCP) problem or the multiconstrained optimal path problem in a heuristic and approximate way with polynomial and pseudo-polynomial-time complexities, paying little attention to the parameters/costs used and their effects on network performance. The multiconstrained problem has been less studied in the context of wireless ad hoc networks, even though these networks have important reliability, energy, and capacity constraints that are not present in wired networks. In [23], the authors propose a probabilistic modeling of the link state for wireless networks as well as an approximation of a local multipath routing algorithm to provide soft QoS under delay and reliability constraints. In [24], a multiconstrained QoS routing algorithm for mobile ad hoc networks is proposed that uses simulated annealing. In [25], the authors present an algorithm based on depth-first search that solves the general *k*-constrained MCP problem with pseudo-polynomial-time complexity. In [26] and [27], well-known routing algorithms for ad hoc networks are extended to support QoS through the usage of multiple constraints. These algorithms focus on the bandwidth- and delay-constrained routing problems. In [28], a QoS routing scheme for ad hoc networks that uses flooding is proposed.

8.2.2 Multicast

The solutions proposed so far for multicast and broadcast communication can be divided into two categories: *augmentation algorithms*, which start with an empty set of nodes that is gradually augmented to a multicast or broadcast tree; and *local search algorithms*, which perform a walk on a multicast or broadcast structure, decreasing its overall energy consumption. Two surveys summarizing much of the related work in the field can be found in [29,30].

A seminal work presenting a series of basic energy-efficient multicasting and broadcasting algorithms, such as minimum spanning tree (MST), shortest-path tree (SPT), and broadcast incremental power (BIP), is [31]. The MST algorithm constructs a minimum-energy spanning tree for broadcasting, while the SPT algorithm uses Dijkstra's algorithm in order to obtain a tree consisting of the minimum-energy unicast paths to a destination. The BIP algorithm maintains a single tree rooted at the source node and adds new nodes to the tree, one by one, on a minimum incremental cost basis. The incremental cost of adding a new node to the tree is the minimum additional power required by a node in the current tree to reach the new node. A variation of BIP is the broadcast average incremental power (BAIP) algorithm [32] where many new nodes can be added at the same step. The nodes added at each step have minimal average incremental cost, which is defined as the ratio of the minimum additional power required by a node in the current tree to reach these nodes to the number of new nodes reached. The greedy perimeter broadcast efficiency (GPBE) algorithm [33] applies the same tree formation procedure as the BIP algorithm, but it is based on another decision metric, called *broadcast efficiency*, defined as the number of newly covered nodes reached per unit transmission power. In [34], two algorithms are proposed, namely, the minimum longest edge (MLE) and the minimum weight incremental arborescence (MWIA) algorithms. The MLE is based on the property that the MST has the MLE among all spanning trees. Therefore, it first computes an MST using as link costs the required transmission powers and then removes the redundant transmissions based on the nature of the wireless broadcast. In MWIA, a broadcast tree is constructed using as criterion a weighted cost that combines the residual energy and the transmission power of each node. Afterward, the unnecessary edges are removed in a manner similar to the MLE algorithm. All the aforementioned works assume adjustable node transmission power. One of the few papers that assume preconfigured power levels for each node is [35], where two heuristics for the minimum-energy broadcast problem are proposed: a greedy one, where the criterion for adding a new node in the tree is the ratio of the expended power over the number of nodes covered by the transmission, and a node-weighted Steiner tree algorithm. Multicasting is also considered in [36], where an algorithm based on the directed Steiner tree and two heuristic energy-efficient multicasting algorithms are proposed. The criterion used in the heuristics for selecting the nodes to transmit is the transmission power of the nodes averaged over the number of neighbors reached. In [37], the authors present a protocol that uses directional antennas and considers jointly interference and energy efficiency to enhance the network performance, while [38] proposes two energy-efficient multicast algorithms for multichannel and multiinterface wireless networks.

In [39], a so-called relative neighborhood graph (RNG) topology is used for broadcasting. The RNG contains all edges (u, v) such that the intersection of the two disks centered at nodes u and v with radius D_{uv} contains no other nodes, where D_{uv} is the distance between u and v. In [40], each node starts by building a one-hop MST. A link is included in the final graph, called *local minimum spanning tree* (LMST), if it was selected in the LMSTs of both of its edge nodes. In [41], a localized version of the BIP algorithm is presented. Each node uses knowledge about its two-hop neighborhood to apply the BIP algorithm and communicates the decisions taken using a forward packet.

Local search algorithms perform a walk on broadcast forwarding structures. The walk starts from an initial broadcast topology obtained by some algorithm, for example, a spanning tree algorithm. At each step, a local search algorithm obtains a new broadcast topology that maintains the necessary connectivity properties. An energy-related rule is used at each step for selecting the next topology, and the algorithm terminates when no further improvement can be obtained. In [31], the sweep heuristic algorithm was proposed to improve the performance of BIP by removing transmissions that are unnecessary, due to the wireless broadcast advantage. Sweep can be applied to all sorts of broadcast trees and not only to BIP broadcast trees. Iterative maximum branch minimization [42] starts with a trivial broadcast tree where the source transmits directly to all other nodes and, at each step, replaces the longest link with a two-hop path that consumes less energy. In [43], Embedded Wireless Multicast Advantage (EWMA) is proposed that modifies an MST by checking whether increasing a node's power, so as to cover a child of one of its children, would lead to power savings. The *r*-shrink heuristic [44] is applied to every transmitting node and shrinks its transmission radius so that less than *r* nodes hear each transmission. One node is examined at each step, and the new tree formation is kept if there are savings in energy consumption. The Largest Expanding Sweep Search (LESS) heuristic [45] extends the EWMA algorithm. While EWMA removes a transmitting node at each step, the LESS algorithm generalizes its operation by permitting a slight increase in the transmission power of a node if multiple other nodes can stop transmitting or reduce their transmission power.

8.3 Multicost Approach for Unicast Traffic

For the case of unicast communication, we apply multicost routing by assigning to each link of the network a cost vector consisting of several cost parameters. The cost vector of a path is obtained from the cost vectors of the links that comprise it by applying, component-wise, a monotonic associative operator to each cost vector parameter. The parameters that may be included in the path cost vector are categorized by the way they are obtained from the link cost vectors, i.e., by the associative operator used for each cost vector component, and by the criterion applied to them (maximization or minimization) to select the optimal path. To be more specific, we denote by $V_l = (v_{1l}, v_{2l}, ..., v_{kl})$, where k is the number of cost parameters, the link cost vector of link l; by $V(P) = (V_1, V_2, ..., V_L)$ the cost vector of the path P that consists of links $l = 1, 2, ..., L$; and by $f(V(P))$ the optimization function that has to be minimized in order to select the optimal path. The cost vector $V(P) = (V_1, V_2, ..., V_L)$ of a path P is obtained from the cost vectors of the links that comprise it by applying, component-wise, a monotonic associative operator $\oplus$ to each cost vector parameter $V_m = \oplus_{l=1, ..., L} \Rightarrow v_{ml}$. The associative operator $\oplus$ may be different for different cost vector components. For example, the mth parameter of the cost vector may be of one of the following types:

- Additive cumulative, where $V_m = \sum_{l=1}^{L} v_{ml}$, $v_{ml} \geq 0$ and f is monotonically increasing in V_m (so our objective is to minimize V_m)
- Restrictive, where $V_m = \min_{l=1...L} v_{ml}$ and f is monotonically decreasing in V_m (so our objective is to maximize V_m
- Maximum representative, where $V_m = \max_{l=1...L} v_{ml}$ and f is monotonically increasing in V_m (so our objective is to minimize V_m)

Additive cumulative parameters include several important cost measures used in practice. For example, if v_{ml} is the delay on link l, then V_m represents the delay of the path, which, in most practical situations, has to be minimized. If $v_{ml} = 1$ for all links l, then V_m corresponds to the number of hops on the path. Since paths that use a small number of links are more economical in terms of resource utilization, it is natural to assume that the cost function f is an increasing function of V_m. If v_{ml} is the energy consumed on link l of a wireless network, then V_m represents the energy consumed for transmitting a packet on the path, which has to be minimized. Another interesting case arises when $v_{ml} \in [0,1]$, represents the probability that link l is operational, and $V_m = \prod_{l=1}^{L}, v_{ml} \geq 0$ is the probability that all links on a path are operational (assuming links fail independently of each other). For the routing algorithm to favor reliable paths, the cost function f should be a decreasing function in V_m. This problem can be reduced to a problem involving cumulative additive components by defining new cost components $v'_{m1} = -\ln v_{m1}$, $v'_{m2} - \ln v_{m2}, \ldots, v'_{ml} = -\ln v_{ml}$, where $v'_{ml} = 0$. Then maximizing the reliability V_m of a path is equivalent to minimizing $V'_m = \sum_{l=1}^{L} v'_{ml}$.

Restrictive cost parameters appear in routing problems when capacities or transmission rates are considered. In particular, if v_{ml} is the available capacity on link l, then $V_m = \min_{l=1 \ldots L} v_{ml}$ represents the capacity of a path, defined as the minimum of the capacities available on the links of the path. For the routing algorithm to favor less congested paths, the cost function f should be a decreasing function of V_m. Another interesting case arises when v_{ml} represents the remaining energy at the transmitting node of link l, in which case V_m represents the minimum energy available over all nodes of a path, which, in most practical cases, we want to maximize.

An example of a maximum representative parameter is the case where v_{ml} is the energy consumed for transmitting a packet on link l, in which case $V_m = \max_{l=1 \ldots L} v_{ml}$ represents the most energy-expensive transmission on the path. Another example is the case where v_{ml} is the BER on link l, in which case $V_m = \max_{l=1 \ldots L} v_{ml}$ represents the link with the highest BER on the path, which is often a good approximation of (or at least of the same order of magnitude with) the path BER.

It is important to note that the path that optimizes $f(V_1, V_2, \ldots, V_k)$ is generally different than the path that optimizes $\sum_{l=1}^{L} f(v_{1l}, f_{2l}, \ldots, v_{kl})$, indicating that multicost routing is a generalization of single-cost (shortest-path) routing. Also, in contrast to single-cost routing, multicost routing is not always compatible with distributed routing, since for some choices of the cost function f, the optimal paths do not have the inclusion property that shortest paths have; subpaths of shortest paths are also shortest paths, but this is not generally the case with optimal paths found by multicost routing for specific choices of the optimization function. These show that multicost routing is very different from single-cost routing in terms of the decision it takes, its properties, and the way it is implemented.

A multicost routing algorithm consists of two phases [47]. In the first phase, an enumeration of an appropriate set of candidate paths for a given source–destination pair is performed. This can be viewed as a generalization of Dijkstra's algorithm. The basic difference of this algorithm with Dijkstra's algorithm is that a set of paths between a source node and a destination node is obtained, instead of a single path. Also, a destination node for which a path has already been found may have to be considered again later. The set of candidate paths that a multicost routing algorithm produces at the end of the first phase consists of the so-called nondominated paths. These are paths for which it is impossible to find other paths that are better with respect to one cost parameter (of their cost vectors) without being worse with respect to some other cost parameter. This reduces to a large extent the algorithm's computational effort, since the optimization function does not need to be applied to every possible path between a certain source–destination pair. An example of the enumeration of the nondominated paths is given in Figure 8.1, where an additive cumulative parameter h and a restrictive parameter R are assumed. In the second phase, the optimal path is chosen from this set according to the optimization function $f(V)$ used.

A formal description of the multicost routing algorithm is presented next. The algorithm obtains the optimal path for a given source S–destination E pair. Without the loss of generality, let the cost vector of each link have k cost parameters, the first s of which are additive and have to be minimized and the rest $k-s$ are restrictive and have to be maximized. We assume that each

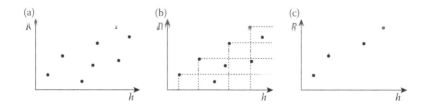

Figure 8.1 Enumeration of the set of nondominated paths for the case of two cost parameters. Each dot represents the cost vector of a path. The parameters of this vector are an additive cumulative parameter h (representing, e.g., the number of hops or the delay of the path) and a restrictive parameter R (representing, e.g., the minimum residual energy on the nodes of the path or the available capacity on the path). (a) The set of all paths. (b) Obtaining the nondominated paths. (c) The set of nondominated paths; paths that have both larger R and smaller h than some of the other paths that have been discarded.

path is represented by a label that includes the cost vector associated with it and the first hop to the source using that path. The source that serves the connection is taken to be node S. We let W_i be the set of labels of the paths from node S to a node n_i, and $W = \bigcup_{n_i \neq s} W_i$ be the set of all labels.

Initially, every node has a single label corresponding to the link (if any) that connects it directly to the origin node. In each subsequent step, the algorithm marks labels (equivalently, paths) from the set W as final. We let $W^f \subseteq W$ be the subset of all final labels for all the nodes and let $W_i^f \subseteq W_i$ be the set of final labels for node n_i. We also let T be the set of nodes with at least one final label. The algorithm can now be described as follows:

Phase 1: Enumeration of a Set of Nondominated Paths

Step 0: Initialization. $W = \{V_{p1}, V_{p2} \ldots, V_{pN}\}$, $W_f = \{\ \}$. $T = \{\ \}$, where V_{pi} is the label of the path p_i (if any) leading directly from node S to node n_i.

Step 1: Choosing the optimum label. The label of path p whose cost vector minimizes the additive component is chosen. In the case of a tie, we look at the second component, which is the binary capacity availability vector, and a dominant one is chosen. If V_{pi} is the cost vector of the chosen label and n_i is the node to which it leads, then the following updates are performed: $W_f^i = W_f^i \cup \{V_{pi}\}$, $W_f = W_f \cup \{V_{pi}\}$, $T = T \cup n_i$.

Step 2: Obtaining the new labels. The neighbors of node n_i, which may or may not belong to the set T, are now considered and are given new labels (except for the origin node and the node specified as the previous node in the label). The new label for the path p_j leading to the neighbor n_j of node n_i by extending the path p_i through the link $l = (n_i, n_j)$ is then computed as follows. The new cost vector is updated according to $V'_{pi} = V_{pi} \oplus V_l$, where V_l is the label of link $l = (n_i, n_j)$ and $\oplus$ represents the monotonic associative operator described earlier.

Step 3: Discarding dominated paths. Each neighbor considered in Step 2 compares its new label with its previous labels using the domination relation. Let n_j be one of the neighbors of node n_i, V'_{pi} be the new label obtained from Step 2, and W_j be the set of labels for this node. The new label has to be compared with the labels $V_{pj} \in W_j$ (both final and nonfinal). If any cost vector in W_j dominates V'_{pi}, then V'_{pi} is discarded and W_j and W do not change. If the new cost vector V'_{pi} is not dominated by any of the vectors in W_j, then V'_{pi} is added to the set W_j and W so that $W_j = W_j \cup V'_{pi}$ and $W = W \cup V'_{pi}$. If the new vector dominates one of the vectors in W_j, then W_j and W are updated by eliminating the dominated vectors and adding the new vector V'_{p_i}. Note that it is not possible for the new vector to dominate an existing vector and be dominated by another one at the same time.

Step 4: Termination. If after an iteration the set W' is equal to W, the algorithm is completed. Otherwise (when there are still some labels to be chosen), we go back to Step 1. The set P_{nd} of nondominated paths from the given source S to the given destination node E is the final set W'_E.

Phase 2: Selection of the Optimal Path

In the second phase, the optimal path is chosen from this set according to the optimization function $f(V)$ used. The final cost of a path P is given by a function $f(V(P))$ of its cost vector $V(P)$, and the routing algorithm selects the path with the minimum cost from the set of nondominated paths.

The multicost algorithms we propose for energy-aware routing in wireless ad hoc networks use three link cost parameters: the hop count, equal to 1 for all links; the residual energy R_i; and the transmission power T_i at the transmitting node i of a link (i, j). The number of hops h of a path is obtained by counting the links that belong to it and is an additive cost parameter. The minimum residual energy R of a path represents the minimum residual energy left on the nodes of the path and is a restrictive cost parameter. Finally, the transmitted power parameter T of a path is defined as the sum of the transmission powers T_i of the path's nodes and is an additive cost parameter.

The proposed optimization functions $f(h, T, R)$ are listed below. Given that we focus on energy efficiency, we combine the aforementioned energy-related cost parameters in several ways. Each optimization corresponds to a different routing algorithm since it produces a distinct set of routing decisions.

- Minimum hop: $f_1(h, T, R) = h$
- SUM/MIN energy: $f_2(h, T, R) = \sum_{i \in p} T / \sum_{i \in p} R_i$
- SUM/MIN energy–hop: $f_3(h, T, R) = h \sum_{i \in p} T_i / \sum_{i \in p} R_i$
- SUM/MIN energy–half-hop: $f_3(h, T, R) = \sqrt{h} \sum_{i \in p} T_i / \sum_{i \in p} R_i$

In all cases, the algorithms first find the set of nondominated paths with cost parameters (h, T, R), and therefore they have the same (or similar, if the parameters are not the same) phase 1, and then use the corresponding optimization function $f(h, T, R)$ in phase 2 to select the optimal path. In other words, the computation of the set of nondominated paths is common to all algorithms, and the selection of the optimal path is done at the end in a way that is different for each of the algorithms proposed. The function to be optimized at the last step may depend on the QoS requirements of the user. The optimization functions considered penalize paths that use a large number of hops, consume a large amount of energy, or pass through nodes that have little energy left, differentiating, however, from each other by giving different importance to each of these factors.

The optimization functions presented, except for the first one, cannot be optimized over all paths by using a single-cost routing algorithm, which shows that multicost routing is a strict generalization of single-cost routing. Finally, based on [22], the complexity of any multicriteria algorithm using at least two additive parameters is exponential, except in the case where one of the two is the hop metric. Also, when one additive and one restrictive or maximum representative parameter are used, the complexity of the corresponding algorithm (optimization function) is polynomial. As a result, all the proposed algorithms have polynomial complexity.

8.4 Multicost Approach for Multicast Traffic

In this section, we describe an energy-efficient algorithm for performing multicasting based on the multicost concept. The objective of the algorithm is to find, for a given source node and a desired multicast group, an optimal sequence of nodes for transmitting so as to implement multicasting in an energy-efficient way. In particular, it selects a transmission schedule that optimizes any desired function of the total power T consumed by the multicasting task and the minimum R of the residual energies of the nodes, provided that the optimization function used is monotonic in each of these parameters, T and R.

Similar to the unicast case, the algorithm's operation consists of two phases. In the first phase, the source node u calculates a set of candidate node transmission sequences $S_{u,M}$, called a *set of nondominated schedules*, which can send to all nodes in the multicast group M any packet originating at that source. In the second phase, the optimal sequence of nodes for multicasting is selected based on the desired optimization function. Below, a formal description of the optimal multicast algorithm is given.

In the first phase of the algorithm, every source node u maintains at each time a set of candidate multicast schedules S_u.* A multicast schedule $S \in S_u$ is defined as $S = ((u_1 = u, u_2,\ldots, u_h), V_s)$, where $(u_1, u_2,\ldots,u_h)$ is the ordered sequence of nodes used for transmission and Vs (R_S, T_S, P_S) is the cost vector of the schedule, consisting of (i) the minimum residual energy R_S of the sequence of nodes $(u_1, u_2,\ldots,u_h)$, (ii) the total power T_S consumed when these nodes are used for transmission, and (iii) the set P_S of network nodes covered when nodes $(u_1, u_2,\ldots,u_h)$ transmit a packet.

When node u_i transmits a packet at distance r_i, the energy expended is taken to be proportional to r_i^a, where a is a parameter that takes values between 2 and 4. Because of the broadcast nature of the medium and assuming omnidirectional antennas, a packet being forwarded by a node can be correctly received by any node within range r_i of the transmitting node u_i. Therefore, multicast communication in these networks corresponds to finding a sequence of transmitting nodes instead of a sequence of links as is common in the wireline world.

The formal description of the algorithm is given as follows.

Phase 1: Enumeration of the Candidate Multicast Schedules

　　Step 0: Initialization. Initially, each source node u has only one multicast schedule $(\varnothing,(\infty,0, \{u\}))$ with no transmitting nodes, infinite node residual energy, and zero total power consumption, while the set of covered nodes contains only the source.

　　Step 1: Extending the multicast schedules. Each multicast schedule $S = ((u_1, u_2,\ldots,u_{i-1}), R_S, T_S, P_S)$ in the set of nondominated schedules S_u is extended by adding to its sequence of transmitting nodes a node $u \in P_S$ that can transmit to some node u_j not contained in P_S. If no such nodes u_i and u_j exist, we proceed to step 4.

　　A schedule S is extended to schedule S' as follows:

　　　　■ Node u_i is added to the sequence $u_1, u_2,\ldots,u_{i-1}$ of transmitting nodes.
　　　　■ $R_S' = \min (R_i, Rs)$, where R_i is the residual energy of node u_i.

---

* The schedules in S_u are not only for multicasting to the desired multicast group M but to any set of nodes.

- $T_S' = T_i + T_S$, where T_i is the (fixed) transmission power of node u_i.
- The set of nodes $D(u_i)$ that are within the transmission range from u_i are added to P_S.
- The extended schedule $S' = ((u_1, u_2, ..., u_i),$ min $(R_i, R_s),\ T_i + T_S,\ D(u_i) \cup P_S)$ obtained in the way described above is added to the set S_u of candidate schedules.

Step 2: Discarding dominated schedules. Next, the *domination relation* between the various multicast schedules of source node u is applied, and the schedules found to be dominated are discarded. In particular, a schedule S_1 is said to *dominate* a schedule S_2 when $T_1 < T_2$, $R_1 > R_2$, and $P_1 \supset P_2$ In other words, schedule S_1 dominates schedule S_2 if it covers a superset of the nodes covered by S_2, using less total transmission power and with larger minimum residual energy on the nodes it uses. All the schedules found to be dominated by another schedule are discarded from the set S_u.

Step 3: Checking for end of loop. The procedure is repeated, starting from step 1, for all multicast schedules in S_u that meet the above conditions. If no schedule $S \in S_u$ can be extended further, we proceed to Step 4.

Step 4: Selection of the schedules matching the multicast group. Among the schedules in S_u, we form the subset of schedules S for which $P_S \supset M$. This subset is called the *set of nondominated schedules* for transmitting from source node u to *multicast group M* and is denoted by $S_{u,M}$.

Phase 2: Selection of the Optimal Multicast Schedule

An optimization function $f(V_S)$ is applied to the cost vector V_S of every nondominated schedule $S \in S_{u,M}$ with source node u produced in the first phase. The optimization function combines the cost vector parameters in order to produce a scalar metric representing the cost of using the corresponding sequence of nodes for multicasting. The schedule with the minimum cost is selected.

The node residual energy is used as a restrictive cost parameter, since its minimum value on the nodes of a schedule defines the schedule's residual energy, while the node transmission power is used as an additive cost parameter. This is because the residual energy on a set of nodes is more accurately characterized by its minimum value among all nodes in the set, while the power consumed by a set of nodes is described by the sum of their transmission powers. We also observe that Steps 1–3 of the algorithm do not depend on the target multicast group M, and the set of S_u schedules produced in these steps can be used for multicasting to any set of nodes. Therefore, if the desired multicast group M changes, only Step 4 of the algorithm is recalculated. This makes the algorithm particularly useful for cases where a node belongs to multiple multicast groups or where the multicast group membership changes with time.

In the performance results described in Section 8.5, the optimization function used is $f(S) = T_S / R_S$ for $S \in S_{u,M}$, which favors, among the schedules that cover all nodes in the multicast group M, those that consume less total energy T_S and whose residual energy R_S is large. Other optimization functions could also be used, depending on the interests of the network, and different functions could be used for different multicast groups. Also, our algorithm, with straightforward modifications, can include parameters other than the transmission power and the residual energies of the nodes. For example, we can include as a cost parameter the number h_S of transmissions required by schedule S to complete the multicast (which is also related to the multicast delay in the absence of other traffic and queuing delays in the network), and the optimization function used could incorporate this parameter when deciding the optimal schedule. The only requirement is that the optimization function has to be monotonic in each of its parameters (e.g., an increasing function of T_S and h_S and a decreasing function of R_S). In this case, the proposed algorithm calculates the optimal schedule. The following theorem proves the previous statement.

Theorem 8.1: If the optimization function $f(V_S)$ is monotonic in each of the parameters involved, the algorithm finds the optimal multicast schedule.

PROOF: Since $f(V_S)$ is monotonic in each of its parameters, the optimal schedule has to belong to the set of nondominated schedules (a schedule S_1 that is dominated by a schedule S_2, meaning that it is worse than S_2 with respect to all the parameters, cannot optimize f). Therefore, it is enough to show that the set S_u computed in steps 1–3 includes all the nondominated schedules for multicasting from node u. We let $S = ((u_1, u_2,...,u_h), R_S, T_S, P_S)$ be a nondominated schedule that has a minimal number of transmissions h among the schedules not produced by the algorithm. Then, for the schedule $S' = ((u_1, u_2,...,u_{h-1}), R_{S'}, T_{S'}, P_{S'})$, we have that $R_S = \min (R_h, R_{S'})$, $T_S = T_h + T_{s'}$, and $P_S = D(u_h) \cup P_{S'})$. The fact that S is nondominated and was not produced by the algorithm implies that S' was not produced by the algorithm either. Since S is a nondominated schedule with a minimal number of transmissions among those not produced by the algorithm and S' was not produced by the algorithm and uses less transmissions, we conclude that S' is dominated. However, S is nondominated, implying that S' is also nondominated (otherwise, the schedule S'' that dominates S', in the sense that $T_{S'} < T_{S'}$, $R_{S'} > R_S$, and $P_{S''} \supset P_S$, extended by the transmission from node u_h would dominate S), which is a contradiction.

The algorithm presented above solves the multicasting problem optimally but has nonpolynomial complexity. This is because the number of different nondominated schedules S produced by the first phase of the algorithm increases exponentially with the number n of wireless nodes since the set P_S of covered nodes parameter can take 2^n different alternatives. On the basis of work in [22], we know that the complexity of any multicost algorithm using one additive (T) and one restrictive (R) parameter is polynomial. So, there are cases where, in the second phase of the optimal algorithm, the optimization function $f(V_S)$ is applied to an exponentially large number of schedules S. This leads to exponential execution time and to nonpolynomial worst-case complexity.

In order to obtain a polynomial time algorithm, we relax the domination condition so as to obtain a smaller number of candidate schedules. In particular, we define a *pseudo-domination* relation among schedules, according to which schedule S_1 *pseudo-dominates* schedule S_2 if $T_1 < T_2$, $R_1 > R_2$, and $|P_1| > |P_2|$, where T_i, R_i, and $|P_i|$ are the total transmission power, the residual energy of the transmitting nodes, and the cardinality of the set P_i of nodes covered by schedule S_i ($i = 1, 2$), respectively. When this pseudo-domination relationship is used in step 2 of the algorithm, it results in more schedules being pruned (not considered further) and smaller algorithmic complexity. In fact, by weakening the definition of the domination relationship, the complexity of the algorithm becomes polynomial, as can be seen by arguing that T_i, R_i, and $|P_i|$ can take a finite number of values, namely, at most as many as the number of nodes. The decrease in time complexity, however, comes at the price of losing the optimality of the solution.

To highlight the differences between the proposed optimal and near-optimal algorithms, let us consider the network given in Figure 8.2, where node 1 broadcasts a packet across the network. Both algorithms first find the schedule, $S_0 = \{(1),(1,0.6,(1,2,3))\}$ where node 1 transmits and covers nodes 2 and 3. When this schedule is expanded, both algorithms obtain the schedule

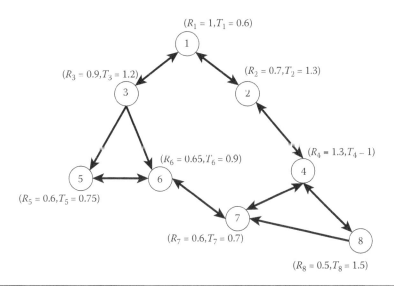

Figure 8.2   An example of a broadcast illustrating the differences between the optimal and the near-optimal algorithm for a multicast. Node 1 broadcasts a packet across the network. Each node's cost vector is in the form (R, T), where R is its residual energy in Joules and T is its transmission power in watts.

S_1 = {(1,2),(0.7,1.85,(1,2,3,4))} or S_2 = {(1,3),(0.9,1.8,(1,2,3,4,5,6))}by expanding S_0 with the transmission of node 2 or 3, respectively. In the case that the near-optimal algorithm is used, schedule S_1 is dominated by schedule S_2 since S_2 in comparison to S_1 covers more nodes with higher minimum residual energy and less total transmission power. This leads the near-optimal algorithm to finally select the schedule $S_{near\ optimal}$ = {(1,3,6,7,4),(0.6,4.4,(1,2,...,8))} as the node schedule to implement the desired broadcast. On the contrary, when the optimal algorithm is used, schedule S_1 is not dominated by schedule S_2 since it covers a node (node 4) not covered by S_2. Thus, the optimal algorithm selects the schedule $S_{optimal}$ = {(1,2,3,4),(0.7,4.1,(1,2,...,8))}.

8.5 Performance Results

In this section, we present an extensive simulation study of the proposed algorithms for unicast and multicast communication. The algorithms were implemented and tested using the network simulator [46].

8.5.1 Unicast Communication

We first evaluate the performance of the multicost unicast algorithms presented in Section 8.3. The algorithms are studied in the context of two different models, namely, the network evacuation model and the infinite time horizon model. In the network evacuation model, the network starts with a certain number of packets that have to be routed and a certain amount of energy per node, and the objective is to serve the packets in the smallest number of steps or to serve as many packets as possible before the energy at the nodes is depleted. In the infinite time horizon model, packets with uniformly distributed destinations are generated at each network node according to

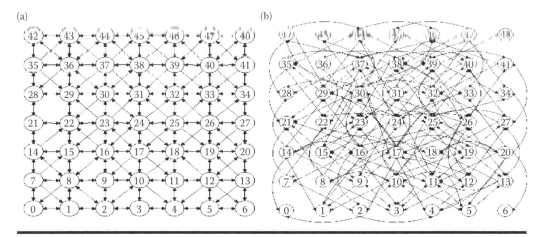

Figure 8.3 **(a) Network topology where the transmission range of the nodes varies between 50 and 100 m. (b) Network topology where the transmission range of the nodes varies between 50 and 150 m [we omit the edges already shown in topology (a)].**

a random process and energy is also added at each node at a given recharging rate over an infinite time horizon.

The wireless ad hoc network simulated consists of 49 stationary nodes placed along a 7×7 grid. The distance between neighboring grid points is set at 50 m. The topologies studied in the experiments are either a regular grid topology (where the transmission range of the nodes is fixed at 50 m) or a random topology (where the transmission range varies from node to node and is uniformly distributed between 50 and 100 m in one set of experiments and between 50 and 150 m in another set of experiments) (Figure 8.3). For the case of randomly produced network topologies, each measured value represented in the graphs corresponds to the average of a set of 10 experiments.

The initial energy of the nodes was taken to be either 100 or 2 J. The former case represents a scenario of essentially unlimited energy reserves ("infinite" energy), while in the latter case some nodes run out of energy during the experiments (finite energy), depending on the amount of traffic they end up serving. The number of packets per node that have to be delivered to their destinations ("evacuated" from the network) in our experiments varies from 100 to 1,000 (at steps of 100) packets per source node. All packets have equal length that is taken to be 500 bytes. Packet destinations are uniformly distributed over all remaining network nodes and the packet generation rate at each node is equal to 0.1 packets/s. The interval between nondominated path recalculations is equal to 1 s.

8.5.1.1 Network Evacuation Model

Figure 8.4a illustrates the average residual energy in the network at the end of an evacuation experiment as a function of the number of packets evacuated per node. The minimum-hop algorithm results in a higher average residual energy E at the end of the evacuation experiments than the other routing algorithms examined. However, the minimum-hop algorithm also results in less uniform energy consumption in the network and in smaller energy depletion times than the other algorithms, indicated by Figures 8.4b and 8.4c, respectively. As a result, the minimum-hop algorithm also achieves lower throughput and a higher dropping ratio. Note in Figure 8.4a that when more than 400 packets are generated per source node, the average residual energy stops

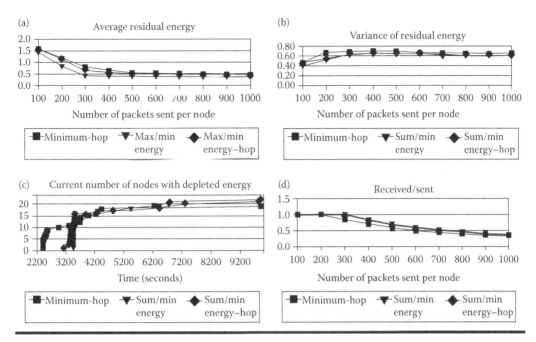

Figure 8.4 Illustration of (a) the average residual energy at the end of the evacuation experiment, (b) the variance of the residual energy, (c) the current number of nodes with depleted energy at the end of the evacuation experiment, and (d) the received-to-sent packets ratio for the minimum-hop, SUM/MIN energy, and SUM/MIN energy–hop algorithms. The results were obtained for the case of finite energy and the topology of Figure 8.3a.

decreasing for all algorithms examined. This happens because many nodes run out of energy, and consequently the network becomes disconnected and no more packets are served.

The minimum-hop algorithm uses the same path for the entire duration of a session or until the energy of a node on the path is depleted. As a result, a relatively small subset of nodes has more active participation in the transmission of packets than other nodes. The SUM/MIN energy and SUM/MIN energy–hop algorithms, on the other hand, are based on parameters R_i that change over time, and the path selected may not remain the same for all the packets of a session. In this way, the traffic forwarding and the resulting energy consumption are spread more uniformly over a larger number of nodes, leading to smaller average residual energy E and smaller variance σ^2 of the residual energy than the minimum-hop algorithm. Regarding the time in which the energy of the nodes is depleted, the SUM/MIN energy–hop algorithm exhibits the best performance in all the experiments, while the minimum-hop algorithm seems to result in the worst depletion time. Note that with the SUM/MIN energy and SUM/MIN energy–hop algorithms, when nodes start running out of energy, this happens almost simultaneously for all nodes. This is because these algorithms spread the energy consumption uniformly in the network so that when one node is at the point of first running out of energy, most other nodes are at the same energy-critical situation.

In most of the experiments conducted, we found that the performance of the SUM/MIN energy–hop algorithm was between that of the minimum-hop algorithm and that of the SUM/MIN energy algorithm, and actually in most cases, it was closer to that of the minimum-hop algorithm. The SUM/MIN energy–half-hop algorithm was found to behave very similarly to the SUM/MIN energy algorithm in all cases considered. It seems that the 0.5 exponent on the number of hops in the former algorithms effectively eliminates the impact of the hop term on the cost function. This is the reason we chose not to present in great detail the results on the SUM/MIN energy–half-hop algorithm.

Figure 8.4d shows the received-to-sent packets ratio for various algorithms examined. When the initial energy of the nodes is infinite, all the packets are delivered to their destinations, but when the initial energy is finite, the fraction of packets delivered to their destinations decreases after a certain number of packets have been inserted in the network. The reason is that nodes run out of energy, limiting the ability of the network to route packets. We observed that the SUM/MIN energy–hop and SUM/MIN energy algorithms achieve the best received-to-sent packets ratio in almost all the experiments, since with these algorithms the network nodes remain alive for longer periods of time.

Figure 8.5 shows the average number of hops of the path. When the nodes' energy is infinite or the number of packets exchanged is small, the SUM/MIN energy algorithms select paths that are on average longer than the minimum-hop paths. However, when the nodes have finite energy, there are cases where the SUM/MIN energy algorithm achieves a similar average number of hops to that of the minimum-hop paths. This is because some of the nodes run out of energy and the minimum-hop algorithm eventually has to use longer paths. The SUM/MIN energy-hop algorithms are dominated by the hop parameter and give results that are similar to those of the minimum-hop.

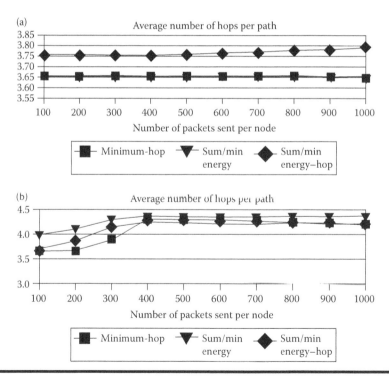

Figure 8.5 Illustration of the average number of hops of the paths constructed by the minimum-hop, SUM/MIN energy, and SUM/MIN energy–hop algorithms. The results were obtained for the case of (a) infinite (very large) and (b) finite energy as well as the topology of Figure 8.3a.

8.5.1.2 Infinite Time Horizon Model

In this section, we compare the performance of the SUM/MIN energy algorithm to that of the minimum-hop algorithm in the context of the infinite time horizon model. All packets are assumed to have equal length and require one slot in order to be transmitted over a link. Time is slotted, and a new packet is generated at each node with probability p during a slot. In our experiments, the duration of the slot is 0.08 s while the packet transmission time is 0.016576 s for the 2,000 byte-sized packets we use in our experiments. We chose this slot time in order for the RTS/CTS handshake mechanism to have been completed by the time the next packet is generated. Packet destinations are uniformly distributed over all nodes.

In addition to the usual capacity and interference constraints, the network is also assumed to be energy-constrained. More specifically, we assume that energy is generated at each node at a recharging rate of X units of energy per slot. Initially, the network is without energy. Each packet transmission consumes an equal amount of energy E. Furthermore, we define a threshold on the residual energy of a node, and when the energy at a node falls below this threshold, the node stops forwarding packets and starts storing them in its queue. The same happens when the receiver's residual energy is below this threshold. Each node periodically checks its energy reserves and those of its neighbors, and if they both exceed the threshold, the node starts forwarding its queued packets.

We are interested in the steady-state performance of the proposed schemes for varying recharging rates and packet generation probabilities. The network is assumed to have reached the steady state when the variance in the packet delivery delay is below some threshold. The network topology in our experiments is that of Figure 8.3a, where the transmission range of the network nodes is uniformly distributed between 50 and 100 m. The performance metrics of interest are the largest packet generation probability p_{max} for which the network remains stable (maximum throughput) and the average packet delivery delay for a given packet generation probability $p < p_{max}$. By stability, we mean that the incoming traffic can be served appropriately, with small average packet delay and high packet delivery ratio. When either of these conditions is not fulfilled, the network is assumed to enter the unstable region, and then there is no point in studying it further. Each measured value, represented in the graphs, corresponds to the average of a set of 10 experiments.

Figure 8.6 shows the average packet delay as a function of the packet generation rate p for recharging rates of $X = 5 \times 10^{-3}$ and $X = 9 \times 10^{-3}$ J/slot* for both the minimum-hop and the

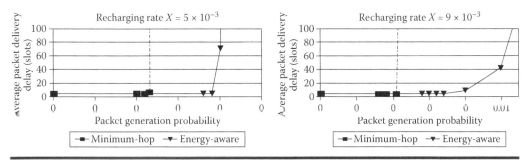

Figure 8.6 Illustration of the packet delay (in slots) as a function of the packet generation rate p for the minimum-hop and the SUM/MIN energy (energy-aware) algorithms for energy recharging rates of (a) $X = 5 \times 10^{-3}$ J/slot and (b) $X = 9 \times 10^{-3}$ J/slot.

---

* To be more specific, energy equal to 0.005 Joules and 0.009 Joules was offered every 10 s in the experiments.

SUM/MIN energy routing algorithms. The SUM/MIN energy algorithm outperforms the minimum-hop algorithm by enabling the network to remain stable for heavier traffic loads. For the topology considered, the traffic generation probabilities p that the SUM/MIN energy algorithm is able to handle, with adequately small packet delivery delay, are nearly twice those of the minimum-hop algorithm for both recharging rates considered. The transition of the network to the unstable region, as indicated by the rise in the average packet delay in Figure 8.6, is very steep for the minimum-hop algorithm for both recharging rates $X = 5 \times 10^{-3}$ and $X = 9 \times 10^{-3}$ J/slot; from values of the delay around 4 or 5 slots in the stable region, there is an almost instant increase to large (practically infinite) values above 100 slots. This is because when the minimum-hop algorithm (which is not energy-efficient) is used, the network for both the values of the recharging rate X is energy-constrained. When the energy at some nodes gets depleted, the energy of many other nodes also starts getting depleted soon afterward, and the rise in the delay is very abrupt. In this state, the delivery of the incoming packets becomes difficult (large delays) or impossible (dropping of packets) due to the weakened connectivity of the network. When the SUM/MIN energy algorithm is used and for the low recharging rate $X = 5 \times 10^{-3}$ J/slot, the network is again energy-constrained, but because it uses energy more efficiently, the rise in the delay is less abrupt than with the minimum-hop algorithm. When the SUM/MIN energy algorithm is used and the recharging rate is relatively high, $X = 9 \times 10^{-3}$ J/slot, the network is mainly capacity-constrained and the rise in the delay is rather smooth.

Figure 8.7 shows the number of packets received with respect to the number of packets sent for recharging rates $X = 9 \times 10^{-3}$ and $X = 15 \times 10^{-3}$ J/slot. It can be observed that the SUM/MIN energy algorithm achieves a higher throughput than does the minimum-hop algorithm since the

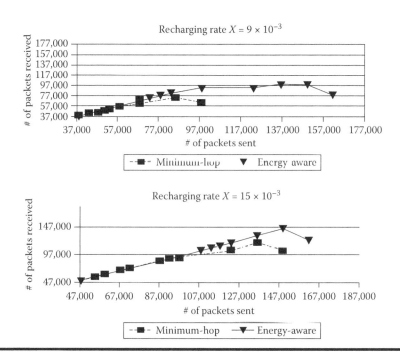

Figure 8.7 The number of the packets received versus the number of packets sent for the minimum-hop and the SUM/MIN energy (energy-aware) algorithms for energy recharging rates of (a) $X = 9 \times 10^{-3}$ J/slot and (b) $X = 15 \times 10^{-3}$ J/slot.

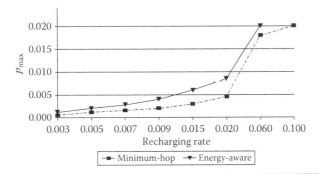

Figure 8.8 **The maximum throughput (maximum packet generation probability) p_{max} for which the network remains stable as a function of the recharging rate X (J/10 s) at the network nodes for the minimum-hop and the SUM/MIN energy (energy-aware) algorithms.**

degradation of the received-to-sent packets ratio begins later relative to the minimum-hop algorithm. For both algorithms, the number of packets delivered to their destinations grows linearly, initially, with the number of packets that enter the network, since for light traffic, few packets are dropped. For packet generation probabilities greater than p_{max}, however, there is a steep decline in the packet delivery ratio. The number of packets successfully delivered to their destinations not only stops increasing as the number of incoming packets grows, but it even declines after the network enters the unstable region.

Figure 8.8 shows the maximum throughput (maximum packet generation probability) p_{max} for which the network remains stable as a function of the recharging rate X at the network nodes for both the minimum-hop and the SUM/MIN energy routing algorithm. The maximum throughput p_{max} achieved by the network is taken to be the highest packet generation probability for which the network manages to serve the incoming traffic appropriately, meaning with small average packet delivery delay and high packet delivery ratio. The thresholds set for these two metrics used for detecting experimentally when the network enters the unstable region (above 100 slots for the average packet delivery delay and under 80% for the delivery ratio) are not important qualitatively for the results obtained, since we found that a different setting of the thresholds causes only a small shift in the values presented without altering any of the conclusions drawn.

Figure 8.8 shows that the SUM/MIN energy algorithm outperforms the minimum-hop algorithm, achieving significantly larger p_{max} for all recharging rates considered. The maximum throughput p_{max} seems to depend on the recharging rate almost linearly until the very end for both routing algorithms. This linear increase verifies that the network in this region of recharging rates is energy-constrained. When the recharging rate increases beyond some point, the network starts getting constrained by capacity/interference limitations, and the rate at which p_{max} grows with respect to the recharging rate is slowed down until it reaches a plateau, indicating that the capacity/interference limitation has been reached. The performance difference between the SUM/MIN energy algorithm and the minimum-hop algorithm is larger for low-energy recharging rates at the nodes, and the difference is gradually reduced as the limitation posed by the network capacity is approached. The maximum throughput p_{max} achieved by the SUM/MIN energy algorithm is nearly twice that of the minimum-hop algorithm. This is because the further away the network is from the capacity-constrained region, the more important becomes the efficiency in the use of the energy resources. In summary, when energy is the factor defining the ability of the network to serve incoming traffic, the SUM/MIN energy algorithm performs better. However, as energy becomes abundant and capacity becomes the main limitation on network performance, the performance gap between the SUM/MIN energy and the minimum hop algorithm is narrowed.

8.5.2 Multicast Communication

We evaluate the performance of the multicost multicast algorithms presented in Section 8.4. Apart from the standard case in multicasting where a node transmits to a subset of the network nodes, we also include in our simulation study the special case where the multicast group contains *all* the network nodes, which is referred to as broadcasting. In what follows, we will refer to the optimal and near-optimal algorithms for multicasting as optimal total and residual energy multicost multicast (OTREMM) and near-optimal total and residual energy multicost multicast (NOTREMM) algorithms, respectively.

In the case of multicasting, the proposed algorithms are compared against the multicast version of the BIP algorithm [31], to be referred to as multicast incremental power (MIP) algorithm [31], the MWIA algorithm [34], and the node-join-tree (NJT) heuristic [36]. BIP finds an MST using as link cost the required transmission power and then removes the unnecessary transmissions by taking into account the broadcast nature of the wireless medium. The MIP algorithm prunes, from the broadcast tree, the transmissions to nodes not belonging to the multicast group in order to construct the multicast tree. The MWIA algorithm uses as criterion, for the selection of a node as transmitter in the multicasting schedule, the ratio of the node's transmission power over the node's residual energy. The NJT algorithm builds a multicast tree using as criterion, for the selection of a transmitting node, the node's transmission power over the number nodes covered by the transmission. Note that the MIP and NJT algorithms try to minimize the network's energy consumption, while the MWIA algorithm considers both the minimization of the network's energy consumption and the maximization of its lifetime. However, MWIA considers these two objectives using the single-cost approach by defining a corresponding link cost metric, while our algorithms consider them using the multicost approach, where the two parameters are treated separately until the very end.

For the special case where the multicast group contains the entire set of the network nodes, the proposed algorithms are compared against established solutions for energy-efficient broadcasting. The first algorithm to be included in our performance comparison is the BIP algorithm [31], which operates equivalently to the MLE algorithm [34] when the node transmission power is fixed. We also consider the broadcast version of the MWIA algorithm [34] described above and the BAIP heuristic, which uses as criterion, for the addition of a node in the tree, the power consumed when using the corresponding link over the number of newly covered nodes. The BAIP heuristic corresponds to the greedy-h heuristic [35], which, to the best of our knowledge, is the best solution proposed so far for energy-efficient broadcasting when the nodes' transmission power is fixed. As with other algorithms originally proposed under the assumption of adjustable transmission power, the BAIP heuristic used in our comparisons corresponds to the fixed-power versions of BAIP [32] and GPBE [33] algorithms.

The proposed multicasting algorithms are evaluated under both the network evacuation model and the infinite time horizon model. In the network evacuation model, each node starts with a certain amount of initial energy and a given number of packets to be transmitted to the nodes belonging to the multicast group. The experiments run until all packets are successfully received by their destinations or until no more transmissions can take place due to the lack of energy reserves. In the infinite time horizon setting, packets and energy are generated over time according to a round-based scenario. At the beginning of each round, the node energy reserves are restored to a certain level, and an equal number N of packets to be multicasted is generated at every node. A round terminates when the residual energy of at least half of the network nodes falls below a certain safety limit. Packets that did not successfully reach their destinations during a round continue from the point they stopped in the following round(s). The succession of rounds continues until the network reaches the steady state or until it becomes inoperable (unstable).

In our simulations, we use a 4×4 two-dimensional grid network topology of 16 stationary nodes with a distance of 50 m between neighboring nodes. Each node's transmission radius is fixed at a value uniformly distributed between 50 and 100 m. The initial energy E_0 is taken to be the same for all nodes (equal to 5, 10, and 100 J). Each node multicasts 200–1,000 packets (at steps of 200). Finally, a node belongs to the multicast group M with a probability q that takes values 0.25, 0.5, and 0.75. In the case of a broadcast, the multicast group M contains all nodes in the network.

8.5.2.1 Network Evacuation Model

8.5.2.1.1 Multicasting

Figure 8.9a illustrates the average number of transmissions h undertaken by a packet in order to reach all destinations in its multicast group for different values of the number of packets multicasted per source. It can be seen that the MIP and MWIA algorithms exhibit the worst performance among the algorithms examined. This can be explained by the fact that both algorithms select at each step a transmission with low required power, resulting in more transmissions needed per packet in order to reach all destinations in its multicast group. The OTREMM and NOTREMM algorithms perform better than the MIP and MWIA algorithms but seem to be slightly worse than the NJT algorithm. However, this is misleading since, as can be seen in Figure 8.9c, NJT manages to complete fewer multicasts than do OTREMM and NOTREMM due to energy depletion.

In Figure 8.9b, we illustrate the average node residual energy R at the end of an evacuation period assuming that the nodes' initial energy $E_0 = 5$ J. The best performance is achieved by the OTREMM algorithm and is closely followed by that of the NOTREMM algorithm. It is important to note that NOTREMM performs comparably to the optimal algorithm but has considerably smaller computational overhead. When the number of packets evacuated is small, the NJT algorithm performs similarly to the OTREMM and NOTREMM algorithms, but when this number increases, the NJT performance deteriorates. The MIP and MWIA algorithms are the worst performers among the algorithms examined; even though they both use transmissions that are less energy-consuming, the total energy expended is larger due to the larger number of transmissions h required.

In Figure 8.9c, the proposed OTREMM and NOTREMM algorithms outperform all other strategies with respect to the multicast success ratio p they achieve. The success ratio p of MIP and NJT starts falling, compared with that of OTREMM and NOTREMM, even for relatively light inserted traffic. The MWIA algorithm performs significantly better than do MIP and NJT algorithms since it takes into account both the node transmission powers and residual energies in selecting the multicast schedule. Again, NOTREMM achieves very good performance, which is only marginally inferior to that of OTREMM.

Figure 8.10 illustrates the current number of nodes L with depleted energy reserves as a function of time. We can see that the OTREMM and NOTREMM algorithms do not only result

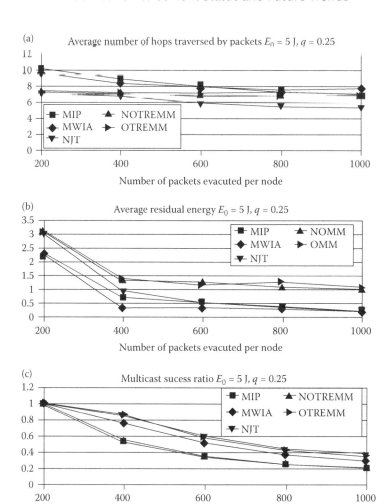

Figure 8.9 (a) The average number of transmissions *h* required by each multicast packet. (b) The average node residual energy *R* at the end of an evacuation period. (c) The multicast success ratio *p*. Results were obtained for $q = 0.5$ and the node's initial energy $E_0 = 5$ J.

in fewer nodes running out of energy but also result in node energy depletions occurring later in the experiment when compared with the MIP, MWIA, and NJT algorithms. When the node's initial energy E_0 is set at 5 J, NJT seems to perform slightly better than MIP, but these differences decrease when the initial energy is 10 J. The MWIA algorithm outperforms both MIP and NJT, since when using MWIA, the first node energy depletions tend to happen later than when using MIP and NJT. This explains the higher multicast success ratio *p* of MWIA in comparison to that of MIP and NJT. As mentioned earlier, when discussing the results of Figure 8.9b, if even one node of the multicast group *M* runs out of energy, no more successful multicasts can be completed. Even though NJT consumes, under some scenarios, similar energy as that of OTREMM and NOTREMM, it spreads energy consumption less uniformly, with many nodes running out of energy rather soon while other nodes still have plenty of available energy. The performance of the NOTREMM algorithm is again almost identical to that of the optimal OTREMM

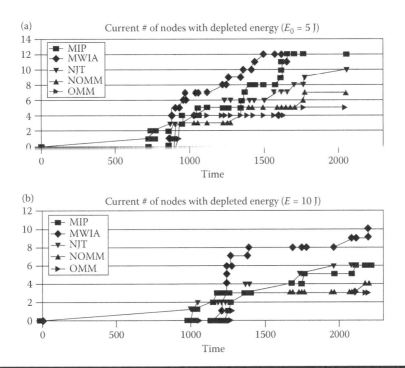

Figure 8.10   The current number of nodes *L* with depleted energy reserves for the case *q* = 0.5 and the nodes' initial energy equal to (a) 5 J and (b) 10 J.

algorithm, verifying that it can obtain most of the benefits of OTREMM at a fraction of its computational cost.

8.5.2.1.2 Broadcasting

The average number of transmissions *h* required to complete a broadcast is depicted in Figure 8.11a. OTREMM outperforms all other algorithms, with NOTREMM achieving only slightly larger *h*. BAIP also seems to perform similarly to or even better than OTREMM and NOTREMM, but this is rather misleading since BAIP does not successfully complete the same number of broadcasts with these schemes (Figure 8.11c).

The overall energy expenditure of the algorithms is depicted in Figure 8.11b, where the average node residual energy *R* at the end of an experiment is shown. Generally, the average energy reserves of the nodes decrease as the number of packets broadcasted increases, but the rate of decrease is reduced when a certain number of packets (400–600 packets per node, depending on the scheme used) are broadcasted in the network. This happens because, beyond this point, many nodes run out of energy, the network gets disconnected, and no more packet transmissions can take place regardless of the increase in the incoming traffic. The OTREMM and NOTREMM algorithms utilize the node energy reserves more efficiently, compared with the rest of the algorithms, resulting in a higher average residual energy at the end of an evacuation period. Even though OTREMM selects the most energy-efficient set of nodes for broadcasting, NOTREMM achieves comparable results.

Figure 8.11c depicts the broadcast success ratio *p* for all the algorithms considered as a function of the number of packets evacuated per node. Clearly, OTREMM outperforms all the other

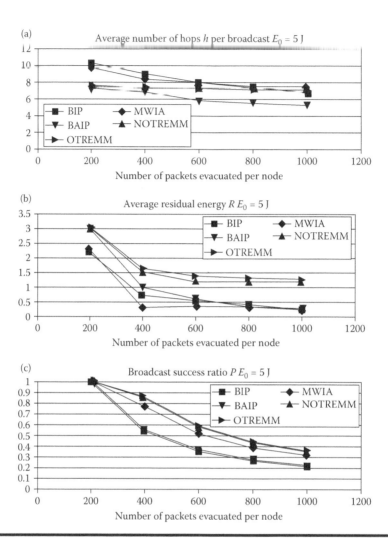

Figure 8.11 (a) The average number of transmissions *h* per broadcast. (b) The average residual energy *R* at the end of the experiment. (c) The broadcast success ratio *p* for all the algorithms evaluated when the nodes' initial energy is equal to 5 J.

algorithms, with NOTREMM being only slightly inferior. The performance of the BIP and BAIP algorithms starts degrading even for small traffic load, and only the MWIA algorithm stays close to the OTREMM and NOTREMM algorithms. The characteristic enabling MWIA to also perform rather well is that it is dynamic since its selection criterion involves the time-varying current residual energy of the nodes. In contrast, the BIP and BAIP algorithms do not change their broadcast paths and they quickly exhaust the node energy reserves. The OTREMM and NOTREMM algorithms spread traffic more uniformly across the network, with nodes remaining operational and able to broadcast packets for longer times.

The longer network lifetime achieved by the OTREMM and NOTREMM algorithms can be observed in Figure 8.12, where the current number of nodes *L* with depleted energy reserves is presented as a function of time. At any time, the OTREMM and NOTREMM algorithms have fewer nodes with depleted energy than the other algorithms. The BIP algorithm seems to have

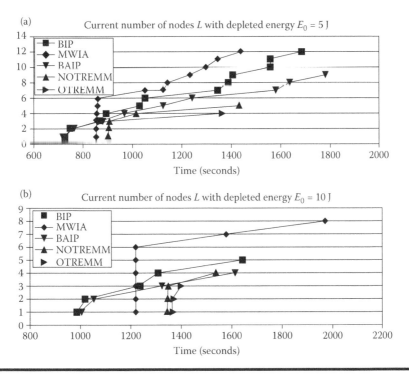

Figure 8.12 **The current number of nodes *L* with depleted energy for all the algorithms evaluated when the nodes' initial energy is equal to (a) 5 J and (b) 10 J.**

the worst performance. MWIA tends to spread energy consumption uniformly across a *subset* of network nodes, and therefore when these nodes run out of energy, they do so almost simultaneously. The BAIP algorithm performs better than MWIA and BIP, but significantly worse than OTREMM and NOTREMM with respect to network's lifetime.

8.5.2.2 Infinite Time Horizon Model

Figure 8.13a shows the broadcast success ratio p in each round for all the algorithms evaluated; for the case where $N = 23$, broadcast packets per round are generated at each node. The broadcast success ratio p of all the algorithms stabilizes after several rounds, with the OTREMM and NOTREMM algorithms yielding the highest broadcast success ratio p.

Figure 8.13b shows the average broadcast delay D, measured in packet times, for the same traffic scenario. A *packet time* is defined as the time needed for the successful transmission of a packet between two neighboring nodes. We observe that when the BIP, MWIA, and BAIP algorithms are used, the delay D increases over consecutive rounds, indicating that the insertion rate of $N = 23$ packets per round corresponds to unstable operation for these schemes. In other words, the maximum stable broadcast throughput of these schemes is less than 23 packets per node per round. On the contrary, the OTREMM and NOTREMM algorithms reach steady state and achieve quite smaller broadcast delays D, indicating that this packet insertion rate is sustainable and within the stability region of these schemes.

Figure 8.14a presents the broadcast success ratio p at steady state for a different number of broadcast packets N inserted at each node per round. We observe that even for relatively light

(a)

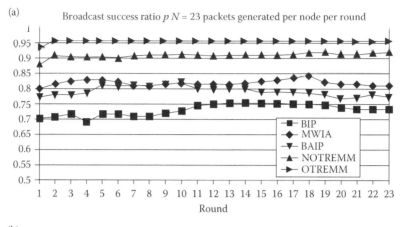

(b)

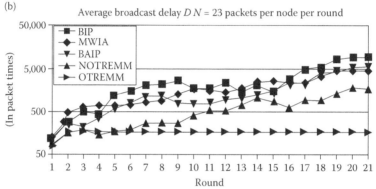

Figure 8.13 (a) The broadcast success ratio *p* in each round and (b) the average broadcast delay *D*, measured in packet times, in each round for all the algorithms evaluated when *N* = 23 packets are generated at each node per round.

traffic, the BIP, MWIA, and BAIP algorithms are not able to successfully broadcast all the packets generated. In particular, the BIP scheme remains stable for loads up to $N = 15$ packets per node per round and the BAIP scheme remains stable for loads up to $N = 17$ packets per node per round. The MWIA scheme performs slightly better, remaining stable for up to $N = 19$ packets per node per round. The OTREMM and NOTREMM schemes have the maximum stability region (maximum broadcast throughput) and remain stable for up to $N = 21$ packets per node per round. By taking into account energy-related cost parameters and switching through multiple energy-efficient paths, OTREMM and NOTREMM spread energy consumption more evenly across the nodes, increasing the volume of broadcast traffic that can be successfully served. NOTREMM performs comparably to OTREMM and only for the heavier traffic loads does its success ratio *p* fall significantly below that of the optimal algorithm. This is important, considering the great gain in computational effort achieved when using the NOTREMM instead of the OTREMM algorithm.

Figure 8.14b shows the average broadcast delay *D*, measured in packet times, at steady state of all the algorithms evaluated, as a function of the number of broadcast packets inserted per node and round. Recall that the broadcast delay *D* includes the delays incurred by a packet during all the rounds that elapse from the time it is generated until the time it reaches all nodes in

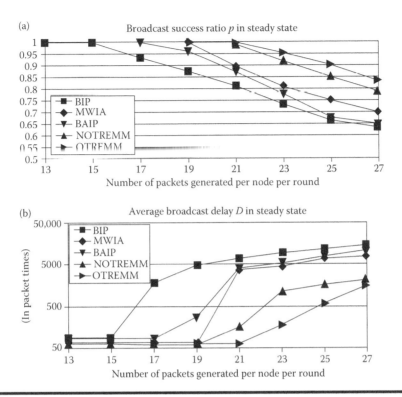

Figure 8.14 (a) The broadcast success ratio *p* and (b) the average broadcast delay *D* in the steady state of the algorithms evaluated for a different number of broadcast packets *N* inserted in the network.

the network. As can be seen from Figure 8.14b, the delay versus traffic load curves of the BIP, MWIA, and BAIP algorithms are above those of the OTREMM and NOTREMM algorithms. Since packets whose broadcast is not completed during a round fill the node queues and congest the network, the average delay of the BIP, MWIA, and BAIP algorithms quickly becomes very large. Naturally, when the traffic load inserted increases beyond each scheme's maximum stable throughput, the delays will become unbounded and the success ratio *p* will start falling. The OTREMM and NOTREMM algorithms have smaller average delay *D* and remain stable for higher loads than the other schemes considered. Again, NOTREMM manages to keep up with OTREMM, and only at the end does its performance deteriorate.

8.6 Conclusions

In this chapter, we overviewed the basic trends in unicast and multicast routing for wireless ad hoc networks and then presented a novel class of routing algorithms based on multiple criteria. The experiments conducted show that in the unicast case, the proposed multicost energy-aware routing algorithms distribute the traffic more uniformly across the network, prolonging its life-time and improving its performance. Similarly, in the multicasting case, our results illustrate that the proposed multicost algorithms outperform the other heuristic algorithms considered, consuming less energy and successfully multicasting more packets to their destination under both the packet evacuation and the infinite time horizon model. An interesting conclusion drawn from our simulations is that the near-optimal multicast algorithm has a performance similar to that of the optimal algorithm while having considerably smaller execution time, indicating that the computation overhead of the optimal algorithm is not justified by its performance superiority.

References

1. D. Bertsekas and R. Gallager. *Data Networks*. New Jersey: Prentice Hall, 1992.
2. S. Singh, M. Woo, and C.S. Raghavendra. Power-aware routing in mobile ad hoc networks. In *Proceedings of Annual International Conference on Mobile Computing and Networking (MOBICOM)*, 1998, pp.181–190.
3. J.-H. Chang and L. Tassiulas. Maximum lifetime routing in wireless sensor networks. *IEEE/ACM Transactions on Networking*, vol. 12, 2004, pp. 609–619.
4. C.-K. Toh. Maximum battery life routing to support ubiquitous mobile computing in wireless ad hoc networks. *IEEE Communications Magazine*, vol. 39, no. 6, 2001, pp. 138–147.
5. D.F. MacEdo, A.L. Dos Santos, L.H.A. Correia, J.M. Nogueira, and G. Pujolle. Transmission power and data rate aware routing on wireless networks. *Computer Networks*, vol. 54, no. 17, 2010, pp. 2979–2990.
6. M.K. Jakobsen, J. Madsen, and M.R. Hansen. DEHAR: a distributed energy harvesting aware routing algorithm for ad-hoc multi-hop wireless sensor networks. In *IEEE International Symposium on A World of Wireless, Mobile and Multimedia Networks (WoWMoM)*, 2010.
7. W. Yang, W. Liang, J. Luo, and W. Dou. Energy-aware online routing with QoS constraints in multi-rate wireless ad hoc networks. In *Proceedings of the 6th ACM International Wireless Communications and Mobile Computing Conference (IWCMC)*, 2010, pp. 721–725.
8. S. Mahfoudh and P. Minet. Energy-aware routing in wireless ad hoc and sensor networks. In *ACM International Wireless Communications and Mobile Computing Conference (IWCMC)*, 2010, pp. 1126–1130.
9. N. Gupta and S. Das. Energy-aware on-demand routing for mobile ad hoc networks. In *Proceedings of the 4th International Workshop on Distributed Computing, Mobile and Wireless Computing (IWDC)*, 2002, pp. 164–173.
10. M. Younis, M. Youssef, and K. Arisha. Energy-aware management for cluster-based sensor networks. *Computer Networks*, vol. 43, no. 5, 2003, pp. 649–668.
11. S. Banerjee and A. Misra. Minimum energy paths for reliable communication in multi-hop wireless networks. In *Proceedings of the ACM International Symposium on Mobile Ad Hoc Networking and Computing (MobiHoc)*, 2002.
12. K. Woo, C. Yu, D. Lee, H.-Y. Youn, and B. Lee. Non-blocking, localized routing algorithm for balanced energy consumption in mobile ad hoc network. In *Proceedings of the Ninth International Symposium in Modeling, Analysis and Simulation of Computer and Telecommunication Systems (MASCOTS)*, 2001, pp. 117–124.
13. V. Rodoplu and T. Meng. Minimum energy mobile wireless networks. In *Proceedings of the IEEE International Conference on Communications (ICC)*, 1998, pp. 1633–1639.
14. R. Ramanathan and R. Rosales-Hain. Topology control of multihop wireless networks using transmit power adjustment. In *Proceedings of IEEE INFOCOM*, 2000, pp. 404–413.
15. J.-H. Chang and L. Tassiulas. Energy conserving routing in wireless ad-hoc networks. In *Proceedings of the IEEE INFOCOM*, 2000, pp. 22–31.
16. B. Chen, K. Jamieson, H. Balakrishnan, and R. Morris. Span: an energy-efficient coordination algorithm for topology maintenance in ad hoc wireless networks. In *Proceedings of the 7th ACM International Conference on Mobile Computing and Networking (MobiCom)*, 2001, pp. 85–96.
17. S. Chen and K. Nahrstedt. On finding multi-constrained paths. In *Proceedings of the IEEE International Conference on Communications (ICC)*, 1998.

18. T. Korkmaz and M. Krunz. Multi-constrained optimal path selection. In *Proceedings of the IEEE INFOCOM*, 2001, pp. 834–843.
19. P. Mieghem and F. Kuipers. Concepts of exact QoS routing algorithms. *IEEE/ACM Transactions on Networking*, vol. 12, no. 5, 2004, pp. 851–864.
20. X. Yuan and X. Liu. Heuristic algorithms for multiconstrained quality-of-service routing. *IEEE/ACM Transactions on Networking*, vol. 10, no. 2, 2002, pp. 244–256.
21. P. Mieghem and F. Kuipers. On the complexity of QoS routing. *Computer Communications*, vol. 26, no. 4, 2003, pp. 376–387.
22. Z. Wang and J. Crowcroft. Quality-of-service routing for supporting multimedia applications. *Journal of Selected Areas in Communications*, vol. 14, no. 7, 1996, pp. 1228–1234.
23. X. Huang and Y. Fang. Multi-constrained soft-QoS provisioning in wireless sensor networks. In *Proceedings of the 3rd International Conference on Quality of Service in Heterogeneous Wired/Wireless Network (QSHINE)*, 2006.
24. L. Liu and G. Feng. Simulated annealing based multi-constrained QoS routing in mobile ad hoc networks. *Wireless Personal Communications*, vol. 41, no. 3, 2007, pp. 393–405.
25. Z. Li and J. Garcia-Luna-Aceves. Finding multi-constrained feasible paths by using depth-first search. *Wireless Networks*, vol. 13, no. 3, 2007, pp. 323–334.
26. H. Badis and K. Al Agha. Quality of service for ad hoc optimized link state routing protocol (QOLSR), http://tools.ietf.org/html/draft-badis-manet-qolsr-03, Accessed on 20th June 2011.
27. C. Perkins and E.M. Belding-Royer. Quality of service in ad-hoc on-demand distance vector routing, http://tools.ietf.org/id/draft-perkins-manet-aodvqos-00.txt, Accessed on 20th June 2011.
28. S. Chen and K. Nahrstedt. Distributed quality-of-service routing in ad-hoc networks. *Journal on Selected Areas in Communications*, vol. 17, no. 8, 1999, pp. 1488–1505.
29. S. Athanassopoulos, I. Caragiannis, C. Kaklamanis, and P. Kanellopoulos. Experimental comparison of algorithms for energy-efficient multicasting in ad hoc networks. In *Proceedings of ADHOC-NOW*, 2004, pp. 183–196.
30. S. Guo and O.W.W. Yang. Energy-aware multicasting in wireless ad hoc networks: a survey and discussion. *Journal of Computer Communications*, vol. 30, no. 9, 2007, pp. 2129–2148.
31. J.E. Wieselthier, G.D. Nguyen, and A. Ephremides. On the construction of energy-efficient broadcast and multicast trees in wireless networks. In *Proceedings of IEEE INFOCOM*, 2000, pp. 585–594.
32. P.-J. Wan, G. Calinescu, X. Li, and O. Frieder. Minimum-energy broadcast routing in static ad hoc wireless networks. In *Proceedings of IEEE INFOCOM*, vol. 2, 2001, pp. 1162–1171.
33. I. Kang and R. Poovendran. A novel power-efficient broadcast routing algorithm exploiting broadcast efficiency. In *Proceedings of IEEE Vehicular Technology Conference*, vol. 5, 2003, pp. 2926–2930.
34. M.X. Cheng, J. Sun, M. Min, Y. Li, and W. Wu. Energy-efficient broadcast and multicast routing in multihop ad hoc wireless networks. *Journal of Wireless Communications and Mobile Computing*, vol. 6, no. 2, 2006, pp. 213–223.
35. D. Li, X. Jia, and H. Liu. Energy efficient broadcast routing in static ad hoc wireless networks. *Journal of IEEE Transactions on Mobile Computing*, vol. 3, no. 2, 2004, pp. 144–151.
36. D. Li, Q. Liu, X. Hu, and X. Jia. Energy efficient multicast routing in ad hoc wireless networks. *Journal of Computer Communication*, vol. 30, no. 18, 2007, pp. 3746–3756.
37. Z. Li, D. Li, and M. Liu. Interference and power constrained broadcast and multicast routing in wireless ad hoc networks using directional antennas. *Computer Communications*, vol. 33, no. 12, 2010, pp. 1420–1439.
38. G. Zeng, B. Wang, Y. Ding, L. Xiao, and M. Mutka. Efficient multicast algorithms for multichannel wireless mesh networks. *IEEE Transactions on Parallel and Distributed Systems*, vol. 21, no. 1, 2010, pp. 86–99.
39. J. Cartigny, D. Simplot, and I. Stojmenovic. Localized minimum-energy broadcasting in ad-hoc networks. In *Proceedings of the IEEE INFOCOM*, vol. 3, 2003, pp. 2210–2217.
40. N. Li, J. Hou, and L. Sha. Design and analysis of an MST-based topology control algorithm. In *Proceedings of IEEE INFOCOM*, vol. 3, 2003, pp. 1702–1712.
41. F. Ingelrest and D. Simplot-Ryl. Localized broadcast incremental power protocol for wireless ad hoc networks. *Journal of Wireless Networks*, vol. 14, no. 3, 2008, pp. 309–319.

42. F. Li and I. Nikolaidis. On minimum-energy broadcasting in all-wireless networks. In *Proceedings of Local Computer Networks*, 2001, pp. 193–202.

43. M. Cagalj, J.-P. Hubaux, and C. Enz. Minimum-energy broadcast in all-wireless networks: NP-completeness and distribution issues. In *Proceedings of ACM MobiCom*, 2002, pp. 172–182.

44. A.K. Das, R.J. Marks, M. El-Sharkawi, P. Arabshahi, and A. Gray. r-Shrink: a heuristic for improving minimum power broadcast trees in wireless networks. In *Proceedings of IEEE GLOBECOM*, vol. 1, 2003, pp. 523–527.

45. I. Kang and R. Poovendran. Broadcast with heterogeneous node capability. In *Proceedings of IEEE GLOBECOM*, vol. 6, 2004, pp. 4114–4119.

46. The Network Simulator NS-2, http://www.isi.edu/nsnam/ns/, Accessed on 20th June 2011.

47. F.J. Gutierrez, E. Varvarigos, and S. Vassiliadis. Multi-cost routing in max–min fair share networks. In *Proceedings of Annual Allerton Conference on Communication, Control and Computing*, vol. 2, 2000, pp. 1294–1304.

Chapter 9

Security Issues in FHAMIPv6

Jesús Hamilton Ortiz, Jorge Luís Perea Ramos, Julio
Cesar Rodríguez Ribon, and Juan Carlos López

Contents

FHAMIPv6 was designed to extend the features of F-HMIPv6 to mobile ad hoc networks (MANETs). To do so, it uses a small set of messages to make the registrations with the ad hoc mobile anchor point (AMAP) and the ad hoc home agent (AHA). Other messages are used by the ad hoc previous access router (APAR) or the ad hoc new access router (ANAR) to announce their presence to the ad hoc mobile nodes (AMNs) that approach the network. FHAMIPv6, like other MANET protocols, assumes that the nodes of the network are always willing to cooperate in traffic forwarding. Moreover, FHAMIPv6 does not implement authentication mechanisms for either base stations (APAR, ANAR, AHA) or mobile (AMN) and does not have tools to ensure the integrity of data and control messages. That is why this protocol can suffer many attacks, some of which can cause either just a simple delay in the network or its complete collapse.

In addition the implicit security issues in FHAMIPv6, those of the medium access protocol used by it (IEEE 802.11), so the risks for the information are even higher, including two other types of attacks that can be performed on FHAMIPv6. These are flooding attacks and association attacks, which are not based on the weaknesses of the medium access layer or the lack of security mechanisms for FHAMIPv6 messages.

9.1 Introduction

In this chapter, the security issues related to the FHAMIPv6 protocol [1] will be studied, including the way it works and how its messages are presented. Subsequently, the safety of ad hoc mobile networks, the security problems they suffer from, and some countermeasures to mitigate them will be studied. Then, some safety issues of the medium access protocol used by FHAMIPv6 will be taken into account, which present some security problems, many of which are also present in FHAMIPv6. We will also consider some ways of compromising security in the FHAMIPv6 by manipulating its diverse messages. Finally, we will consider some high-risk security issues that are unrelated to the messages used by FHAMIPv6.

9.2 FHAMIPv6 Operation

In the following sections, FHAMIPv6 will be introduced. A scenario and the most relevant message sequences used in FHAMIPv6 will be given. This will be the starting point for understanding how FHAMIPv6 security is compromised by manipulating its messages.

9.2.1 FHAMIPv6

The IPv4 protocol was enough for a long period of time to satisfy the needs of Internet users with regard to the network layer. However, given the current massive use of wireless technologies and the rise of mobile computing, this protocol began to be insufficient for the new demands of the users, mostly due to the necessity of staying connected in a mobile environment. In order to solve this problem, MIPv4 [2] appeared to provide the mobile capacity that users were beginning to demand. Still, this protocol produced a very high delay when a mobile node changed from an access point to another in an external network. To amend this problem, some extensions for the protocol were designed. The first one, known as *HMIP* [3], tried to decrease the home network overload, introducing a hierarchical scheme. The second proposal, known as *F-MIP*, sought to reduce the transfer delay through methods that are well defined in [4]. In the same way, a third extension was created, merging the best of HMIP and F-MIP: F-HMIPv6 [5], which delivered a low delay transfer hierarchical scheme that supports mobility in infrastructure networks. FHAMIPv6 [1] then comes up as an extension of F-HMIPv6 for MANETs.

9.2.2 FHAMIPv6 Scenario

Figure 9.1 shows the FHAMIPv6 scenario. Letter A at the beginning of each abbreviation refers to the ad hoc condition of FHAMIPv6.

Initially, the AMN is located in the area of the AHA, which means that all communications with the ad hoc correspondent node (ACN) go via AHA. Then, the AMN moves toward where the APAR is located and completes a registration process with the AMAP and the AHA. In addition, it obtains a temporary network address (Care of Address (CoA)). From then on, all communications toward the AMN will proceed as follows: ACN > AHA > AN1 > AMAP > AN2 > APAR > AMN. Finally, the AMN moves toward the ANAR area, and after registering its new position with the AMAP and receiving a new CoA, the ACN will be able to contact the AMN through the ANAR using the following route: ACN > AHA > AN1 > AMAP > AN3 > ANAR > AMN (see Figure 9.1).

9.2.3 Message Sequence of FHAMIPv6

In this section, the sequence of messages of FHAMIPv6 is described. For further details, see [1]. Figure 9.2 illustrates the exchange of messages for the registration process.

When the AMN is about to reach the APAR area, it receives a warning message (MIPT_ADS), which contains the CoA address of the APAR as well as the address of the AMAP. Then, the AMN is registered with the AMAP sending a MAP_REG_REQUEST message via the route AMN > APAR > AN2 > AMAP. Next, the AMAP returns a MAP_REG_REPLY message, and when it reaches the AN2, it is changed for a new REPLY_AMAP message and sent to the APAR, which reconstructs the original MAP_REG_REPLY and sends it to the AMN. When the latter receives the message, the registration process is complete. Registration with the AHA works the same way, so it will not be explained.

Figure 9.3 shows the message sequence process when the AMN moves from the APAR toward the ANAR area.

First, the AMN sends a start message of fast handover to the APAR (FAST_RTSOLPR). Next, the APAR replies with a FAST_PRRTADV message where the ANAR address is indicated. Then, the registration with the AMAP is fulfilled, as described below. The only difference is that the new route to the AMAP is now AMN > ANAR > AN3 > AMAP. Finally, MIPT_NA and MIPT_NACK messages are exchanged between the AMN and the ANAR to indicate that the AMN is fully located in the ANAR area (layer 2 handover completed).

Figure 9.4 illustrates the registration process between the AMN and its AHA from the ANAR area.

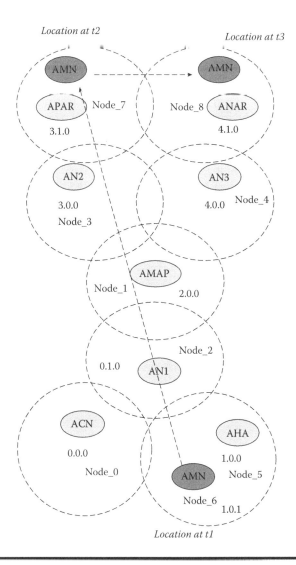

Figure 9.1   Simulation scenario for FHAMIPv6.

As shown in Figure 9.4, the process is similar to that made from the APAR area. This time the route is AMN > ANAR > AN3 > AMAP > AN1 > AHA.

9.2.4 Allocation of Functions to the Intermediate Nodes

Since the APAR, the ANAR, and the AHA are not always located in the coverage area of the AMAP, the presence of intermediate nodes between the nodes mentioned earlier and the AMAP is needed. FHAMIPv6 implements a simple proceeding to detect these intermediate nodes. Figure 9.5 illustrates the process.

Initially, the AMAP sends DISCOVERY_AN messages in the broadcast mode. These messages are received by the intermediate nodes, which in turn spread DISCOVERY_BS messages. When a base station (APAR, ANAR, or AHA) detects these messages, they return a DISCOVERY_REPLY

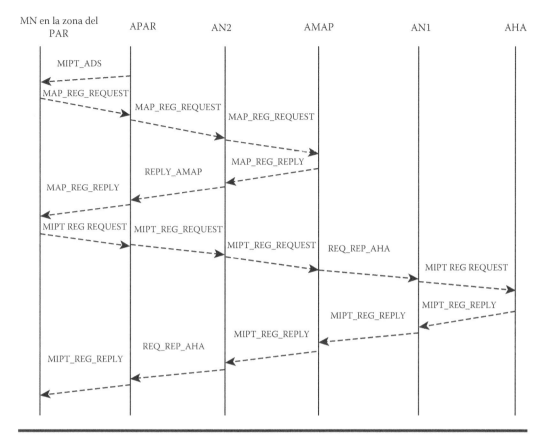

Figure 9.2 Sequence of exchange during register process in a mobile node for FHAMIPv6 agent.

message. If this message reaches the AMAP, the intermediate node is discarded (given the fact that the arrival of the last message confirms that it is not necessary to use an intermediate node between the AMAP and the base station since these are close enough). If the DISCOVERY_REPLY message does not reach the AMAP, the intermediate node will be used as an FHAMIPv6 node. That means it supports FHAMIPv6 and can be used as a bridge between the AMAP and a base station. Figure 9.6 illustrates all the above.

9.3 Security Issues in MANETS

9.3.1 Fundamental Concepts of MANETs

A MANET is defined as a wireless network of nodes that communicate with each other in a multihop mode without the support of a fixed infrastructure, such as an access point or a base station [6].

The MANETs are meant to be used where no wired or cellular networks exist or where the cost of that infrastructure would be very high. MANETs have no central coordination, which makes the operation more complex than in other wireless networks such as IEEE 802.11 or cellular networks. MANETs are formed by a collection of mobile nodes that have no preestablished

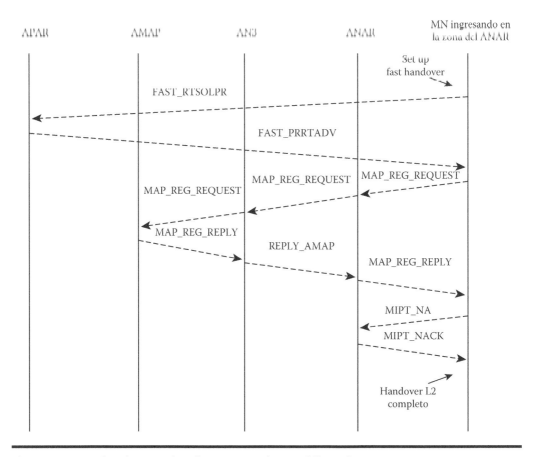

Figure 9.3   Fast handover and register process in a mobile node (FHAMIPv6).

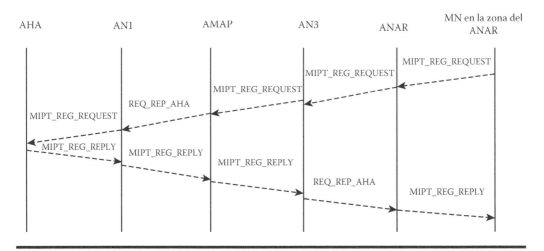

Figure 9.4   AMN moving to the ANAR area.

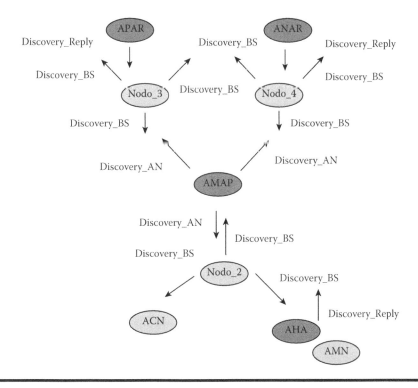

Figure 9.5 Allocation functions to the intermediate nodes. (AN1, AN2, AN3).

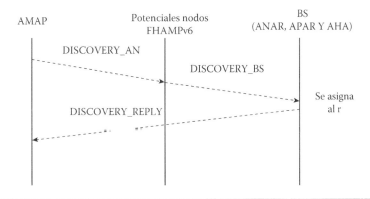

Figure 9.6 Acknowledgement of the intermediate nodes. (AN1, AN2, AN3).

infrastructure, creating a temporary network for specific purposes [7]—hence the ad hoc idea. Each node has a wireless interface to contact others by either radio or infrared. Laptops and PDAs are examples of these network nodes.

In general, the bandwidth of these networks is rather limited, and given its reduced scope, it is required to establish a multihop scheme so that a node can reach others. MANETs can be formed and transformed dynamically without a central node [8]. This decentralized nature is useful for scenarios that require spreading out a network quickly. In addition, since a preexisting physical infrastructure is not necessary, MANETs can be used, for instance, to communicate across a group of soldiers in enemy territory, in distributed computing environments, such as in classrooms where students need to interact and have meetings; in a conference room, where scientists share information; or in emergency situations, when the physical network has been destroyed by a war or a natural disaster such as hurricanes or earthquakes. Thus, the prompt deployment of a MANET will be extremely useful for the proper coordination of search and rescue works [6].

9.3.2 Routing in MANETs

Given the limited power of the wireless interfaces of the nodes forming a MANET, a sender may need to use one or more intermediary nodes to send a packet to a distant destination. The fact that a packet must take several hops to reach its destination justifies the need for routing algorithms [7].

Traditional routing algorithms such as link state or distance vector (DV) are not appropriate because they were designed for environments where the topology does not change or changes only very slightly. Another disadvantage of traditional routing algorithms is the constant delivery of routing messages between different nodes. Such an exchange of information in a MANET scenario would cause a consumption of resources of bandwidth, battery, and processing that are usually scarce in such networks. Plus, traditional routing algorithms assume that links are bidirectional, which cannot be taken for granted in MANETs because the transmission power of the nodes may vary, so a node may have enough power to transmit to another node, but the latter cannot send back or forward packages. For this reason, extensive research was done to find routing algorithms more suitable for MANETs. These are the main proposals [9]:

- Proactive routing protocols: These protocols react to all changes in the network topology.
- Reactive routing protocols (on-demand protocols): Routes between two nodes are established only if there is data flow between them.
- Hybrid routing protocols: These are a combination of reactive and proactive protocols, and they seek to improve the overall performance of the network.
- Position-based routing protocols: They are based on the geographical coordinates of the nodes within the network to improve the throughput.

9.3.2.1 Traditional Routing Protocols

Since some ad hoc routing protocols are based on traditional routing protocols, a brief overview of the latter is presented to make understanding of the former easier.

9.3.2.1.1 Link State

This routing algorithm operates on the fact that each node maintains a view of the complete topology and stores it in a table that is broadcasted regularly to all its neighbors to keep it updated. Each node receiving this information updates its own table and calculates the shortest route to set up the next hop to any destination. This protocol is not exempt from routing loops or inconsistent information topology due to propagation delays through the network, but the problem of loops is usually short-lived.

9.3.2.1.2 Distance Vector

Here, each node is interested only in the links coming out of itself (not the links of the entire network) and only sends information about the shortest paths of the network's nodes to its neighbors, which receive this information and recalculate their routing tables using the shortest-path algorithms.

9.3.2.1.3 Source Routing

With this routing algorithm, packets contain all the information hops from the source to the destination. That is, the source node includes in the packet full information of the path to follow in order to reach a particular destination. The advantage of this routing algorithm is that it avoids the loops, and the disadvantage is that it slightly increases the network load because the packets are bigger.

9.3.2.1.4 Flooding

The activity of this algorithm is based on the delivery of broadcast messages. Here, a source node sends information to all its neighbors, which in turn send it to their neighbors and so on until the packet reaches all nodes in the network.

9.3.2.2 Ad Hoc Routing Protocols

The study of security in MANETs is based on the two ad hoc routing protocols described below.

9.3.2.2.1 AODV

The ad hoc on demand distance vector (AODV) routing protocol is a protocol that provides multihop routing between mobile nodes in a MANET. This protocol is based on the DV algorithm. AODV is reactive, while DV is proactive. This means that AODV requests routes only when needed, while DV continuously sends routing messages to discover and update routes. AODV operates as follows.

When a node wants to find a route to another node, it broadcasts a route request (RREQ) to all its neighbors. This message is spread all over the network until it reaches the destination node or a node that has a path to it. The route discovered is enabled by sending route error (RRER) messages back to the source. Furthermore, AODV uses Hello messages [a special type of route reply (RREP)] that it continually sends to its neighbors to confirm its location. If a node stops sending Hello messages, its neighbors may assume that it has left the network and they will consider the link broken. Then, the affected nodes will be notified.

9.3.2.2.1.1 Route Discovery — Route discovery refers to the fact that if node A wants to send packets to node B, it must previously obtain a route. When a node needs to learn a route to a destination and does not have one available (it does not know a route or its former route has expired), it broadcasts RREQ messages and waits for an RREP message. If after some time there is no answer, it will keep on sending RREQ messages or will assume that there is no route to the destination.

RREQ-type-message forwarding occurs when a node receiving this message knows no route to the destination, so the node will send the message to its neighbors and maintain a temporary reverse path to the source. The next hop would be neighbor's IP that sends the RREQ. This temporary route is created in order to eventually send back through it an RREP from the destination. This route is considered temporary because its expiration time is much shorter than for normal routes.

When an RREQ-type message reaches its destination or a node that has a valid route to it, an RREP message is created and sent to the node that produced the RREQ request. In the forwarding process of the RREP from the destination to the source, and via intermediaries, a path to the destination is also created, so when the RREP reaches the source, a complete route between it and the destination is established.

9.3.2.2.1.2 Route Keeping

— This refers to the mechanism used by the issuer node to detect whether the network topology has changed, making it impossible to send packets along routes previously discovered. This occurs when a node moves out of the transmission radius of others or when a node is turned off.

When a node detects that a route to another node is no longer valid, it will remove it from its routing information and will send a link-failure-type message. This message is directed to the neighbors that could be constantly using that route to inform them that it is no longer valid. These neighbors will also send the message to their neighbors. To fulfill this task, each AODV node maintains a record of the routes used by its neighbors. When the link failure notification reaches any affected node, each AODV node can choose between sending information via this route and sending RREQ requests to discover a new route.

9.3.2.2.2 DSR

The Dynamic source routing (DSR) protocol is a reactive routing protocol that allows nodes to dynamically discover routes that involve multiple hops to any destination. The term *source routing* means that each packet contains in its header the complete path information from the source to the destination: the header contains the addresses of all nodes between the source and the destination. However, DSR uses no periodic routing messages to reduce network overload, battery consumption, CPU usage, and bandwidth.

9.3.2.2.2.1 Route Discovery

— In DSR route discovery, a source node broadcasts RREQ-type messages. When a node receives this message, it will seek an entry in its route cache toward the requested destination. If not found, it will forward the message, adding its own address to the packet's hop sequence record. Such a message will travel through the network until it finds the destination node or, alternatively, a node with a route toward it. When this happens, it will send an RREP to the source. This package contains the sequence of hops required from the source to the destination.

The routing of this reply message back to the source can be accomplished in several ways. The simple way is to get the RREQ packet routing information and send it via that route, but this is based on the assumption that the route links are bidirectional. Another way is to find the path to the source through another RREQ message. This makes the DSR protocol appropriate even when unidirectional links are present in the network.

9.3.2.2.2.2 Route Keeping — When detecting a problem with a route in use, a DSR RERR message is sent to the source node. At the time the error message is received, the hop in error is removed from the host route cache and all routes containing the hop are truncated.

9.3.3 Security in MANETs

In MANETs, security is one of the most important issues because they are more vulnerable to attacks than are wired networks or other wireless networks. The design of a secure protocol for MANETs is a challenging task mainly because MANETs share an unprotected radio channel, they lack a central authority and partnership mechanisms for users, and they have limited resources and physical vulnerabilities [6].

A comprehensive analysis of security in MANETs involves a detailed security analysis for each layer of this architecture. This chapter will list the security features of each layer. Only the network layer security will be closely studied.

The following are the security issues relating to the protocol pile of MANETs [10]:

- Application layer: In this layer, the topics of interest are based on vulnerable applications running on the user's machine and the execution of unwanted viruses, worms, or other malicious codes.
- Transport layer: This layer authenticates and secures end-to-end communication via encryption means.
- Network layer: This layer protects routing protocols and packet forwarding.
- Link layer: The important issue of this layer is to protect the Media Access Protocol (MAC) and provide security support for linking protocols.
- Physical layer: This layer prevents jamming attacks that seek to cause denial or deterioration of MANET services.

The security issues in the network layer are for the routing protection protocol and forwarding mechanisms. Below, there is a description of the security mechanisms of these two components.

9.3.3.1 Routing Protocol Security in MANETs

One of the most relevant studies in the field of MANETs is to provide security mechanisms for routing protocols in MANETs. The attacks against the security of MANETs can be classified as actives and passives:

- Active attacks: This type of attack is based on the modification of packets or introducing false information. These attacks can be internal or external: performed by a node inside the network or by a node outside of it.
- Passive attacks: A passive attack does not impair the normal operation of the routing protocol. It simply obtains valuable information by listening to the routing traffic. This makes them difficult to detect.

The following are the security problems of the AODV and DSR routing algorithms. The description will focus on the introduction of advertising routing update actions that do not follow the routing protocol specifications, causing serious damage to the network's security.

9.3.3.2 Security in AODV

9.3.3.2.1 Attacks against AODV

Several attacks against the normal operation of the AODV protocol messages have been described [11]:

- Black hole attack: An attacker can trick the other nodes into believing that it is possible to obtain the shortest route to a specific destination through it. The purpose is to intercept packets between the source and the destination, so the source sends data packets through it. To achieve this, when a malicious node receives an RREQ, it may respond with an RREP. When the source receives this message, it will define a route to the destination through the malicious node, so all data packets between the source and the destination that pass through the malicious node are discarded by this node to cause the complete loss of the data.

- Gray hole attack: Once the attacker manages to pass the traffic to a certain destination through itself, this malicious node can selectively forward all RREQ and RREP messages, sending only a few data messages and eliminating all others. This attack belongs to the category of internal and active attacks.

- Wormhole attack: This attack is performed by two or more nodes that belong to the ad hoc network and have a private connection between them. Assuming that nodes A and B form a wormhole, all packets arriving at A will be forwarded to B to route them and the ones arriving at B will be sent to A to do the same. Clearly, this action affects the routing protocol by altering the short routes that it uses.

- Denial of service (DoS) attack: In this attack, the malicious node consumes the entire bandwidth of the network. To do so, it uses either of the following techniques: flooding the network with very frequent RREQ messages, which may exhaust the resources of the network for other nodes, or the malicious node can send notifications of broken links to other nodes, so they remove from their routing table the entry of the node that the attacker chooses.

- Routing table overflow attack: If a malicious node frequently sends RREQ-type messages to the network, the nodes will enter in their routing tables information about the route to the destination contained in the RREQ messages (for a destination that does not actually exist in the network) in order to send via this same route an eventual RREP or RERR. Since the size of the routing table of the nodes is limited, the malicious node can fill it with useless information and when a genuine RREQ arrives at a node, it will not be added to the routing table and, therefore, no route discovery will be successful.

- Power consummation attack: Power is a critical parameter in MANETs. Nodes try to save energy by transmitting only the necessary information. A malicious node send frequent and unnecessary RREQ messages and forward irrelevant packets to cause other nodes to rapidly consume their energy, making them process more information than is actually needed.

9.3.3.2.2 Security Mechanism in AODV

Secure AODV (SAODV) [11] is a protocol that seeks to solve some security problems in AODV. It fights external active attacks by avoiding forwarding RREQ messages to external nodes. This can be achieved by the authentication of all nodes in the network with a password. Thus, before forwarding an RREQ message to its neighbors, nodes first check the authenticity of the sender and verify the password. Only if it is correct then is the RREQ forwarded. As a result of this process, external nodes will not be able to enter the MANET.

To solve the routing table overflow attack, the table has to be updated at regular intervals of 70 ms. On the contrary, SAODV faces the black hole attack, disabling the RREP send function of the intermediate nodes and enabling the generation of these RREP messages via the destination node only. After receiving an RREP from an intermediate node, the source node checks that there is a route between the intermediate node and the destination. If there is one, the source node relies on the intermediate node and sends its data packets through it. Otherwise, the source node discards the RREP sent by the intermediate node and broadcasts a warning message to the network. Consequently, the rest of the nodes will isolate the intermediate node of the network, and a new route discovery process will start. Finally, SAODV encrypts data packets, so the attacker cannot read them and the reliability improves.

In addition, some security schemes developed for ad hoc routing should be underlined, such as secure routing basic scheme, trust-based routing scheme, initiative-based schemes, and schemes that use detection and isolation mechanisms [8].

9.3.3.3 Security Problems in DSR

DoS attack: DSR builds the full route from a source to a specific destination depending on the contribution of the intermediate nodes. The RREP message returns from the destination to the source through the intermediate nodes, and a malicious node could cause serious problems of availability. Assuming that node A discovers that the route to E is A > B > C > D > E and B is a malicious node, B could change the route before A completes it and could leave it as A > B > C > E, removing node D from the route. When the RREP arrives at node A, it would send data packets through the route manipulated by B. Since C has no direct communication toward E, it would return an RERR-type message to A through B, which would eliminate that message, and A would not learn that its data are not being delivered to E.

Some of the attacks against AODV, such as the black hole attack and the energy consumption attack, can also affect DSR.

9.3.3.4 Security in Forwarding Mechanisms

Forwarding mechanisms refer to the fact that a node forwards packets with a different destination than itself.

In the design of the network layer functions of MANETs, it has to be assumed that mobile nodes work together to perform those functions. However, this collaborative model of mobile nodes is not the only one in a MANET. There are other models known as "selfish" and "malicious." The first one uses the network, but does not cooperate in forwarding packets or routing information, so the battery consumption is low. These types of nodes do not attempt direct damage to the network. On the contrary, malicious nodes do try to damage other nodes in the network.

Selfish nodes can be of three different types: Those of type 1 do not forward data packets, those of type 2 do not participate in the DSR route discovery phase; and type 3 are a combination of types 1 and 2, in the sense that they adopt one or the other behavior depending on the level of the battery [12].

The most common types of attacks against MANETs [6] are as follows:

- Rushing attack: When a node sends requests for route discovery, the malicious node takes advantage of the fact that duplicate responses are removed and shall promptly send a reply to this message; thus, it becomes part of the group of nodes in charge of the forwarding.

■ Neighbor attack: This attack is based on the fact that each intermediate node puts its address in the header of the RREP packets received. A malicious node could choose not to put its address in the header of those packets, pretending that they are within its radius transmission when in reality they are not. This event leads to the loss of links in the network.

■ Jellyfish attack: In this attack, the malicious node significantly slows down the process of forwarding packets and so the end-to-end delay increases.

9.4 Safety Issues of FHAMIPv6 Inherited from IEEE 802.11

Since FHAMIPv6 uses IEEE 802.11 as an access protocol to the medium, some of its security problems are adopted by FHAMIPv6. The following sections illustrate the most relevant inconveniences.

9.4.1 Weak Management of CTS Frames

Initially, the IEEE 802.11 medium access standard [13] presented the problem of hidden and exposed stations [14]. This means that station A could send data to station B without realizing that the latter was in communication with a third station, C, which causes the collision of A frames and C frames when approaching B (hidden station problem: in this case, C is not in the scope of A, so the latter has no way of knowing that C is transmitting to B), which is not convenient to the wireless network performance. The issue of the exposed station was evident when a station, A, for example, wrongly assumed that the medium was being used by another station, C, so A refrained from sending any frames to the medium. Both issues were solved when two new types of frames were introduced: clear to send (CTS) and request to send (RTS). When a wireless station wishes to start a transmission to another, first of all it sends an RTS. If the second station replies with a CTS, the transmission can begin. All other stations that received the RTS or the CTS refrain from transmitting for some time.

Although RTS and CTS frames solved the difficulties related to hidden and exposed stations already mentioned, the IEEE 802.11 protocol cannot check their security status. That is why attackers could introduce hundreds of CTS frames into the network. As a result of this, the other wireless stations will wrongly believe that the medium is busy and will stop transmitting data for a long period of time. An issue like this seriously degrades the wireless network's efficiency.

This type of attack can be performed against FHAMIPv6, since it does not implement any additional control over CTS and RTS frames.

9.4.2 Nonencrypted Packet Transmission

The wireless nature of IEEE 802.11 is itself a security problem. Given that the messages (electromagnetic emissions from wireless stations or access points) travel through the air without any encryption, attackers could read the content just by putting the network card in the monitor mode. For this reason, different protocols have been developed to protect data that is transported in IEEE 802.11 networks. Some well-known protocols are Wireless Equivalent Privacy (WEP) [15], Wi-fi Protected Access (WPA), and, especially, WPA2, which is the most effective. These considerations were not taken into account in the design of FHAMIPv6, so both control and date messages of this protocol cross the air without any encryption, which is a serious security problem (more specifically, a confidentiality problem) of FHAMIPv6, since all the messages in the coverage area of the attackers' wireless network card can be read.

9.4.3 Physical Layer Weaknesses

The IEEE 801.11 standard owns physical layer specifications that are implicitly used by certain networks that implement some version of the standard properly. Next, the most relevant security issues of the specification will be introduced.

A massive attack of unlimited resources can be performed on an IEEE 801.11-based network sending high-power signals to mobile stations in a wide frequency spectrum that could definitely disrupt all communications. A variation of this attack could be the transmission of high-power signals randomly (not constantly as in the first case), which would make locating the attackers a difficult task to achieve. Since the delivery of high-power signals could consume the energy resources of the intruders, they may introduce a variation to previous attacks and carry out a reactive one, waiting for some station to transmit data, so they can activate interference signals. The effect of this would be similar to those above, in the sense that the network performance would be seriously damaged and could even cause the collapse of the entire network [16].

On the contrary, the IEEE 801.11 standard does not provide security mechanisms to manage the Synchronization (SYNC) field of the physical layer protocol header. This field informs the wireless stations if a signal is acceptable or if it is just noise or void. In addition, it supports time synchronization with the packets received [13]. Attackers could use this field to make the wireless stations believe that the signals they receive are just noise and thus be discarded. This could also add synchronization difficulties between stations [16].

FHAMIPv6 does not implement any improvement to these security problems of the physical layer, which makes it vulnerable to the attacks described earlier.

9.5 Compromising FHAMIPv6 Security by Message Manipulation

In the following sections, FHAMIPv6 security issues will be revealed by manipulating some of its most important messages.

9.5.1 Warning Messages Manipulation (MIPT_ADS)

In normal conditions, a MIPT_ADS warning message is used by the APAR to announce itself to the AMN. This message contains both the APAR and the AMAP addresses. From then on, the AMN will send all data via these two agents.

Since FHAMIPv6 does not implement security mechanisms for this type of message, attackers located close to the APAR area could send MIPT_ADS messages (using their own address pretending to be the APAR and keeping the AMAP address unmodified) to an AMN that has just arrived in that area. The AMN would then send all data through the attackers (man in the middle attack) assuming that they are a genuine APAR. As a result, attackers could manipulate all the information of the AMN. In addition, the attackers may decide not to forward AMN traffic toward the AMAP, causing a DoS to the AMN.

9.5.2 Manipulation of Registration Messages with the MAP (MAP_REG_REQUEST and MAP_REG_REPLY)

As explained in Section 9.2, the AMN uses a MAP_REG_REQUEST message type to start the registration with an AMAP, which will reply with a MAP_REG_REPLY message. The process will be completed when the message finally reaches the AMN.

FHAMIPv6 cannot validate that the arrival of a message MAP_REG_REPLY truly comes from a previous MAP_REG_REQUEST. For this reason, attackers could be able to send MAP_REG_REPLY messages to the AMN, which will process them as valid. Therefore, if the attacker assigns a CoA (other than that originally assigned by the AMAP) to the AMN through one of these messages, the AMN will take it to replace the old one. Of course, all current connections will be lost, given the change of address of the network layer. In addition, the AMN will be unable to send or receive data in the future because the AMAP has not recorded its new CoA.

9.5.3 Manipulation of Registration Messages with the AHA (MIPT_REG_REQUEST and MIPT_REG_REPLY)

Ideally, MIPT_REG_REPLY and MIPT_REG_REQUEST messages are used by FHAMIPv6 so an AMN can notify its AHA of its new CoA and the current address of its AMAP. However, because no security mechanisms are implemented to protect these messages, several problems may occur. When the AMN wants to set up the registration with the AMAP by sending a MAP_REG_REQUEST message, this process is performed through the APAR. Once it receives the message from the AMN, the APAR sends a broadcast request, expecting some node in the scope of the AMAP to forward the data. At this point, attackers could respond to the request in order to settle in between the APAR and the AMAP.

When the AMN is registered with the AMAP, it will try to notify the AHA of the new location (CoA and AMAP address). Since all traffic passes through the attackers, they will have the ability to change the AMAP address contained in the MIPT_REG_REQUEST message when the message passes through the AMAP. The AMAP will not consider whether the address in the message is the right one, and it will end up sending the message to the AHA. When this message reaches the AHA, the real AMAP address will be changed for the address contained in the message (the one amended by the attackers). Later, when the AHA sends packets to the AMN, the packets will never reach the destination because they are actually directed toward an incorrect or nonexistent AMAP, causing a DoS to the AMN. On the contrary, attackers need to generate a return REQ_REP_AHA message to the AMN so that it will assume that the registration with its AHA has been completed successfully.

9.5.4 Manipulation of Fast Handover Messages (FASTPRRTADV)

When an AMN approaches the ANAR area, it actually sends a request to start a fast handover (FAST_RTSOLPR) to its respective APAR, which receives the request and sends back a FASTPRRTADV to the AMN. That message indicates the address of the ANAR. Then, the AMN begins the registration process with the AMAP through that ANAR. Next, it sets the registration with its AHA by the same route. Since FHAMIPv6 does not implement mechanisms to validate the origin of a FASTPRRTADV, attackers could send this type of message to the AMN on behalf of the APAR at any time, so the AMN may change from an ANAR to another at the attackers' will. Also, attackers could send a specially crafted FASTPRRTADV message containing the address of the APAR as the origin and the attackers' own address as the genuine address of the ANAR. Thereafter, this would cause the AMN communications to pass through the attackers (alleged ANAR), which can manipulate them at their will.

9.5.5 Manipulation of Function Allocation Messages to Intermediate Nodes (DISCOVERY_REPLY)

In Section 9.2.4, it was already described how DISCOVERY_AN, DISCOVERY_BS, and DISCOVERY_REPLY messages operate. Basically, the AMAP sends a broadcast DISCOVERY_AN message to the broadcast. If an intermediate node receives this message, it will transmit a DISCOVERY_BS message. Such messages can only be replied to by a base station (APAR, ANAR, or AHA). If the response message (DISCOVERY_REPLY) arrives at the AMAP, it is presumed that the base station is close enough to the AMAP, so an intermediate node is not necessary. Otherwise, an intermediate node will be used to send data to the base station.

Again, FHAMIPv6 does not implement any mechanism to validate that a DISCOVERY_REPLY message comes from the base station that it claims to be from. Likewise, no verification is performed to ensure that the message comes in response to a previous request. Therefore, attackers, when receiving a DISCOVERY_BS message, will be able to reply to the AMAP with a DISCOVERY_REPLY message on behalf of any base station. This will result in the AMAP not using intermediate nodes to reach that particular base station and not being able to contact it. However, attackers could send a DISCOVERY_REPLY message at any time using the address of a base station as the source, which will actually cause the collapse of the network.

9.6 Other Security Issues

The problems mentioned previously are due to the lack of security mechanisms of FHAMIPv6 for its various types of messages. In the following sections, other security issues of this protocol will be observed.

9.6.1 Flooding Attacks

In most current networks, it is really difficult to defend against attacks based on packet flooding that produces a DoS to a host or an entire network. Many solutions have been developed based on firewalls for both hardware and software. However, crackers always create new ways to collapse the networks or their central servers.

A flooding attack involves sending a huge volume of legitimate traffic to certain hosts of a network in order to exhaust its resources to bring it out of order. Traditionally, these types of attacks have been carried out using User Datagram Protocol (UDP) packets, but other protocols have been used too.

FHAMIPv6 does not implement any security mechanisms to protect itself from flooding attacks, so it is not a complex challenge to organize an attack on this type of a network. Attackers could easily send a massive volume of traffic to the base stations (APAR, ANAR, AHA) or to the AMAP in order to avoid that the traffic flows through the normal paths. In the event that attackers send many MIPT_REG_REQUEST messages to an AMAP, this one (AMAP) will try to register with the AMN home network repeatedly. As a result, an excess of MIPT_REG_REPLY and REQ_REP_AHA return messages will travel through the network. Even if all FHAMIPv6 stations have sufficient resources to handle such a huge volume of traffic, the active nodes will quickly lose all power, leading the entire network to collapse.

9.6.2 Attack via AMAP Association

In the field of computer networks, it is well known that computational resources are always limited. Buffer size, network bandwidth, and the processing capability of hosts and routers are some of the most important resources. In FHAMIPv6-based networks, the issues are not different and therefore some security problems clearly emerge.

When an AMN arrives in an external network, it has to register with the AMAP, which allocates a CoA and keeps a record of it. On this basis, attackers could carry out an association attack. This involves sending hundreds of MAP_REG_REQUEST messages to the AMAP where all of them change their address of origin. As a result, the AMAP will assume that they are new AMNs, so a CoA will be allocated and recorded for every MAP_REG_REQUEST received. If the attackers fill all the space reserved to keep the connections, a new, legitimate AMN will not be able to perform the registration with the AMAP due to a DoS.

References

1. S. Renan, J. López, and J. Ortiz. FHAMIPv6: Una propuesta para extender el protocolo hierarchical mobile IPv6 with fast-handover a escenarios de redes móviles Ad-Hoc. Tesis de Master, Universidad de Castilla-La Mancha, 2009.
2. Perkins, C. IP mobility support. RFC 2002, October 1996.
3. H. Soliman, C. Castelluccia, K. El-Malki, and L. Bellier. Hierarchical mobile IPv6 (HMIPv6) mobility management. RFC 5380, October 2008.
4. R. Koodli. Fast handover for mobile IPv6. RFC 4068, July 2005.
5. H. Jung, H. Soliman, S. Koh and J. Lee. Fast handover for hierarchical MIPv6 (F-HMIPv6). Internet Draft, IETF, April 2005.
6. L. Nguyena. Study of security attacks on multicast in mobile ad hoc networks. Master Thesis, York University, 2006.
7. T. Larsson and N. Hedman. Routing protocols in wireless ad-hoc networks-a simulation study. Master Thesis, Lluleå University of Technology, 1998.
8. R. Carlton. Security protocols for mobile ad hoc networks. PhD Thesis, McGill University, 2006.
9. C. Schwingenschlögl. A framework for secure and efficient communication in mobile ad-hoc networks. Doktors der Naturwissenschaften genehmigten dissertation, Technische Universität Munchen, 2004.
10. H. Yang. Security in mobile ad hoc networks: challenges and solutions. *IEEE Wireless Communications*, vol. X, 2004, pp. 38–47.
11. S. Deswal and S. Singh. Implementation of routing security aspects in AODV. *International Journal of Computer Theory and Engineering*, vol. 2, no. 1, 2010, pp. 135–138.
12. P. Michiardi and R. Molva. *Simulation-Based Analysis of Security Exposures in Mobile Ad-Hoc Networks*. France: Institut Eurécom, 2002.
13. IEEE Computer Society. *IEEE Standard for Information Technology—Telecommunications and Information Exchange between Systems. Local and Metropolitan Area Networks—Specific Requirements: Wireless LAN Medium Access Control (MAC) and Physical Layer (PHY) Specifications*. New York: IEEE Computer Society, 2007.

14. A. Tanenbaum. *Redes de computadoras*, Cuarta edición. Mexico: Pearson Education, 2003.

15. K. Scarfone. *Guide to Securing Legacy IEEE 802.11 Wireless Networks*. National Institute of Standards and Technology, Gaithersburg, 2008.

16. K. Bicakci and B. Tavli. Denial-of-Service Attacks and Countermeasures in IEEE 802.11 Wireless Network. *Computer Standards & Interfaces*, vol. 31, 2009, pp. 931–941.

Chapter 10

Channel Assignment in Wireless Mobile Ad Hoc Networks

Javad Akbari Torkestani and Mohammad Reza Meybodi

Contents

Due to the tremendous growth of the mobile ad hoc networks (MANETs), the efficient use of the scarce bandwidth allocated to the wireless communications is mandatory. In current wireless networks, a fixed spectrum assignment strategy is used to regulate the radio systems. In this strategy, the whole radio spectrum is subdivided into a fixed number of radio ranges, each exclusively assigned to a specific user. Such a spectrum assignment strategy leads to an undesirable condition under which some systems use only a small portion of the allocated spectrum while the others have very serious spectrum insufficiency. In addition to the current inefficient channel division policies, the channel interferences caused by the unconstrained simultaneous transmissions are the main origin of the problems with the radio spectrum assignment techniques. Cognitive radio is a highly potential technology to address the spectrum scarcity challenges in wireless ad hoc networks. This technology is the future trend of the MANETs in the field of radio spectrum management. The aim of a channel assignment scheme is to assign a minimum number of channels to the radio nodes in such a way that no interference happens. Although a MANET does not have the infrastructure of the base stations, the channel assignment in these networks can be efficiently conducted in a way very similar to that in infrastructure-based cellular systems, specifically when the network is clustered. This chapter aims to exhibit the current status of the channel assignment in MANET and the problems of the existing techniques by providing an in-depth survey of the most recent channel assignment protocols designed and developed for mobile wireless ad hoc networks. In this chapter, depending on the nature of ad hoc environments, the channel assignment schemes are generally categorized as contention-free and contention-based channel assignment schemes. Contention-free schemes are further subdivided into fixed and on-demand channel assignment (ODCA) schemes as the ad hoc networks require. This chapter summarizes the key design issues, objectives, and performances of each category and argues that due to the explosive growth of using mobile devices and ad hoc networks and the scarceness of the channel bandwidth in ad hoc networks, cognitive radios and intelligent bandwidth allocation schemes are the future trends of the MANET researchers to design efficient radio channel assignment protocols.

10.1 Introduction

Dynamic network topology changes; network mobility; severe constraints on network resources such as communication channel bandwidth, processing power, and battery life; and the lack of a fixed infrastructure or centralized administration are the major challenging issues from which the ad hoc networking protocols suffer, while the traditional networking does not consider them. In spite of the essential differences between the traditional and wireless mobile ad hoc networking, the channel assignment schemes reported in literature show that the protocols tailored for wired or cellular networks are also employed in MANET environments. In current wireless networks, the radio spectrum is partitioned into a fixed number of radios each of the same range or the channel bandwidth is evenly allocated to different users. This significantly degrades the channel utilization and network performance since different users have different requirements and so need different portions of the channel. Recent advances in wireless communication technology have provided an opportunity to develop a new approach of intelligent or cognitive radios in which the radio frequency spectrum can be adaptively distributed among the users proportional to their needs. In other words, the cognitive radio is an emergent paradigm to address the spectrum allocation strategy issues in wireless networks in which the wireless nodes are capable of changing their transmission or reception parameters to communicate efficiently without interference. A cognitive radio system supports a very dynamic medium access control (MAC) layer adaptation based on the active monitoring of available channel bandwidth. The following are the main functions of a cognitive radio [1]:

1. Exploring the unused ranges of the radio spectrum and sharing them with the other users avoiding interference and collision.
2. Selecting the best available spectrum to meet the system constraints and user requirements.
3. Supporting the user connection requests even if it exchanges the frequency of operation.
4. Providing the fair collision-free spectrum scheduling method.

However, cognitive radio is still in the very early stage of the research and development, and certainly will be the future trend of wireless ad hoc networking [1,2].

In wireless ad hoc networks, bandwidth is a scarce resource. The tremendous growth of wireless ad hoc networks requires an efficient use of the scarce bandwidth (radio spectrum) allocated to the wireless communications. However, the main difficulty against the efficient use of the bandwidth arises from interferences, caused by unconstrained simultaneous transmissions, which result in damaged communications that need to be retransmitted, leading to a higher cost of the service. Indeed, the aim of channel assignment is to assign a required number of channels to each host such that efficient bandwidth utilization is provided and interference effects are minimized. Two sorts of interferences must be avoided by an effective channel assignment algorithm. The first one occurs when a host simultaneously transmits and receives signals over the same channel, and the second one occurs when a node simultaneously receives more than one signal over the same channel. To prevent the first group of interferences, two hosts can be assigned the same channel if and only if none of them is within the transmission range of the other. Similarly, to prevent the second group of interferences, two hosts can be assigned the same channel if and only if by no means another host is located in the intersection of their transmission ranges. Interferences can be eliminated (or at least reduced) by means of suitable channel assignment schemes. Indeed, co-channel interference caused by channel reuse is one of the most critical factors on the overall system capacity in wireless networks. Channel assignment schemes partition the given bandwidth into a set of disjoint channels that can be used simultaneously by the hosts while maintaining acceptable radio signals. The purpose of channel assignment algorithms is to assign the channels to the hosts in such a way that the co-channel reuse distance and channel separation constraints are verified as well as the difference between the highest and lowest channels assigned is kept as small as possible. By taking advantage of the physical characteristics of the radio environment, the same channel can be reused by two or more hosts at the same time without interferences (co-channel stations), provided that the hosts are spaced sufficiently apart. The minimum distance at which co-channels can be reused with no interferences is called *co-channel reuse distance*. The interference phenomena may be so strong that even different channels used at near hosts may interfere if the channels are too close [3]. Figure 10.1 illustratively represents the concepts of the co-channel and co-channel reuse distance.

The conflict-free channel assignment problem is equivalent to the vertex coloring of a special class of geometric graphs, which is an NP-hard problem in graph theory. The conflict-free channel assignment problem seeks an assignment of the fewest channels to a given set of radio nodes with specified transmission ranges without any interference. It is a classic and fundamental problem in wireless ad hoc networks. It is shown that the channel assignment problem is NP-hard even if all the nodes are located in a plane and have the same transmission radii [3]. Since perfect filters are not available, interference between the close frequencies is a serious problem, which can be handled either by adding guard frequencies between adjacent channels or by imposing channel

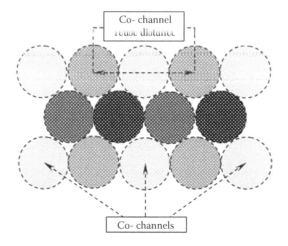

Figure 10.1   Co-channel and co-channel reuse distance.

separation. In the latter approach, the channels assigned to the near hosts must be separated by a gap on the radio spectrum—counted in a certain number of channels—which is inversely proportional to the distance between the hosts [4].

Roughly speaking, channel assignment schemes in wireless ad hoc networks are generally classified into fixed assignment, on-demand assignment, and contention-access algorithms. In fixed assignment, as the name implies, the channels are nominally assigned to the hosts in advance according to the predetermined estimated traffic intensity. In fixed assignment approach, the further transmission requirements are not taken into consideration. In on-demand assignment, channels are dynamically assigned as the calls arrive. Therefore, on-demand assignment is also called *dynamic channel assignment.* The dynamic channel assignment schemes schedule the channel access based on the demand of the hosts for packet transmission. The latter approach makes ad hoc environments more efficient, particularly if the traffic distribution is unknown or changes with time. On the contrary, the on-demand assignment algorithms are generally time-consuming and require more complex control. In both fixed and on-demand (dynamic) channel assignment algorithms, the MAC algorithms are collision-free. Various extensions or combinations of the above two schemes have been discussed in the literature. These channel assignment schemes are also referred to as *contention-free approaches,* since the hosts do not compete to seize the channel. The third class of the channel assignment algorithms comprises contention-access (or random access) algorithms, in which the hosts contend for channel access, and the hosts that lose it try again later. Since the collisions are not prohibited by the contention-access algorithms, they require a method for detecting and recovering the collisions. The fixed channel assignment (FCA) approach is the simplest off-line allocation scheme that outperforms the other schemes under uniform and heavy traffic loads. Furthermore, FCA problems can serve as bounds for the performance of the other schemes. For these reasons, FCA schemes constitute a significant research subject for the operations research, artificial intelligence, and mobile communication fields [3]. Frequency division multiple access (FDMA) [5], code division multiple access (CDMA) [6], and time division multiple access (TDMA) [7] schemes are some fixed assignment MAC layer protocols. Among the controlled access MAC protocols, TDMA is the most commonly used in wireless ad hoc networks. Combinations of FDMA and CDMA schemes with TDMA have been also proposed for ad hoc networks in the literature [8]. Polling [9], trunking [10], and reservation [11] are some well-known MAC layer protocols for supporting on-demand assignment approach. ALOHA [12] and carrier sense multiple access (CSMA) [13] are two famous medium access protocols proposed for wireless ad hoc networks.

This chapter presents a new classification of the channel assignment schemes for MANET, in which the allocation strategies are classified in two main categories: contention-free and contention-based channel assignment schemes. This chapter studies and compares the most recent effective schemes of each category, with the emphasis on the key design issues, objectives, performances, and costs. The rest of this chapter is organized as follows. The channel assignment problem is defined in Section 10.2. Section 10.3 presents a new classification of the channel assignment schemes and gives an in-depth overview of the well-known channel allocation strategies based on the proposed classification. Section 10.4 concludes the paper.

10.2 Channel Assignment Problem

In wireless communication systems, the radio spectrum is a limited resource. However, the efficient use of available channels has been shown to improve the system capacity. The role of a channel assignment scheme is to allocate channels to the users in such a way as to minimize call blocking or call dropping probabilities and also to maximize the quality of service (QoS). The channel assignment problem is an NP-hard problem that can be modeled as an appropriate bandwidth multicoloring problem. The channel assignment problem aims at finding an assignment of the fewest channels to a given set of radio nodes with specified transmission ranges without any interference. An efficient channel assignment scheme in a multihop ad hoc network should not only guarantee successful data transmissions without any interference but also enhance the channel spatial reuse to maximize the system throughput. In this section, the channel assignment problem in ad hoc networks is described in detail.

Definition 1.1: A wireless ad hoc network can be modeled as an undirected graph $G < V, E >$, where the vertex set V represents the individual hosts and an edge connects two hosts if the corresponding hosts are within the transmission range of each other.

Definition 1.2: The demand vector $\Psi = \{\psi_i \mid v_i \in V\}$ is a function $\Psi: V \to \mathbb{N}$, where ψ_i is the required number of connections that must be simultaneously supported for host H_i.

Definition 1.3: The channel separation vector $\Lambda = \{\lambda_{ij} \mid \forall e_{i,j} \in E\}$ is a function $\Lambda: E \to \mathbb{N}$, where $\lambda_{i,j}$ denotes the minimum (valid) distance between the channels assigned to host H_i and H_j to avoid interference.

Definition 1.4: The channel assignment problem can be formally modeled by a quintuple $< G, \Psi, \Lambda, \mathbb{C}, \mathbb{F} >$, where $G \langle V, E \rangle$ denotes an undirected graph representing the wireless ad hoc network topology whose vertices in V correspond to the hosts and edges in E correspond to the pairs of hosts that can hear each other's transmission; $\Psi = \{\psi_i \mid v_i \in V\}$ denotes the demand vector associated with the vertex set V; $\Lambda = \{\lambda_{i,j} \mid \forall e_{i,j} \in E\}$ denotes the channel separation vector (or weight constraints) imposed to the edges of graph G; and $\mathbb{C}$ denotes the channel set. The channel assignment problem is to find a function $\mathbb{F}: V \to 2^{\mathbb{C}}$ from the vertex set to the channel set such that

$$|\mathbb{F}(v_i)| = \psi_i \text{ for all } v_i \in V,$$
$$\mathbb{F}(v_i) \cap \mathbb{F}(v_j) = \emptyset \text{ for all } e_{(i,j)} \in E, \text{ and}$$
$$|c_i - c_j| \geq \lambda_{i,j} \text{ for all } e_{(i,j)} \in E, c_i \in \mathbb{F}(v_i), \text{ and } c_j \in \mathbb{F}(v_j).$$

From the above definition, it can be seen that the channel assignment problem can be modeled as an appropriate bandwidth multicoloring problem. Each channel assignment algorithm aims at finding an assignment function F. A channel assignment algorithm is optimal if it minimizes the number of channels assigned to the hosts.

10.3 Channel Assignment Schemes

Optimal channel assignment in an arbitrary wireless ad hoc backbone is an NP-hard problem (similar to the graph coloring problem). An optimal channel assignment algorithm looks for finding an allocation strategy that maximizes the channel reuse without violating the constraints. Such an optimal channel strategy minimizes the blocking rate. The constraints of the channel assignment can be classified into three categories: first, the channel constraint that specifies the number of available channels (frequencies) in the radio spectrum. This constraint is imposed by the national and international regulations; second, the traffic constraint that specifies the minimum number of frequencies required by each host; and third, the interference constraint that is further classified as the co-channel constraint and the adjacent channel constraint. The co-channel constraint states that the same channel cannot be assigned to certain pairs of wireless hosts simultaneously, and the adjacent channel constraint states that the adjacent channels in the frequency domain cannot be assigned to the adjacent wireless hosts at the same time. As described earlier, the existing channel assignment schemes in the literature are mostly heuristic based. These schemes can be classified as the contention-free (FCA schemes, ODCA schemes) and contention-based channel assignment schemes. Figure 10.2 shows the classification of channel assignment schemes presented in this chapter. In the remaining of this chapter, the above-mentioned classification of the channel assignment schemes is described in more detail and the existing channel assignment schemes are briefly discussed in each class.

10.3.1 Fixed Channel Assignment Schemes

The FCA approach is the simplest off-line channel allocation scheme in which the channels are assigned to the hosts either permanently or for a long time interval. In an FCA strategy, each host is allocated a predetermined set of channels. These schemes do not consider the further variations in communication requirements. This makes the fixed assignment schemes often the easiest to implement, but also the most inflexible in response to the changing network conditions.

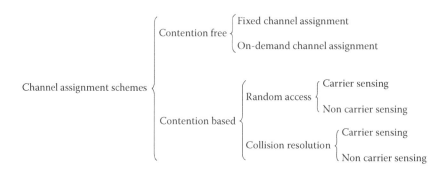

Figure 10.2   Classification of channel assignment schemes.

This characteristic is a large disadvantage for wireless ad hoc networks due to their ad hoc, self-organizing, and self-maintaining nature. Therefore, FCA schemes are rarely proposed for ad hoc networks. FCA protocols assure the hosts that their transmitting messages will not collide with the messages from the other hosts. That is, in FCA schemes the MAC algorithms are collision-free. TDMA, FDMA, and CDMA are typical protocols that belong to this class of channel assignment schemes. The following is a brief review of the most popular FCA protocols in the MAC layer.

10.3.1.1 Time Division Multiple Access

In the TDMA scheme, a single channel is time-shared. That is, the use of the channel is divided among several hosts by allowing each host to access the channel periodically, but only for a small period of time referred to as *time slot*. A set of such periodically repeating time slots is known as the *TDMA frame*. During a time slot, the entire bandwidth is available and then the host must relinquish the channel. A given host may be assigned more than one time slot in each frame. Since the channel is available only to one of the hosts at a (fraction of) time, TDMA is a collision-free scheme. Difficulties with TDMA largely center on the problem of synchronizing a number of independent hosts. To cope with this problem a perfect synchronization between the hosts is required, and a guard band (or guard interval) is proposed as a solution to relieve the impact of synchronization errors, clock drift during the slot, and differences in propagation delay between the hosts. Indeed, a guard band is a period of time during which the channel is assigned to no host. Due to the small size of the time slots, guard bands results in a significant overhead for the system [14]. Although the TDMA scheme is essentially a half-duplex mechanism in which only one of the two communicating hosts is able to transmit at a time, the small duration of the time slots gives the illusion of a two-way simultaneous communication. Among the controlled access MAC protocols, TDMA is the most commonly used in wireless ad hoc networks. Figure 10.3 shows the structure of a TDMA scheme.

In [14], a dynamic TDMA frame length expansion and recovery method called *dynamic frame length channel assignment* (DFLCA) was proposed. The proposed method, taking advantage of the channel spatial reuse concept, efficiently utilizes the channel bandwidth by assigning the unused slots to the new nodes as well as enlarging the frame length when the number of slots in the frame is insufficient to support the nodes. DFLCA controls the expansion and recovery of unassigned time slots by dynamically changing the frame length according to the traffic load and the number of mobile nodes in the contention area. For this purpose, the nodes are allowed to release the unused slots and shrink their channel tables when the frame is inefficient. An adaptive time slot

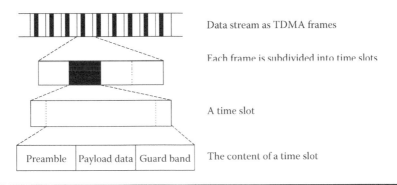

Data stream as TDMA frames

Each frame is subdivided into time slots

A time slot

| Preamble | Payload data | Guard band |

The content of a time slot

Figure 10.3 A simple structure of a TDMA scheme.

assignment algorithm was proposed in [15] for variable bandwidth switching systems. Kim and Khu [15] claim that the amount of change between the consecutive switching configurations is small, and so it is unnecessary to schedule all traffic for each configuration. The adaptive time slot assignment algorithm proposed in [15] employs such an idea and reduces the computational burden of finding switching configurations. Mo and Chew [16] proposed a new TDMA scheme in which the time slot duration is not fixed and vary with time for each user. In this method, the time slot duration is independently adjusted for each user proportional to its transmission requirements. In [16], it is shown that the proposed variable frame length TDMA scheme improves the bandwidth utilization. Xin et al. [17] proposed a simple and efficient time allocation scheme called *MES-ESRPT* for delay-sensitive Variable Bit Rate (VBR) traffic in accordance with IEEE 802.15.3 standard. In this scheme, the coordinator allocates one MCTA (management channel time allocation) for each stream, which is the process of communication at the end of superframe. During the MCTA period, each transmitter should report the fragment number of the first and remainder MAC service data unit to the coordinator. In the next superframe, the coordinator allocates the channel time based on these by a shortest remaining processing time technique.

Akbari Torkestani and Meybodi [1] proposed a cognitive radio for clustered wireless MANETs based on learning automata. In this method, the wireless hosts are grouped into clusters and each cluster head takes the responsibility of a collision-free channel access scheduling within the cluster. That is, Akbari Torkestani and Meybodi [1] design an adaptive TDMA scheme for slot assignment in each cluster of a clustered ad hoc network. To do this, each cluster head is equipped with a learning automaton whose action set includes an action for each of its cluster members. At each stage, cluster head randomly chooses one of its actions according to its action probability vector. Then, the cluster member corresponding to the selected action is permitted to transmit its packets during the current time slot. If the selected member has a packet to transmit, the cluster head rewards the selected action and penalizes it otherwise. As the proposed algorithm proceeds, the probability of choosing a given host converges to the proportion of time it has a packet to transmit. This probability specifies that the fraction of TDMA frame must be assigned to the host. Akbari Torkestani and Meybodi [1] argue that the proposed channel assignment scheme performs well under conditions that the input traffic parameters are unknown and time variable.

10.3.1.2 Frequency Division Multiple Access

In an FDMA scheme, the available spectrum is divided into a number of equal frequency channels, and one or more channel is assigned to each host for its own exclusive use. These frequency subchannels are sufficiently separated (via guard bands) to prevent co-channel interference. A significant portion of the effective channel bandwidth is usually wasted by the guard bands. FDMA is able to accommodate simultaneous packet transmissions (one on each subchannel) without collision. To receive packets from a particular sender, the destination host must be listening on the proper subchannel. Assigning frequencies for all potential users of the spectrum, when far fewer users will be present most of the time, leads to the poor spectrum utilization. These results in a significant fraction of the available spectrum will be unoccupied. While this is a minor or nonexistent problem for some services such as broadcasting that occupy the assigned frequencies almost continuously, it is a significant problem for other services, such as public safety communications that seldom require use of their frequencies. The FDMA scheme is shown in Figure 10.4.

OFDMA (orthogonal frequency division multiple access) is a multiuser version of the popular orthogonal frequency division multiplexing (OFDM) digital modulation scheme. OFDM is a

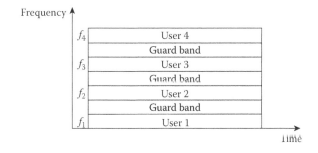

Figure 10.4   The FDMA scheme.

frequency division multiplexing scheme utilized as a digital multicarrier modulation method. In OFDM, the subcarrier frequencies are selected so that they are orthogonal to each other. By this, the cross talk between the subchannels does not occur and so intercarrier guard bands are not required. In this method, a large number of closely spaced orthogonal subcarriers are exploited to carry the data. OFDM divides the data into several data streams, each sending over a subcarrier. In OFDM, each subcarrier is modulated by using modulation. Multiple access is achieved in OFDMA by assigning subsets of subcarriers to individual users. This allows simultaneous low data rate transmission from several users. Heikkinen and Hottinen [18] designed a distributed subchannel allocation scheme for an OFMDA system. To this end, they first showed that the centralized subchannel assignment problem in a two-hop multiuser relay network can be formulated as a three-dimensional assignment problem. The three-dimensional assignment problem is known to be *NP-hard*. Therefore, the centralized solutions perform well only in small-sized networks, and so the distributed subchannel allocation seems to be a promising approach. The distributed subchannel allocation was studied by using the noncooperative game theory. In the proposed scheme, a sequential greedy channel allocation mechanism (sequential game model) is used to select the first link that implies a good performance (without pricing) in a relay network relative to a per-hop optimal assignment. An auction-based game model is also used to do the per-hop optimal assignment, which imposes an additional updating price to the system. Therefore, there is a trade-off between efficiency and signaling costs in the proposed distributed resource allocation scheme [18]. A joint channel assignment, routing, and scheduling scheme called *JARS* was proposed in [19] for wireless ad hoc networks in which the nodes are equipped with multiple radios. JARS shows the benefits of integrating the routing, scheduling, and channel assignment procedures, which is achieved by using the multiple radios at each node to transmit and receive simultaneously on different orthogonal channels.

San Jose-Revuelta [20] proposed a genetic algorithm-based heuristic to solve the frequency reuse problem in cellular radio communication systems. The proposed algorithm aims to minimize the required number of channels (or the minimum theoretical number of channels) to handle a certain number of connections. In this technique, the probabilities of mutation and crossover are on-line adjusted based on the diversity of the population. Huang et al. [21] proposed a new 802.11-like multichannel MAC protocol, called *self-adjustable multichannel MAC* (SAM-MAC). In the proposed scheme, one common channel and two half-duplex transceivers are used for each network node. Due to the overhead caused by the channel assignment process, the current 802.11-like schemes of the multichannel MAC are not able to efficiently use the available bandwidth of the multiple channels. On different channels, SAM-MAC uses a self-adjustment mechanism so as to do allocation and reallocation of the channels and to balance the traffic. Huang et al. [21] claim that due to less contention in common channel and smaller channel assignment overhead, their protocol has a higher throughput in comparison with the previous approaches.

10.3.1.3 Code Division Multiple Access

A CDMA is a spread spectrum multiple access scheme in which a transmitter spreads the information signal in a wide frequency band by using a spreading code. A receiver uses the same code to retrieve the received signal as well. This approach provides multiple accesses by allowing the simultaneous transmission by different nodes and is employed to reuse the bandwidth and to reduce the interferences. In the CDMA scheme, each group of nodes can be given a shared code. Many codes occupy the same channel, but only nodes associated with a particular code can understand each other. If the codes are orthogonal, or nearly so, so that any bit errors caused by co-channel interference can be handled by forward error correction, multiple nodes may occupy the same band. In the spread spectrum CDMA system, simultaneous transmissions can be isolated by using different spreading codes. Therefore, a unique code must be assigned to each transmitter and the receiver should be set to the same code as the designated transmitter. Assigning a unique code to each transmitter, which is called the *code assignment problem,* is a trivial problem if the network size is small. But when we employ the CDMA scheme in a large multihop ad hoc network, the code assignment becomes an intractable problem. It is impossible to assign a unique code to each transmitter or receiver since the number of available codes is limited. Therefore, the concept of the code spatial reuse seems to be promising. This means that two or more nonneighboring hosts can be assigned the same code if they are sufficiently separated. An interference-free code assignment problem is similar to the vertex coloring problem in which the neighboring nodes (hosts) are refrained from choosing the same colors (codes). Figure 10.5 provides a good comparison of TDMA, FDMA, and CDMA schemes.

10.3.1.4 Hybrid Multiple Access Scheme

The above-mentioned multiple access schemes can be combined in a single application to improve the network throughput. CDMA/TDMA [8,22], CDMA/FDMA [23], and FDMA/TDMA [24] are three possible hybrid multiple access schemes made of basic multiple access schemes. Among the hybrid multiple access schemes, CDMA/TDMA has received more attention. CDMA/TDMA is a combinational scheme in which the CDMA scheme is overlaid on top of the TDMA. It is a promising approach to solve the code assignment problem in wireless ad hoc networks [8,25]. In the CDMA/TDMA scheme, the networks must be initially partitioned into several groups

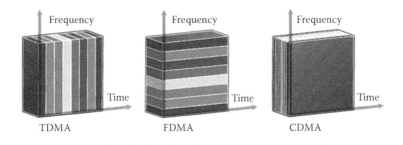

Figure 10.5   TDMA, FDMA, and CDMA schemes.

(clusters). In this scheme, the collision-free intracluster communications are organized by the cluster heads using a TDMA scheme, and a CDMA scheme is overlaid on the TDMA to organize the interference-free intercluster communications. A CDMA/TDMA scheme reduces the required number of codes to the number of clusters, which is significantly less than that of a CDMA scheme. To design a CDMA/TDMA scheme for clustered wireless ad hoc networks, we encounter the following three intricate problems. The first problem is dividing the network into a minimum number of nonoverlapping clusters. The second problem is to assign an interference-free code to each cluster considering the concept of the maximum code spatial reuse. That is, this problem is assigning the minimum number of codes to the clusters so that no two neighboring clusters are assigned the same codes. This problem is similar to the graph (vertex) coloring problem in graph theory, which is known to be NP-hard. In CDMA/TDMA networks, the TDMA scheme is used to schedule the intracluster communications. By this scheme, the code assigned to each cluster is divided into several time slots. The third problem we encounter is to assign a proper portion of the code (or TDMA frame) to each cluster member so that the maximum number of connections can be served during a TDMA frame.

Many studies have been carried out on the CDMA/TDMA scheme in cellular networks [26–29], while in ad hoc networks it has not received the attention it deserves. Gerla and Tsai [30] proposed an overlaid CDMA/TDMA channel access scheduling scheme for a clustered multihop wireless network. Gerla and Tsai [30] also proposed a distributed cluster formation algorithm. The cluster heads act as local coordinators to resolve channel scheduling, perform power measurement/control, maintain time division frame synchronization, and enhance the spatial reuse of time slots and codes. Using a CDMA scheme, an interference-free code is assigned to each cluster, and a TDMA scheme is used within the clusters. Richard and Gerla [31] also proposed a CDMA/TDMA-based scheme for multimedia support in a self-organizing multihop mobile network. They introduced a network architecture in which the nodes are organized into nonoverlapping clusters. In this method, the clusters are independently controlled and are dynamically reconfigured as the nodes move. In [31], an interference-free channel access scheduling method is proposed to handle the intercluster communications based on graph coloring problem. Due to the node clustering, the proposed method provides spatial reuse of the bandwidth. Furthermore, the bandwidth can be shared or reserved in a controlled fashion in each cluster. Wu [8] proposed a CDMA/TDMA structure for clustered wireless ad hoc networks. He designed a dynamic channel assignment algorithm called *hybrid-DCA* to make the best use of available channels by taking advantage of the spatial reuse concept. In this approach, the TDMA scheme is overlaid on top of the CDMA scheme to divide the bandwidth into smaller chunks. The proposed DCA algorithm forms the channel as a particular time slot of a particular code.

Akbari Torkestani and Meybodi [2] designed a learning automata-based dynamic frame length CDMA/TDMA scheme for clustered wireless ad hoc networks with unknown traffic parameters. In the proposed scheme, intracluster communications are scheduled by the cluster head using a TDMA scheme, and a CDMA scheme is overlaid on the TDMA to organize the interference-free intercluster communications. Therefore, to design the proposed scheme, the following three important problems were considered: cluster formation, code assignment (in CDMA scheme), and slot assignment (in TDMA scheme) problems. The authors proposed three learning automata-based algorithms to solve the addressed problems. By the proposed cluster formation algorithm, the network is partitioned into a number of clusters, each with a cluster head and a number of cluster members. Intercluster connections are handled by a CDMA scheme, in which an interference-free code must be assigned to each cluster. To do so, the authors proposed a code assignment algorithm based on the vertex coloring problem in which the neighboring clusters are refrained from choosing the same codes. Intracluster channel access scheduling is based on the TDMA scheme, and the cluster heads are responsible for the collision free slot assignments within the clusters. In [2], after each cluster is assigned a code, the cluster head divides the code into time slots and assigns them to the cluster members. To optimally assign the time slots, the authors proposed a dynamic frame length slot assignment algorithm in which each cluster member receives a portion of TDMA frame proportional to its traffic load. By the proposed scheme, the authors claim that the following advantages can be achieved. This scheme organizes the channel access in groups and so can be effectively used in scalable multihop ad hoc networks. The number of required codes decreases to at most the number of groups, and exploiting the code spatial reuse concept, it can be minimized. In each group, using the TDMA scheme, a large number of connections can be served in a time-efficient manner. The experimental results show that the channel assignment scheme proposed in [2] significantly outperforms CS-DCA [25] and hybridDCA [8] in terms of number of clusters, code and channel spatial reuse, blocking rate, waiting time for packet transmission, control overhead, and throughput.

Jovanovic and Djordjevic [32] proposed a hybrid multiple access scheme called *TFMAC* in which the timing feature of TDMA is combined with the multiple channels of FDMA. TFMAC comes from time frequency MAC that exploits the existence of multiple channels and the ability of transceivers to switch between them quickly to increase network throughput. In order to provide conflict-free communication for data packets, TFMAC divides time into a fixed number of time slots and allows each node to use different frequencies within different time slots to send data packets to its neighbors. The slot assignment is accomplished in a distributed way through exchange of a limited number of control messages during the contention slot at the beginning of each time frame. Trifonov et al. [23] proposed a hybrid CDMA/FDMA scheme for 3G mobile communications. The proposed scheme uses a novel adaptive coding approach. In this method, the transmission bandwidth is subdivided into a number of subbands, each allocated to a group of users (FDMA), which transmit in a CDMA fashion. The proposed method exploits an efficient adaptive subband allocation (ASBA) approach. The ASBA has been shown to provide a significant gain in the uncoded system performance as compared with the usual fixed frequency mapping, based on the interleaving of the carriers assigned to different user groups.

As noted earlier, the FCA problem models the task of assigning radio spectrum to a set of hosts on a permanent basis. The formulation of this problem as a combinatorial optimization problem in the beginning of the 1980s led a number of computer scientists and operations research scientists to try and find optimal solutions. Heuristic methods are capable of finding near-optimal solutions in a reasonable computational cost to such a time-consuming problem. An overview of the most basic heuristics for fixed-channel assignment in the literature is the subject of the study in the remaining of this section. Li et al. [33] proposed an evolutionary-dynamic TDMA slot assignment protocol for ad hoc networks. In this slot management method, the frame length and transmission schedule are dynamically updated according to the topology density of network and bandwidth requirement. This protocol allows the transmitter to reserve one or more unscheduled slots from the set of unassigned slots. San Jose-Revuelta [34] proposed a new genetic algorithm with good convergence properties and remarkable low computational load to solve the channel reuse problem. The proposed method tunes up the probabilities of mutation and crossover on the basis of the analysis of the individuals' fitness entropy. The proposed algorithm aims at obtaining a conflict-free channel assignment in such a way that the resulting bandwidth is close to the minimum channel span required for the whole network. Vidyarthi et al. [35] developed an evolutionary strategy that optimizes the channel assignment. The proposed evolutionary strategy approach uses an efficient problem representation as well as an appropriate fitness function. In this approach, a novel way of generating the initial population is proposed that generates a possibly better initial parent. The new method of generating the initial population reduces the number of channels reassignments and therefore yields a faster running time. Smith and Palaniswami [36] apply the neural networks for minimizing the co-channel interference in FCA approaches. They first introduce a new nonlinear integer programming representation to formulate the FCA problem. Then, they propose two different neural networks for solving this problem. The first is an improved Hopfield neural network that resolves the issues of infeasibility and poor solution quality which have plagued the reputation of the Hopfield network. The second approach is a new self-organizing neural network that is able to solve the FCA problem and many other practical optimization problems due to its generalizing ability. Perez et al. [37] proposed a fuzzy-based interference canceller for CDMA schemes. The proposed interference canceller exploits a fuzzy filter to remove the interference signal from the received one.

10.3.2 On-Demand Channel Assignment Schemes

The economic viability of a wireless service is often a strong function of how efficiently it uses the available spectrum [38]. Spectral efficiency, in turn, is greatly affected by the employed channel access method. As mentioned earlier, FCA protocols cannot efficiently allocate the available channels. ODCA protocols attempt to improve on the channel inefficiencies of FCA schemes by reassigning the unused channels to the users, if they need. In ODCA schemes, the channels are not preallocated to any user and dynamically assigned as the calls arrive. Therefore, ODCA is also called *DCA*. DCA schemes attempt to optimize the system performance by adapting to the traffic variations. Unlike the fixed assignment, in ODCA schemes all the channels can be used by all users as long as the co-channel constraints are satisfied. The benefit of dynamic assignment is the ability to switch an interface to any channel, thereby offering the potential to use many channels with few interfaces. Furthermore, in ODCA schemes, the MCA algorithms are collision-free. However, it should be noted that the ODCA algorithms are generally time-consuming and require more complex control. In the rest of this section, we review the well-known ODCA algorithms proposed in the literature.

10.3.2.1 Polling Scheme

The polling is an ODCA scheme in which a centralized controller queries the hosts, in a cyclic predetermined order, whether they have data to transmit or not. Due to the recent advances in communication systems, some other variations of the polling scheme have also been considered. These variations deal with noncyclic allocation policies, which include random, Markovian, or, more generally, nondeterministic allocation policies. In a polling scheme, controller polls (one by one) the hosts to give them an opportunity to access the medium. The hosts that have no packet to be transmitted (or do not need the channel access) decline, and the other hosts begin the packet transmission upon receiving the query. In polling scheme, the centralized controller is responsible for coordinating the transmissions, and so polling is a collision-free scheme. In this scheme, the entire bandwidth is available for the host that is permitted to transmit data. Although in realistic scenarios, traffic load of the different hosts is not the same, the major drawback of the basic polling scheme is to give the same importance (or equal access to the channel) to all hosts. A prioritized polling system may provide better results. Furthermore, the polling scheme suffers from the substantial overhead caused by the large number of messages generated by the controller to query the communicating hosts. As mentioned earlier, polling is based on a centralized control system. Therefore, in ad hoc networks, due to the lack of fixed infrastructures and centralized administrations, polling cannot be a practical channel assignment policy. Clustering the ad hoc networks in which the network is subdivided into several nonoverlapping groups is a promising approach to solve the above-mentioned problem. In clustered multihop ad hoc networks, the cluster head assumes the role of a centralized controller.

Polling systems have been extensively studied for the last three decades because of the applicability to the computer networks and communication systems. Grillo [39] provided a survey on applications of polling scheme in communication systems. Wang et al. [40] proposed an efficient distributed scheduling algorithm based on a prioritized polling policy for multihop wireless networks. The proposed algorithm maximizes the spatial and time reuse with an interference-based network model. Lye and Seah [41] also studied a priority-based random polling scheme. A QoS supportive adaptive polling scheme was proposed by Lagkas et al. [42] for wireless networks. In this scheme, an access point polls the wireless nodes in order to grant them permission to transmit. The polled node sends its data directly to the destination node. The proposed polling scheme is based on an adaptive algorithm by which the active nodes are polled with a higher probability. This results in a higher throughput and lower packet delays. Yang and Liu [43] also proposed a QoS support bandwidth polling scheme called *BBP*. In this scheme, to allocate a proper portion of bandwidth to each node, a coordinator defines a framing structure of time slots. A coordinator is allowed to poll a node more than once, and this causes it to allocate a proper number of slots (or a proper bandwidth portion) to each active node. However, in ad hoc networks, due to the lack of centralized coordination, the polling scheme has not received the attention it deserves. Dimitriadis and Pavlidou [44] proposed a polling access scheme for clustered multihop ad hoc networks called *two-hop polling* (2HP). 2HP is a revised version of the polling scheme tailored for the clustered environments. The authors claim that by this scheme it is possible to utilize intercluster links (distributed gateways) without adding much to the complexity of polling. 2HP changes the medium access by giving more liberty to the non-cluster-head hosts. In clustered networks, the hosts that belong to the different clusters must communicate through the cluster heads. This results in many potent links between the hosts not to be used. In the proposed scheme, the members of the neighboring clusters can directly communicate by the intercluster connections they have in between. Tseng and Chen [45] proposed a priority-based polling scheme with reservation for QoS guarantee in wireless ad hoc networks. The proposed scheme combines the priority-based and randomly addressed polling schemes to guarantee QoS constraints.

10.3.2.2 Reservation Method

The basic idea behind the reservation-based schemes is to set some time slots for carrying reservation messages. Figure 10.6 shows a reservation-based channel access scheme. As the name suggests, reservation methods require a controller device to reserve a communication channel prior to transmission. In this method, the time is subdivided into superframes and each superframe is then further divided into a reservation period and a data-transmission period. The reservation period is also divided into frames, with one frame assigned to each host in the network. During the reservation time frame, the host transmits a code word, indicating whether or not it has message traffic to send, and the number of data transmission slots it requires. All the other hosts do the same in turn. When the controller receives the reservation request, it computes a transmission schedule and announces the schedule to the hosts. At the end of the reservation period, all hosts know which hosts will be transmitting during the data-transmission period. This method avoids the collisions since each host sends only in its assigned time slot. This form of reservation-based methods is not

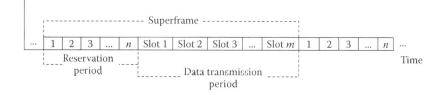

Figure 10.6   A reservation-based channel access scheme.

fair, since there are a finite number of available data-transmission slots and hosts request them in a preferred order. Therefore, the initial hosts (in predefined order) are always able to transmit, while the last ones find slots available only if the other hosts have left some for them. One way to solve this problem is to define an aging value for each host in its reservation code word, which indicates how long it has been waiting for transmission. In this method, each message is assigned a priority on the basis of its aging value. The messages with higher priorities are then selected for transmission during the data-transmission period.

Goodman et al. [11] presented a reservation method called *PRMA* that uses the concept of packet reservation. PRMA is a reservation protocol with features of TDMA and ALOHA. In PRMA, a star network is assumed. Time is divided into frames, each of which has many numbered slots. The network controller transmits an acknowledgment message at the end of each slot, which identifies that slot as being "reserved" or "unavailable." When a host has message traffic, it uses the ALOHA protocol to contend for an available slot. When the controller successfully receives the message, it replies with a "reserved" acknowledgment message, indicating receipt and indicating that the host has reserved that slot for future frames. The host now has an assigned slot in the frame (similar to TDMA), and so the transmission is collision-free in that slot in all future frames since the other neighboring hosts also detect the acknowledgment message. When the host has completed its traffic, the slot reservation is released by the simple expedient of not transmitting in it. The network controller then transmits an "available" acknowledgment message at the end of the slot.

Hou and Papavassiliou [46] presented a dynamic channel reservation scheme to provide the QoS guarantees in wireless mobile networks. Providing QoS is one of the most critical issues in MANETs. QoS is considered in routing, channel access and call admission control, resource reservation, and mobility management in MANET. Due to frequent topology changes and formation of dynamic connections, improvement of QoS through minimizing the interferences and collisions is a challenging issue in mobile ad hoc networking [47]. The proposed method in [46] is based on a concept called *influence curve*. The concept of influence curve is defined on the basis of the handoff probabilities [46] and directional factors [46]. The basic idea behind this scheme is that besides the requirements in the current cell, a mobile node exerts some influence on the channel allocation process within the neighboring cells. In [46], it is shown that due to the handoffs, traffics in the various cells are not necessarily independent of each other. That is, when a call enters a cell, not only it does need to consume a channel in the current cell, but the channels of the neighboring cells are probably also affected by this call. Such an influence is greatly related to the moving pattern of the node and can be calculated statistically. Through simulation experiments, the authors show that the proposed reservation scheme outperforms traditional channel reservation methods and can effectively adapt to the real-time network conditions.

Menache and Shimkin [48] presented a distributed channel access control mechanism in which a reservation scheme is applied to efficiently use the shared medium. The proposed technique adopts the virtual carrier sense mechanism of 802.11 (CSMA scheme will be discussed later in this chapter). That is, the channel access scheme proposed in [48] is a carrier sense reservation method. In this scheme, as a mobile station gets ready to transmit data, it first sends an RTS (request to send) control packet including the source, destination, and requested duration of the data transmission period. Then, the base station responds the RTS by sending back a CTS (clear to send) control packet with the same information as above.

10.3.2.3 Trunking

Trunking is a multiple access scheme that dynamically assigns the available logical channels to the communication requests. The earliest trunked systems were wired telephone systems, in which multiple lines between points were installed and calls were routed to a line with available capacity. This greatly increased the reliability of the network, since a single line outage would be unlikely to result in a loss of service, and also improved infrastructure economy, since any peaks in call volume could be rerouted over other, less busy, lines, and every line did not have to be designed for the peak call volume requested over that route. The first wireless systems to employ trunking methods (other than the microwave systems employed by the telephone system itself) were FDMA land-mobile radio (LMR) systems. These systems employed repeaters to provide communication links among mobile devices, normally organized into groups, and between mobile devices and wireline infrastructure such as telephone interconnect. Devices on LMR systems typically have very low average data throughput, but very high peak throughput—the worst-case scenario for channel efficiency using FDMA. To install an FDMA trunking system, the system operator amasses a collection of 5–40 FDMA frequency pairs (inbound/outbound) by using a repeater for each of them. One repeater is designated the control channel, while the others are used for message traffic. When not engaged in sending or receiving message traffic, network devices monitor the outbound control channel. When a device has a message for a particular group, it sends the request on the inbound control channel. The trunking system identifies an available repeater (channel) and transmits a command on the outbound control channel for all devices in the requested group to change to the available channel, where the requesting device transmits and the rest of the group receives [10]. Since a much larger number of users can be served with the existing spectrum allocation, this scheme greatly improves the economics of LMR. In the United States, FDMA trunking has since been expanded into the Associated Public Safety Communications Officials Project 25 Advanced Narrowband Digital Communications standard [49]. Trunking principles can also be applied to TDMA systems. The terrestrial trunked radio (formerly the trans-European trunked radio) (TETRA) standard [50] employs TDMA with four slots per frame. The control frame is the last frame in a series of 18 consecutive frames, called a TETRA "multiframe." Operation is analogous to FDMA trunked systems; mobile devices monitor the (outbound) control frame transmitted by the base station and are assigned communication resources, in the form of identified slots in identified frames, to communicate. TETRA is designed to transmit both voice and data as separate services; the use of TDMA engenders great flexibility in channel access for this purpose, since voice services can be assigned frequent, repetitive slots, while data transfers can be assigned larger blocks of time and interrupted for the more latency-critical voice transmissions.

Lima et al. [51] proposed two adaptive genetic algorithms called *GALC* and *GASC* for dynamic channel assignment in mobile communication systems. GALC locks in the channel assigned throughout the call holding time, and GASC switch the assigned call from a channel to another one during the call holding time. The proposed algorithms aim at minimizing the blocking probability of new calls and the dropping probability of handoff calls in channelized systems. To improve the efficiency and the convergence speed of algorithms, a number of mechanisms are added to the canonical genetic algorithm. These mechanisms are adaptive parameters, random immigrants, a greedy policy, a reservoir to assist the initial population, a truncation selection scheme, and a three-point crossover. Wang et al. [52] proposed a genetic channel assignment algorithm to minimize the effects of interferences where the number of available channels is substantially less than the minimum number of channels required for interference free assignment. The proposed channel assignment algorithm takes advantage of genetic algorithms to minimize the interference between the calls while demands for channels are satisfied. Vidyarthi et al. [35] also proposed an evolutionary channel assignment algorithm that exploits a new method for generating the initial population to reduce the number of channel reassignments.

Tseng et al. [53] considered the channel assignment problem in a MANET that has access to multiple channels. They proposed a new location-aware channel assignment algorithm called *GRID-B,* which exploits the concept of channel borrowing. Several channel borrowing strategies are proposed to dynamically assign the available channels to the mobile hosts to improve the channel reuse and to resolve the unbalanced traffic loads. The proposed protocols assign channels to mobile hosts based on the location information of mobile hosts that might be available from the positioning device. Wu and Yang [54] proposed a novel channel assignment scheme for improving the channel reuse efficiency. The authors believe that by overhearing the control packets of one-hop neighbors, a host can easily know the channel condition within the range of two-hop neighbors and so can select a suitable transmitting/receiving data channel to form the better reuse pattern. To enhance the probability of forming channel reuse pattern, they propose a back-off counter adjustment scheme such that a host with more channel information can transmit control packets earlier than those with less channel information. Gong et al. [55] presented three distributed channel assignment protocols for multichannel MANETs. They first proposed a new channel assignment protocol called *CA-AODV,* in which a channel assignment algorithm is combined with the AODV routing protocol. Then, they also presented two extensions to the CA-AODV protocol, namely, the enhanced 2-hop CA-AODV (E2-CA-AODV) protocol and the enhanced k-hop CA-AODV (Ek-CA-AODV) protocol. The proposed protocols combine channel assignment with distributed on-demand routing. In these protocols, the available channels are only assigned to the active nodes. They are shown to require fewer channels and exhibit lower communication, computation, and storage complexity, compared with existing approaches.

Wu et al. [56] proposed a protocol that assigns channels dynamically in an on-demand style. This protocol, called *DCA,* requires one dedicated channel for control messages and other channels are for data transmission. Each host has two transceivers so that it can listen on both the control channel and the traffic channel simultaneously. DCA follows an "on-demand" style to assign channels to mobile hosts and does not require clock synchronization. This kind of scheme does not perform well when the number of channels is large because all the negotiations are fulfilled on the control channel and too much contention will cause the saturation problem over the control channel. Sallent et al. [57] presented an integrated framework for heterogeneous wireless networks in order to design efficient dynamic and decentralized spectrum and radio management schemes. The proposed framework is defined as a layered model in which the radio resource allocation and spectrum management mechanisms are identified at both the intraoperator and interoperator levels. Such a framework needs to be strongly supported by the multiple radio access technologies and flexible spectrum capabilities. Therefore, it can be fully accomplished solely based on intelligent and cognitive radio as the new generation of radio technology [57]. Sallent et al. [57] propose an on-demand cognitive pilot channel called *CPC* to enable the radio for decision making and decentralized operation at the mobile terminal side. The authors show the superiority of the proposed on-demand CPC over the broadcast CPC in terms of the delay to retrieve the information.

10.3.3 Contention-Based Channel Assignment Schemes

As we have seen, ODCA protocols can improve the channel efficiency of FCA protocols. However, most ODCA schemes require a centralized coordinator to assign the channels. In many networks, for example, wireless ad hoc networks, such a controller does not exist. Multihop ad hoc networks, in which the network architecture (and even the order of the network) is not known *a priori*, are another difficult application for both FCA and ODCA schemes. To make matters worse, many types of multihop ad hoc networks generate traffic patterns that have a low average message rate, but a high peak rate, a difficult type of traffic pattern for a channel access protocol to support. The solution to this dilemma is the third class of channel assignment schemes: contention (random)-based channel assignment schemes. In these protocols, the hosts contend (compete) among each other for channel access; the hosts that lose access to the channel merely try again later. Contention-based channel access strategies require no coordination among the nodes accessing the channel. They do not exercise any control to determine which communicating node can access the medium next. In these protocols, the colliding nodes back off for a random duration of time before again attempting to access the channel. Furthermore, these strategies do not assign any predictable or scheduled time for any node to transmit. All backlogged nodes must contend to access the transmission medium. If only one neighbor tries its luck, the packet goes through the channel. If two or more neighbors try their luck, these have to compete with each other, and in unlucky cases, for example, due to hidden-terminal situations, a collision might occur, wasting energy of both transmitter and receiver. Collision occurs when more than one node attempts to transmit simultaneously. To deal with collisions, the protocol must include a mechanism to detect collisions and a scheme to schedule colliding packets for subsequent retransmissions. Contention-based protocols were first developed for long radio links and for satellite communications. The ALOHA protocol, also referred to as *pure ALOHA,* was one of the first such media access protocols. ALOHA simply allows nodes to transmit whenever they have data to transmit. Efforts to improve the performance of pure ALOHA lead to the development of several schemes, including slotted ALOHA, CSMA, carrier sense multiple access with collision detection (CSMA/CD), and carrier sense multiple access with collision avoidance (CSMA/CA). The following is an overview of the contention-based channel assignment schemes.

10.3.3.1 Random Access

ALOHA [12] is the simplest type of contention-based channel access schemes developed to regulate access to a shared transmission medium among uncoordinated contending hosts. The ALOHA communication system was part of a wireless time-sharing system used to connect a mainframe computer near Honolulu with remote users on other Hawaiian islands. Channel access in pure ALOHA is completely asynchronous and independent of the current activity on the transmission medium. A host is simply allowed to transmit data whenever it is ready to do so. Upon completing the data transmission, the communicating node listens for a period of time equal to the longest possible round-trip propagation time on the network. This is typically the time it takes for the signal to travel between the two most distant nodes in the network. If the node receives an acknowledgment for data transmitted before this period of time elapses, the transmission is considered successful. The acknowledgment is issued by the receiver after it determines the correctness of the data received by examining the error check sum. In the absence of an acknowledgment, however, the communicating host assumes that the data are lost due to errors caused by noise on the communication channel or because of collision and retransmits the data. If the number of transmission attempts exceeds a specified threshold, the host refrains from retransmitting the data and reports a fatal error. ALOHA is a simple protocol that requires no central control, thereby allowing nodes to be added and removed easily. Furthermore, under light-load conditions, hosts can gain access to the channel within short periods of time. The main drawback of ALOHA, however, is the sever network performance degradation as the number of collisions rapidly increases due to increased load. Assuming that message generation follows Poisson statistics, it can be shown that the ALOHA system becomes unstable (i.e., the number of retransmissions grows without bound) when the fraction of time the channel is utilized exceeds $1/2e \approx 0.184$. To improve the performance of pure ALOHA, slotted ALOHA [58] was proposed. As shown in Figure 10.7, in pure ALOHA protocol, each node starts transmitting the data upon its frame gets ready. Such a random access protocol may cause collision. Boxes indicate the frames, and dotted boxes show the frames which have collided.

Slotted ALOHA introduces the synchronized transmission time slots similar to TDMA. In this approach, nodes can transmit only at the beginning of a time slot. The introduction of time slots doubles the throughput as compared with the pure ALOHA scheme, with the cost of necessary time synchronization. In this scheme, all communication nodes are synchronized and all packets have the same length. Furthermore, the communication channel is divided into uniform time slots whose duration is equal to the transmission time of a data packet. Contrary to pure ALOHA, transmission can occur only at a slot boundary. Consequently, collision can occur only in the beginning of a slot, and colliding packets overlap totally in time. Limiting channel access to slot boundaries results in a significant decrease in the length of collision intervals, resulting in increased utilization of the underlying communication channel. Slotted or nonslotted, the ALOHA protocol is quite simple and is often used as a part of more complex medium-access methods such as PRMA proposed by Goodman et al. [11], which is a packet-reservation multiple access protocol with features of TDMA and ALOHA. Despite this performance improvement, however, pure and slotted ALOHA remain inefficient under moderate to heavy load conditions. In communication networks where the propagation delay is much shorter than the transmission time of a data packet, nodes can become aware almost immediately of an ongoing packet transmission. This observation led to the development of a new class of medium access schemes, whereby before a transmission is attempted, a host that has a packet to transmit first senses the carrier by listening to the channel. Carrier sensing forms the basis of the CSMA schemes. The CSMA-based schemes further reduce the possibility of packet collisions and improve the throughput. Figure 10.8 shows the slotted ALOHA protocol for a sample network including five nodes. Boxes show the frame, and dotted boxes indicate the frames that are collided.

CSMA [13] algorithms attempt to improve upon the relatively poor channel capacity of ALOHA by obliging the hosts to sense the channel for any ongoing activity prior to transmission.

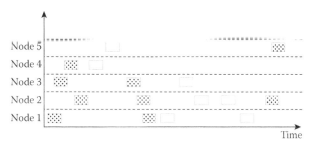

Figure 10.7   Pure ALOHA protocol.

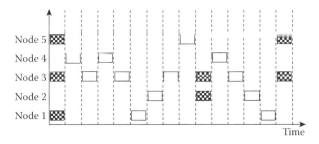

Figure 10.8 Slotted ALOHA protocol.

In a CSMA system, the hosts can be in three possible states: transmitting, idle, or listening. When the propagation delay is small compared to the packet transmission time, the throughput of the CSMA system is significantly better than that of ALOHA. CSMA protocols can be divided into two categories—nonpersistent CSMA and persistent CSMA—depending on the strategy used to acquire a free channel and the strategy used to wait for a busy channel to become free. In nonpersistent CSMA protocol, when a node becomes ready to transmit a packet, it first senses the carrier to determine whether another transmission is in progress. If the channel is idle, the node transmits its packet immediately and waits for an acknowledgment. In setting the acknowledgment timeout value, the node must take into account the round-trip propagation delay and the fact that the receiving node must also contend for the channel to transmit the acknowledgment. In the absence of an acknowledgment, before a timeout occurs, the sending node assumes that the data packet is lost due to collision or noise interference. The node schedules the packet for retransmission. If the channel is busy, the transmitting node "backs off" for a random period of time after which it senses the channel again. Depending on the status of the channel, the station transmits its packet if the channel is idle or enters the back-off mode if the channel is busy. This process is repeated until the data packet is transmitted successfully.

The nonpersistent CSMA protocol minimizes the interference between packet transmissions as it requires the hosts that find the channel busy to reschedule their transmissions randomly. The major drawback of the nonpersistent CSMA scheme, however, results from the fact that a channel may become idle during the back-off time of a contending host. The unnecessary waste of channel capacity can reduce significantly the overall network throughput. The need to address the shortcomings of nonpersistent CSMA led to the development of a class of *p*-persistent CSMA schemes. These schemes differ in the algorithm they use to acquire a free channel. The *1-persistent* scheme never allows the channel to remain idle if a node is ready to transmit. Based on this scheme, a node ready to transmit a data packet first senses the channel. If the channel is free, the node transmits its message immediately. If the channel is busy, however, the node persistently continues to listen until the channel becomes idle. Transmission is attempted immediately after the channel is sensed idle. The *p*-persistent algorithm represents a compromise between the nonpersistent and 1-persistent schemes. Based on this algorithm, a node that senses the channel idle transmits its packet with probability *p*. With probability $(1 - p)$, the node waits for a specific period of time before attempting to transmit the packet again. At the end of the waiting period, the node senses the channel again. If the channel is busy, the node continues to listen until the channel becomes idle. When the channel becomes idle, the node repeats the foregoing *p*-persistent channel acquisition algorithm. This process continues until the data packet is transmitted successfully. The optimal value of *p* for maximum throughput depends on the offered traffic rate. A drawback to all CSMA protocols is the so-called hidden terminal problem. Nonpersistent and persistent (1-persistent and *p*-persistent) CSMAs belong to the category of CSMA protocols with collision detection.

Pathmasuntharam et al. [59] identified a new type of exposed terminal problem, known as the *critically exposed terminal problem*, which causes severe throughput degradation in ad hoc networks. They proposed two possible methods: via scheduling and channel assignment to solve this problem with further elaboration of the latter scheme. They also analyzed the proper method of channel assignment and worked out the minimal channel assignment required to eliminate the critically exposed terminal problem. They suggest use of certain clustering approach to achieve the minimal assignment.

10.3.3.2 Collision Resolution

Busy tone is the first solution for the hidden-terminal problem in CSMA proposed by Tobagi and Kleinrock [60]. This solution rests on the realization that the hidden-terminal problem, and frame collisions in general, occurs at the receiving device while the CSMA algorithm is being performed at the transmitting device. The busy-tone solution requires each network device receiving a frame to simultaneously transmit a "busy tone" on another signaling channel, indicating that its receiver is busy. Devices desiring to transmit are required to check for the presence of busy tones prior to transmission. If present, they delay transmission since the channel (at the receiving device, where it matters) is busy.

Karn [61] proposed another influential single-channel solution to the hidden-terminal problem of CSMA, called *CSMA/CA*. CSMA/CA is also called *MACA*. The MACA protocol introduces the use of two control messages that can (in principle) solve the hidden and exposed terminal problems. The control messages are called *RTS* and *CTS*. The essence of the scheme is that when a node wishes to send a message, it issues an RTS packet to its intended recipient. If the recipient is able to receive the packet, it issues a CTS packet. When the sender receives the CTS, it begins to transmit the packet. When a nearby node hears an RTS addressed to another node, it inhibits its own transmission for a while, waiting for a CTS response. If a CTS is not heard, the node can begin its data transmission. If a CTS is received, regardless of whether or not an RTS is heard before, a node inhibits its own transmission for a sufficient time to allow the corresponding data communication to complete. Figure 10.9 illustrates the function of the CSMA/CA MAC layer protocol.

CSMA/CD is a modification of pure CSMA. CSMA/CD is used to improve the performance of CSMA by terminating the simultaneous transmissions upon detection of a collision. The CSMA/CD protocol is generally composed of a carrier sensing scheme and a collision detection algorithm. CSMA/CD can be described as follows: If some node has a data for transmitting, it initially assembles the data frame. Then, it checks to see if the channel is busy or not. If so, the transmitting node waits for a random time interval and tries again. Otherwise, if the channel is idle, it starts transmitting the data. During the transmission of the data frame, transmitting node checks for a collision. If no collision occurs, it continues sending the data until the frame is thoroughly transmitted. Upon detecting a collision, it terminates the current transmission and calls the collision recovery procedure. This procedure sends the jam signal to inform the other transmitting nodes that there has been a collision, waits for a random (back-off) period, and initiates a new transmission. In this mechanism, by waiting for a random period of time before retransmission, the probability of the second collision on retransmission significantly decreases. This is because the choice probability of the same random time by different nodes is negligible. The function of CSMA/CD is illustrated in Figure 10.10.

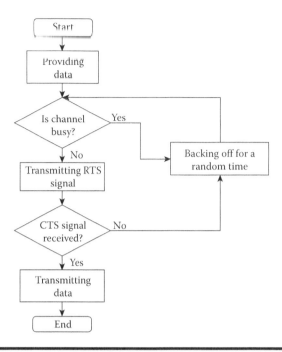

Figure 10.9 CSMA/CA MAC layer protocol.

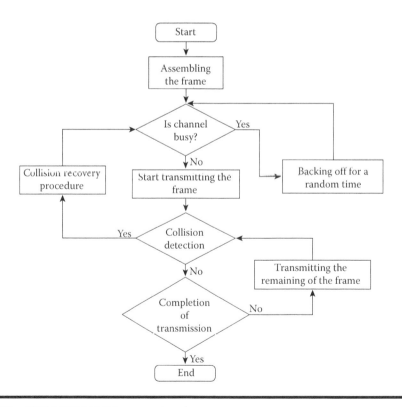

Figure 10.10 CSMA/CD MAC layer protocol.

MACA was very influential and led to many variants, including MACAW [62], floor acquisition multiple access (FAMA) [63], and the MAC methods used by wireless local area network standards such as IEEE 802.11 [64]. Most variants attempted to address identified weaknesses in MACA, such as the still-non-zero probability of frame collisions [65] and its back-off algorithm. MACA proposed the use of a simple binary exponential back-off, in which the back-off time is doubled after every collision and returned to the minimal value after a successful RTS/CTS exchange. This algorithm was shown [62] to be unfair in that over time one network device would "win" the channel and have a low back-off value (with frequent channel access), while all remaining network devices would have very large back-off values and be effectively frozen out of the network. MACA's back-off problem has been addressed in a number of ways; MACAW, for example, shares back-off values between network devices. The MACA and MACAW [61,62] protocols perform poorly since the time periods of RTS contentions can be very long. Several other protocols have been proposed that are based on RTS/CTS exchanges; they differ in the methods used to resolve the collisions of RTSs. FAMA proposed by Garcia-Luna-Aceves and Fullmer [66] is another MACA-based scheme that requires every transmitting station to acquire control of the floor (i.e., the wireless channel) before it actually sends any data packet. Unlike MACA or MACAW, FAMA requires that collision avoidance be performed at both the sender and receiver nodes. To "acquire the floor," the sending node sends out an RTS using either nonpersistent packet sensing or nonpersistent carrier sensing. The receiver responds with a CTS packet, which contains the address of the sending node. Any station overhearing this CTS packet knows about the station that has acquired the floor. The CTS packets are repeated long enough for the benefit of any hidden sender that did not register another sending node's RTS. The authors recommend the nonpersistent carrier sensing variant for ad hoc networks because it addresses the hidden-terminal problem effectively.

Nicopolitidis et al. [67] proposed a learning automata-based carrier-sense-assisted adaptive learning MAC protocol for wireless LANs, capable of operating efficiently in bursty traffic wireless networks with unreliable channel feedback. According to the proposed protocol, the mobile station that is granted the permission to transmit is selected by means of learning automata. At each station, the learning automaton takes into account the network feedback information in order to update the choice probability of each mobile station. The proposed protocol utilizes carrier sensing in order to reduce the collisions that are caused by different decisions at the various mobile stations due to the unreliable channel feedback.

10.4 Conclusions

Over the past few years, due to the rapid development of the ubiquitous computing techniques and proliferation of the mobile communication devices, the wireless mobile ad hoc networking has been experiencing exponential improvements. A MANET is a self-organizing and self-configuring multihop wireless network that can be instantly developed in situations where either a fixed infrastructure is unavailable or a fixed infrastructure is difficult to install. Because of many applications of ad hoc networks in personal area networking, military environments, civilian environments, medical and emergency operations, and so on, these networks received much attention from academic and industrial researchers during last years. Due to ease and speed of implementation and deployment of ad hoc networks in different situations as well as ease of joining the mobile hosts to these networks, the size of ad hoc network rapidly and incredibly grows. Therefore, the radio bandwidth is a scarce resource in wireless channels that must be economically assigned to the users. The channel assignment techniques are used to allocate a (minimum) required number of channels to the hosts in such a way that the interference effects are minimized and the available bandwidth can be effectively used. The conflict-free channel assignment problem can be modeled by the graph coloring problem of a special class of geometric graphs, which is an NP-hard problem. Therefore, the channel assignment problem is known to be NP-hard even if all the nodes are located in a plane and have the same radio transmission range. Two types of interferences might happen in a wireless ad hoc network. The former interference happens if a host decides to transmit and receive signals over the same channel simultaneously, and the second one takes place if a host receives more than one signal over the same channel at the same time. Since in wireless ad hoc networks the radio spectrum is a very limited resource, it must be profitably assigned to the radio nodes avoiding both interferences. Many studies have been carried out on channel assignment in wireless and cellular networks, while in ad hoc networks it has not received the attention it deserves. This chapter provided an in-depth survey of the most recent channel assignment schemes for wireless ad hoc networks. Depending on the nature of ad hoc networks, this chapter generally classified the channel assignment techniques into two main categories as contention-free and contention-based techniques. This chapter investigated and compared the well-known schemes of each category, with the emphasis on the key design issues, objectives, performances, and drawbacks. This chapter also discussed that because of bandwidth scarceness and rapid growth of using wireless mobile devices and ad hoc networks, cognitive radios and intelligent bandwidth allocation schemes are the promising approaches that are opening the new horizons to the researchers to design efficient radio channel assignment protocols.

References

1. J. Akbari Torkestani and M.R. Meybodi. A learning automata-based cognitive radio for clustered wireless ad-hoc networks. *Journal of Network and Systems Management*, vol. 19, 2011, pp. 278–297.
2. J. Akbari Torkestani and M.R. Meybodi. An efficient cluster-based CDMA/TDMA scheme for wireless mobile ad-hoc networks: a learning automata approach. *Journal of Network and Computer applications*, vol. 33, 2010, pp. 477–490.
3. A. Sen and M.L. Huson. A new model for scheduling packet radio networks. *ACM/Baltzer Journal Wireless Networks*, vol. 3, 1997, pp. 371–382.
4. C.-W. Yi. Maximum scan statistics and channel assignment problems in homogeneous wireless networks. *Theoretical Computer Science*, vol. 410, 2009, pp. 2223–2233.
5. A.M. Saleh. Inter-modulation analysis of FDMA satellite systems employing compensated and uncompensated TWTs. *IEEE Transactions on Communications*, vol. 30, no. 5, 1982, pp. 1233–1242.
6. W.C.Y. Lee. Overview of cellular CDMA. *IEEE Transactions on Vehicular Technology*, vol. 40, no. 2, 1991, pp. 291–302.
7. T. Sekimoto and J.G. Puente. A satellite time-division multiple-access experiment. *IEEE Transactions on Communications*, vol. 16, no. 4, 1968, pp. 581–588.
8. C.-M. Wu. Hybrid dynamic channel assignment in clustered wireless multihop CDMA/TDMA ad hoc networks. *Wireless Personal Communications*, vol. 42, 2007, pp. 85–105.
9. A. Capone, M. Gerla, and R. Kapoor. Efficient polling schemes for Bluetooth picocells. In *Proceeding of the IEEE International Conference on Communications (ICC2001), Finland*, vol. 7, 2001, pp. 1990–1994.
10. A. Chrapkowski and G. Grube. Mobile trunked radio system design and simulation. In *Proceedings of the IEEE Vehicular Technology Conference*, 1991, pp. 245–250.
11. D.J. Goodman, R.A. Valenzuela, K.T. Gayliard, and B. Ramamurthi. Packet reservation multiple-access for local wireless communications. *IEEE Transactions on Communications*, vol. 37, 1989, pp. 885–890.

12. N. Abramson. The ALOHA system—another alternative for computer communications. In *Proceedings of the AFIPS Fall Joint Computer Conference*, vol. 37, 1970, pp. 281–285.
13. L. Kleinrock and F.A. Tobagi. Packet switching in radio channels. Part I:Carrier sense multiple-access modes and their throughput-delay characteristics. *IEEE Transactions on Communications*, vol. 23, no. 12, 1975, pp. 1400–1416.
14. M. Wu. Dynamic frame length channel assignment in wireless multihop ad hoc networks. *Journal of Computer Communications*, vol. 30, 2007, pp. 3832–3840.
15. S. Kim and J.I. Kim. An adaptive time slot assignment algorithm for variable bandwidth switching systems. *Computers & Operations Research*, vol. 27, 2000, pp. 423–435.
16. R. Mo and Y.H. Chew. System throughput analysis of rate adaptive TDMA system supporting two class services. *Wireless Networks*, vol. 11, 2005, pp. 687–695.
17. L. Xin, D. Qionghia, and W. Qiu-feng. Time allocation scheme in IEEE 802.15.3 TDMA mechanism. *Journal of Zhejiang University Science A*, vol. 7, 2006, pp. 159–164.
18. T. Heikkinen and A. Hottinen. Distributed subchannel assignment in a two-hop network. *Computer Networks*, vol. 55, 2011, pp. 33–44.
19. X. Wang and J.J. Garcia-Luna-Aceves. Distributed joint channel assignment, routing and scheduling for wireless mesh networks. *Computer Communications*, vol. 31, 2008, pp. 1436–1446.
20. L.M. San Jose-Revuelta. A heuristic search technique for fixed frequency assignment in non-homogeneous demand systems. *Signal Processing*, vol. 88, 2008, pp. 1461–1476.
21. R. Huang, H. Zhai, C. Zhang, and Y. Fang. SAM-MAC: an efficient channel assignment scheme for multi-channel ad hoc networks. *Computer Networks*, vol. 52, 2008, pp. 1634–1646.
22. J.T. Wang. Throughput analysis for interference limited TDMA and TDMA/CDMA wireless ad hoc networks. *International Journal of Communication Systems*, vol. 22, 2009, pp. 365–372.
23. P. Trifonov, E. Costa, and A. Filippi. *Adaptive Coding in MC-CDMA/FDMA Systems with Adaptive Sub-Band Allocation*. XXXX: Kluwer Academic Publishers, 2004.
24. I. Horikawa and M. Hirono. A digital FDMA/TDMA microcell system for the next generation cordless telephones. In *Proceedings of the Fourth Nordic Seminar on Digital Mobile Radio Communications*, Oslo, 1990.
25. Y. Akaiwa and H. Andoh. Channel segregation—a self-organized dynamic channel allocation method: application to TDMA/FDMA microcellular system. *IEEE Journal of Selected Areas in Communications*, vol. 11, no. 6, 1993, pp. 949–954.
26. J. Perez-Romero, O. Sallent, and R. Agusti. On the optimum traffic allocation in heterogeneous CDMA/TDMA networks. *IEEE Transaction on Wireless Communications*, vol. 6, no. 9, 2007, pp. 3170–3174.
27. K. Navaie and H. Yanikomeroglu. Optimal downlink resource allocation for non-real time traffic in cellular CDMA/TDMA networks. *IEEE Communication Letters*, vol. 10, no. 4, 2006, pp. 278–280.
28. K. Navaie and H. Yanikomeroglu. Downlink joint base station assignment and packet scheduling algorithm for cellular CDMA/TDMA networks. In *IEEE International Conference on Communications*, 2006, pp. 4339–4344.
29. R. Vannithamby and E.S. Sousa. An optimum rate/power allocation scheme for downlink in hybrid CDMA/TDMA cellular system. In *52nd IEEE Conference on Vehicular Technology*, 2000, pp. 1734–1738.
30. M. Gerla and J. Tsai. Multicluster, mobile, multimedia radio network. *ACM/Baltzer Journal on Wireless Networks*, vol. 1, no. 3, 1995, pp. 255–265.
31. C.R. Richard Lin and M. Gerla. Adaptive clustering for mobile wireless networks. *IEEE Journal of Selected Areas in Communications*, vol. 15, no. 7, 1997, pp. 1265–1275.
32. M.D. Jovanovic and G.L. Djordjevic. TFMAC: multi-channel MAC protocol for wireless sensor networks. In *IEEE – Telsiks*, 2007, pp. 23–27.
33. W. Li, J.B. Wei, and S. Wang. An evolutionary-dynamic TDMA slot assignment protocol for ad hoc networks. *Wireless Communications and Networking Conference*, 2007, pp. 138–142.
34. L.M. San Jose-Revuelta. A new adaptive genetic algorithm for fixed channel assignment. *Information Sciences*, vol. 177, 2007, pp. 2655–2678.

35. G. Vidyarthi, A. Ngom, and I. Stojmenovic. A hybrid channel assignment approach using an efficient evolutionary strategy in wireless mobile networks. *IEEE Transactions on Vehicular Technology*, vol. 54, 2005, pp. 1887–1895.

36. K. Smith and M. Palaniswami. Static and dynamic channel assignment using neural networks. *IEEE Journal on Selected Areas in Communications*, vol. 15, no. 2, 1997, pp. 238–249.

37. A.I. Perez, J. Bas, and M.A. Lagunas. A neuro-fuzzy system for source location and tracking in wireless communications. In P. Stavroulakis (Ed.) *Neuro-Fuzzy and Fuzzy-Neural Applications in Telecommunications*. Springer, 2005, pp. 119–148.

38. S. Bandyopadhyay and E.J. Coyle. An energy efficient hierarchical clustering algorithm for wireless sensor networks. In *Proceedings of INFOCOM 2003*, San Francisco, 2003.

39. D. Grillo. Polling mechanism models in communication systems—some application examples. In H. Takagi, ed., *Stochastic Analysis of Computer and Communication Systems*. Amsterdam: Elsevier, 1990, pp. 659–698.

40. K. Wang, M.-G. Peng, and W.-B. Wang. Distributed scheduling based on polling policy with maximal spatial reuse in multi-hop WMNs. *The Journal of China Universities of Posts and Telecommunications*, vol. 14, 2007, pp. 22–27.

41. K. Lye and K. Seah. Random polling scheme with priority. *Electronics Letters*, vol. 28, 1992, pp. 1290–1291.

42. T.D. Lagkas, G.I. Papadimitriou, and A.S. Pomportsis. QAP: a QoS supportive adaptive polling protocol for wireless LANs. *Computer Communications*, vol. 29, 2006, pp. 618–633.

43. C. Yang and C.F. Liu. A bandwidth-based polling scheme for QoS support in Bluetooth. *Computer Communications*, vol. 27, 2004, pp. 1236–1247.

44. G. Dimitriadis and F.N. Pavlidou. Two-hop polling: an access scheme for clustered, multihop ad hoc networks. *International Journal of Wireless Information Networks*, vol. 10, no. 3, 2003, pp. 149–158.

45. C. Tseng and K.-C. Chen Priority polling with reservation wireless access protocol for multimedia ad hoc networks. In *Proceedings of Vehicular Technology Conference*, vol. 2, pp. 899–903, 2002.

46. J. Hou and S. Papavassiliou. A dynamic reservation-based call admission control algorithm for wireless networks using the concept of influence curve. *Telecommunication Systems*, vol. 22, no. 1–4, 2003, pp. 299–319.

47. K. Kim. A distributed channel assignment control for QoS support in mobile ad hoc networks. *Journal of Parallel and Distributed Computing*, vol. 71, 2011, pp. 335–342.

48. I. Menache and N. Shimkin. Reservation-based distributed medium access in wireless collision channels. In *Proceedings of the 3rd International Conference on Performance Evaluation Methodologies and Tools*, Institute for Computer Sciences, Social-Informatics and Telecommunications Engineering, Brussels, Belgium, 2008.

49. G.M. Stone and K. Bluitt. Advance digital communications system design considerations for law enforcement and internal security purposes. In *Proceedings of the IEEE 29th Annual International Carnahan Conference on Security Technology*, 1995, pp. 402–408.

50. European Telecommunication Standards Institute. Terrestrial trunked radio (TETRA), voice plus data (V + D). Part 2: Air Interface (AI). Document ETSI EN 300 392-2 V2.4.2 (2004-02). France: European Telecommunication Standards Institute, 2004.

51. M.A.C. Lima, A.F.R. Araújo, and A.C. Cesar. Adaptive genetic algorithms for dynamic channel assignment in mobile cellular communication systems. *IEEE Transaction on Vehicular Technology*, vol. 56, no. 5, 2007, pp. 2685–2696.

52. L. Wang, S. Arunkumaar, and W. Gu. Genetic algorithms for optimal channel assignment in mobile communications. In *Proceedings of the 9th International Conference on Neural Information Processing (ICONIP'02)*, vol. 3, 2002, pp. 1221–1225.

53. Y.C. Tseng, C.M. Chao, S.L. Wu, and J.P. Sheu. Dynamic channel allocation with location awareness for multi-hop mobile ad hoc networks. *Computer Communications*, vol. 25, 2002, pp. 676–688.

54. S.L. Wu and J.Y. Yang. A novel channel assignment scheme for improving channel reuse efficiency in multi-channel ad hoc wireless networks. *Computer Communications*, vol. 30, no. 17, 2007, pp. 3416–3424.

55. M.X. Gong, S.F. Midkiff, and S. Mao. On-demand routing and channel assignment in multi-channel mobile ad hoc networks. *Ad hoc Networks*, vol. 7, 2009, pp. 63–78.
56. S. Wu, C. Lin, Y. Tseng, and J. Sheu. A new multi-channel MAC protocol with on-demand channel assignment for multi-hop mobile ad hoc networks. In *Proceedings of ISPAN'00*, USA, 2000.
57. O. Sallent, R. Agustí, J. Pérez-Romero, and L. Giupponi. Decentralized spectrum and radio resource management enabled by an on-demand cognitive pilot channel. *Annual Telecommunication*, vol. 63, 2008, pp. 281–294.
58. L.G. Roberts. ALOHA packet system with and without slots and capture. *Computer Communications Review*, vol. 5, no. 2, 1978, pp. 28–42.
59. J.S. Pathmasuntharam, A. Das, and A.K. Gupta. *Channel Assignment for Nullifying the Critically Exposed Node Problem in Ad hoc Wireless Networks*. IEEE Conference on Sensor and Ad hoc Communications and Networks, Santa Clara, USA, 2004.
60. A. Tobagi and L. Kleinrock. Packet switching in radio channels. Part II: The hidden terminal problem in carrier sense multiple-access and the busy tone solution. *IEEE Transactions on Communications*, vol. COM-23, no. 12, 1975, pp. 1417–1433.
61. P. Karn. MACA—a new channel access method for packet radio. In *Proceedings of the ARRL/CRRL Amateur Radio 9th Computer Networking Conference*, 1990, pp. 134–140.
62. V. Bharghavan, A. Demers, S. Shenker, and L. Zhang. MACAW: a medium access protocol for wireless LANs. *ACM SIGCOMM Computer Communication Review*, vol. 24, no. 4, 1994, pp. 212–225.
63. L. Fullmer and J.J. Garcia-Luna-Aceves. Solutions to hidden terminal problems in wireless networks. *ACM SIGCOMM Computer Communications Review*, vol. 27, no. 4, 1997, pp. 39–49.
64. Institute of Electrical and Electronics Engineers, Inc. *IEEE Standard for Information Technology—Telecommunications and Information Exchange between Systems—Local and Metropolitan Area Networks—Specific Requirements. Part 11: Wireless LAN Medium Access Control (MAC) and Physical Layer (PHY) Specifications, IEEE Standard 802.11-1999 (ISO/IEC 8802-11: 1999)*. IEEE Press, 1999.
65. S. Singh and C.S. Raghavendra. PAMAS—power aware multi-access protocol with signalling for ad hoc networks. *ACM SIGCOMM Computer Communication Review*, vol. 28, no. 3, 1998, pp. 5–26.
66. J.J. Garcia-Luna-Aceves and C. Fullmer. Floor acquisition multiple access (FAMA) in single-channel wireless networks. *ACM Mobile Networks and Application (MONET)*, vol. 4, no. 3 (Special Issue on Ad hoc Networks), 1999, pp. 157–174.
67. P. Nicopolitidis, G.I. Papadimitriou, M.S. Obaidat, and A.S. Pomportsis. Carrier-sense-assisted adaptive learning MAC protocols for distributed wireless LANs. *International Journal of Communication Systems*, vol. 18, 2005, pp. 657–669.

Quality-of-service State Information-Based Solutions in Wireless Mobile Ad Hoc Networks: A Survey and a Proposal

Lyes Khoukhi, Ali El Masri, and Dominique Gaiti

Contents

Ubiquitous access to information anytime and anywhere will characterize whole new kinds of information systems in the future. This is being made possible by rapidly emerging communication systems that exploit wireless technologies. These systems have the potential to change how societies will evolve, as people are no longer constrained by information location or communication mechanisms. Wireless mobile ad hoc networks, also called *MANET*s, are a key component of these future wireless networks. MANETs are likely to expand their presence in future applications. The emergence of real-time applications and their widespread usage in communication have generated the need to provide quality-of-service (QoS) support in wireless and mobile networking environments. The classical QoS architectures proposed for wired networks are not readily applicable to the dynamic nature of ad hoc networks. Although challenging, it is quite interesting to design and develop QoS support mechanisms for MANETs. Two broad QoS approaches have been suggested by academic and industry experts: a *stateful approach*, requiring all nodes to create and maintain state information for each flow passing through them, and a *stateless approach*, where the nodes differentiate traffic according to the class they belong to without maintaining any state information. Both the approaches have their merits and drawbacks. To support more reliable and scalable communications, it is critical to reduce the states to be maintained by the network and make the routing not significantly impacted by topology changes. This chapter presents a novel hybrid stateless QoS model for multimedia services in wireless MANETs.

This chapter describes both stateful and stateless approaches and their advantages and disadvantages. This chapter proves that in a large network with the highest mobility scenarios, these approaches suffer from problems of scalability and false admission. Then, a hybrid stateless QoS model with service differentiation for wireless ad hoc networks is presented. The proposed model named *HybQoS* maintains some advantages and overcome some problems of both stateful and stateless approaches in wireless ad hoc networks.

11.1 Introduction

Wireless and mobile communication technologies are becoming an essential part of our daily lives. These technologies are expected to provide a wide variety of services, from high-quality voice to high-definition video through high-data-rate wireless channels anytime and anywhere in the world. Wireless ad hoc networks represent one trend in the future generation of wireless communications.

A MANET can be defined as self-organizing and a rapidly deployable network in which neither a wired backbone nor a centralized control exists (Figure 11.1). The mobile devices in MANETs are free to move randomly and organize themselves in arbitrary fashion. These features make them ideal for average users, for Internet service providers, and while reacting to emergency situations in which normal communication is impossible. MANETs can be used with success in disaster areas (e.g., flood, hurricane, and earthquake), military training grounds, schools, conference rooms,

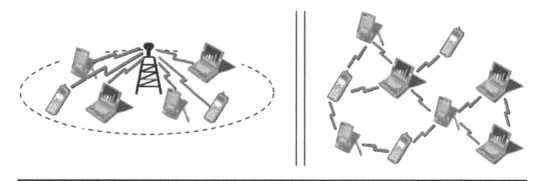

Figure 11.1   Infrastructure-based wireless network and ad hoc network.

hotels, airports, houses, and so on. MANETs are a new paradigm of wireless wearable devices enabling person-to-person, person-to-machine, or machine-to-person communications immediately and easily. This kind of network is the best alternative for developing countries and places where a communication infrastructure does not exist [1].

Ad hoc networks are likely to expand their presence in future applications. The emergence of real-time applications and their widespread usage in communication have generated the need to provide QoS support in wireless and mobile networking environments. The support of multimedia services over ad hoc networks is one of the hottest challenges facing today's industry and research community working on this area. The QoS support means a guarantee by the network to satisfy a set of predetermined service performance constraints for the user in terms of end-to-end delay statistics, available bandwidth, probability of packet loss, and so on. The cost of transport and total network throughput may be included as parameters [2]. The QoS architectures proposed for the Internet are not readily applicable to the dynamic nature of ad hoc networks. Although challenging, it is quite interesting to design and develop QoS support mechanisms for MANETs.

Most existing works on QoS in ad hoc networks have been carried out under the assumption that the underlying QoS architecture is reservation based. In such architecture, mobile nodes maintain per-flow *state information* (e.g., flow identity, and priorities) and source nodes use explicit reservation and control messages to indicate their QoS requirements. On the contrary, in the stateless approach, there is no flow or session state information maintained at intermediate nodes in support of end-to-end communications between the source and destination pairs.

The rest of this chapter is structured as follows. First, the environment of wireless MANETs is reviewed. Then, some stateful and stateless QoS solutions in ad hoc networks are presented, and their merits and drawbacks are discussed. Finally, a hybrid solution to support multimedia services is presented that maintains some advantages and overcomes some problems of both stateful and stateless approaches in wireless ad hoc networks.

11.2 Ad Hoc Environment and QoS Paradigm

MANET technology includes an autonomous collection of wireless mobile devices that communicate over bandwidth-constrained wireless links. Interactions between only such mobile devices are used to provide the necessary control and administration functions for such networks. The deployment of MANETs can be performed instantly and efficiently in situations where infrastructure is unavailable. This technology has several potentialities that are not available with classical

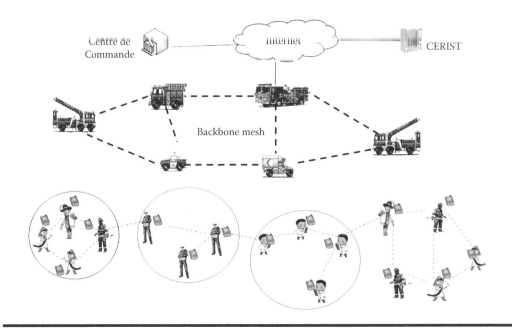

Figure 11.2   MANET for emergency/rescue operations.

wired or wireless networks technologies. Hence, MANET can be useful for different environments. Significant examples include establishing survivable and efficient communication for emergency/rescue operations (Figure 11.2), disaster relief efforts, and military networks (Figure 11.3). In recent years, home and small office networking and collaborative computing with laptop computers in small areas (e.g., conference rooms or classrooms) have emerged as other major areas of potential applications of MANETs. In such setups, some potential applications cannot rely on centralized and organized connectivity but can naturally be conceived as MANET applications [3,4].

The last few years have witnessed a wealth of research ideas on ad hoc networking that are moving rapidly toward implemented standards. Although ad hoc networks research is a relatively new field, it is gaining more popularity for various new applications. For instance, multimedia applications that open up for converged services and new applications is quickly becoming a key focus area for wireless communications. With the increase in both the bandwidth of wireless channels and the computing power of mobile devices, it is expected that video and audio services will be offered over ad hoc networks in the future [5,18,27].

However, enabling multimedia communications over such networks is remaining a challenging task for both academic and industrial communities. Video and audio services typically require stringent bandwidth and delay guarantees. This makes the deployment of QoS mechanisms a vital need for the satisfaction of users' requirements. These kinds of applications generate traffic at varying rates and usually require the network to be able to support such a changing rate. Besides, in a network consisting of mobile nodes, the connection path between a source and a destination is constantly broken and has to be frequently updated.

Therefore, providing QoS guarantees is crucial for supporting disparate services envisioned for the future wireless ad hoc networks. Despite the efforts made to alleviate this issue, there still exist a number of barriers to the widespread deployment of real-time applications. The most prominent one is how to ensure the quality delivery for real-time traffic in the case of MANET device

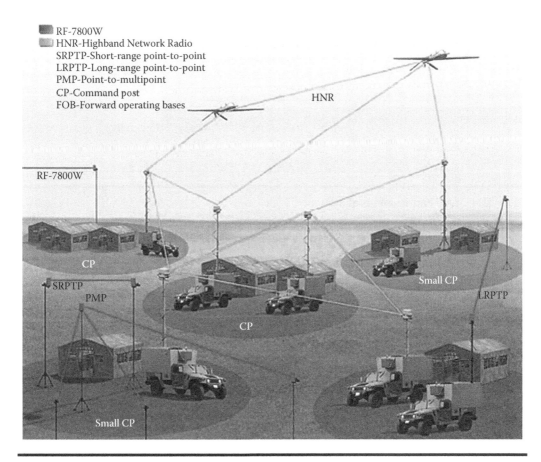

RF-7800W
HNR-Highband Network Radio
SRPTP-Short-range point-to-point
LRPTP-Long-range point-to-point
PMP-Point-to-multipoint
CP-Command post
FOB-Forward operating bases

Figure 11.3 Example of MANET application in military applications. (From Meyer, S., *Harris Corporation Technology Supports Streaming Video over Airborne Tactical Network during Demo for Australian Forces*, Harris Government Communications Systems, 2009. With Permission.)

mobility. It is important to note that the existing solutions developed for wired networks cannot be deployed directly within ad hoc networks. Difficulties with these solutions lie in the fact that they are not adapted to different node states and resource variation, as in ad hoc environments the available bandwidth for each node varies with time since the nodes are mobile and the medium is shared. In the rest of the chapter, some QoS concepts in MANET are discussed in more detail, focusing especially on the notion of "state information."

11.3 QoS in Ad Hoc Networks

Future generation of wireless networks will carry diverse kinds of multimedia applications characterized by their high exigency level of quality delivery (Figure 11.4 and Table 11.1). Then, the need arises in the wireless technology for QoS mechanisms to support real-time multimedia services. However, this is a challenging task in ad hoc architecture characterized by the mobility of nodes and the absence of infrastructure.

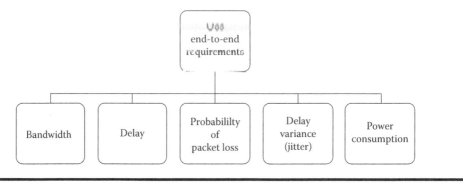

Figure 11.4   QoS end-to-end requirements.

Table 11.1   Applications QoS Requirements

Performance Dimensions				
			Sensitivity to	
Applications	Bandwidth	Delay	Jitter	Loss
VoIP	Low	High	High	Medium
Videoconference	High	High	High	Medium
Streaming video on-demand	High	Medium	Medium	Medium
Streaming audio	Low	Medium	Medium	Medium
Client/server transactions	Medium	Medium	Low	High
E-mail	Low	Low	Low	High
File transfer	Medium	Low	Low	High

11.3.1 QoS Challenges in MANETs

To be able to support the real-time voice and video communications over MANETs, novel QoS provisioning mechanisms need to be developed. However, there are many problems in providing QoS in MANETs in comparison with static wired networks due to many reasons [4–6,56]:

■ Interconnection between source-destination devices relies on peer wireless and mobile nodes that operate as routers. Then, rerouting among mobile nodes causes topology and network load conditions to change dynamically, which may complicate the support of real-time applications with an appropriate QoS.
■ Scarcity of resources in the ad hoc nodes requires that QoS mechanisms and signaling should not exhaust these resources.
■ Flows relying on a preestablished path and resource reservation along the path will suffer traffic interruptions due to frequent path changes.

- QoS requirements such as latency are time-variant and can considerably change in a short time due to the mobility of nodes and radio interferences.
- The dynamic nature of wireless ad hoc networks makes difficult the task of the dynamic assignment of a central controller to maintain connection state and reservations.
- The performance of most wired routing algorithms relies on the availability of precise state information. However, the mobility of nodes in an ad hoc network makes the available state information inherently imprecise.

It is clear that classical wired QoS models (e.g., IntServ, DiffServ) cannot entirely cope with the needs of QoS provisioning in the wireless ad hoc networks. The IntServ/Resources reSerVation protocol (RSVP) model is not suitable for ad hoc networks because of the following [7]:

- The necessity to maintain and establish the QoS information for each mobile node: this IntServ/RSVP feature requires a high capacity of processing and storage, which is undesirable for power- and resource-constrained ad hoc networks.
- RSVP is an "out-of-band" signalization protocol (i.e., RSVP signalization is not included in the data packets). This may cause a problem of competition between signalization and data packets in terms of resources.

On the contrary, DiffServ (Figure 11.5) seems to be an interesting model for the implementation in the internal routers (core routers) because various flows will be in an aggregate, which facilitate the routing inside the network. However, the dynamic topology of ad hoc networks disturbs the definition of the DiffServ architecture, in particular the notion of ingress, egress, and core routers in the network.

QoS support architecture in MANETs should have two main attributes [8,9]:

- Flexibility: This is useful for the heterogeneity of the physical and medium access control (MAC) layers, as well as multiple routing protocols.
- Efficiency: This is useful for the limited processing power and storage capabilities of nodes, as well as the scarce bandwidth available in such networks.

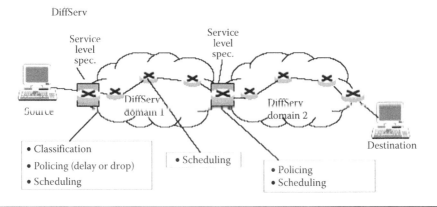

Figure 11.5   DiffServ network.

11.3.2 QoS Solutions

A survey of the existing works reveals that there are many subproblems that are often brought up when attempting to provide QoS in a wireless ad hoc environment. Some of the aspects that researchers tend to deal are as follows:

- QoS routing addresses the best and simple way to find a path through the network that is capable of supporting the requested level of QoS.
- The QoS model is a framework for achieving QoS over one or multiple layers.
- The MAC layer offers to the upper layers the mechanisms to ensure QoS, such as the service differentiation.
- Signalization-based models define control messages that deal with the changes in the resource availability of the network.

Other topics such as QoS maintenance and fairness issues are also interesting problems. The majority of the solutions proposed in the literature till now, regarding the aspects described above, can be classified in two broad approaches: stateful and stateless solutions.

11.4 Stateful and Stateless Approaches

The following is a summary of major QoS solutions proposed in the literature regarding the *state information* in MANETs.

11.4.1 Stateful Approach

This section considers stateful models and protocols that store information concerning the state of both nodes and traffic. A number of stateful QoS approaches have been discussed in the literature. Most of them are related to mechanisms under which paths are determined on the basis of some knowledge of resource availability in the network and the QoS requirements of the flows or connections. Figure 11.6 illustrates the relationship between the functional blocks typically involved in such a stateful approach. Hereafter, we proceed to detail some relevant works on each of QoS subareas as presented in Figure 11.7.

11.4.1.1 QoS Routing

This is probably the most active of the QoS for ad hoc network subareas. The goals of QoS routing are twofold: selecting paths that can satisfy given QoS requirements of arriving communication requests, and achieving global efficiency in resource utilization [7]. A brief description of some works is presented in what follows. Sinha et al. [10] proposed a core-extraction distributed ad hoc routing (CEDAR) algorithm that can identify a group of nodes called *the core of the network*, which can help in proving routes to applications with minimum bandwidth requirements. Two methods are used: the first one aims to maintain an approximate dominating set called the *core* to reduce the routing complexity. The second method uses an algorithm that performs QoS routing by local propagation of state control messages. CEDAR algorithm includes three key components: (1) core extraction, (2) link state propagation, and (3) route computation. CEDAR algorithm is designed for small- to medium-sized networks with tens to hundreds of nodes. Liao et al. [11] proposed a distributed (i.e.,

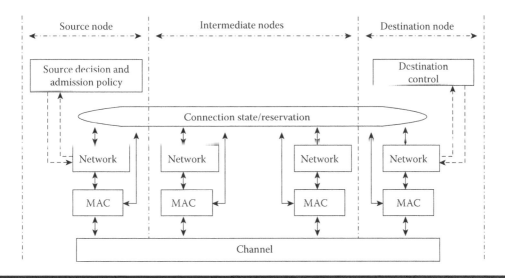

Figure 11.6   Functional diagram of a stateful approach.

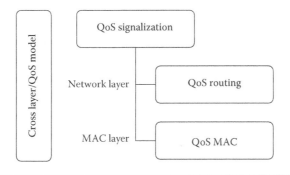

Figure 11.7   Stateful approaches classification.

hop-by-hop) multipath QoS routing scheme for MANETs called *ticket-based probing*. This protocol selects a network path with sufficient resources to satisfy a certain delay (or bandwidth) requirement. Similar to CEDAR algorithm, ticket-based probing does not use a flooding-based route discovery technique. Instead of randomly selecting many potential routes to search for acceptable one, ticket-based probing attempts to search only the best possible routes [12]. Imprecise state information can be tolerated, and multiple paths are searched simultaneously to find the most feasible path.

Adaptive distance vector (ADV) [13] is a stateful distance vector routing algorithm that exhibits some on-demand characteristics by varying the frequency and the size of the routing updates in response to the network load and mobility conditions. The authors have shown that ADV outperforms ad hoc on-demand distance vector (AODV) [14] and dynamic source routing (DSR) [15] by having significantly higher peak throughput and lower delay even at higher node mobility. On the basis of the same idea, Das et al. [16] proposed a protocol that focuses more on traffic load issues with the idea of combating congestion. The authors developed a mechanism for adaptive computation of multiple paths in order to transmit a large volume of data packets from a source to a destination. Two features are considered: the first one is to perform preemptive route rediscoveries before the occurrence of route errors while transmitting a large volume of data. This aspect be able to find out dynamically a series of multiple paths in temporal domain to complete the data transfer. The second aspect is to select multiple paths in spatial domain for data transfer at any instant of time and to distribute the data packets in sequential blocks over those paths in order to reduce congestion and end-to-end delay [16].

Lin and Liu [17] proposed a protocol for QoS support in a multihop environment based on the destination sequenced distance vector (DSDV) routing scheme. This protocol provides QoS support via separate stateful bandwidth computation and allocation mechanisms. The proposed bandwidth scheme depends on the use of a code division multiple access (CDMA) over a time division multiple access (TDMA) medium access scheme in which the wireless channel is time-slotted, the transmission scale is organized as frames (each containing a fixed number of time slots), and a global clock or time-synchronization mechanism is utilized [12]. That is, the entire network is synchronized on a frame-and-slot basis. This protocol can be applied to two important scenarios: multimedia ad hoc wireless networks and multihop extension wireless Asynchronous Transfer Mode (ATM) networks. The fast rerouting at the path's failure is ensured by maintaining a secondary path. When the primary path fails, the secondary route is used as the primary route and then another secondary route will be discovered.

The proposed work in [7] focuses on providing delay-constrained routes for data sessions. The key features of this protocol are as follows. First, a stateful proactive distance vector algorithm is used to establish and maintain routing tables containing the distance and next hop along the shortest path to each destination node. When a delay-constrained path is required, this information is used to send a probe to the destination along the shortest path to test its suitability. If this path satisfies the maximum delay constraint, the destination returns an Acknowledgment (ACK) packet to the source, which reserves resources. For this purpose, a resource reserving MAC protocol is assumed [3].

A QoS routing protocol based on optimized link state routing (OLSR) [19] was presented in [20]. OLSR is a proactive protocol in which information about 1-hop and 2-hop neighbors is maintained in each node's routing table. This information is disseminated via periodically broadcasted Hello messages. OLSR minimizes the control overhead involved in flooding routing information by employing only a subset of nodes, termed *multipoint relays* (*MPRs*), to rebroadcast it. As a consequence, only MPRs are discovered during route discovery, and thus, only they are used as intermediate nodes on routes. Also, calculating the optimal MPR set to reach all 2-hop neighbors is an NP-complete problem; and therefore, heuristics are applied. QoS Optimized Link State Routing (QOLSR) appears to be a promising proactive QoS routing protocol for finding and maintaining the shortest-widest paths in terms of delay and throughput. While QOLSR does not rely on the use of lower-layer information directly, it does require notifications to be sent by the MAC protocol in order to calculate QoS metrics [3].

11.4.1.2 MAC Layer

It is important to note that the ability to provide QoS also depends on how well the resources are managed at the MAC layer. Most of the proposed MAC QoS solutions use either a contention-free, scheduled access control such as TDMA or a contention-based MAC. TDMA MAC protocols with QoS support include GSM [21] for cellular telephony, satellite communications [2,31], and Bluetooth in personal area networking [23]. Wang et al. [24] proposed a stateful token-based scheduling scheme for a fully connected wireless local area network (WLAN) that supports both voice and data traffic. The proposed scheme can provide guaranteed priority access to voice traffic and, at the same time, provide more quantitative service differentiation for data traffic, which provides flexibility to the network service provider for service class management. Compared with a contention-based scheme and a centralized polling scheme, the proposed scheme improves the channel utilization by avoiding collisions (in the contention-based scheme) and the polling overhead (in the polling scheme).

In [25,26], two similar stateful proactive protocols (called *Coop-MAC* and *Relay-Enabled Medium Access (rDCF)*) based on the IEEE 802.11 DCF are proposed to mitigate the throughput bottleneck caused by low-data-rate nodes. A high-rate node is allowed to help a low-rate node through 2-hop transmission. With joint routing and cooperation, a cross-layer approach is introduced in [6]. Clusters of nodes near each transmitter form virtual multiple-input single-output link to a receiver on the routing table and as far as possible to the transmitter. Space–time codes are utilized to support transmission over a long distance, thus reducing the number of transmission hops and improving the communication reliability.

Most recent QoS solutions for contention-based MAC are extensions to IEEE 802.11 [28]. The IEEE 802.11e standard [29] attempts to propose QoS support for WLANs with a new enhanced distributed coordination function and a hybrid coordination function polling scheme. The authors in [31] suggested a stateful QoS extension to IEEE 802.11 to include a distributed priority scheduling technique and a multihop coordination scheme that enables downstream nodes to adjust the priority of packets in transit to compensate for upstream delays. The work presented in [30] proposed modification of the collision avoidance algorithm of IEEE 802.11 by implementing a sophisticated bandwidth allocation mechanism, which supports both Variable Bit Rate (VBR) and Constant Bit Rate (CBR) traffic.

While the standard IEEE 802.11 provides only best-effort service and makes no provisions for QoS support, other works have attempted to provide service differentiation at the MAC layer by manipulating the contention window [33] associated with the backoff algorithm. The modification of backoff algorithm proposed in [34] is capable of producing several service classes. To demonstrate the performances of this approach, the authors used three service classes, each with different channel access priorities.

Lin and Gerla [35] described a stateful periodical packet reservation technique *MACA/PR* (multiple access collision avoidance with piggyback reservations), which uses Carrier-Sense Multiple Access/Collision Avoidance (CSMA/CA) as the MAC layer and a bandwidth reservation technique at the network layer. MACA/PR combines between RSVP and QoS routing algorithm to provide end-to-end (i.e., source to destination) QoS capabilities in ad hoc networks. Collectively, these components are called the MACA/PR architecture and are based on CSMA/CA and TDM (e.g., time-slotted bandwidth reservations). All nodes maintain a reservation table to check when and who is transmitting. The basic access scheme in MACA/PR for non-real-time packets requires Request to Send/Clear to Send (RTS/CTS) dialogue followed by the transmission of the data packet. Upon reception of data packet by the destination, an ACK is sent to the source, providing a timeout mechanism for fast recovery in case of packet collisions. For real-time traffic, real-time scheduling information is carried (i.e., piggybacked) in the headers of data packets and ACK messages [12].

11.4.1.3 QoS Models

Some researches have presented mechanisms that enable QoS support autonomous of the routing protocols. Usually, a QoS model does not rely on specific protocol or layer. Instead, it defines a framework by which some services (e.g., per-flow or class-based) can be offered over one or multiple layers in the network.

FQMM (flexible quality-of-service model for mobile ad hoc network) [36] was the first QoS model proposed for wireless ad hoc networks. This model combines the advantages of both solutions implemented on the Internet: IntServ and DiffServ. FQMM tries to preserve the per-flow granularity for a small portion of traffic in MANETs, given that a large amount of the traffic belongs to per-aggregate of flows, i.e., per-class granularity. FQMM defines three types of nodes as in DiffServ [36] (Figure 11.8): *ingress node* (mobile node that sends data), *interior nodes* (nodes that forward data and control messages), and *egress node* (destination node). Since nodes are free to move that result in topology changes, a single host may have multiple roles. An adaptive traffic conditioner is placed at the ingress nodes where the traffic originates. It includes several components: a traffic profile, meter, marker, and dropper. The traffic conditioner is responsible for marking the traffic streams and discarding packets according to the traffic profile.

FQMM offers a good solution for small- and medium-sized ad hoc network with fewer than 50 nodes, but it is not suitable for large networks, which means that it suffers from the problem of scalability. Due to bandwidth limitation, FQMM tries to preserve the per-flow granularity for a small portion of traffic types in MANETs, given that these form a small percentage of the total traffic load [12]. Since the states of per-flow granularity come from only small portion of the traffic, the scalability problem as in IntServ is expected to be significant. On the contrary, this model does not consider large mobility scenarios.

Convenient stateful frameworks based on the cluster approach have been developed [37–40] to satisfy the requirements of efficient network resource control and multimedia traffic support. These works considered important features such as channel access, bandwidth allocation, virtual circuit support, and power control. In the clustering approach, all nodes are grouped into clusters. A *cluster* is a subset of nodes delegated to act as local coordinators to resolve channel scheduling; to maintain time division frame synchronization, channel access, routing, and bandwidth allocation; to enhance the spatial reuse of time slots and codes (among clusters); and to perform power measurement/control. A node that can hear two or more cluster heads represents a gateway. Within a cluster, time-division scheduling is enforced [41].

In [42], we proposed IntelliGent Quality Of Service model (GQOS), an intelligent stateful QoS model with service differentiation based on neural networks in MANETs. GQOS aims to satisfy some QoS requirements, especially the reduction of end-to-end delay, in networks whose topologies change at a low to medium rate. GQOS is composed of a kernel plan, which ensures basic functions of routing and QoS support control, and an intelligent learning plan, which ensures the training of GQOS kernel operations by using multilayered feedforward neural network (MFNN). The advantages of using a neural network algorithm in GQOS are the fast learning of different operations performed by the kernel and the reduction of time processing in the network. However, the performances obtained by GQOS are not very promising; the gain achieved in terms of delay via the stateful complex training process accomplished by the MFNN is only 10% as compared with that accomplished by IEEE 802.11.

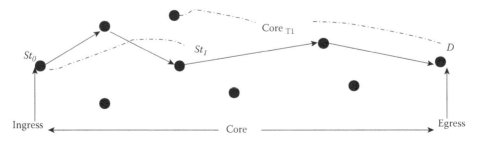

Figure 11.8 FQMM model.

11.4.1.4 QoS Signalization

Signaling is used to negotiate, reserve, maintain, and free up resources and is one of the challenging aspects of ad hoc network because it should be performed reliably with minimum overhead even when the topology changes. The two commonly used approaches are "out-of-band" and "in-band" signaling. *In-band signaling* refers to the fact that the control information is encapsulated into the data packets, making the signaling approach easy and "lightweight." In contrast, explicit control packets are used in the out-of-band signaling approaches. The latter are characterized as "heavyweight" because additional information in the network can consume more bandwidth [43].

INSIGNIA (in-band signaling support for QoS in mobile ad hoc network) [44] is one of the noteworthy QoS frameworks with per-flow granularity and reasonable treatment for mobility. The main goal of INSIGNIA is to provide adaptive QoS guarantees for real-time traffic. INSIGNIA supports fast reservation, restoration, and adaptation algorithms. For that aim, INSIGNIA employs an in-band approach by encapsulating some control signals in the Internet Protocol (IP) option of every data packet, which is called INSIGNIA option. Similar to RSVP, INSIGNIA allows a management of traffic in per-flow fashion. The flow state information is maintained for real-time traffic on an end-to-end basis, informing the source nodes about the status of their flows.

INSIGNIA is composed of a set of modules: the flow state information and its periodic update are managed by a soft-state module. In coordination with the module of admission control, INSIGNIA module performs the allocation of bandwidth to flows when the required resource can be satisfied, otherwise the packets will be degraded to the best-effort service. INSIGNIA is only a signalization protocol; therefore, it is necessary to cooperate with other protocols, in particular a routing protocol (e.g., AODV and DSR). The latter detects the change in node positions and ensures the update of routing tables in each node.

Two QoS levels are considered in INSIGNIA: *base QoS* and *enhanced QoS*. A flow in the network carries MIN/MAX bandwidth requests in the packet headers. At a bottleneck (i.e., node can support only MIN or best-effort QoS), all the nodes preceding the bottleneck will adjust their reservation to no more than the bottleneck's QoS. The source node sends either *base QoS* or *enhanced QoS* traffic when it receives a QoS report from the receiver, indicating the total bandwidth reserved along the path [6].

INSIGNIA is an adaptive in-band signaling protocol that respects the constraint of fast reestablishment of QoS reservation for the new route appeared in the network after a change in topology due to the network mobility. However, INSIGNIA requires the maintaining of per-flow state information at each mobile node, which can be seen as compromising the scalability of a whole network. On the contrary, the bandwidth is the only QoS parameter used in INSIGNIA.

11.4.2 Stateless Approach

This section deals with stateless approach; the models developed within category do not store any information regarding data sessions at intermediate nodes (Figure 11.9). A number of stateless works found in the literature have proposed techniques that build on a combination of well-established algorithms to provide efficient stateless distributed traffic control in wireless ad hoc networks. For instance, additive increase multiplicative decrease (AIMD) [17], fair queuing [45], and explicit congestion notification (ECN) [46] have proven to be efficient components to achieve a distributed traffic control. AIMD algorithm is implemented to control the sending rates of competing sources. While many of the proposals can provide some level of QoS support, they are

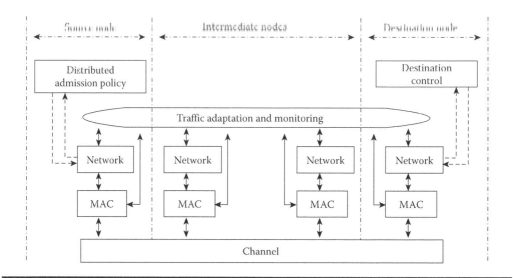

Figure 11.9 Functional diagram of a stateless approach.

based on a set of architectural assumptions where all nodes in the network implement a certain set of end-to-end control algorithms or require the support of QoS-capable MAC at each node along the path.

SWAN (service differentiation in stateless wireless ad hoc network) [47] is the best example of stateless approaches. SWAN proposes a service differentiation in wireless ad hoc networks by using distributed control algorithms. SWAN distinguishes between two traffic classes: real-time and best-effort. A classifier decides which type an incoming packet belongs to. In the case of real-time data packet, it is processed as soon as possible. Otherwise, the packet is considered as low priority and should wait until all real-time packets are dispatched. A shaper limits the relay of low-priority traffic to reduce the contention between neighboring stations [47].

SWAN cooperates with almost all routing protocols. When a source station wants to send a real-time traffic to another station, it probes the path to the destination station to identify the bandwidth available for real-time traffic. The source station is responsible only for admitting or denying its own session. If the available bandwidth is sufficient, the source station launches the session, else the session is refused. Intermediate stations along the routing path do not charge of evaluating the session admission, and they do not keep any per-flow information and thus avoid complex signaling and state control mechanisms. SWAN relies on feedback information received from the MAC layer as a measure of congestion in the network by using mechanisms of rate control and source-based admission control. The AIMD rate control algorithm is used at each node in order to perform the control of the best-effort traffic. The rate control restricts the bandwidth usage of best-effort traffic so that real-time applications can exploit the required bandwidth. The bandwidth not used by real-time applications can be exploited by the best-effort traffic. On the contrary, SWAN uses sender-based admission control in order to perform the real-time traffic control [48].

One of the drawbacks of SWAN is how to calculate the threshold rate, limiting any excessive delay that might be experienced. SWAN adopts engineering techniques that attempt to set the admission threshold rate at mobile nodes to operate under the saturation level of the wireless channel [49]. Also, it uses merely two levels of services: real-time and best-effort traffic. However,

SWAN remains the best example of stateless distributed QoS framework developed for wireless ad hoc networks.

Chen and Kassler [50] proposed E-SWAN, an extension to SWAN that facilitates the QoS-enabled integration of ad hoc and infrastructure networks, providing stateless QoS support for communication between ad hoc nodes and Internet network's routers. E-SWAN assumes that the infrastructure and access networks are connected through an access router or gateway. The ad hoc path between an ad hoc node and the gateway is abstracted as a virtual link to the infrastructure network. Admission control is performed at the gateway with collaboration of the MANET nodes, as in SWAN it takes into account the admission control for the core network. More specifically, an ad hoc node wishing to establish a real-time session through the infrastructure network gathers information about the availability of resources in the virtual link to the gateway by using a normal SWAN probing process. However, the authors did not study the impact of nodes mobility on the performance in terms of end-to-end delay and throughput. In addition, E-SWAN did not support multiple gateways that allow bringing additional benefit of load balancing and enable traffic engineering as well as multiple paths between the fixed network and the MANET extensions.

Domingo and Remondo [51] proposed DS-SWAN (differentiated services–stateless wireless ad hoc network) that enable ad hoc networks connected to fixed DiffServ domains to cooperate to ensure the QoS support. DS-SWAN warns nodes in the ad hoc network when congestion is excessive for the accurate functioning of real-time applications. These wireless nodes react by slowing down best-effort traffic. DS-SWAN uses sequence numbers in warning messages to prevent neighboring nodes from overreacting. The authors claim that DS-SWAN significantly improves the end-to-end delay for real-time applications without starvation of background traffic, but other QoS parameters such as jitter and throughput have not been deeply studied.

Xiang et al. [52] proposed a RSGM (robust and scalable geographic multicast), which can scale to a large group size and network size and provide multicast packet transmissions in a dynamic mobile ad hoc network environment. The protocol is designed to be simple; thus, it can operate more efficiently and reliably. RSGM includes several virtual architectures for more robust and scalable membership management and packet forwarding in the presence of high network dynamics due to unstable wireless channels and frequent node movements. Both the data packets and control messages are transmitted along tree-like paths; however, different from other tree-based protocols, there is no need to explicitly create and maintain a tree structure. RSGM makes use of position information to support reliable packet forwarding. Instead of addressing only a specific part of the problem, RSGM introduces a zone-based scheme to handle the group membership management and takes advantage of the membership management structure to efficiently track the locations of all the group members without resorting to any external location server [53].

A stateless distributed admission control for MANET environment (DACME) was presented in [54,55]. The DACME protocol assesses the achievable QoS on a given route by means of a set of back-to-back probe packets. In [55], techniques are explained for estimating route capacity, end-to-end delay, and delay jitter when DACME is operating with a single-path routing protocol. In [54], the emphasis is on throughput-constrained applications only, but a multipath routing protocol is considered. The authors state that the optimal operating environment for DACME is based on an Enhanced Distributed Channel Access (EDCA) MAC scheme, but it can also operate with a non-QoS-aware MAC [58]. The estimation of delay in DACME is determined to be half of a probe/probe reply's round-trip time. The source device explicitly notifies the destination of its packet sending rate, so that the expected interpacket interval can be calculated. A disadvantage of the DACME approach is that until the traffic is actually being carried on the two routes, there is no way to predict the effect of the interroute interference on the achievable QoS.

In [2], we proposed an integrated stateless cross-layer QoS protocol FuzzyQoS based on fuzzy logic for wireless MANETs. The choice of using fuzzy logic is justified by the fact that fuzzy logic is well adapted to systems characterized by imprecise states as in the case of ad hoc networks. The fuzzy approach aims to improve the control of traffic regulation rate and congestion control of multimedia applications. FuzzyQoS uses fuzzy thresholds to adapt the traffic transmission rate to the dynamic conditions. The performance evaluation has shown that our fuzzy stateless QoS model can achieve low and stable end-to-end delay and high throughput under different network conditions.

11.4.3 Discussion

Table 11.2 presents a comparison of the previously described QoS state-information protocols. Most existing stateful solutions (e.g., INSIGNIA, MACA/PR, FQMM, CEDAR, ADV, and GQOS) in ad hoc networks have been carried out under the assumption that the underlying QoS

Table 11.2 Classification of QoS Models Based on State Information

Protocol/ Algorithm	States Category		Performance			QoS Parameters		Layer		
	STAT	STLS	CO	SC	ST	DE	TH	NET	MAC	CL
ADV	X					X	X	X		
CEDAR	X		H	H	L	X	X	X		
INSIGNIA	X		H	L	L		X		X	
SWAN		X	L	H	H	X	X			X
CoopMAC	X		L	H	H	X			X	
RSGM		X	H	H	H	X	X			X
MACA/PR	X		H	L	H	X	X		X	
FQMM	X		L	L	H	X	X			X
E-SWAN		X	L	H	L	X	X			X
DSDV	X		H	L	H	X	X	X		
QOLSR	X		L	L	H	X		X		
GQOS	X		H	L	H	X	X			X
DS-SWAN		X	H	H	L		X			X
VMAC	X		L	L	H	X	X		X	
OLSR	X		L	H	L	X	X	X		
DACME		X		H						

STAT, stateful; STLS, stateless; CO, control overhead; NET, network layer; SC, scalability; MAC, MAC layer; ST, storage capability; CL, cross layer.

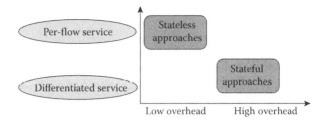

Figure 11.10   QoS state information-based approaches.

architecture is reservation-based. In such an assumption, mobile nodes maintain per-flow state information (e.g., flow identity, priorities, and the amount of allocated bandwidth) and source nodes use explicit control and reservation notifications (e.g., route reservation messages) to indicate their QoS requirements. The advantage of these state information-based models is that they allow the source nodes to guarantee the requested QoS on an end-to-end basis with a high degree of accuracy. However, this kind of model needs complex and signaling control mechanisms to refresh and update per-flow state information.

State information maintained at mobile devices, as proposed by the protocols described above, is difficult and problematic to manage in the face of devices' mobility and limits the scalability as the number of mobile device grows (Figure 11.10). This kind of approach may not be feasible for typical mobile devices, which are limited in terms of battery life and computational power.

In the stateless approach, there is no flow or session *state information* maintained at intermediate devices in support of end-to-end communications between source and destination pairs. Note that the state information related to data sessions is still stored at source nodes and sometimes also destination nodes. This approach has the advantage of high adaptability and scalability as shown earlier with SWAN, DACME, RSGM, DS-SWAN, E-SWAN, and others. This is largely because of the light processing achieved by the intermediate mobile devices. However, in a large network with highest mobility scenarios, the stateless-based models suffer from the problem of *illusory readings*. This means that source nodes may have an old view about the real state of network resources (i.e., bandwidth) because of the dynamic nature of ad hoc network. This problem may occur when multiple source nodes check the state of network simultaneously, which leads to the admission of more traffic than what an intermediate node can support. Since there is no centralized QoS controller in a MANET, multiple sessions in a network may send probe messages to check current bandwidth availability at the same time. This may result in a situation where multiple sessions (sharing common mobile devices) send probe messages simultaneously and observe that the QoS resources are available, but in reality they are not.

11.5 Proposed Hybrid Stateless Model

This section presents a hybrid QoS stateless model, named *HybQoS*, for service differentiation in wireless MANETs. The HybQoS model makes resource reservation in advance before the flow uses it unlike other models that make the resource exclusively reserved for the flow and no additional traffic is allowed to use the reserved resource. HybQoS aims essentially to minimize the end-to-end delay of multimedia traffic by using distributed control mechanisms to support real-time traffic and service differentiation delivery. For that purpose, an admission control with a

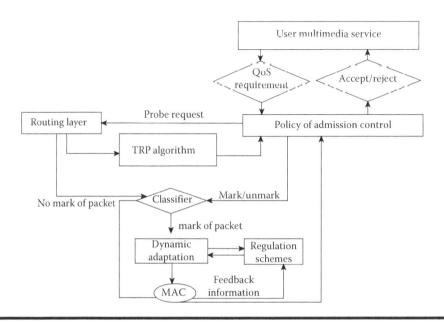

Figure 11.11 HybQoS model.

temporary reservation process (TRP) is used for User Datagram Protocol (UDP) real-time traffic, and two regulation schemes are proposed in order to ensure the control of best-effort traffic. The objective of the regulation techniques is to adjust dynamically the transmission of traffic according to the network conditions in order to ensure a good utilization of resources. The response to the fluctuations is performed by the source nodes, which adjust consequently the transmissions. HybQoS tries to take some advantages and overcome some problems of both stateless and stateful approaches. In order to refresh per-flow state information, the HybQoS model uses minimal information available on the network nodes without relying on complex mechanisms.

A schematic diagram of the HybQoS model is illustrated in Figure 11.11. It shows a number of mechanisms used to support both real-time and best-effort traffic and to adapt the traffic rate according to the dynamic changes of ad hoc network. HybQoS assumes that there will always be best-effort flows present that can be rapidly rate controlled by using the regulation schemes in an independent manner to yield the necessary low delays and stable throughput for real-time flows. The routing scheme and the TRP perform the discovery of routes and bandwidth reservation (more details in the next section). The admission controller efficiently estimates the local available bandwidth at each node. Many multimedia applications such as Voice over IP (VoIP) are delay-sensitive or bandwidth-sensitive applications. Hence, providing information about the real network state can be useful for the decision of acceptance of a new flow in the network. The decision to admit a new flow is done by the admission control mechanism. The classifier is able to differentiate between flows in terms of QoS requirements: best-effort flows and real-time flows. The packets classified are regulated by using the regulation schemes according to the application requirements and network conditions. As the case of stateless solutions, the HybQoS model does not require the support of a QoS-capable MAC to deliver service differentiation. Rather, real-time services are built by using existing best-effort wireless MAC technology. The regulation schemes dynamically adjust the traffic rate according to the feedback information that concerns the network state. More details about the mechanisms of HybQoS are given in the following sections.

11.5.1 Admission Control Based on TRP

The aim of the admission control mechanism is to determine whether a new flow should be admitted or not. This function is conducted together by the source node and other intermediate nodes. Note that the source node has a final decision to accept or reject the user QoS requirements based on the feedback information state received from the network. This feedback measurement represents the packet delay measured by the MAC layer, which is calculated by the difference between the time of receiving an ACK packet (from the next hop) and the time of sending a packet to the MAC layer (from the upper layer). The admission control also measures the rate (bits/s) of the real-time traffic by listening to the packet sent within the radio transmission range. The best-effort traffic is regulated by using the proposed regulation schemes. Hence, the cooperation between the admission control mechanism and the regulation technique ensures that the total traffic (i.e., best-effort and real-time traffic) is below a certain threshold rate that would cause congestion and trigger an excessive delay.

HybQoS attempts to use a multilevel admission control, unlike SWAN, which uses only a source-based admission control. In a mediate or large network with high mobility, the stateless source-based admission control may fail or get an old view about the real state of resources (i.e., bandwidth) because of the dynamic nature of ad hoc network. In addition, the problem of illusory readings, in which multiple source nodes read simultaneously the state of the network (via request/response probes), may lead to the admission of more traffic than what an intermediate node can support. This problem may be aggravated in the stateless approaches where intermediate nodes do not maintain the state information, and the admission control is conducted only at the source node in a decentralized manner. This problem may appear in stateless models when multiple source nodes simultaneously initiate the admission control process at the same moment and share common routing paths between source-destination pairs. The illusory readings may also appear during the period of network exploration (i.e., the time between sending the probe request and receiving the probe response at the source node). The source nodes may receive response to their request, indicating that resources are available when in fact they are not. Under the assumption that the available resources can satisfy the flow requirements, the source node may falsely admit the flow.

The following example illustrates the problem of illusory readings. Consider two voice flows at a rate of 32 Kbps and one video flow at a rate of 200 Kbps. If we consider that the available bandwidth at a common intermediate node is 220 Kbps, then only one single flow video could be supported. However, all flows are accepted in the case of illusory readings; this results in an aggregation of 264 Kbps at the common node. To overcome this problem, which may cause a local congestion and excessive delay for multimedia applications, some stateless approaches such as SWAN use AIMD rate control. However, this technique does not predict the illusory readings; rather it lets them occur and then performs the regulation; this solution may complicate the task of the rate controller in the case of a large number of flows. In HybQoS solution, we resolve this problem as follows.

The estimation of the end-to-end available bandwidth is performed by sending request from the source node toward the destination. For that purpose, a UDP control packet is exploited by using an additional field B_{bott} that initially contains the value of the requested bandwidth B_{req}. At each intermediate node, a comparison is performed between the value of B_{bott} and the available bandwidth B_{avai} of the current node. The value of the field B_{bott} is updated if it is greater than the value of B_{avai} of the current node. When the destination receives a UDP packet, the B_{bott} value represents the minimum bandwidth available along the path, and it is copied from UDP to a newly generated short message replay (SMR). The latter packet is transmitted back to the source node at the same time as the execution of the TRP.

The TRP mechanism exploits additional fields in the node's routing table to specify the "temporary reservation status" and "status duration" associated with intermediate nodes. The field "temporary reservation status" is set to 1 to indicate that the reservation of connection is active, and the "status duration" is set to a certain value TU. TU indicates the period of time within which the temporary reservation is performed; i.e., during this period, the intermediate node reserves temporarily the available resources to flows traversing the node. This is especially useful in the case of a large network in which the processes of path discovery or session recovery may take a long time. During the TU period, the available resources at an intermediate node, originally affected to a session CS, can be allocated to other sessions, for instance, during the interruption period of CS or during the path reservation process executed from destination to source nodes. Note that the reserved bandwidth should be released after the expiration of TU duration, i.e., when TU = 0, or at the reception of data packets belonging to the original flow belonging to CS (e.g., interrupted session and session concerned by the path discovery). The algorithms permitting the flow management within the TRP mechanism are presented later in this chapter.

Upon the reception of an SMR message by the source node, the latter compares the end-to-end available bandwidth and the required bandwidth for the new real-time session. Note that the probe request is exploited to carry, initially, the requested bandwidth and to perform the intermediate nodes control and then to start the temporary reservation. This permits, on the one hand, to overcome the illusory reading problem since the bandwidth state of an intermediate node is checked before the transmission of data packets and, on the other hand, to exploit efficiently the utilization of network resources by temporarily allocating the available bandwidth to other flows traversing the node. Note that in the existing stateful and stateless models, the available bandwidth in intermediate nodes is not exploited during TU.

The TRP mechanism is described in more detail in what follows. Let us consider $\Delta t = $ TU (i.e., the time given to a temporary reservation interval of flow in a given intermediate node). Other flows originated from other source nodes can use the available bandwidth during Δt. Let μ be the target satisfaction, which defines the desired percentage of packets to be sent within the QoS constraint, where $\mu = 1$ corresponds to the best QoS guaranty and $\mu = 0$ corresponds to the best-effort transmission. Then, (11.1) verifies the probability that Δt is less than a given time value δ and the flow request to be accepted:

$$P(\Delta t \leq \delta) \geq \mu \qquad (11.1)$$

The exact evaluation of (11.1) requires the destination to be acquainted with the statistical descriptions of delay of each node along the path. However, in many cases, the statistical distribution of such parameter can be approximated by a Gaussian distribution. Under this hypothesis and assuming independency among node statistics, the temporary reservation time among the nodes turns out to be a Gaussian variable. If we consider m_{T_r} and $\sigma_{T_r}^2$ the statistical average and variance of the random variable T_r, respectively (T_t is the temporary reservation time in a given node), then the temporary reservation interval statistics can be expressed as

$$P(\Delta t \leq \delta) = 1 - Q\left(\frac{\delta - m_{T_r}}{\sigma_{T_r}}\right) \qquad (11.2)$$

where Q represents the complementary distribution function of a Gaussian variable with mean 0 and variance 1.

Let us consider δ and μ to be, respectively, the maximum time and satisfaction target for an accepted flow request. Let υ be the actual time satisfaction provided by the intermediate node as given by (11.2). Hence, the flow request would be satisfied even the average temporary reservation time was increased to the value m_{T_r} given by

$$m_{T_r} = \delta - \sigma_{T_r} Q^{-1}(1 - \upsilon) \tag{11.3}$$

The violation of the TRP is detected at a given node if the temporary reservation time is greater than m_{T_r} (m_{T_r} is the time bound of the temporary reservation interval). Upon the detection of violation, the reserved resources are released in order to be used by other flows; this ensures a good utilization of resources.

11.5.2 Dynamic Traffic Adaptation

The periodic measurements of real-time traffic at an intermediate node permit to estimate the local available bandwidth. Each intermediate node can detect anomalies (i.e., buffer overload/congestion) by using these periodic measurements. In the HybQoS model, when a congestion is detected the congested node sends a "congestion notification message" (CNM) to the source node. Note that there is no need to mark the ECN bits in the IP header of the traffic packets such as in the case of stateless approaches. For instance, in SWAN, the destination node monitors the ECN bits and notifies the source node by using an additional message. Figure 11.12 shows the theoretical gain that can be obtained in comparison with SWAN.

Let us consider n to be the number of hops between the source node S and the destination D, k the number of hops between the congested nodes P and D, and t the transmission time between two hops. The time required by SWAN and HybQoS to notify S is given, respectively, by $T_{\text{SWAN}} = (2n - k)t$ and $T_{\text{HYBQoS}} = kt$. Hence, the gain T_{GAIN} of HybQoS is obtained as shown in

$$T_{\text{GAIN}} = T_{\text{SWAN}} - T_{\text{HYBQoS}} = (2n - k)t - kt = 2(n - k)t \tag{11.4}$$

The achieved gain in terms of delay permits to reduce the waiting time at source nodes to adjust the traffic rate after the detection of anomalies (e.g., congestion) and then to minimize the end-to-end delay. Upon the reception of the CNM packet by a source node, the latter tries to reestablish the session and adapt the traffic rate, taking into account both the minimum requirement of the original requested bandwidth and the new state of the connection, using one of the following proposed regulation schemes.

The regulation schemes aim to adjust dynamically the traffic transmission rate according to the network conditions in order to ensure a good utilization of resources. The source nodes respond to

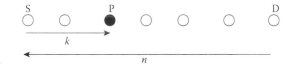

Figure 11.12   Notification message.

the fluctuations and adjust consequently the transmissions. Some issues considered by our regulation schemes are the capability of these schemes to accommodate multiple transmitters that may or may not differ in their connectivity to the network, the rate control in response to feedback, and the quality of received multimedia streams.

Here are presented in what follows two schemes to address the traffic regulation problem. Note that by adopting the feedback mechanism used in some stateless models, each mobile node increases its transmission rate gradually (this rate is noted in the simulation results by the "increasing rate" parameter) every T s until the packet delays become excessive. In the case of congestion, which is generally the main cause of any excess in packet delays, one of the following regulation schemes is performed by the source nodes.

11.5.2.1 Scheme 1: Flow-Based Adaptation

In this scheme, the intermediate node, which detects the congestion, sends a CNM packet to all active source nodes. CNM includes the traffic regulation rate and the moment of connection reestablishment (i.e., the convenience time to reactivate the current session). Some stateless models send a notification without specifying the rate of flow regulation and use a random starting time for a session reestablishment. Therefore, the source nodes may initiate the regulation at the same time, which may produce a second situation of congestion. HybQoS computes the new rate as follows:

$$
\begin{aligned}
Cf_1^N &= Cf_1 &- h \\
Cf_2^N &= Cf_2 &- h \\
&\vdots &\vdots \\
Cf_r^N &= Cf_r &- h
\end{aligned}
\tag{11.5}
$$

$$
Cf_1 + Cf_2 + \cdots + Cf_r > Th
$$

where Cf_1, Cf_2, ..., Cf_r are the old rates of the congested sessions and Cf_1^N, Cf_2^N, ..., Cf_r^N are the new ones. The variables r and Th represent, respectively, the number of congested sessions and the accepted threshold admission rate. The distribution of the excess bandwidth rate over the congested sessions is performed equitably by a factor h calculated in

$$
h = \frac{\sum_{i=1}^{r} Cf_i - Th}{r}
\tag{11.6}
$$

The second step is to compute the starting time TS_i, which determines the moment of starting the regulation of a congested session i, in order to avoid a simultaneous starting transmission. Assuming that t is the transmission time between two hops and k is the number of hops between the congested node and the source node, Equations 11.7 and 11.8 compute the time between the CNM packet sent by the congested node and the reception of this packet by the appropriate source node:

$$
TS_1 = kt
\tag{11.7}
$$

$$TS_2 = TS_1 + \Delta\varphi_1 \tag{11.8}$$

The factor $\Delta\varphi_j$ is the congestion avoidance time of the session j, which represents the time required to start sending the packets of the flow without causing congestion or overload at an intermediate node. The optimization of this factor would be useful for the minimization of the end-to-end delay constraint of real-time traffic:

$$\Delta\varphi_1 = \frac{h}{Cf_1} \tag{11.9}$$

where h is the regulation factor calculated in (11.6) and Cf_1 is the traffic rate of the first flow arrived at the congested node. The starting time and the congestion avoidance time of the second flow are calculated, respectively, in

$$TS_3 = TS_2 + \Delta\varphi_2 \tag{11.10}$$

$$\Delta\varphi_2 = \frac{h}{Cf_2} \tag{11.11}$$

Similarly, the starting time and the congestion avoidance time for a given session i are computed, respectively, as shown in

$$TS_i = TS_{i-1} + \Delta\varphi_{i-1} \tag{11.12}$$

$$\Delta\varphi_{i-1} = \frac{h}{Cf_{i-1}} \tag{11.13}$$

11.5.2.2 Scheme 2: Priority-Based Adaptation

Rather than considering that all congested sessions have the same priority, the regulation scheme considers different priorities for the source nodes. In this case, the priority of the source nodes and the priority of the flow within the flow sets are taken into account. The notion of priority is very interesting for many applications and can be useful in several areas. For instance, in the battlefield, the message sent by the leaders troop should have more priority than that sent by the soldiers. Therefore, it is preferable in some cases that the traffic regulation should take into account the priority of the source nodes. To the best of our knowledge, this kind of regulation has been little considered in the existing stateless models, where all the congested flows are treated in the same manner. This regulation scheme is described hereafter.

$$
\begin{aligned}
Cf_1^N &= Cf_1 &- \gamma_1 h \\
Cf_2^N &= Cf_2 &- \gamma_2 h \\
& \cdot &\cdot \\
& \cdot &\cdot \\
Cf_r^N &= Cf_r &- \gamma_r h
\end{aligned}
\tag{11.14}
$$

$$\frac{\sum_{i=1}^{r} \gamma_i}{r} \le 1 \tag{11.15}$$

where $h = \left(\sum_{i=1}^{r} Cf_i - \text{Th}\right)/r$ and $\gamma_1, \gamma_2, \ldots, \gamma_r$ are the factors of priority of r different traffic sources. The source node that has a great priority is the one that has a small value. Note that the packets belonging to lower priority flows are selectively dropped prior to the packets of higher priority flows. The starting time TS_i of the congested session i is determined as shown in

$$\text{TS}_1 = kt \tag{11.16}$$

$$\text{TS}_2 = \text{TS}_1 + \Delta\varphi_1 \tag{11.17}$$

$$\Delta\varphi_1 = \frac{1}{\gamma_1} \frac{h}{Cf_1} \tag{11.18}$$

where γ_1 represents the priority factor of the first source node. The computation of the starting time and the congestion avoidance time of a given session r are performed, respectively, in

$$\text{TS}_r = \text{TS}_{r-1} + \Delta\varphi_{r-1} \tag{11.19}$$

$$\Delta\varphi_{r-1} = \frac{1}{\gamma_{r-1}} \frac{h}{Cf_{r-1}} = \frac{1}{r-1} \frac{\sum_{i=1}^{r-1} Cf_i - \text{Th}}{\gamma_{r-1}Cf_{r-1}} \tag{11.20}$$

11.5.3 Simulation

In the following, the performance evaluation and simulation results of HybQoS using the NS-2 simulator are presented. Throughout the simulation, the first scheme of regulation is used, and each wireless mobile host has a transmission range of 250 m and shares an 11 Mbps radio channel with its neighboring nodes. The source and destination nodes associated with flows are distributed among the mobile nodes in the wireless ad hoc network. The simulated environment has a square shape of 150 m × 150 m, where all wireless ad hoc mobile nodes share a single radio channel of 11 Mbps.

A scenario of 32 Transmission Control Protocol (TCP) best-effort traffic, 4 voice flows, and 4 video flows is considered with the aim to better understand the properties of the HybQoS regulation. In the simulation, the TCP traffic is modeled as a mixture of File Transfer Protocol (FTP) and Web traffic. Web traffic represents microflows, whereas FTP traffic corresponds to macroflows. TCP flows are greedy FTP type of traffic with a packet size of 512 bytes. Web traffic is modeled as short TCP file transfers with random file size and random silent period between transfers. The file size is driven from a Pareto distribution, with a mean file size of 10 Kbytes and a shape parameter of 1.2. The length of the silent period between two transfers is also Pareto in distribution with the same shape parameter with a mean of 10 s.

During the simulation, 4 voice and 4 video flows are active and monitored for a duration of 100 s representing real-time traffic. Voice traffic is modeled as 32-Kbps constant rate traffic with a packet size of 80 bytes. Video traffic is modeled as 200-Kbps constant rate traffic with a packet size of 512 bytes. The simulation investigates the performance of HybQoS compared with both the

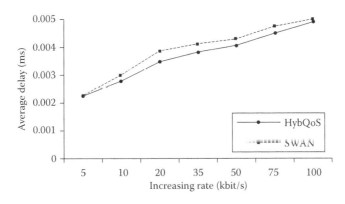

Figure 11.13   Average delay versus increasing rate.

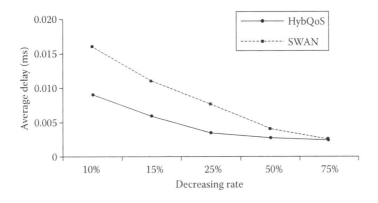

Figure 11.14   Average delay versus decreasing rate.

stateless SWAN model described in [49] and the stateful IEEE 802.11 wireless networks without HybQoS mechanisms.

Figures 11.13 and 11.14 show the average MAC delay of real-time traffic, and Figures 11.15 and 11.16 show the total throughput of best-effort traffic. The value of the increasing rate parameter is represented by the *x*-axis in Figures 11.13 and 11.15. In Figures 11.14 and 11.16, the *x*-axis represents the value of the decreasing rate parameter, which represents the factor *h* expressed as percentage (see Section 11.5.2.1). The increasing rate parameter does not have much impact on the average delay, as shown in Figure 11.13. The average delay grows from 2 to 4.5 ms in both HybQoS and SWAN models (the average delay in SWAN is slightly greater that in HybQoS). However, as shown in Figure 11.14, the increasing rate has more impact on the total throughput of best-effort traffic. It is remarkable that when a small value of increasing rate is chosen, the total throughput is decreased. Thus, when the increasing rate is set to 5 Kbps, the throughput is reduced by about 8% in comparison when the increasing rate equals 75 Kbps. For the 32 TCP flow scenario, our model outperforms the SWAN model by about 10% for different increasing rate values.

The impact of the decreasing rate parameter on the average delay is shown in Figure 11.14. The average delay of the real-time traffic is noticeably reduced when a great value of decreasing rate is chosen. When the decreased rate is less than 25%, the average delay of HybQoS is reduced by

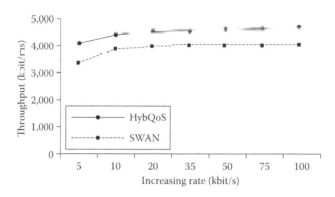

Figure 11.15 Total throughput versus increasing rate.

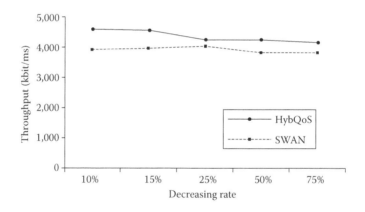

Figure 11.16 Total throughput versus decreasing rate.

about 15–20% in comparison to the SWAN model. The average delay is reduced slowly when the decreasing rate is greater than 25%. Figure 11.16 shows the impact of decreasing rate on the total throughput of the best-effort TCP flows when the increasing rate is set to 35 Kbps. Throughput in HybQoS becomes almost as larger as throughput in SWAN when the deceasing rate is less than 25%. For decreasing rate values superior to 25%, a small impact on the total throughput is observed in both models. These results demonstrate that by using HybQoS mechanisms, a reduction of 15–20% in the average delay of real-time traffic and about 5% increase of TCP throughput can be achieved as compared with SWAN.

Figure 11.17 shows the average delay of real-time traffic with a growing number of Web microflows in both the original and HybQoS models. The last one refers to wireless ad hoc networks without the HybQoS mechanisms. The real-time traffic is modeled by using 4 voice flows of 32 Kbps and 4 video flows of 200 Kbps. It is observed in Figure 11.10 that the average delay of real-time traffic is the same in both the HybQoS and original models without Web microflows. The average delay in our model remains less than 3 ms in contrast to the original model where

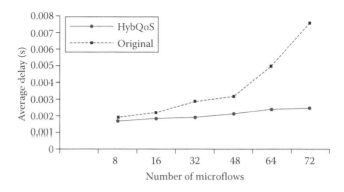

Figure 11.17   Average delay versus number of web microflows.

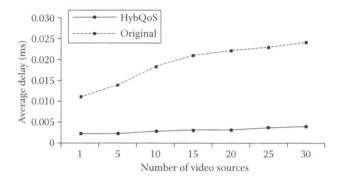

Figure 11.18   Average delay versus number of video sources.

the average delay grows linearly from 1.8 to 7 ms when the number of Web microflows increases from 8 to 72.

The average delay of real-time traffic for a growing number of UDP video sources is shown in Figure 11.18. The TCP best-effort traffic is modeled as a mixture of Web and FTP traffic. It is observed that the original model shows an average delay of larger than 12 ms with only 5 video flows and over 20 ms with 15 or more video flows. HybQoS shows delays inferior to 2 ms with 5 video flows and less than 3 ms with 20 video flows. These results show the efficiency of HybQoS mechanisms.

11.6   Conclusion and Future Trends

The growing popularity of real-time and multimedia applications over wireless networks has stimulated strong interest in extending QoS support to existing ad hoc protocols. Several recent studies have acknowledged the need for scalable QoS solutions and have given the stimulus to a number of proposals on how to conceive a stateless QoS models. So far, most stateful QoS works have been carried out under the concept of per-flow granularity, so the amount of *state informa-tion* increases proportionally with the number of flows. This results in a processing and storage overhead on mobile nodes, which is a well-known scalability problem of stateful approach. The scalability problem is less likely to occur in small-scale MANETs considering the small number of both flows and nodes and the bandwidth of the wireless links. However, as the quality of wireless technology increases rapidly, high-speed and large-sized MANETs may be a matter of fact some day. Although one could argue that whenever large high-performance MANETs will be developed in future, processing capabilities will increase as well, which can alleviate some challenges of stateful models. On the contrary, the stateless approach has the advantages it offers the scalability, since no session-state information is maintained at intermediate nodes. Even though stateless models ensure good QoS provision, they suffer from the problem of illusory reading of network resources because of the dynamic nature of the ad hoc network.

After having described these two approaches and their merits and drawbacks, this chapter has proposed a hybrid stateless QoS support model for wireless ad hoc networks named "HybQoS." This chapter has also presented the main mechanisms of HybQoS: admission control, temporary reservation process, algorithms of flow management, and regulation schemes. The hybrid stateless solution discussed in this chapter looks promising in terms of the performance results that show the benefits of our model under diverse mobility, traffic, and channel conditions. The use of HybQoS mechanisms proved to be efficient, robust, and scalable. The extensive simulations conducted by using the NS-2 simulator have shown that real-time traffic experiences low delays under various mobility, traffic, and multihop conditions.

In line with the ultimate aims of current protocols, the goals of future work on QoS state information in MANET are twofold. First, it is to make the initial resource reservation process more flexible and "lighter," leading to reduce the amount of information stored at intermediate nodes. Second, it is to make the QoS stateless approach more robust in the face of network dynamics. As this chapter has highlighted, the maintenance of flow state information can be performed on-demand or proactively. However, all of these models have weaknesses. Future work on more intelligent hybrid stateless models that adapt the transmission traffic rate to the state of the network resources would be of practical significance. The ideal plan for the QoS provisioning would be able to predict the changes in MANET's topology and the availability of resources before they happen and then to maintain state information according to the resource available in the network.

References

1. M. Natkaniec. Ad hoc mobile wireless networks: principles, protocols, and applications. *IEEE Communications Magazine*, vol. 47, no. 5, 2009, pp. 12–14.
2. L. Khoukhi, S. Cherkaoui. Intelligent QoS management for multimedia services support in wireless mobile ad hoc networks. *Journal of Computer Networks*, vol. 54, no. 10, 2010.
3. L. Hanzo, R. Tafazolli. A survey of QoS routing solutions for mobile ad hoc networks. *IEEE Communications Surveys and Tutorials*, vol. 9, no. 1–4, 2007.
4. L. Khoukhi, S. Cherkaoui, and D. Gaïti. Managing rescue and relief operations using wireless mobile ad hoc technology: the best way? In *IEEE LCN ON-MOVE 2009*, Switzerland, October 20–23, 2009.
5. S. Chakrabarti and A. Mishra. Quality of service challenges for wireless mobile ad hoc networks. *Wireless Communications and Mobile Computing*, vol. 4, 2004, pp. 129–53.
6. Q. Xue and A. Ganz. Ad hoc QoS on-demand routing (AQOR) in mobile ad hoc networks. *Journal of Parallel and Distributed Computing*, vol. 63, no. 2, 2003, pp. 154–165.
7. B. Zhang and H.T. Mouftah. QoS routing for wireless ad hoc networks: problems, algorithms, and protocols. *IEEE Communications Magazine*, vol. 43, no. 10, 2005, pp. 110–117.
8. C.T. Calafate, M.P. Malumbres, J. Oliver, J.C. Cano, and P. Manzoni. QoS support in MANETs: a modular architecture based on the IEEE 802.11e technology. *IEEE Transactions on Circuits and Systems for Video Technology*, vol. 19, no. 5, 2009, pp. 678–692.

9. M.S. Corson and A.T. Campbell. Towards supporting quality of service in mobile ad hoc networks. In: *Proceedings of the IEEE Conference on Open Architecture and Network Programming*, San Francisco, CA, April 1998.

10. P. Sinha, R. Sivakumar, and V. Bharghavan. CEDAR: a core extraction distributed ad hoc routing algorithm. *IEEE Journal on Selected Areas in Communications*, vol. 17, no. 8, 1999, pp. 1454–1465.

11. W.-H. Liao, Y.-C. Tseng, J.-P. Sheu, and S.-L. Wang. A multi-path QoS routing protocol in a wireless mobile ad hoc network. In: *Proceedings of the IEEE International Conference on Networking*, Part II, July 2001, pp. 158–167.

12. D.D. Perkins and H.D. Hughes. A survey on quality-of-service support for mobile ad hoc networks. *Wireless Communications and Mobile Computing*, vol. 2, no. 5, 2002, pp. 503–513.

13. R. Boppana and S. Konduru. An adaptive distance vector routing algorithm for mobile ad hoc networks. In: *Proceedings of the IEEE INFOCOM Conference*, Anchorage, AK, April 2001.

14. C.E. Perkins and E.M. Royer. Ad-hoc on-demand distance vector routing. In: *Proceedings of the IEEE Workshop on Mobile Computing Systems and Applications*, New Orleans, LA, February 1999, pp. 90–100.

15. D. Johnson and D. Maltz. Dynamic source routing in ad-hoc wireless networks. In: *Mobile Computing*. Kluwer Academic Publishers, 1996, pp. 153–181.

16. S.R. Das, A. Mukherjee, B. Bandyopadhay, K. Paul, and D. Saha. Improving quality-of-service in ad hoc wireless networks with adaptive multi-path routing. In: *Proceedings of the IEEE GLOBECOM Conference*, San Francisco, CA, November 2000.

17. C.R. Lin and J.S. Liu. QoS routing in ad hoc wireless networks. *IEEE Journal on Selected Areas in Communications*, vol. 17, no. 8, 1999, pp. 1426–1438.

18. O. Younis, L. Kant, A. Mcauley, K. Manousakis, D. Shallcross, and K. Sinkar. Cognitive tactical network models. *IEEE Communications Magazine*, vol. 48, no. 10 (Special Issue on Military Communications), October 2010, pp. 70–77.

19. P. Jacquet, P. Muhlethaler, T. Clausen, A. Laouiti, A. Qayyum, and L. Viennot. Optimized link state routing protocol for ad hoc networking. In: *Proceeding of the IEEE International Multi Topic Conference*, Lahore, Pakistan, December 2001, pp. 62–68.

20. H. Badis and K. Al Agha. QOLSR, QoS routing for ad hoc wireless networks using OLSR. *European Transactions on Telecommunications*, vol. 15, no. 4, 2005, pp. 427–442.

21. J. Eberspaecher, H. J. Voegel, and C. Bettstetter. *GSM Switching, Services, and Protocols*. John Wiley & Sons, 2001.

22. G. Maral and M. Bousquet. *Satellite Communications Systems: Systems, Techniques and Technology*. Singapore: John Wiley & Sons, 2002.

23. I. Cardei and S. Kazi. MAC layer QoS support for wireless networks of unmanned air vehicles. In: *Proceedings of the Hawaii International Conference on System Sciences (HICSS)*, Big Island, HI, January 2004.

24. P. Wang and W. Zhuang. A token-based scheduling scheme for WLANs supporting voice/data traffic and its performance analysis. *IEEE Transactions on Wireless Communications*, vol. 7, no. 4, 2008, pp. 1708–1718.

25. P. Liu et al. CoopMAC: a cooperative MAC for wireless LANs. *IEEE Journal on Selected Areas in Communications*, vol. 25, no. 2, 2007, pp. 340–354.

26. H. Zhu and G. Cao. rDCF: a relay-enabled medium access control protocol for wireless ad hoc networks. *IEEE Transactions on Mobile Computing*, vol. 5, no. 9, 2006, pp. 1201–1214.

27. G. Jakllari et al. A framework for distributed spatio-temporal communications in mobile ad hoc networks. In: *Proceedings of IEEE INFOCOM*, Orlando, FL, 2006.

28. H.T. Cheng and W. Zhuang. QoS-driven MAC-layer resource allocation for wireless mesh networks with non-altruistic node cooperation and service differentiation. *IEEE Transactions on Wireless Communications*, vol. 8, no. 12, 2009, pp. 6089–6103.

29. IEEE Standard 802.11. Medium Access Control (MAC) Enhancements for Quality of Service (QoS). IEEE Draft Standard 802.11e, November 2001.

30. S.T. Sheu and T. F. Sheu. A bandwidth allocation/sharing/extension protocol for multimedia over IEEE 802.11 ad hoc wireless LANs. *IEEE Journal on Selected Areas in Communications*, vol. 19, no. 10, 2001, pp. 2065–2080.

31. V. Kanodia, C. Li, A. Sabharwal, B. Sadeghi, and E. Knightly. Distributed multi-hop scheduling and medium access with delay and throughput constraints. In: *Proceedings of the ACM MOBICOM Conference,* Rome, Italy, July 2001

32. H. Luo, S. Lu, V. Bharghavan, J. Cheng, and G. Zhong. A packet scheduling approach to QoS support in multihop wireless networks. *ACM Journal of Mobile Networks and Applications (MONET)* vol. 9, no. 3 (Special Issue on QoS in Heterogeneous Wireless Networks), 2004, pp. 193–206.

33. M. Barry, A. Campbell, and A. Veres. Distributed control algorithms for service differentiation in wireless packet networks. In: *Proceedings of the IEEE INFOCOM Conference,* Anchorage, AK, April 2001.

34. S.S. Kang and M.W. Mutka. Provisioning service differentiation in ad hoc networks by the modification of backoff algorithm. In: *Proceedings of the IEEE ICCCN Conference,* Scottsdale, AZ, October 2001.

35. C.R. Lin, M. Gerla. MACA/PR: an asynchronous multimedia multihop wireless network. In: *Proceedings of the IEEE INFOCOM Conference,* Kobe, Japan, April 1997.

36. H. Xiao, W.K.G. Seah, A. Lo, and K. Chaing. Flexible QoS model for mobile ad-hoc networks. *IEEE Transactions on Vehicular Technology,* vol. 1, 2000, pp. 445–449.

37. H.T. Cheng and W. Zhuang. Pareto optimal resource management for wireless mesh networks with QoS assurance: joint node clustering and subcarrier allocation. *IEEE Transactions on Wireless Communications,* vol. 8, no. 3, 2009, pp. 1573–1583.

38. C. Lin and M. Gerla. Adaptive clustering for mobile wireless networks. *IEEE Journal on Selected Areas in Communications,* vol. 15, no. 7, 1997, pp. 1265–1275.

39. H.T. Cheng and W. Zhuang. Novel packet-level resource allocation with effective QoS provisioning for wireless mesh networks. *IEEE Transactions on Wireless Communications,* vol. 8, no. 2, 2009, pp. 694–700.

40. H.K. Wu. Multimedia mobile: multihop networks in channel fading environment. PhD dissertation. University of California at Los Angeles, Los Angeles, 1999.

41. G.N. Aggélou. An integrated platform for quality-of-service support mobile multimedia clustered ad hoc networks. In: *The Handbook of Ad Hoc Wireless Networks.* Boca Raton, FL: CRC Press, 2003.

42. L. Khoukhi and S. Cherkaoui. Toward neural networks solution for multimedia support in wireless mobile ad hoc networks. *Journal of Networks,* vol. 4, no. 2, 2009, pp. 148–161.

43. Y.D. Zeinalipour. A glance at quality of services in mobile ad-hoc networks [online]. Available at http://www.cs.ucr.edu/csyiazti/courses/cs260/manetqos.pdf, November 2001.

44. S. Lee, G. Ahn, X. Zhang, and A. Campbell. INSIGNIA: an IP-based quality of service framework for mobile ad hoc networks. *Journal of Parallel and Distributed Computing,* vol. 60, no. 4 (Special Issue on Wireless and Mobile Computing and Communications), 2000, pp. 374–406.

45. H. Luo, P. Medvedev, J. Cheng, and S. Lu. A self-coordinating approach to distributed air queueing in ad hoc wireless networks. In: *Proceedings of INFOCOM,* vol. 3, Anchorage, AK, April 2001, pp. 1370–1379.

46. S. Lee, G. Ahn, and A. Campbell. Improving UDP and TCP performance in mobile ad hoc networks with INSIGNIA. *IEEE Communications Magazine,* vol. 39, no. 6, 2001, pp. 156–165.

47. G. Ahn, A.T. Campbell, A. Veres, and L.H. Sun. SWAN: service differentiation in stateless wireless ad hoc networks. In: *Proceedings of IEEE INFOCOM,* New York City, NY, June 2002.

48. M. Hejmo, B. Mark, C. Zouridaki, and R. Thomas. Design and analysis of a denial-of-service-resistant quality-of-service signaling protocol for MANETs. *IEEE Transactions on Vehicular Technology,* vol. 55, no. 3, 2006, pp. 743–751.

49. Y.L. Morgan and T. Kunz. PYLON: an architectural framework for ad-hoc QoS interconnectivity with access domains. In: *Proceedings of the Hawaii International Conference on System Sciences (HICSS),* Big Island, HI, January 2003.

50. S.S. Chen and A. Kassler. Extending SWAN to provide QoS for MANETs connected to the Internet. In: *Proceedings of the IEEE International Symposium on Wireless Communication Systems (ISWCS),* Siena, Italy, September 2005.

51. M. Domingo and D. Remondo. An interaction model for QoS support in ad hoc networks connected to fixed IP networks. *International Journal of Wireless and Mobile Computing,* vol. 4, no. 3, 2010, pp. 179–187.

52. X. Xiang, X. Wang, and Y. Yang. Stateless multicasting in mobile ad hoc networks. *IEEE Transactions on Computers,* vol. 59, no. 8, 2010, pp. 1076–1090.

53. S. Meyer. *Harris Corporation Technology Supports Streaming Video over Airborne Tactical Network during Demo for Australian Forces*. Harris Government Communications Systems, July 2009.
54. C. Calafate, J. Oliver, J. Cano, P. Manzoni, and M. Malumbres. A distributed admission control system for MANET environments supporting multipath routing protocols. *Microprocessors & Microsystems*, vol. 31, no. 4, 2007, pp. 236–251.
55. C. Calafate, J. Cano, P. Manzoni, and M. Malumbres. A QoS architecture for MANETs supporting real-time peer-to-peer multimedia applications. In: *Proceedings of the IEEE International Symposium on Multimedia*, Irvine, CA, December 2005, pp. 193–200.
56. J. Khoukhi, A. El Masri, and D. Gaïti. A hybrid stateless QoS approach for wireless mobile ad hoc networks. In: *Proceedings of the IFIP International Conference on New Technologies, Mobility and Security (NTMS)*, Paris, France, February 2011.

FUTURE NETWORKS INSPIRED BY MANET

Chapter 12

Connecting Moving Smart Objects to the Internet: Potentialities and Issues When Using Mobile Ad Hoc Network Technologies

Bernardo Leal and Luigi Atzori

Contents

The Internet of Things is evolving to enable the seamless communication of moving smart objects with nodes on the Internet. When these objects move away from structured infrastructures, the mobile ad hoc network (MANET) becomes one of the most appropriate technologies to connect them to the Internet. In this chapter, issues about MANET integration with the Internet are considered, with particular attention to the handover performance in the scenario where moving objects roam between different multihomed hybrid ad hoc networks. We first review the mechanisms that mainly affect the management of the handover procedure, which are the IP mobility, the external route computation, the ad hoc routing, and the gateway discovery. We then provide a performance evaluation of the handover when different adjacent MANET subnetworks are connected to the Internet by means of their own gateway and mobile IP agent. We conclude that the adoption of reactive routing protocols combined with a proactive gateway discovery procedure is highly recommended and that the use of multiple gateway routes and early agent preregistration is crucial for seamless handovers.

12.1 Introduction

The expression "Internet of Things" is used to refer to the idea of a global infrastructure of interconnected physical objects [1]. This concept is mainly motivated by the growing adoption of the Radio Frequency Identification (RFID) technologies, which have been widely used for tracking objects, people, and animals, making use of an architecture that combines the use of simple RFID tags and extensive and complicated interconnection of RFID readers. This architecture optimally supports tracking physical objects within well-defined areas (such as stores), but it limits the sensing capabilities and deployment flexibility that other challenging application scenarios may require.

An alternative architectural model for the Internet of Things may be a more loosely coupled, decentralized system of smart objects with sensing, processing, and networking capabilities. In contrast to simple RFID tags, smart objects may carry segments of application logic that may let them evaluate their local environment, and by means of a unique addressing scheme, probably IP, they interact with each other and with human users, wherever they are.

Several wireless technologies allow mobile smart objects to increase their pervasive presence around us. Wi-Fi, WiMAX (Worldwide Interoperability for Microwave Access), and sensing and cellular networks are examples of technologies that may support object interconnection, but when they move away from network structures, MANETs may be the recommended way to interconnect them to the Internet. In fourth-generation (4G) wireless systems, we consider ubiquitous computing and universal access for mobile users (or objects) that wish to connect to the Internet through heterogeneous technologies and that wish to maintain connectivity globally without interrupting their ongoing communications even when they cross from one type of network to another or when their connection paths change the gateways their packets go through [2]. Figure 12.1 shows an integration of different communication technologies, including including MANETs, Universal Mobil Telecommunication Systems (UMTS), and Code Division Multiplexing (CDMA) systems, which permit object's ubiquitous communication. In such a type of scenario, a farmer may receive real-time information in his mobile phone about data directly coming from Sensors on his caws in a different country.

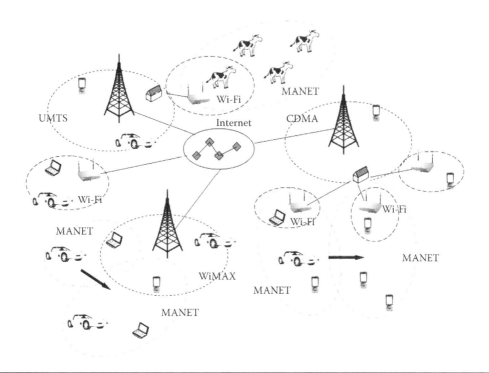

Figure 12.1   Fourth-generation heterogeneous wireless systems.

In a way different to traditional fixed IP networks, members of MANETs communicate over multihop relays by equally participating in the routing information distribution and maintenance by using the same ad hoc routing protocol, which must be adaptive to topological changes and traffic demands. To interconnect a MANET with other types of networks, such as Internet, any MANET object (or router) that has connectivity with both types of networks may effectively become a gateway between the ad hoc domain and the Internet.

The integration of MANETs with fixed infrastructures must be carefully studied to evaluate how it performs. In such an integrated scenario, commonly known as *hybrid ad hoc network*, a MANET can be seen as an extension to the existing infrastructure, whose mobile objects may seamlessly communicate with nodes on the fixed network, forwarding packets throughout the gateways found on the edge that join both types of networks. If during an ongoing communication between an object on a wireless network and a node on a fixed network, a gateway change occurs on their traffic path, some packets may get lost, or their delay variation may increase, which may affect communication performance.

MANET integration with fixed networks is a research topic that has received great attention in recent years, but not so much has been argued about seamless handover between hybrid MANET subnetworks, which is an important topic when we think about object ubiquitous and universal communication in 4G systems. In this chapter, issues about MANET integration with the Internet are considered, with particular attention to the performance when moving objects roam between different MANET subnetworks in a multihomed hybrid ad hoc network. After presenting the considered scenario, we review the mechanisms that mainly affect the management of the handover procedure, which are the IP mobility, the external route computation, the ad hoc routing, and the gateway discovery. We then provide a performance evaluation of the handovers when different adjacent MANET subnetworks are connected to the Internet by means of their own gateways.

12.2 Smart Objects on MANETs

Hybrid ad hoc networks are composed of three different parts: (1) the fixed Internet, where traditional internal gateway protocols are set to find suitable routes; (2) the MANET, where mobile objects running a routing protocol are grouped in different subnetworks and normally share a common address prefix; (3) the gateways, which are special routers that interconnect the MANET subnetworks to the fixed network, allowing data packets to traverse from one network type to the other, and the propagation of routing protocol information.

Gateways are required to have at least one interface associated with the fixed network and one interface associated with the MANET. Objects in a MANET subnetwork normally set its IP address according to the MANET gateway interface prefix [3]. When two or more gateways connect the MANET to the fixed network, it is referred to as multihomed hybrid ad hoc networks, and in this case, if these gateways use different address prefixes, independent MANET subnetworks are formed.

Prior to the establishment of a connection with a node on the fixed network, a moving object on a MANET must first find the closest gateway, set its IP address, find and register to a mobile IP agent [4], and find the correspondent node external route. To have ubiquitous and transparent connectivity in this nomadic integrated scenario, there should not be any communication interruptions when moving objects change their registrations from one subnetwork gateway to another during ongoing connections with hosts on a fixed network (handover). Figure 12.2 shows a scenario where an object engaged in communication with an Internet correspondent node moves from one MANET subnetwork to another and is compelled to change its agent/gateway registration from one subnetwork to another to maintain connection.

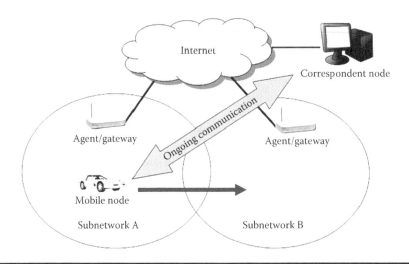

Figure 12.2 Hybrid MANET handover.

12.3 Global Routing

The communication between any pairs of objects inside a MANET is handled by ad hoc routing protocols, which were originally created to permit communication between MANET members away from network infrastructures. To permit communication between MANET objects and fixed nodes, gateways are used to join both types of networks, to help share routing information between them, and then to find the appropriate external route.

To allow moving objects to communicate with hosts on a fixed network, other than using ad hoc protocols and gateways, a MANET object must also use an IP mobility management protocol, such as mobile IP, to maintain the same IP address even when they roam between different MANET subnetworks [5,6]. To operate, mobile IP needs the use of agents on each subnetwork, which normally run on the same subnetwork gateways that connect MANET subnetworks to fixed networks.

12.3.1 IP Mobility

Most existing architectures for connecting MANETs to the Internet are based on mobile IP [7]. Mobile IP was originally designed as a mechanism that allows users to roam among IP networks without changing their IP addresses. Three functional entities are necessary in mobile IP: MN (mobile node), FA (foreign agent), and HA (home agent).

HAs and FAs broadcast agent advertisements via one-hop link to advertise their presence. The MN checks these received agent advertisements to detect whether it is on its home network or has roamed to a foreign network. If the MN has roamed to a foreign network, it can obtain a care of address (CoA) from the FA's network and send a registration request to its HA. When the HA receives the request, it updates its binding list and sends a registration reply to the MN. All packets destined to the MN's home address is tunneled to the MN's CoA by the HA. One disadvantage of this technique is that packets may take longer paths to arrive to the destination than when the MN remains in its home network.

Mobile IP was designed for nomadic hosts on traditional fixed IP networks, and it assumes that the mobile nodes have direct link connectivity with agents. In a MANET, a mobile node (or object) generally does not have direct connectivity to agents. To solve this problem, mobile IP must be modified to work over MANETs to meet the following requirements [2,4]:

■ Objects should be able to use the agent's address prefix even if multiple hops away from them.
■ Agents should be able to forward mobile IP signaling packets using multihop routes.

A version of mobile IP was also developed for IPv6 [7]. For this case, every time an object roams to a different MANET, it autoconfigures a new stateless IP address according to the network access router's address prefix. This version of mobile IP does not manage well handovers, because the object does not keep the same IPv6 address when continuing forwarding packets to its correspondent node during handovers.

In a multihomed hybrid ad hoc network, a common architecture is composed of different MANET subnetworks, each one including a different agent. By evaluating agent advertisements, objects may register to the nearest one. The agent to which an object registers first becomes its home agent, whereas agents on visited networks perform as foreign agents.

12.3.2 External Routes

If an object wishes to communicate to another one on the same MANET, it uses ad hoc routing protocols to find suitable routes to reach it. But if it wants to communicate with hosts on the Internet, it must first find external routes throughout its current gateway. Objects may recognize external routes from internal routes in different ways [2]. The most popular is by verifying if the destination address prefix belongs to the same MANET subnetwork. This is the recommended way when implementing mobile IP since it helps to recognize them when they roam into different subnetworks. One alternative method requires that the local gateway stores the registrations from all the objects in the same subnetwork. Under this circumstance, an object that wants to communicate must first inquire the local gateway to find out about the destination presence in the same subnetwork. This mechanism does not guarantee the tracking of every possible object placed in the same subnetwork at every possible time.

Once it has been identified that the destination is outside the subnetwork, forwarding packets is mainly done in two different ways:

- Default route: Source objects use the default route entry to the gateway to forward packets toward the Internet. Neighbor objects are responsible for delivering packets toward the gateway.
- Tunneling: It is also called *source routing*. The source object adds a new IP header to the original IP packet, with its current gateway address as the new destination. Upon receiving the tunneled packet, the gateway decapsulates it and forwards it to the Internet.

Packet delivery reliability is low with the default route method, whereas traffic congestion is incremented with the tunneling method.

12.4 Local Routing

Communications between objects inside a MANET is handled by ad hoc routing protocols. The route finding time, whether for external, for internal, or for gateway nodes, depends on the type of protocol used. This time is important especially in highly nomadic scenarios, where routes are frequently lost and recomputed, as it occurs in MANETs.

12.4.1 Routing Protocols

In general, ad hoc routing protocols can be mainly divided into proactive and reactive. Reactive routing protocols discover routing paths only when traffic demands it, trading off longer packet delays in the interest of lower protocol overhead. Proactive protocols maintain and regularly update full sets of routing information, trading off greater protocol overhead and higher convergence time in the interest of smaller packet delays [8,9].

12.4.1.1 Proactive Protocols

In proactive protocols, each object updates its routing table by continually exchanging routing information with other objects. Routing information is obtained by propagating link state information throughout the whole network. Destination sequenced distance vector (DSDV) and optimized link state routing (OLSR) are two popular proactive routing protocols.

Mobile IP agents normally run on MANET gateways, and objects are compelled to register with the closest agent from which they receive advertisements. The proactive routing protocol is responsible for relaying agent advertisements and registration messages between the ad hoc objects and the agent via multihop paths.

With proactive protocols, routes (internal, external, or agent/gateway) are normally available when needed. However, proactive protocols suffer from high routing traffic and long route recovering time.

12.4.1.2 Reactive Protocols

Reactive routing protocols have been developed to decrease control overhead and to preserve bandwidth. Routing information is obtained only when needed, and two main phases are involved: route discovery and route maintenance. During the first phase, route requests are broadcast until an answer is received from an object that has the requested destination information. During the second phase, active routes are preserved via periodic Hello messages. Ad hoc on-demand distance vector (AODV) and dynamic source routing (DSR) are currently the most popular and mature reactive routing protocols.

One method for managing IP mobility is MIP-MANET, which provides AODV-based MANETs with access to the Internet using mobile IP [10]. In MIP-MANET, a gateway periodically broadcasts agent advertisements, which are rebroadcasted by MANET objects. Later, objects discover gateway routes on-demand, so the routing traffic is significantly decreased.

With reactive protocols, routes (internal, external, to agent/gateway) are not available when needed; thus, a variable amount of time is required to find them. However, the time to recover lost routes is normally shorter than in proactive protocols, because they take less time to declare lost routes as broken [11].

12.4.2 Agent/Gateway Discovery

Agent/gateway discovery is a key component for providing Internet connectivity to MANETs. Objects must discover gateways before communicating with Internet nodes and discover agents to maintain their home address while roaming into different MANET subnetworks. Agents normally run on subnetwork gateways, and their advertisements are used for setting objects IP addresses and allowing objects know their presence. The agent/gateway discovery approaches can be broadly divided into two main categories: proactive and reactive.

12.4.2.1 Proactive Approach

The proactive approach is a popular way to advertise the presence of agents. Mobile IP agents broadcast agent advertisements periodically, which are later rebroadcasted by other objects, to be received by those beyond the gateway transmission range. One broadcast can satisfy the registration requirements of all objects in a MANET. Upon receiving agent advertisements, the mobile IP registration process begins when objects send registration requests to the selected agent.

To reduce the congestion caused by frequent broadcast advertisements, periodic unicast advertisements may be implemented for objects that have already been registered to maintain the registration alive. In addition, to avoid frequent broadcast of agent requests, objects may use agent lists to record the agent's addresses from which advertisements have been previously received. Alternatively, mechanisms for piggybacking routing information on mobile IP signaling messages may be implemented. This helps objects to avoid requiring for gateway routes and sending the answers. It is also possible to define a gateway's service scope by defining a time to live (TTL) value on the advertisement packet header equal to a predefined value *N*, so that the rebroadcasting of advertisements does not extend beyond *N* hops. Finally, the registering process may also contribute with congestion reduction by using unicast forwarding mechanisms instead of broadcast mechanisms.

The proactive approach allows objects to detect when they are closer to other subnetwork agents. If this occurs, objects may be triggered to initiate a handover. This type of handover is called *proactive gateway handovers*.

12.4.2.2 Reactive Approach

With the reactive discovery approach, a mobile object is allowed to actively discover an agent when necessary. The mobile object may broadcast a solicitation that is relayed by other objects to reach the agent. In response, an agent may broadcast an advertisement. At this moment, the object may initiate its registration process.

The trigger for sending solicitations is a key issue in the operation of the reactive discovery approach. One possible trigger may be the advertisement lifetime expiration. This triggering mechanism does not help to notice the loss of connectivity with gateways, which frequently occurs in changing topology scenarios. To address this problem, gateway route expiration triggers may be set to detect invalid gateway routes. A different type of triggering may be implemented for roaming objects when they are set to listen for route requests and answers occurring on visited subnetworks. This type of triggering is useful for handover initiations. After receiving agent requests or answers, the roaming object may proceed to register with the FA.

To reduce registration congestion, unicast solicitations and advertisements may be used instead of broadcast ones. It is also possible to use multicast addresses for requesting valid routes to an FA. Another contribution is the integration of gateway discovery requests within the external route discovery cycle.

12.5 Current Status of Object Mobility

12.5.1 Inter-MANET Handover Approaches

Initially, when an object is connected, it has set its IP address according to the closest gateway address prefix from which it receives the advertisements and is registered to the collocated mobile IP agent, which becomes the object's HA. In this scenario, objects are able to communicate with any host on the Internet using multihop routes and are able to roam between different subnetworks using the same global IP address.

When the object moves around, it may find itself closer to a different gateway than that to which it is currently registered. Normally, hop count is the metric used to estimate the gateway distance. Under this circumstance, it initiates a handover process (see Figure 12.3) to change its registration to the FA to continue communicating with hosts on the Internet. A roaming object may notice that it is entering a different MANET segment when it begins either receiving periodic advertisements from other subnetwork agents (proactive discovery approach) or hearing of other object's registration requests and answers. This corresponds to the first step shown in Figure 12.3. With the reactive discovery approach, the presence of an FA is not detected, but agent discovery

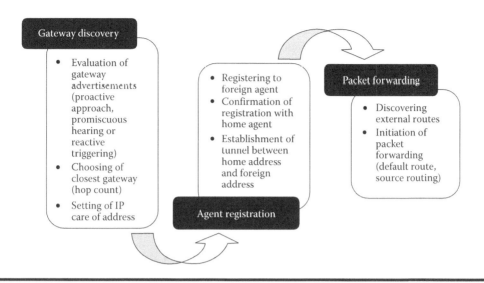

Figure 12.3   Inter-MANET handover process.

may be initiated when the object is triggered to do so if gateway routes or agent advertisements expire. Next, after setting its CoA according to the visited subnetwork address prefix, the moving object may register to the FA. From this moment, communication between the object and the Internet correspondent node will be made through the tunnel that is established between the HA and the FA.

From the moment an object starts to handover, packet delivery will be interrupted until the registration to the FA is completed and the external route is found. Once an object has reestablished the connection with its correspondent node, it may begin forwarding its packets throughout the foreign agent/gateway. Forwarded packets will tunnel to the home gateway before arriving to the correspondent node on the Internet. Packets returning to the mobile object will follow the opposite path. This path is normally longer than the one used before the handover occurred, and thus the delivery delay will increase. The forwarding mechanism implemented and the tunneling between agents, which is part of the mobile IP protocol, may contribute to the packet delivery delay.

A different handover procedure occurs when the reactive discovery approach is used. In this case, objects will be triggered to initiate a handover when they realize that a connection to its current gateway is lost or if the agent advertisement life expires. In either case, objects are compelled to initiate a new registering process by searching for a new agent/gateway. If a different gateway is found, a handover process is executed, but if advertisements are received from the same agent/gateway, when the registering process is completed, it must not be considered a handover. In the reactive approach, the handover procedure requires more time, because it needs to wait for the triggering to occur, and the additional time is wasted to find available agents.

To achieve seamless handovers, every ongoing communication should be uninterrupted and the packets belonging to these connections should be continuously delivered. But the required operations consume a significant amount of time, which is given by the following main procedures: handover triggering, gateway discovery, agent registration, external route discovery, and implementation of the forwarding mechanism. The time needed to complete these procedures is affected by the used routing protocol, as shown in Table 12.1.

Table 12.1 Aspects Affecting MANET Handovers

Handover Procedures	Proactive Protocol	Reactive Protocol
Handover triggering type	• Other gateway advertisements	• Other gateway advertisements • Advertisement lifetime trigger • Gateway broken route
Gateway discovery	• Proactive approach (mostly) • In some cases, uses reactive approach or hybrid approach	• Main use: reactive approach, but may equally use any type
External route discovery	• Prefix identification (mostly)	• Prefix identification (mostly) • In some cases, search first inside the MANET
Forwarding mechanism	• Every object has a route to every other object and gateway	• Default route • Tunneling

From the table we see that associated with the discovery approach implemented are the possible triggering options. For instance, handover triggering normally uses a proactive discovery approach, independent of the type of routing protocol used, since objects initiate handovers only if they receive FA advertisements. Advertisement lifetime expiration and broken routes are instead normally used with the reactive discovery approach because those are the objects that notice the losses. These triggers are more frequently used to reinitiate gateway discovery processes. We may also notice that the gateway discovery approaches are associated with the type of routing protocol used. As to the gateway discovery, the approach used is usually in accordance with the running routing approach (proactive or reactive), but this is not a fixed rule. Address prefix identification is the preferred method for discovering external routes in both types of routing protocols. This is the easiest way when using mobile IP, since different agents/gateways may be identified by their address prefix. Finally, each type of routing protocol uses a different way to deliver packets to its destination. The forwarding mechanism is simpler when using proactive routing protocols, because like in fixed networks, once the convergence has been reached, every object in the subnetwork has routes to every other object and additionally act knowing that any other object has this information too. Hence, the neighbor object's routing responsibility is only to forward packets through the proper interface. Contrarily, reactive routing protocols, having a limited picture of the MANET subnetwork topology, use default routes or tunneling mechanisms to forward packets versus external destinations. Tunneling-based forwarding, also known as *source routing,* is a more reliable method since intermediate objects only need to forward the packets received to the immediate destination indicated in the packet header.

12.5.2 Proposed Schemes

Having explained details about the inter-MANET handover approaches, we present now a summary of the principal schemes for MANET mobility proposed by researchers grouped according the type of ad hoc routing protocol. Table 12.2 shows a list of these schemes.

Table 12.2 Proposed Schemes for MANET Mobility

Protocol Type	Scheme	Characteristics
Proactive protocol	• Lei and Perkins • Tseng et al. • Ammari and El-Rewini • Benzaid et al. • Ergen	• Usually, only one gateway serves a MANET subnetwork • The proactive agent discovery is the dominant approach • The overhead caused by the proactive routing protocol and the proactive agent discovery increases the network congestion
Reactive protocol	• Jonsson et al. • Sun et al. • Ratanchandani and Kravets • Broch et al.	• The overhead is noticeably reduced with the use of reactive routing protocols • The use of more than one gateway is allowed

In the proactive group, Lei and Perkins [12] proposed the integration of ad hoc networks by using mobile IP and a modified routing information protocol (RIP) as a proactive ad hoc routing protocol. Objects discover agents using a proactive approach. In this architecture, the frequent agent advertisement and the frequent exchange of RIP tables increase the network congestion. Tseng et al. [13] presented a project that uses DSDV routing and mobile IP. Each MANET subnetwork service scope is controlled defining a TTL value equal to the maximum gateway distance N. Objects farther than N are not allowed to register to the gateway. Tseng et al.'s scheme also suffers from high congestion caused by the frequent agent broadcast and DSDV signaling. Another scheme based on DSDV and mobile IP was proposed by Ammari and El-Rewini [14], but in this case, mobile gateways, which are one-hop away from the FAs, are used to serve as an interface between them and the mobile objects. Only mobile gateways use standard mobile IP to register with the FAs. Mobile objects register with the mobile gateway to be able to communicate with the Internet by using DSDV and are not aware of the existence of the FAs. Benzaid et al. [15] designed an architecture based on OLSR and mobile IP. As in the precedent cases, the frequent agent advertisements and the use of a proactive routing protocol increase network congestion. Mobile enriched wireless local area network architecture (MEWLANA) was presented by Ergen and Puri [16]. It uses a combination of DSDV and tree-based bidirectional routing (TBBR) protocols, each one aimed at different network sizes and congestions. In this design, less congestion is generated because it uses the reactive agent discovery approach.

In the reactive group, Jonsson et al. [17], Sun et al. [18], and Ratanchandani and Kravets [19] designed schemes based on mobile IP and AODV. The Jonsson scheme is called MIP-MANET, and it permits the use of up to two gateways for each MANET subnetwork. Agents are discovered using a proactive approach, but the use of a reactive routing protocol reduces noticeably the network congestion. On the contrary, Broch et al. [20] proposes the use of DSR and a single gateway. Each gateway explicitly defines a MANET subnetwork, and objects handover occur when mobile objects change their agent registration.

12.6 Future Trends

12.6.1 Performance Evaluation

The different schemes proposed to manage object mobility in MANETs were classified according to the ad hoc protocol they use. This and the agent discovery approach implemented are the main factors affecting the network performance. Next, a qualitative comparison of the handover performance is presented in Table 12.3 for the different types of ad hoc routing protocols.

From the table, it is possible to make the following observations about MANET handovers:

■ Frequent transmissions of agent advertisements allow the proactive approach to perform a faster handover with respect to the reactive approach. Indeed, this feature permits moving objects to easily recognize when they roam to a different subnetwork. In addition, objects may have ready information about several agents/gateways, which may be immediately used in the event that registration to the current gateway is lost. Differently, with the reactive approach, a longer time is required to wait for the triggers to initiate the handover process and an additional time interval to find a new agent.

■ An important problem to consider is the network congestion produced by the frequent transmission of agents and route advertisements on proactive gateway discovery and routing. This congestion may increase forwarding delays and may affect route and gateway discovery. To reduce congestion, the use of reactive routing protocols is recommended in addition to alternative techniques for reducing congestion, such as unicasting, piggybacking, and TTL range scope.

Some changes that improve MANET performance have been proposed in the past, which are mainly aimed to reduce the disadvantages that characterize each type of protocol [19]. In Table 12.4, for the two routing approaches we summarize the proposals that improve the performance during the handover process. As we can see from the table, both types of routing protocols

Table 12.3 Comparison between Proactive and Reactive Routing Protocols during Handover

Protocol Type	Advantages	Disadvantages
Proactive protocol	• Permanent availability of internal and external routes • Mobile IP prefix identification ready • Permanent availability of agent advertisements (first step for handover)	• Significant congestion • Long time to recover broken routes • Reactive triggers not defined
Reactive protocol	• Low congestion • Low time to recover broken routes (low time to find a gateway) • Possibility of proactive agent advertisements	• Routes not available • Long time to find routes • In some cases, mobile IP prefix identification not ready • In some cases, nonpermanent availability of agent advertisements

Table 12.4 Mechanisms for Improving Handover Performance

Protocol Type	Mechanism for Improving Inter-MANET Handover
Proactive protocol	Mechanisms implemented to reduce congestion: • Unicast advertisement to those specific objects that register with agents • Route piggybacking to avoid route rebroadcasting • TTL rebroadcasting control to cover only the desired physical area • Maintaining agent lists is recommended
Reactive protocol	Mechanisms implemented to reduce congestion (when broadcasting advertisements): • Unicast advertisement to those specific objects that register with agents • Route piggybacking to avoid route rebroadcasting • TTL rebroadcasting control to cover only the desired physical area • It is recommended to configure proactive agent advertisements to maintain object informed about the agent's presence and to maintain agent lists • It is preferable to use tunneling instead of default routes. The latter does not guarantee packet forwarding • Computation of multiple routes to a single destination is recommended to improve the robustness

may use various mechanisms to reduce routing traffic. When using a proactive discovery approach it is very important to keep a list of agents from which advertisements have been received. In addition, it is preferable to use source routing because it facilitates the forwarding job on each intermediate object.

We may see that even with the use of a reactive routing protocol and proactive discovery approach, a finite amount of time is needed to register to an FA before packet forwarding may be resumed. The exact amount of time depends on the route finding and forwarding mechanism, the improving mechanisms implemented, and the particular conditions of the MANET at handover time. The best combination is achieved for reactive routing protocols with the proactive discovery approach, which may be improved with unicasting, piggybacking, and TTL range scope.

We believe that for achieving seamless handovers, three conditions must be accomplished. First, the roaming object should have multiple routes to the current gateway available—at least two. If this is the case, the problem of loosing contact with the current agent will be solved, avoiding the need to reinitiate a new registering cycle. That is, in the event that one of the available gateway routes is lost, the others should be used to reach the gateway, and a new one should immediately be found if the total number of remaining routes is less than 2. This may be easily implemented with both types of routing protocols, being already available in some cases [21], but the use of reactive types is recommended to reduce congestion. Second, proactive gateway advertisement should be implemented to permit objects realize agent/gateway presence anywhere and at anytime. Third, as soon as an object considers that it has to handover according to the mentioned triggering mechanisms, it should first register to the new agent through two or more routes before deregistering from the current one. This new registration could be marked as a transitional registration, which may become a definitive registration as soon the object finishes the deregistration process from its current gateway. This would permit the object to continue forwarding its packets without any interruption.

12.6.2 Future Improvements

Potentialities offered by the Internet of Things (IoT) make possible the development of a large number of applications, of which only a very small part is currently available. New applications based on the IoT would likely improve the quality of our lives, and these environments will be equipped with objects with constantly growing intelligence. Applications in the transportation, in the logistics, and in the healthcare domains, for example, will require mobile objects that are permanently reachable on the Internet, giving them the possibility to communicate between each other and to elaborate the information perceived from the surroundings.

MANET objects will have to handover between different subnetworks while they move, and the communication interruption time has to be minimized in order to support sensible applications. It is clear that to reduce the network congestion and to reduce the handover time, the future trend will guide us to use reactive ad hoc routing protocols and a proactive agent advertisement approach. But this would not be enough. The idea is to get a "seamless handover" scenario, and for this, work has still to be done. Some of the issues that must be attended are the following:

- Reduction of route holding time on reactive protocols to decrease the time for declaring a broken route.
- Management of mobile object IP addresses when a handover occurs. Even with the use of mobile IPv6, there are some characteristics such as address collision discovery that do not permit seamless handover.
- Evaluate the use of multiple agent registration in order to have a backup registration when the actual one is lost. Mobile Ipv6 gives a first step by keeping user data in the old agent while the mobile object executes a handover.
- An alternative way to reduce the handover time is to maintain multiple routes to the actual gateway. Thus, broken routes will not cause agent loss.

As stated before, in 4G wireless systems, MANET is known to be the preferred type of network for nodes (objects) that are far from structured networks, and its performance during handovers must still be investigated.

12.7 Conclusions

In this chapter we have analyzed the scenario of moving objects connected to hybrid MANETs and performing handovers from one subnetwork to another. In this scenario, objects must first be triggered to register to a different agent/gateway; second, the object must register and set a new CoA; and third, they must reroute packets directed to nodes on the fixed network using the new agent/gateway. Accomplishing these steps require a finite amount of time during which the ongoing communications may be interrupted. The time to reestablish fluid communication depends on the type of routing protocol used, the gateway discovery approach, and the packet forwarding mechanism implemented.

None of the current schemes for MANET handover being used nowadays are completely conditioned to handle MANET handovers in a seamless manner. We propose that multiple gateway routes and early agent preregistration are the conditions needed to achieve handovers without communication interruptions. In additional, we recommend the use of a reactive routing protocol with proactive agent advertisement, modified with mechanisms such as unicasting, piggybacking, and TTL range scope.

References

1. G. Kortuem, F. Kawsar, D. Fitton, and V. Sundramoorthy. Smart objects as building blocks for the Internet of things. *IEEE Internet Computing*, vol. 14, no. 1, 2010, pp. 44–51.
2. S. Ding. A survey on integrating MANETs with the Internet: challenges and designs. *Computer Communications*, vol. 31, 2008, pp. 3537–3551.
3. F. Ros, P. Ruiz, and A. Gomez-Skarmeta. Performance evaluation of interconnection mechanisms for ad hoc networks across mobility models. *Journal of Networks*, vol. 1, no. 2, 2006, pp. 9–17.
4. F.M. Abduljalil and S.K. Bodhe. A survey of integrating IP mobility protocols and mobile ad hoc networks. *IEEE Communication Surveys and Tutorials*, vol. 9, no. 1, 2007, pp. XX–XX.
5. K. Ur, R. Khan, R. Zaman, A, Venu, and G. Reddy. Integrating mobile ad hoc networks and the Internet: challenges and a review of strategies. In *Communication Systems Software and Middleware and Workshops (COMSWARE 2008)*, 2008, pp. 536–543.
6. S. Ding. Mobile IP handoffs among multiple Internet gateways in mobile ad hoc networks. *IET Communication*, vol. 3, no. 5, 2009, pp. 752–763.
7. T. Wu, C. Huang, and H. Chao. A survey of mobile IP in cellular and mobile ad-hoc network environments. *Ad Hoc Networks*, vol. 3, 2005, pp. 351–370.
8. A. Derhab and N. Badache. Data replication protocols for mobile ad-hoc networks: a survey and taxonomy. *IEEE Communications Surveys & Tutorials*, vol. 11, no. 2, 2009, pp. 33–51.
9. P. Engelstad, A. Tønnesen, A. Hafslund, and G. Egeland. Internet connectivity for multi-homed proactive ad hoc networks. *IEEE International Conference on Communications*, vol. 7, 2004, pp. 4050–4056.
10. Y. Bin and S. Bin. Modify AODV for MANET/INTERNET connection through multiple mobile gateways. In *The 11th International Conference on Advanced Communication Technology (ICACT 2009)*, 2009, pp. 1519–1523.
11. D.D. Giusto, A. Iera, G. Morabito, and L. Atzori (eds). The internet of things: a survey. *Computer Networks: The International Journal of Computer and Telecommunications Networking*, vol. 54, no. 15, 2010. pp. 2787–2805.
12. H. Lei and C. Perkins. Ad hoc networking with Mobile IP. In *Proceedings of the 2nd European Personal Mobile Communication Conference*, 1997.
13. Y.-C. Tseng, C.-C. Shen, and W.-T. Chen. Integrating mobile IP with ad hoc networks. *Computer*, vol. 36, no. 5, 2003, pp. 48–55.
14. H. Ammari and H. El-Rewini. Integration of mobile ad hoc networks and the internet using mobile gateways. In *Proceedings of the 18th International Parallel and Distributed Processing Symposium (IPDPS04)*. IEEE Computer Society, 2004, p. 218.
15. M. Benzaid, P. Minet, K. Agha, C. Adjih, and G.Allard. Integration of mobile-IP and OLSR for a universal mobility. *Journal of Wireless Networks*, vol. 10, no. 4, 2004, pp. 377–388.
16. M. Ergen and A. Puri. MEWLANA-mobile IP enriched wireless local area network architecture. In *2002 IEEE 56th Proceedings of the Vehicular Technology Conference, 2002 (VTC 2002-Fall)*, vol. 4, 2002, pp. 2449–2453.
17. U. Jonsson, F. Alriksson, T. Larsson, P. Johansson, and G.Q. Maguire. MIPMANET mobile IP for mobile ad hoc networks. In *First Annual Workshop on Proceedings of the Mobile and Ad Hoc Networking and Computing, 2000 (MobiHOC, 2000)*, 2000, pp. 75–85.
18. Y. Sun, E. Royer, and C.E. Perkins. Internet connectivity for ad hoc mobile networks. *International Journal of Wireless Information Networks*, vol. 9, no. 2 (Special Issue on 'MANETs: Standards, Research, Applications), 2002, pp. 75–88.

19. P. Ratanchandani and R. Kravets. A hybrid approach to Internet connectivity for mobile ad hoc networks. In *IEEE Wireless Communications and Networking Conference (WCNC 2003)*, vol. 3, 2003, pp. 1522–1527.

20. J. Broch, D.A. Maltz, and D.B. Johnson. Supporting hierarchy and heterogeneous interfaces in multi-hop wireless ad hoc networks. In *ISPAN'99: Proceedings of the 1999 International Symposium on Parallel Architectures, Algorithms and Networks (ISPAN'99)*. Washington, DC: IEEE Computer Society, 1999, p. 370.

21. T. Kim, S. Yeo, J. Park, and B. Vaidya. Performance evaluation of hybrid multipath mobile ad hoc network. In *IEEE Consumer Communications and Networking Conference (CCNC 2009)*, 2009, pp. 1–5.

Chapter 13

Vehicular Ad Hoc Networks: Current Issues and Future Challenges

Saleh Yousefi, Mahmood Fathy, and Saeed Bastani

Contents

13.1 Introduction to Vehicular Ad Hoc Networks

Vehicular ad hoc networks (VANETs) are a special type of mobile ad hoc networks (MANETs) where wireless-equipped vehicles form a network spontaneously while traveling along the road. Direct wireless transmission from vehicle to vehicle makes it possible to communicate even where there is no telecommunication infrastructure, such as the base stations of cellular phone systems or the access points of wireless dedicated access networks. This new way of communication has been attracting much attention in the recent years in academic and industry communities. The US Federal Communications Commission (FCC) has allocated seven 10-MHz channels in the 5.9-GHz band for dedicated short range communication (DSRC) to enhance the safety and productivity of the transportation system [1]. The FCC's DSRC ruling has permitted both safety and nonsafety (commercial) applications, provided safety is assigned priority. The IEEE has taken up working on a new standard for VANETs, which is called IEEE 802.11p [2].

The communication pattern in VANETs includes two forms: vehicle to vehicle (V2V) communications and vehicle to infrastructure (V2I) communications. The former leads to a pure MANETs, while the latter can be viewed as a hybrid network. Although VANETs can be seen as a special case of MANETs, there are several distinctive characteristics that dictate special treatment of VANETs. The most important distinctive characteristics of VANETs are as follows:

■ Specific mobility patterns: As any MANET, mobility pattern has a noticeable effect on the behavior and performance of the network. In the context of MANETs, many mobility patterns are proposed. However, almost none of the previous mobility models can be applied to VANETs mostly due to distinctive mobility pattern of vehicles on roads. On the one hand, the mobility pattern of VANETs is one-dimensional or stripelike. On the other hand, the mobility of vehicles is affected by sophisticated interactions between individual vehicles. The optimistic point here is that there is very rich literature in the field of transportation engineering about modeling mobility patterns. Furthermore, there are many good common traffic simulators that can be of advantage in VANET simulation. Indeed, the current trend in VANET simulation is to find ways for exploiting analytical and simulation tools of transportation engineering in the simulation process of VANETs.

■ Highly dynamic topology: The topology of the network is a burden of high variability mainly due to the mobility of vehicles (in particular vehicles of the opposite direction) in highway scenario and rural environments as well as traffic lights and junctions in the urban environments. Thus, the performance of protocols' application in such unstable environments may be degraded and thus needs specific treatment.

■ Intermittent connectivity: Due to the reason mentioned above, vehicles may not be able to last their communication for a long time and they may encounter several pauses during the communication period. This is the main reason why many common network protocols and their applications should be tailored for VANETs to address this kind of connectivity.

■ Strict quality of service (QoS) requirements: The most important motivation of VANETs is safety applications that address life safety of people (drivers and passengers). Therefore, in spite of ordinary applications, very tough QoS metrics should be guaranteed by the network. Any violation of QoS metrics may be at the expense of people's life. This issue necessitates new performance evaluation metrics as well as protocol design techniques.

The goal of this chapter is to provide a basic knowledge about VANETs and to review state-of-the-art methods tackling the challenges as well as to present some directions for future research. To fulfill these goals, in the first four sections of this chapter, basic discussions have been presented, which are intended to give the reader a solid knowledge about VANETs and their applications and challenges. Then in the remainder of the chapter some important and advanced topics have been selected. In each case, we first state the problem in hand and the challenges, and then some example solutions are provided; finally, some open problems are presented for future research.

13.1.1 Motivation

Each year, many people suffer from different traffic causalities around the world. This issue followed by the huge economic burden of accidents urges governments to improve the level of safety on roads. In 1970s, passive safety systems such as safety belts and airbags were introduced. Although they have decreased the severity of accidents and the number of deaths and injuries, passive safety systems have not been able to decrease the number of accidents, and therefore active safety systems, such as antilock brake systems and electronic stability program systems, have been invented. The statistics published in [3] show that the active safety systems have stopped a rising trend of accidents, but the number of accidents remains almost the same for many years. In other words, it seems that the current safety systems are no longer able to decrease the number of accidents. To tackle the aforementioned problem, many investigations have been performed in order to understand the effective factors that cause accidents. The results of studies in many countries (e.g., [4]) show that the information error has the highest impact on accidents. By the information error, we mean that either the driver receives the critical information too late or the process has failed. The key factor here is the reaction time of the driver, which is relatively high (more than 1.0 s [5]) such that he or she cannot react promptly. As a result, if we could somehow increase the information horizon of the driver such that he or she could be informed about distant events earlier, it might lead to a noticeable decrease in the number of accidents as well as the number of deaths and injuries. However, recent advances in wireless technology make the idea of "communication anytime and anywhere" more reachable. Inspired from this idea, there is a growing belief that embedding wireless radios into vehicles may be quite beneficial from safety aspects. The ultimate goal is to provide new technologies that are able to improve safety and efficiency of road transport. It should be noted that telecommunication has been invoked in intelligent transportation systems (ITSs) for many years but the previous systems are centralized and include either cellular or infrastructure-based roadside to vehicle communications [6,7].

13.2 Roads Traffic Theory Basics

In order to clarify the challenges of communication in VANETs, we invoke some basic concepts from traffic theory. From the traffic theory [5] we know that there are three macroscopic parameters—speed (km/h), density (veh/km/lane), and flow (veh/h/lane)—that describe the traffic state on a road. Over the years, many models have been proposed for speed–flow–density relationships (see [5]). Simply, these parameters are related by the so-called fundamental traffic theory equation given as

$$F = S \times D \qquad (13.1)$$

where F, S, and D are the traffic flow, average speed, and traffic density, respectively. As shown in Figure 13.1, the general relationship between the above basic parameters can be studied in two different phases: First, when the density is low, the flow entering and leaving a section of the highway is the same and no queues of vehicles are forming within the section. This state holds until the density reaches a threshold called critical value. This phase is called stable-flow and is shown by the solid line in the figure. The peak of the flow–density curve is the maximum rate of flow or the capacity of the highway. Beyond this density, some breakdown locations appear on the highways, which lead to forming some queues of vehicles. This phase is called forced-flow and is shown by the dashed line in the figure. If the density increases further, the traffic reaches to the jam state where vehicles have to stop completely. In the stable-flow phase, when the density is sufficiently low, the speed of vehicles and the flow are independent and thus drivers can drive as fast as they want to. This state is called free-flow state.

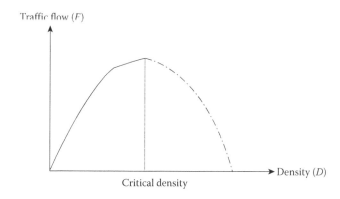

Figure 13.1 **Relationship between the basic parameters in the traffic theory. (From Yousefi, S., Altman, E., El-Azouzi, R., and Fathy, M.,** *IEEE Trans. Veh. Technol.,* **57(6), 3341–3356, 2008. With permission. © (2008) IEEE.)**

The above-mentioned traffic states determine the VANETs' challenges, which should be addressed. From the communication point of view that we peruse in VANETs, different challenges should be addressed in each traffic state. Obviously, connectivity is satisfactory in the forced-flow state while it deteriorates at light load corresponding to the free-flow state in which it might not be possible to transfer messages to other vehicles because of disconnections. However, since the network is sparse, collision between simultaneous transmissions is trivial in the free-flow state while it is one of the main communication challenges that should be addressed in the forced-flow traffic state.

13.3 VANETs' Applications and Messages

According to the FCC frequency allocation, one can categorize applications of VANETs into two main classes. The first class aims to improve the safety level in roads, i.e., safety applications. In this case, VANETs can be seen as a complementary to the current ITSs [6,7] in order to enhance the coverage and performance. The second class of applications, which is predicted to grow very fast in the near future, is commercial services, i.e., comfort applications.

13.3.1 Safety Applications

In safety applications, the goal is to improve the life safety level of passengers by exchanging safety relevant information between vehicles. The information is either presented to the driver or used by the automatic active safety system. Some examples are cooperative forward collision warning, left/right turn assistant, lane changing warning, stop sign movement assistant, and road-condition warning. Due to the stringent delay requirements, applications of this class may demand direct vehicle-to-vehicle communication.

With respect to QoS requirements, these applications are characterized by being delay and loss *intolerant*. In safety applications, usually data should be disseminated to a set of candidate vehicles; thus, broadcast (in its special form) is the dominant pattern of data dissemination. In VANETs' terminology this type of data dissemination is sometimes called geocast or roadcast, aiming at concentrating on the fact that the candidate vehicles are limited to a geographical area and/or a specific part of a road.

In the last decade, much effort has been devoted to substantiate VANETs in the real world. These efforts have brought together industrial and academic bodies in order to realize the idea. Some most important projects in this area include COMeSAFETY [8], COM2REACT [9], COOPERS [10], CVIS [11], and CICAS [12].

Any safety application demands exchange of related messages between vehicles. These messages can be classified into two categories: alarm and beacon, which have different dissemination policies and roles in safety improvement. In the following subsections, we provide a more detailed explanation.

13.3.1.1 Alarm-Based and Beacon-Based Safety Applications

Alarm messages are issued by all vehicles to announce to other vehicles about the previously happened events at a specific location of a road, such as car crash and icy surface, while beacon messages are issued periodically. Using the received beacons, vehicles try to inhibit possible events (not previously occurred) such as erroneous lane changing, forward collisions, and wrong left/right turning.

Besides, beacon messages might be used by other applications (e.g., routing protocols). Note that messages mentioned above are complementary to each other. While alarm messages may be able to inform the driver in time about already happened events in order to prevent more incidents, beacon messages can prevent many incidents before they take place. Moreover, since alarm messages announce events, they are more critical and should be disseminated with a higher priority.

13.3.2 Comfort Applications

In comfort applications, the goal is to improve passenger comfort and traffic efficiency. Examples for this category are traffic-information system, route optimization (navigation), electronic toll collection, map download, video download, and Internet on the roads. These applications are predicted to grow very fast in the near future due to business motivations. Although some of the projects introduced in Section 13.3.1 also have some subtasks related to comfort applications, one may name NOW [13] as a project focusing on data (Internet) access on roads using V2V and V2I techniques.

In this category of applications, both unicast and broadcast communications are justifiable. On the one hand, legacy data applications (e.g., FTP and HTTP) mostly require unicast communications, and on the other hand, many other applications such as map download demand broadcasting of data to all or a subclass of vehicles. Geocast or roadcast is sometimes defined as broadcasting data to a specific geographical location or to a part of the road.

13.4 Performance Evaluation Metrics for VANETs

Due to their QoS requirements, the treatment of comfort applications is similar to that of common networking applications; however, safety applications require a different treatment as explained in the following. In order to deploy safety applications in VANETs, there should be effective ways to evaluate their degree of success in providing safety. For this purpose, inspired from networking literature, researchers utilize common evaluation metrics such as delivery rate and delay. Although these metrics are also valuable for evaluating the performance of safety applications, the following distinctive characteristics necessitate specific treatment of the performance evaluation of safety applications:

■ In safety applications, the lack of fresh information makes each individual vehicle a life threat for the others. Thus, in order to evaluate the performance of a message dissemination protocol or a safety application, the quality of the safety offered to all individual vehicles is critical and should be monitored. While in ordinary networking scenarios, the average values of the metrics of interest are usually evaluated, and in safety scenarios, the average values are no longer useful. Therefore, we propose to monitor the metrics of interest for all vehicles and count on the *worst-case* values.

■ Although the distance (in meters) between senders and receivers may be important in some networking scenarios (e.g., sensor networks), from the safety viewpoint it is the most important metric of performance. To have a solid evaluation in safety scenarios, the coverage property need to be evaluated. In other words, for any given vehicle the range in which the safety applications can provide adequate safety is of ultimate importance. Within this range, vehicles can receive the safety messages of the initiator vehicles with a satisfactory QoS level, i.e., desirable values for delay, delivery rate, etc.

■ The definite goal of any beacon-based safety application deployed in a given vehicle is to inform neighboring vehicles about the vehicle's own status. Hence, evaluating the fact that how well and how fairly the beacon dissemination protocols and/or safety applications are successful in this regard is indispensable from the safety viewpoint.

Note that the above concerns should be taken into account all together. With this in mind, we propose two new metrics, primarily focused on evaluating the performance in safety scenarios. It should be noted that the following metrics may be used along with other common evaluation metrics:

■ Effective range: In any time step (say 1 s), the *effective range* is defined as the range within which the worst case of QoS metrics is satisfied. The satisfaction levels depend on the projected safety application. Although one may consider a different QoS metrics, as a possible way we define the effective range as the range within which (a) the minimum delivery rate is above a predefined threshold and (b) the maximum end-to-end delay is below a predefined threshold. Note that this metric has a value in each time step and the obtained values can be averaged to give an average number during the course of simulation time. This metric can be evaluated for both alarm-based and beacon-based safety applications.
■ Beaconing rate: In any time step (say 1 s), for a given beacon disseminating vehicle the *beaconing rate* is defined as the average value among its worst-case delivery rates to all surrounding vehicles in its transmission range. This metric assists in knowing how well the beacons disseminating from a vehicle reach other vehicles. When we compare the value of this metric for different vehicles, it gives us valuable insights into the capability of the safety application in providing a fair safety performance to each vehicle. Note that this metric has a value in each time step and the obtained values can be averaged to give an average number during the course of simulation time. This metric is suitable for evaluation of safety of beacon-based safety applications.

13.4.1 Application Layer Performance

In Vehicle Safety Communication Project [14], 34 vehicle safety applications, enabled or enhanced by VANETs, were studied. Among them, 8 applications were identified as high priority and selected for extracting their communication requirements. These applications include both alarm-based and beacon-based safety applications. Typically, safety applications have the following communication requirements:

1. Communication ranges from 50 to 300 m (the more the better)
2. Safety message transmission interval from 20 ms to 1 s
3. Safety message size from 200 to 500 bytes
4. End-to-end delays below 150 ms

Note that DSRC is intended to support 1000 m transmission range. However, depending on the channel conditions and environmental obstacles, the practical transmission range may be less than the nominal values. It also uses the IEEE 802.11p MAC (medium access control) protocol, a variant of IEEE 802.11 protocol with Carrier Sense Multiple Access/Collision Avoidance (CSMA/CA) behavior. Therefore, it could be quite reasonable to consider single-hop dissemination as an important type of future intervehicle communication. However, in particular for alarm and

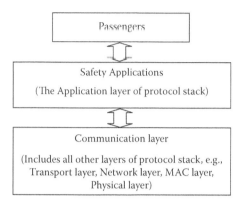

Figure 13.2 Three-layer architecture for safety applications in VANETs.

comfort messages, multihop message propagation is challenging. Note that when we focus on 1-hop beacon dissemination, we get involved in the MAC layer broadcasting, which is quite different from the network layer broadcasting.*

As we know, a network protocol stack is structured in a layered architecture in which each layer offers services to the upper layer. The application layer is the upmost layer in the protocol stack and offers services to the user. To ease explanation, we define a virtual layer, called communication layer, as the combination of all layers under the application layer. This virtual layer normally, as the Open Systems Interconnection (OSI) reference model suggests, includes physical layer, MAC layer, networking layer, and transport layer. However, VANETs may be deployed with a rather different protocol stack and some layers may be absent [15]. For example, normally in beacon dissemination applications, only MAC and physical layers are present. Nevertheless, in the sequel of this section we continue with the general term "communication layer" to deal with more general cases.

Figure 13.2 shows three-layer architecture we consider in this chapter. The communication layer offers services (i.e., sending and receiving data) to the safety applications in the application layer, and, the safety applications are in charge of giving different life safety services to the passengers. In general, in networking literature there are two kinds of relationships between the QoS offered by the communication layer to the application layer and the one that the application layer offers to the users (in our context 'passengers').

1. In reliable data transfer protocols such as HTTP and FTP, the communication layer must guarantee reliability in terms of the delivery rate, even though it may lead to a large delay. Indeed, the role of the transport layer (such as TCP) is to compensate shortcomings of the lower layers, so that the application layer can offer satisfactory services to the user.
2. In real-time applications, normally it is not required that the communication layer offer fully reliable services to the application layer, because it may come at the expense of excessive delay, which is critical in those applications. In other words, the application layer tolerates some unreliability in the services offered by the communication layer.

---

* Beacon-based safety applications normally demand single-hop broadcasting at the MAC layer, while alarm-based safety applications and comfort applications usually demand multihop broadcasting/unicasting at the network layer.

While alarm-based safety applications and comfort applications may be kept in the first category, we believe that beacon-based safety applications can be categorized in the second group. Note that beacon messages do not contain unpredicted information (as the alarm messages do). Thus if the application layer does not receive fresh beacon messages from some vehicles for a short period, it can perform extrapolation to guess their status (e.g., speed, direction, position, and acceleration). It should be emphasized that extrapolation is not possible in the case of alarm messages, because alarm messages announce unpredicted events in which the new status of vehicles does not follow their previous status. Furthermore, when a fresh beacon message is received, the validity of the old one actually expires. Therefore, it is not needed that the communication layer struggle while retransmitting the old one. These considerations convinced us that beacon-based safety applications can get an advantage of some degree of tolerance in QoS offered by the communication layer. In other words, they function properly even though the communication layer is not able to offer a desirable QoS. Recent studies on DSRC show that single-hop beacon dissemination in VANETs is adequate in terms of delay, but the reliability remains defective [16,17]. Hence, in the following we concentrate on reliability as the main performance concern and consider beacon-based safety applications only.

For a given vehicle, we define a time window t by which the safety application must receive at least one fresh beacon from any neighboring vehicles. Let us denote by T the transmission interval of the communication protocol. Let p_{com} be the success rate of the communication protocol, which is the minimum among all delivery rates in a given transmission range of vehicles. Now the beacon-based safety application works adequately if at least one beacon among t/T beacons, from each neighboring vehicle, is received. In other words [16,17],

$$P_{app} = 1 - P(\text{all failed in } N \text{ tries})$$

$$= 1 - (1 - p_{com})^N \tag{13.2}$$

$$= 1 - (1 - p_{com})^{t/T}$$

where P_{app} is the success rate of the safety application. It should be emphasized that the above equation is valid only if the message losses are independent, as we assume here. With Equation 13.2, one can simply relate the reliability of the application layer to that of the communication layer. In Figure 13.3, the relationship between reliability of the application layer and that of the communication layer is shown for different values of t (i.e., the time window as introduced above).

In order to design a safety application, the required attribute of the safety application should be extracted by traffic safety experts and then be used by the VANET protocol and application designers. So far in our model, each application is attributed by two parameters T and t. In addition, we define a vector α that represents the satisfaction levels for each QoS metric (e.g., delivery rate and delay). Table 13.1 shows our proposed attributes for each safety application.

Based on the above attributes, we can consider different classes of safety applications. For instance,

■ Driver-assistant safety applications, which are expected to assist the driver in different maneuvers, such as lane changing and turning left/right. This class of applications can prevent various incidents that may happen because of the driver's fault. One example may be 95% < α < 99% and $N \leq 3$. Note that the sufficient values for α and N should be determined by safety experts.

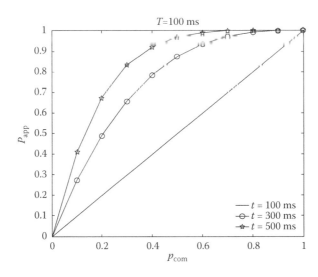

Figure 13.3   The relationship between reliability of the application layer and that of the communication layer.

Table 13.1   Attributes of Beacon-Based Safety Applications

Attributes	Description
T	The time window in which the beacon-based safety application should receive at least one fresh beacon
T	The transmission interval of the beacon communication protocol
α	A vector comprising of satisfaction levels for QoS metrics of interest (e.g., delay and delivery rate)

■ Automatic safety applications, which are expected to control the vehicle as stand-alone systems. Undoubtedly, this class of applications requires restricted QoS satisfaction level. For instance, $\alpha > 99\%$ and $N = 1$. Note that the sufficient values for α and N should be determined by safety experts.

13.5 Simulation of Vehicular Networks

Although there are some efforts made by major car manufactures and research consortiums to establish prototype projects, simulation is still the most tractable and feasible way to evaluate new ideas in large-scale VANETs. Similar to other types of MANETs, there are well-known network simulators that may be invoked to conduct evaluation by simulation (more detailed explanation is provided later). However, it should be stressed that the main difference between VANETs and other types of MANETs is twofold: First, vehicle movement on roads follows sophisticated mobility patterns that are not modeled by means of random mobility patterns as used commonly in many other types of MANETs [18]. Moreover, there is a very mature knowledge on vehicle's mobility due to decades of research in transportation engineering as a field of civil engineering. Therefore, one aiming at simulation of VANETs needs to be familiar with fundamentals of traffic theory or

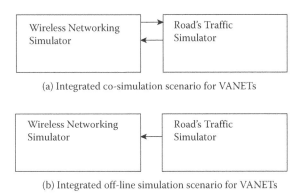

(a) Integrated co-simulation scenario for VANETs

(b) Integrated off-line simulation scenario for VANETs

Figure 13.4   Simulation scenarios in VANETs.

at least be able to make use of plenty of well-known traffic movement pattern generators (see some instances later in this section). Second, in many VANET scenarios the communication between vehicles may affect vehicle's mobility and vice versa. For example, a driver who is informed of a congested junction (through an alarm message) may decide to change his or her path and use an alternative path; thus, delivery of the message causes the change of the vehicle's movement pattern. On the contrary, the change of the vehicle's movement pattern affects road density and message dissemination. This means that we should be able to couple a network simulator to a road's traffic simulator. What is needed in this case is a co-simulation of the wireless network behavior and traffic's movement by which the aforementioned mutual interaction is actualized. However, many current research in VANETs is done by integrated off-line simulation, in which one-sided interaction is held (i.e., the wireless network simulator is fed by mobility patterns generated by the road's traffic simulator). Figure 13.4 shows a block view of the two possible simulation scenarios in VANETs.

13.5.1 Road's Traffic Simulators

Generally, traffic's flow can be viewed in either macroscopic or microscopic perspectives. In the macroscopic view, vehicular traffic is considered as fluid compressible medium. Thus, the basic rules of fluid mechanics are applicable. In particular, there is a basic equation that relates macroscopic metrics such as flow, density, and speed (see Equation 13.1). The macroscopic modeling of vehicular traffic is able to provide only general information such as road's capacity and density and does not consider individual vehicle's movement. On the contrary, the microscopic perspective of vehicle's traffic takes into account each vehicle's movement. Thus, many sophisticated aspects of vehicular traffic such as car-following models (i.e., mutual interaction between vehicles while traveling along roads) can be modeled. As a result, microscopic traffic simulators are more suitable for VANETs research.

There are many proprietary road's traffic simulators such as Paramics [19] and CORSIM [20] that are able to model vehicular traffic in great detail, but without loss of generality, here we focus on free or open source simulators that can be used easily for public research. Nowadays, many of such simulators are presented. At the time of this writing, the most important vehicle's mobility generators include SUMO (simulation of urban mobility) [21], MOVE (mobility model generator for vehicular networks) [22], FreeSim [23], City Mov v.2 [24], VanetMobiSim [25], and STRAW (street random waypoint) [26]. Decision on choosing a particular software should be taken based on the requirements of the simulation scenario in hand. A detailed explanation on a particular simulator is beyond the scope of this book. However, a clarifying and good comparison between these simulators and some others is provided in [27].

13.5.2 Wireless Network Simulators

In this case, all simulators that are usable in other types of MANETs can be of use. The most important proprietary simulators include OPNET [28] and QualNet [29]. However, the most important and commonly used free or open source software are the following: Network Simulator 2 (NS-2) [30], Global Mobile system Simulator (GloMoSim) [31], Objective Modular Network Testbed in C++ (OMNET++) [32], Java in Simulation Time/Scalable Wireless Ad hoc Network Simulator (JiSt/SWANS) [33], SNS (a staged network simulator) [34], and NCTUns (National Chiao Tung University Network Simulator) [35]. It should be noted that GloMoSim is no longer supported and a new version of it is present under the title of the proprietary software QualNet.

13.5.3 Integrated Off-Line Simulation of VANETs

It should be noted that almost all traffic simulators are able to generate mobility patterns for famous wireless network simulators, including NS-2, OMNET++, and GloMoSim (QualNet). Such information can be easily accessible from manual documents of the above-mentioned traffic simulators. Therefore, conducting integrated off-line simulations is a straightforward task. Indeed, most of the simulations in the current literature can be classified in the category of off-line simulations. In other words, a wireless simulator is fed by a mobility pattern generated by a road's traffic simulator. Although in many cases this issue suffices, for many other cases one may need online simulation (co-simulation) to be able to take into account mutual interactions between a traffic simulator and a wireless network simulator.

13.5.4 Integrated Co-simulation of VANETs

Currently, quite a few of these simulators are available and their usage is not that much common in the research community. As mentioned above, due to the special characteristics of VANETs there should be an integrated simulation framework by which both road's traffic and the wireless network aspects of the problem are simulated. Most of the software (except NCTUns) are indeed third-party projects that are developed on top of two other simulators: a road's traffic simulator and a wireless network simulator. In the following, we list the most important ones available at the time of this writing:

a. TraNS (traffic and network simulation environment) [35]: It is written in C++ and Java and facilitates co-simulation between NS-2 and SUMO. It was developed and supported by EPFL, Switzerland.

b. Veins (vehicles in network simulation) [36]: TraCI (traffic control interface) modules for co-simulation of SUMO with OMNeT++ and JiST/SWANS. TraCI is a client/server architecture for connecting to SUMO, in which SUMO behaves like a server that interacts with a client. It was developed by Christoph Sommer et al. in University of Erlangen, Germany.

c. NCTUns [37]: In contrast to the above two software, it has built-in capability to support integrated co-simulation of vehicular traffic and wireless network. NCTUns is a powerful and general-purpose network simulator that has been extended to VANET simulation. The mobility patterns generated by NCTUns are claimed to show a good agreement with those of common traffic simulators.

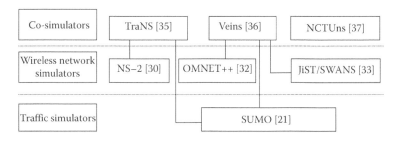

Figure 13.5   Taxonomy of tools for integrated co-simulation of VANETs.

Figure 13.5 shows the most important tools available to conduct integrated co-simulation of VANETs. Note that the case of integrated off-line simulation is not shown here due to plenty of alternatives.

13.6 Transport Layer Protocol for VANETs

VANETs are supposed to be connected to other wireless as well as various wired networks. Therefore, to offer a standard application interface (e.g., socket-like interfaces), one may consider standard protocols such as TCP (transport control protocol) and UDP (user datagram protocol) (and their enhancements) as transport layer protocols of VANETs. TCP is a well-known connection-oriented and reliable protocol operating in the transport layer of the OSI model. This protocol is relied on Acknowledgment (ACK) to be sure about the delivery of each segment of data; thus, it is normally resorted for applications in which reliability is critical. TCP treats data as an ordered sequence of packets and uses retransmissions to guarantee that from the receiver's point of view, all packets are received in order. As a result, a reliable and in-order reception of each piece of data is promised. While UDP is a connectionless and unreliable protocol in which neither reliability nor in-order reception of data is supported. Indeed, the most important feature of UDP is to provide a socket-like interface to the application layer (through defining port numbers in its header). Choosing a particular protocol directly depends on the QoS requirement of the application in hand. In the following we discuss this issue.

13.6.1 Transport Layer Protocol for Safety Application

From Section 13.4 we know that safety applications are characterized by tough timing requirements. In many safety applications, late reception of a message is as bad as nonreception of it. Therefore, it is easy to deduct that TCP does not suit safety applications due to the following reasons: (1) Connection establishment (i.e., three-way handshaking mechanism) is time consuming, which might jeopardize timing requirement of safety applications. (2) TCP uses a closed-loop reliability mechanism (based on ACK) and tries to retransmit lost packets until they are received successfully. However, for many safety applications it might be better to send a fresh message (conditioned on that the message is repeatable) instead of insisting on resending the previously issued one.

Generally speaking, using TCP for safety applications in VANETs is not justified and instead UDP is more suitable. However, for some cases when the safety message is not repeatable, the effort of the application layer is beneficial for compensating the poor reliability of UDP. One example of unrepeatable safety messages may be an alarm message announcing an icy surface. If the vehicle receiving this message is disconnected from other approaching vehicles (due to intermittent connectivity of VANETs), then it should keep the message (in the application layer) or deliver it to an RSU (roadside unit), with the aim of not losing the alarm.

13.6.2 TCP for Comfort Applications in VANETs

On the one hand, many comfort applications are based on bulk data exchange (e.g., file download) or long-lasting data connections (e.g., HTTP). These applications require in-order and reliable exchange of packets. On the other hand, we know that VANETs suffer from intermittent connections due to movement of vehicles [38]; thus, the amount of packets that may be received successfully varies with the traffic pattern, and the order of received packets may not be preserved. Furthermore, an exchange of ACK packets in such intermittently connected network results in increasing traffic load of the network and thus leads to increase of interference and collision level in the MAC layer of the DSRC standard (which is based on IEEE 802.11p). Since IEEE 802.11p is based on the CSMA/CA mechanism, the aforementioned collisions are too restrictive and thus cause poor throughput of TCP connections.

TCP is originally proposed for wired Internet; thus, it suffers from many shortcomings when used for wireless networks. The main problem is that TCP is not able to distinguish between error-prone links and network congestions, which leads to unnecessary slow-start mechanisms. Generally, the previous works tried to modify the congestion control of the TCP (e.g., freeze-TCP [39]) or to use information of intermediate nodes such as Explicit Congestion Notification (ECN) (RFC 3168). There are many works that use TCP for MANETs, which are reviewed in [40] in great detail. Moreover, there are a few works that address TCP challenges for VANETs in particular [41]. However, it seems that TCP has an inherent weakness for such intermittently connected topology as VANETs. Therefore, one may use other approaches for tackling this problem. In the following section we present an example of such methods.

13.6.2.1 Use of UDP Along with Forward Error Correction Techniques

The troublemaking point of TCP in VANETs is its congestion control mechanism through which it reacts whenever the network is disconnected temporarily. If one uses UDP (instead of TCP), the problem related to congestion control is solved, but it is still needed to compensate for the lack of reliability of the UDP. One solution for this problem is to make use of UDP (instead of TCP) along with an application layer protocol based on FEC (forward error correction) to satisfy reliability. In fact this method is based on an open-loop reliability mechanism in which there is no need for any ACK of data packets and retransmission of packets. Instead, probable errors are corrected by taking advantage of some redundant data that have been sent along with the original data.

In a proposed algorithm [42], Yousefi et al. made use of fountain coding as an FEC approach for a category of applications in VANETs that are based on file downloading. The term *fountain* refers to the fact that the only thing that the receiver needs to be able to reconstruct the original input symbols (packets) is to receive a minimum number of *any* output symbols (packets). Let k be the original number of packets that constitute the file to be transmitted, and let n be the total number of packets that need to be received at the receiver so that it can decode the original content. We have $n = k(1 + \varepsilon)$, where ε, termed the decoding efficiency, ranges typically between 10 and 100% (normally in the range of 5–10%), depending on the specific implementation [43]. Please note that when the receiver receives the n symbols, decoding is successful with a probability equal to $(1 - \delta)$, where δ is upper-bounded by $2^{(-k\varepsilon)}$. This means that larger file sizes and/or higher values of ε make the decoding probability even larger. As of complexity, it may be as low as linear in the number of coded symbols.

Among the potential applications of VANETs (safety and comfort applications), we believe that using fountain coding best suits comfort applications. This is because in safety applications we normally encounter a small number of bytes (in the scale of a few kilobytes) to be transmitted and thus there is no need to chop a safety message into pieces. Furthermore, due to restrict delay requirements of the safety message, the time overhead of coding and decoding algorithms might be intolerable by the majority of safety applications.

In the proposed approach of [42], the sender vehicle encodes files using a sample of fountain such as Raptor [44] (or the files may be encoded off-line and stored in the memory of the sender). Then the sender sends a train of encoded packets toward the receiver using the UDP protocol in the transport layer, such as a fountain that spreads water drops. In the beginning of a file transmission, the sender declares the amount of packets of the original file, size of each packet, and the coding algorithm. When the packets arrive, the receiver tries to decode the file using the same coding algorithm. Whenever the amount of received packet is enough, the receiver sends a message to the sender and asks to stop sending packets. Note that the sufficient amount of packets for successful encoding is just slightly larger than the amount of actual file packets. But there is no need to receive packets in a special order and all packets have an equivalent value for the receiver vehicle.

Figure 13.6 shows the number of completely downloaded files of the above-mentioned technique named as fountain as well as the ordinary FTP (which make use of TCP). This metric is important for comfort applications because due to their nature, vehicles can make use of such an application only if the related file is downloaded completely. In the conducted simulation, 30 vehicle pairs are chosen such that their hop count distance is less than 4 (at the initiation of the communication).

As shown in Figure 13.6, the fountain scenario outperforms the FTP scenario. However, as one can conclude from the figure, in most of the cases we observe a poor performance in terms of number of completely downloaded files. Since this is inevitable due to the dynamic nature of the traffic, we need to provide a resume facility. In other words, vehicles could be able to continue their incomplete download from other vehicles somewhere else and/or some time later. The use of fountain coding can best fit this requirement because each neighboring vehicle may have different packets from a file. In such a situation, a given vehicle can reconstruct the file whenever it is able to collect enough distinctive packets from neighboring vehicles. Obviously, this demands some

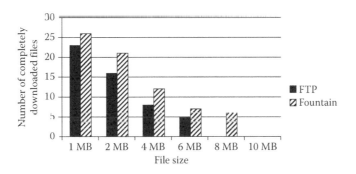

Figure 13.6 Comparison of fountain and FTP scenarios: number of completely downloaded files (out of 30). (From Yousefi, S., Chahed, T., Moosavi, M., and Zayer, K., Comfort applications in vehicular ad hoc networks based on fountain coding, In IEEE WiVec 2010, Taipei, May 2010. © (2010) IEEE. With permission.)

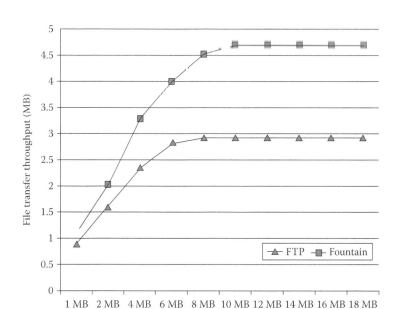

Figure 13.7 Comparison of fountain and FTP scenarios: average byte throughput. (From Yousefi, S., Chahed, T., Moosavi, M., and Zayer, K., Comfort applications in vehicular ad hoc networks based on fountain coding, In IEEE WiVec 2010, Taipei, May 2010. © (2010) IEEE. With permission.)

support from the application layer. On the contrary, in the case of FTP, the vehicle would need to collect specific packets (based on the predefined order) that may be hard to obtain from neighboring vehicles. In order to evaluate this capability of fountain, Figure 13.7 shows byte throughput of the fountain and FTP algorithms. This metric is important since vehicles may be able to resume the download some time later or from some other vehicles. As followed from the figure, the fountain scenario's throughput in terms of byte count is higher than the FTP scenario's. Indeed, by taking advantage of fountain coding we can transfer larger files in comparison to the case when a classic FTP algorithm is used. It is mainly because fountain scenario neither uses retransmission mechanism nor needs in-order packet delivery. In other words, in fountain all file chops have equivalent value, and if one is lost, it can be replaced by another one easily.

13.7 Vehicle to RSU Communications

As mentioned before, vehicular networks exist in two different architectures: V2V and V2I. In the V2V case, which is essential for safety applications, a pure ad hoc network between moving vehicles is established, whereas in the V2I case, vehicles and roadside infrastructure construct a hybrid ad hoc network. The latter is suitable mostly for comfort applications, even though there are also some safety applications that rely on V2I architecture. It is expected that data access from RSUs will become crucial in the near future [45]. Currently, it is assumed that RSUs are equipped with DSRC technology and thus use the IEEE 802.11p MAC layer. However, due to advances in mobile WiMAX (worldwide interoperability for microwave access), it is quite predictable that WiMAX technology is applied in this case (see some more discussion in Section 13.12). Among different applications of VANETs, comfort applications are more significant candidate for vehicle to RSU communications. It is mainly because the delay incurred for transmission from a vehicle to RSU and from the RSU to another vehicle may violate timing requirement of many safety applications. Of course, for some alarm-based safety applications in which an unrepeatable alarm is generated, RSUs can be used to buffer the alarm until an interested vehicle arrives.

Commonly, RSUs can act as a buffer point between vehicles or act as a router for vehicles to access the Internet. Besides, the RSUs may act as servers and thus provide various types of information to vehicles on roads. The following instances are some examples for RSU applications:

1. WEB applications: The passengers can connect to the Internet and make use of various applications such as checking e-mails and browsing Web pages or other Web applications.
2. Real-time traffic: Vehicles can report real-time traffic observations to RSUs. The traffic data then can be transmitted to a traffic center. The result of traffic data analysis then can be accessible to vehicles moving across each RSU.
3. Digital map downloading: When vehicles are driving to a new area, they may hope to update map data locally for travel guidance, such as changing unilateral or deadlock roads.
4. Commercial advertisements: When a vehicle arrives at a new area, it is very helpful to receive local advertisements on hotel reservation, parking places, latest price of petrol, and other stuff. In this case, video and audio advertising files are to be broadcasted by different companies.

In the following we consider an urban environment in which an RSU is established in each junction. Technical challenges in this case can be categorized into intra-RSU and inter-RSU challenges.

13.7.1 Intra-RSU Scheduling

As shown in Figure 13.8, the RSU established in a junction is in charge of serving vehicles that are moving through the cross-road. Actually the RSU responds to the requests submitted by the vehicles. On the one hand, each vehicle stay in the RSU area for a short period of time, and on the other hand, by increasing the number of vehicles (and thus requests), the bandwidth limitation becomes an important challenge. Therefore, it is important to use a scheduling policy to maximize the number of served requests. Besides, some requests that have higher priority should be taken higher priority of service.

In [46], Zhang et al. proposed a scheduling algorithm that is summarized in the following. Each vehicle's request is characterized by a 4-tuple: <*v_id*, *d_id*, *op*, *deadline*>, where *v_id* is the identifier of the vehicle, *d_id* is the identifier of the requested data item, *op* is indicating the operation (upload or download), and *deadline* is the time constraint of the request. If a request is not served within the deadline time limit, it will be dropped from the waiting queue (e.g., the service queues in Figure 13.8) of RSU since the related vehicle is no longer under RSU's coverage. Having this information, the RSU uses scheduling policy to maximize the number of served requests. For this purpose, a scheduling algorithm called D*S has been proposed in which both data size and request deadline are considered. For more detailed discussion and results, please refer to [46].

An open problem and challenge here is to consider multiclass requests. Indeed, in [46] two classes have been mentioned: upload and download. As shown in Figure 13.8, one can consider more sophisticated case in which different QoS classes such as video, audio, and data (text) are distinguished. Therefore, studying different queuing policies such as WFQ and WRR can be taken into account in both simulation and analytical points of view. Furthermore, since many vehicles may have the same requests, invoking multicast approach leads to a higher performance

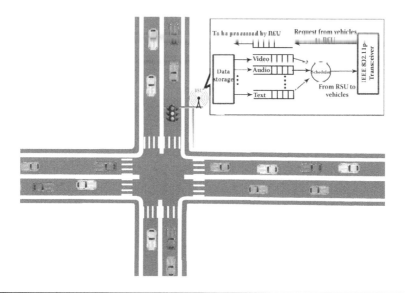

Figure 13.8   Intra-RSU scheduling scenario.

in terms of the number of served requests. In this case the scheduling algorithm should maintain a trade-off between the number of multicast requests and unicast requests which are given service.

13.7.2 Inter-RSU Scenario

If we extend the previous case to an urban environment, an anticipated scenario would be one in which several RSUs are installed in each cross-road. Thus, vehicles moving out from a RSU's range will move into another RSU's range after some time. In other words, the service that is stopped in the first RSU can be resumed in the next RSU. This issue necessitates scheduling algorithm through which RSUs cooperate in order to maximize the number of served requests. Furthermore, reducing delay for individual requests would be another goal of scheduling algorithm. Since the number of vehicles and RSUs as well as the number and size of files are potentially large, scalable scheduling is a challenge. It should be recalled that as a realistic assumption we consider that the RSUs are connected by another network (a wired network or a wireless one such as WiMAX) and managed by a service provider or a group of joint service providers.

In [47], Shahverdy et al. studied a sample of aforementioned problem of file downloading. It is assumed that the files are uploaded by the service provider through the network of RSUs. Therefore, the scenario suits comfort applications only. A vehicle may not be able to finish its download from an RSU; thus, the proposed algorithm allows it to continue its download from the next RSU. In the proposed scheduling algorithm, each RSU implements two separate queues for (1) download requests from the scratch and (2) download requests that are resumed. The data for distribution are chosen from aforementioned queues based on some scheduling policies. A scenario of the problem is depicted in Figure 13.9.

Vehicles retrieve their data from the RSU when they are in the RSU's coverage range. The RSU (server) maintains a service cycle, which is non-preemptive; i.e., one service cannot be interrupted until it is finished. All vehicles can send request to the RSU if they tend to access the data. Each request is characterized by a 5-tuple: <*v-id*, *d-id*, *w-RSU*, *s-rec*, *deadline*>, where *v-id* is the

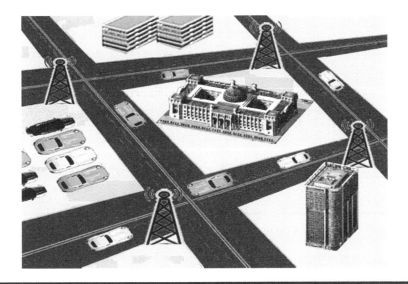

Figure 13.9 A scenario with multiple RSUs that demands inter-RSU scheduling.

identifier of the vehicle, *d-id* is the identifier of the requested data item, *w-RSU* identifies the RSU from which the vehicle has come, *s-rec* is the identifier of the data size that is received until now from the previous RSU (the vehicle now asks for downloading the remaining data from the current RSU), and *deadline* is the critical time constraint of the request, beyond which the vehicle moves out from the RSU area. Each vehicle is equipped with a GPS (global position system); therefore, vehicles know their own geographical position and driving velocity. Therefore, a vehicle can estimate its leaving time, which indeed is the service *deadline*, mentioned above. The scheduling policy adapted here is D*S, which is originally proposed in [46].

Open research problems in this case can be the following. Similar to the case of intra-RSU, one can extend the model to different QoS classes (video, audio, ordinary data). The trade-off between multicast and unicast requests is also challenging in particular because of multiple RSU architecture. Another important issue that is very critical in realizing the idea in real life is to consider the following problem. Take m files (containing l chops) and n RSUs and T be the average request deadline (the average time a vehicle is in the coverage area of an RSU). Then one may face a distributed file download case, which can be solved as a maximization problem. The details of such a problem and objectives are needed to be investigated, but maximizing number of served requests and minimizing average download delay can be considered as objective functions.

13.8 Beacon Based Safety Applications

As mentioned in Section 13.4.1, single-hop beacon dissemination is sufficient for most of the safety applications. However, there is an ongoing debate on whether multihop beacon dissemination would be necessary. Recent results show that multihop dissemination of beacons (periodic safety messages) results in high imposed load on the wireless channel and poor performance of DSRC systems. The main reason here is that DSRC is using IEEE 802.11p, an alternative of IEEE 802.11, which is based on the CSMA/CA paradigm. For a good discussion on this issue, an interested reader may refer to [48]. The following discussion is mainly focused on single-hop message

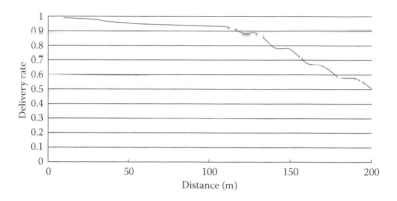

Figure 13.10 Delivery rate of beacons in terms of distance from the sender.

dissemination; however, since multihop dissemination is based on single-hop dissemination, the results may be usable there.

To show the effect of the distance on beacon reception rate, in Figure 13.10, we show the result of a simple beacon dissemination protocol [49] where the vehicles' transmission range is 200 m. The transmission interval is 200 ms and the packet size is 500 bytes. As followed from the figure, the delivery rates are decreasing dramatically by increasing the distance from the sender. This indeed means poorer safety coverage by the beacon-based safety application. We can describe this border effect mainly by a well-known hidden terminal problem. A hidden terminal is one that is within the range of the intended destination but out of the range of the sender. The partial solution to this problem is the use of RTS (ready-to-send)/CTS (clear-to-send) packets. RTS/CTS signaling may solve the problem for unicast communication, but in IEEE 802.11 the RTS/CTS mechanism is not invoked in the broadcast scenarios mainly because CTS messages sent by multiple receivers will result in severe collisions. Hence, the hidden terminal is more troublesome in the broadcast mode of the IEEE 802.11 MAC layer.

Observing such a low delivery rate, the important question that would arise is how to alleviate such an adversity in order to get acceptable QoS for safety applications. The main approach is to control the wireless channel's load. The following factors are the most important ones that should be controlled in order to reduce channel's load:

■ Transmission range: While higher transmission range results in larger awareness distance and is better from the safety point of view, it leads to a larger interference domain. As a result, packets are more likely to collide with each other and throughput degrades more severely. A good example of approaches that are based on controlling power (transmission range) for alleviating channel's load is that of [50].

■ Transmission interval: This parameter is directly related to the requirements of the safety applications and should be determined based on vehicle's speed and driver reaction time, and traffic density. While a smaller transmission interval can prevent unsafe situation in higher speeds and more unsafe conditions, it results in more saturated channels and so it is more likely to cause collision between simultaneous transmissions. An example of works that consider increasing transmission interval (decreasing transmission frequency) is that of [51], which will be discussed briefly in the following subsections.

■ Packet payload size: To estimate the packet size value, we consider that every packet will contain several parameters composing the state of the sender especially location, speed, road hazards, etc. Also, there should be some aggregated information about sender's neighbors. In addition, by including security issues that are very important in intervehicle communication, we can reach packet sizes ranging from 100 to 500 bytes for each message. Generally, safety messages with the size of 1 Kbyte are not far from expectation. An example of works that consider an increase of packet payload size (and simultaneously decreasing transmission frequency) is that of [51], which will be discussed briefly in the following subsections.

■ Control of dissemination pattern: In IEEE 802.11p when the MAC layer is given a packet to disseminate, it starts the process of dissemination (contention phase, etc.) immediately without any knowledge about other vehicle's status. This is inevitable due to MAC's properties. One promising approach is to control the pattern through which the MAC layer is given packets from upper layers (e.g., application layer). In other words, the application layer (i.e., the safety application) can take advantage of a scheduling algorithm using which packets are delivered to the vehicle's MAC layer in a specified order. Therefore, collisions can be avoided and the performance will be improved noticeably. An example of works that consider scheduling algorithm in the application layer is that of [52], which will be discussed briefly in the following subsections.

In the following, we mention two examples of proposed methods that address the problem of controlling wireless channel load, with the aim of decreasing collision level.

13.8.1 Estimation-Based Beacon Dissemination

Although more accurate information (i.e., a larger packet size) could provide safer situation, as argued in the previous subsection, increasing packet size may lead to more saturated channels and as a result more collisions. Nevertheless, due to the nature of CSMA/CA, it could be intuitively understood that the effect of increasing packet size on the performance is not as adverse as the effect of reducing transmission interval. The reason is that acquiring the channel for several transmissions is the bottleneck of CSMA/AC-based MAC protocols. Therefore, one promising idea can be increasing packet size instead of increasing transmission interval (decreasing transmission frequency). To substantiate the idea without hindering the safety level, one possible approach is to estimate several next beacons by using some estimation techniques and send them in advance. In [51,53] a method is proposed that uses the Kalman filter estimation. In the following, we briefly explain the approach suggested in [51].

It is assumed that each vehicle periodically obtains its location, speed, acceleration, etc., through a GPS and/or in-vehicle sensors at every time step. Then the Kalman filter algorithm is implemented in the vehicle to estimate the future longitudinal and lateral location of the vehicle. The Kalman filter [54] is a set of recursive mathematical equations that provide an efficient recursive computational means to estimate the state of a process (here the intention is to estimate future longitudinal and lateral location of the vehicles) in a way that it minimizes the mean of the squared error. As illustrated in the block diagram of Figure 13.11, only one estimator is implemented in each vehicle. The Kalman filter block is in charge of estimating next location information for several future time steps in advance.

Every beacon message contains two categories of information: measured values for the current time step and estimated values for several future time steps. The rationale behind our approach is to prevent dissemination of unnecessary information. The scheduler block in Figure 13.11 is responsible for such a decision: whether any further transmission is necessary or not. In each time step, the location information reported by the GPS is compared with the estimated locations for

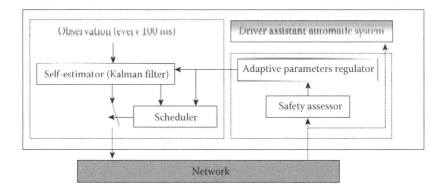

Figure 13.11 Block diagram of the proposed method. (From Armaghan, M., Fathy, M., and Yousefi, S., *FGCN 2009, Korea. Communications in Computer and Information Science* (CCIS Book Series), Berlin: Springer, vol. 56, pp. 74–82, 2009. With permission.)

the same time step (which has been transmitted ahead). When the estimations' error is below a threshold, there is no need of fresh transmissions. Otherwise a new beacon message is composed, containing both present information and estimated information. However, fresh transmission is also done if no estimations are available for a specific time step.

There are two important parameters of the algorithm that play a critical role in the safety level offered by the proposed approach: the threshold of lateral and longitudinal location estimation error and the number of estimated steps. The former is considered as a criterion for initiating fresh beacon transmission as explained above. The latter is indeed the number of time steps for which the estimated data transmitted have been calculated and transmitted ahead. As this number is increased, the consecutive time steps during which no fresh transmissions are performed increased, which causes less crowded wireless medium. However, it should be stressed that there is a trade-off between the safety level and the number of fresh transmissions. On the one hand, disseminating fresh measured information at every time step is obviously very desirable from the safety point of view, but this may lead to increase of the collision level and thus frequent losses take place. On the other hand, relying on estimated data for a large number of time steps may lead to deterioration of the safety level due to limited capability of the estimator as well as the possibility of sending very large packets. Therefore, the value of aforementioned parameters should be determined intelligently. The detailed parameter setting for Kalman estimation as well as other parts of the approach can be a subject of further research but some outlines are provided in [51]. However after being set, the above parameters are saved in the adaptive parameter regulator block, depicted in Figure 13.11.

In Figure 13.11, there is another block (safety assessor) that is in charge of safety assessment through determining whether the current setting of parameters provides enough safety for the vehicle. This block is a critical block and should be implemented based on the knowledge of safety experts and one may use artificial intelligent (AI) decision-making algorithms. Actually, the output of this block is used for setting the values of necessary parameters in the adaptive parameter regulator block.

Design of any block of Figure 13.11 is an open research problem, which is indeed very critical for the success of the safety application. In particular, the use of various estimation techniques can be investigated to reach the better estimation accuracy. Besides, the role of the safety assessor in determining whether a state is safe is a serious challenging problem. In fact, until now there is no clear definition of safe state in the literature of VANETs. Due to the indefinite nature of safety, it seems that one should resort to AI techniques in order to define the level of safety for a given state.

13.8.2 Application Layer Scheduling

As mentioned throughout this chapter, one of the most important challenges with DSRC-enabled services pertains to its IEEE 802.11p MAC layer, which is at risk of extensive collisions in saturated channels. Although these collisions are inevitable due to the nature of the MAC layer, one may think of other alleviating approaches to decrease such collisions. The results could certainly lead to higher beaconing rate and effective range. To fulfill this goal, one possible idea is to do scheduling in the application layer. In other words, vehicles cooperate with each other to disseminate their beacons in a specific order; thus, collisions are removed ideally. In the following, we explain an approach proposed in [52], pursuing the aforementioned idea adapted from the idea of space division multiple access (SDMA) applied in [55].

In the proposed approach, the road is divided into a series of sections (clusters) in which only nonadjacent sections transmit the beacon message simultaneously. Each section is further subdivided into several subsections where only one vehicle can be placed in each subsection. What causes this method to be efficient is the fact that only one subsection (i.e., the vehicle positioned there) can transmit once. Figure 13.12 shows the overall architecture of the proposed method. As followed from the figure, the method is based on a clustering mechanism where all odd-number clusters (e.g., CS1, CS3, and CS5) transmit simultaneously. After a delay that should be computed, the even-number clusters (e.g., CS2, CS4, and CS6) start transmitting at the same time.

During the cluster formation step, a series of contiguous clusters are arranged along the road so that any cluster has a unique cluster-head vehicle (CV). After the end of the cluster formation, each CV broadcasts a Hello message (HM) to announce its current position to the other vehicles in its corresponding CS. Then in order to transmit beacons free of collisions, a kind of SDMA mechanism is employed. To reach this goal, each CS is subdivided into N segments as follows:

$$N = \frac{2R}{L_s} \tag{13.3}$$

where R is the transmission range of the CV and L_s is the minimum allowed distance of two vehicles. Recall that $2R$ is the length of each CS. If the road has M lane, there are B road blocks in each CS as follows:

$$B = M \times N \tag{13.4}$$

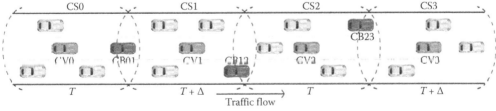

Figure 13.12 Cluster synchronization. (From Sadatpour, V., Fathy, M., Yousefi, S., Rahmani, A.M., Cho, E.-S., and Choi, M.-K., *FGCN 2009, Korea. Communications in Computer and Information Science* (CCIS Book Series), Berlin: Springer, vol. 56, pp. 133–140, 2009. With permission.)

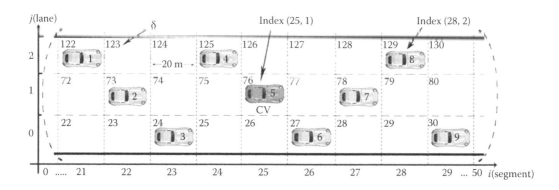

Figure 13.13 Road block partitions and time-slot label assignment for $M = 3$ and $N = 50$. (From Sadatpour, V., Fathy, M., Yousefi, S., Rahmani, A.M., Cho, E.-S., and Choi, M.-K., *FGCN 2009, Korea. Communications in Computer and Information Science* (CCIS Book Series), Berlin: Springer, vol. 56, pp. 133–140, 2009. With permission.)

where only one vehicle can be located in each block. Each road block is identified with index (i, j), where $0 \leq i \leq N - 1$ and $0 \leq j \leq M - 1$. As shown in Figure 13.13, each road block is assigned a time-slot label as

$$\delta = i + j \times N + 1 \tag{13.5}$$

Then each vehicle sends its beacon message in accordance with its own road block; thus, each vehicle should be able to identify its own block. For this purpose, it should be able to determine the index (i, j). It is assumed that each vehicle is equipped with digital maps and it can use its own GPS receiver to recognize its lane; thus, j can be identified for any vehicle. Furthermore, each vehicle can determine i (segment) using the following formula:

$$i = \left\lfloor \frac{R}{L_s} \right\rfloor + \left\lfloor \frac{X_v - X_{cv}}{L_s} \right\rfloor \tag{13.6}$$

where X_v and X_{cv} are the x-coordinate of the vehicle and CV, respectively. For example, Figure 13.13 shows an example for obtaining values of δ for $R = 500$ m, $L_s = 20$ m. The vehicle number 9 can identify index (i, j) using its own GPS receiver and (13.6) (in the figure, $X_{cv} = 25 \times 20 = 500$ m and $X_v = 29 \times 20 = 580$ m). Note that each vehicle is aware of the x-coordinate of its cluster's CV (i.e., X_{cv}) through the HM send by each CV after cluster formation. Thus, it can obtain $\delta = 130$ by using (13.5) and send its message at the $130t + T_{Hello}$, where t is the transmission interval and T_{Hello} is the time when HM was issued. As shown in Figure 13.12, collisions may happen if vehicles belonging to adjacent CSs send beacon messages simultaneously. To avoid this problem, the even-numbered CVs will send HM at T_{Hello} and the odd-numbered CVs will send it at $T_{Hello} + \Delta$, where

$$\Delta = B \times t \tag{13.7}$$

There are several open problems regarding this topic. First, in the proposed approaches, slot allocation is performed based on fix geographical positions, no matter any vehicle exists in the related road block or not. Therefore, one enhancement for this approach is to consider adaptive slot allocation based on traffic's density. In other words, if a given cluster contains fewer vehicles, then the scheduling algorithm can assign more dissemination tickets to those vehicles. Another important problem here is cluster maintenance. This problem is serious due to dynamic topology of VANETs, in particular if the opposite direction traffic is included in the desired safety application. When vehicles leave their home cluster and enter another cluster, the slot allocation of both clusters should be changed. If vehicles disseminate based on the previous allocated slots (in their own home clusters), the disseminated beacons will collide with other dissemination vehicles in the adjacent cluster. The solution here is to initiate the cluster reformation (maintenance) process. One can try to suggest an approach based on which such increasing collisions are detected and, if necessary, reformation of cluster is triggered. In the current approaches, such maintenance is triggered with a fixed and predetermined period.

13.9 Connectivity in VANETs

Connectivity is the primordial condition for message exchange in any network including VANETs. In a mobile network, connectivity is mostly affected by the mobility pattern of nodes. That is, nodes are not able to communicate, since they are not in each other's transmission range. Even when two given nodes are in the transmission range of each other, they may not be able to communicate due to radio interference (e.g., shadowing and fading) and some protocol issues (e.g., collisions due to hidden terminal problem in CSMA-based MAC layers). Therefore, research on connectivity is mostly focused on two things: connectivity impairments due to nodes' mobility and connectivity impairments due to radio and protocol factors (including radio interference and channel's effect such as fading and shadowing and MAC layer problems). In this chapter we focus on the former case.

Although there are many works on the investigation of the effects of mobility on connectivity in other types of MANETs, the obtained results can be hardly applied to VANETs. The reason is that for many other types of MANETs, usually random mobility patterns are taken into account [18]. But vehicular mobility cannot be fallen in the category of common mobility models.

When traffic is in the forced-flow phase, vehicles are closed to each other, and due to a typical transmission range of DSRC (a few hundred meters) the established ad hoc network is always connected. However, here the effect of interference may hinder connectivity. One way to tackle such a problem is to control power by which one can decrease the collision domain of nodes [50,56,57]. However, in such an attempt, safety requirements should be taken into account, according to which the vehicle's transmission range should not be fallen below a threshold (refer to Section 13.4.1). In the free-flow traffic phase, however, vehicles are moving freely due to their distance from each other. Thus, connectivity is affected by road's traffic characteristics as well as vehicle's radio characteristics. In [58], an analytical study has been done on connectivity and the following results are taken. An interested reader may refer to [58] for a more detailed discussion. Hereinafter, we study the connectivity in VANETs by evaluating the probability distribution and expectations of the following metrics:

1. *Platoon size*, which is defined as the number of vehicles in each spatial connected cluster (platoon) or, equivalently, the number of vehicles in the connected path from any given vehicle
2. *Connectivity distance*, which is defined as the length of the connected path from any given vehicle

The former is important because it shows how many vehicles can hear a vehicle in the safety applications and can have data exchange in the comfort applications. The latter metric is important because a larger connectivity distance leads to a larger announcement area for the safety applications and better accessibility to roadside equipment (e.g., Internet gateways) for the comfort applications.

In [58] it is proved that in the free-flow traffic phase, the distribution of intervehicle distance is exponential where its rate can be mapped to the traffic flow parameter (in the traffic theory). This result is also in agreement with the empirical study conducted in [59] for sparse traffic conditions. The interesting point about this work is that it expresses the above-mentioned connectivity metrics based on the fundamental parameters of the traffic theory (flow, speed, and density). Although the analytical discussion that is based on infinite server queues [60] is beyond the scope of this chapter, in the rest of the chapter we will bring forth most important results. In the following equations, R is vehicles' transmission range (m), λ is traffic flow (veh/h/lane), V is a random variable representing vehicles' speed, and N is the random variable representing platoon size.

■ Platoon size: The tail probability of the platoon size (i.e., the probability that at least k vehicles are connected) is

$$P_N(k) = P(N \geq k) = [1 - e^{-\lambda RE(1/V)}]^{k-1} \tag{13.8}$$

The expected value of the number of vehicles in each platoon is given by

$$E(N) = \frac{1}{e^{-\lambda RE(1/V)}} \tag{13.9}$$

■ Connectivity distance: Only the Laplace transform of tail probability of connectivity distance is obtained, which can only be inverted by numerical methods. For more explanation, refer to [58]. However, the average connectivity distance is given by the following explicit expression:

$$E(d) = \frac{1 - e^{-\lambda RE(1/V)}}{\lambda E(1/V)e^{-\lambda RE(1/V)}} \tag{13.10}$$

Using the obtained expression and stochastic ordering tools [61] (the details of which are beyond the scope of this chapter), one can describe the effects of various system parameters, including road traffic parameters (i.e., speed distribution and traffic flow) and the transmission range of vehicles, on the connectivity. From the above equations it is quite easy to see the effects of traffic flow and vehicles' transmission range on the connectivity. However, speed appears to be a random variable; thus, the effect of speed on the connectivity should be studied based on stochastic ordering techniques. In other words, to compare the effects of different speed scenarios on the connectivity, stochastic bounds can be presented. The details of stochastic ordering analysis can be found in [58]. However, in the following, we bring some numerical results aiming at pointing out some important findings.

First it should be noted that the following results hold for low-density traffic, which correspond to the free-flow traffic phase in Figure 13.1. In the free-flow state, the traffic flow is usually

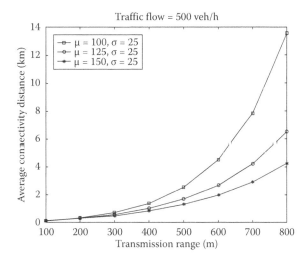

Figure 13.14   Effect of speed scenarios with similar variances and different means on the average connectivity distance. (From Yousefi, S., Altman, E., El-Azouzi, R., and Fathy, M., *IEEE Trans. Veh. Technol.*, 57(6), 3341–3356, 2008. With permission. © (2008) IEEE.)

considered to be below 1,000 veh/h/lane for freeways and below 500 veh/h/lane for other roads [5]. Moreover, although the proposed transmission range for the DSRC standard is 1,000 m [1,2], the current feasible range is about 300 m [14]. In the following figures, we provide results by taking the traffic flow values below 1,000 veh/h/lane and the transmission range values of up to 800 m. Furthermore, we assume that the vehicles' speed is normally distributed, which also holds in the free-flow state [5,62], and use some typical reported values. Furthermore, speed distribution is denoted by $N(\mu, \sigma)$ where μ and σ are the mean and standard deviation values, respectively.

a. If $\mu_1 > \mu_2$ and $\sigma_1 = \sigma_2$: As shown in Figure 13.14, the speed scenario with higher mean leads to a lower expected value of the connectivity distance. This trend also holds for the tail probability of the platoon size, the tail probability of the connectivity distance, and the average platoon size. The related figures are dropped due to space limitation. Consequently if we decrease the mean value of speed distribution, then the connectivity is improved provided that the variance of speed distribution is unchanged.

b. If $\mu_1 = \mu_2$ and $\sigma_1 < \sigma_2$: As shown in Figure 13.15, the traffic's speed with higher variance leads to a higher average connectivity distance. This trend also holds for the tail probability of the platoon size, the tail probability of the connectivity distance, and the average platoon size. The related figures are dropped due to space limitation. Here an interesting result is obtained, which may be in disagreement with our previous belief: If the variance of the speed's distribution is increased, then, provided that the average speed remains fixed, the connectivity is improved. The condition of fixed mean of speed is critical here. One can justify this result by that when the average speed is fixed, an increasing variance that causes more variability may help intermittent connectivity.

13.9.1 Connectivity of VANETs in the Presence of RSUs

In the previous section, we studied connectivity in a pure VANET where all nodes are moving vehicles. However, in real implementations, the pure ad hoc network coexists with fixed RSUs to

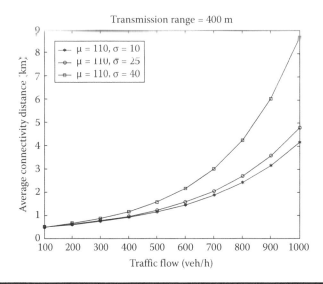

Figure 13.15 Effect of speed scenarios with similar means and different variances on the average connectivity distance. (From Yousefi, S., Altman, E., El-Azouzi, R., and Fathy, M., *IEEE Trans. Veh. Technol.*, 57(6), 3341–3356, 2008. With permission. © (2008) IEEE.)

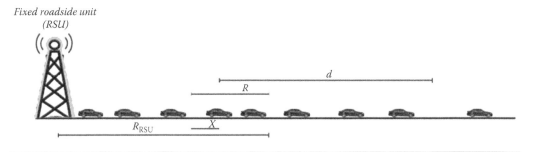

Figure 13.16 Connectivity of VANETs in the presence of RSUs. (From Yousefi, S., Altman, E., El-Azouzi, R., and Fathy. M., *Computer Communications*, vol. 31, no. 9, pp. 1653–1659, 2009. With permission.)

make a hybrid VANET. In particular, the comfort applications are relied on such RSUs to offer services to the vehicles on roads. Therefore, studying connectivity in the presence of RSUs is practically important. It is quite reasonable to assume that an RSU has a larger transmission range than an ordinary vehicle, although the following discussion is not restricted to that assumption. Let R_{RSU} and R_1 be the transmission range of the RSU and of all vehicles, respectively. Let d_{RSU} be the random variable representing the distance through which the RSU can communicate (the connectivity distance from the RSU).

As shown in Figure 13.16, this distance can be obtained by taking into account two independent distances: (1) the distance covered by the RSU with its own transmission range and (2) the distance covered by the pure ad hoc network formed between vehicles. Consider the car whose location is the smallest among all those who are larger than $R_{RSU} - R_1$, and let X be its distance from that point. If $X > R_1$, then this car is not connected to the RSU and the connectivity distance is R_{RSU}. If $X \leq R_1$, then at point $R_{RSU} - R_1 + X$, d is the new connectivity distance. The aforementioned problem has been studied in [63]. Opposed to the case of pure VANETs, no expression for tail probabilities has been obtained. However, for the average connectivity distance, the following expression is presented:

$$\bar{d}_{RSU} = R_{RSU} - R + (\hat{d} + \frac{1}{\xi})(1 - e^{-\xi R}) \tag{13.11}$$

where $\bar{d}_{RSU}$ is the average connectivity distance from the RSU, $\hat{d}$ is the average connectivity distance in a pure ad hoc network (stated in Equation 13.10), and $\xi = \lambda E(1/V)$. Moreover, the following expression has been presented for the average number of vehicles with which the RSU can communicate (in each direction):

$$\bar{N}_{RSU} = R_{RSU}\xi + \hat{N}(1 - e^{-\xi R}) - e^{-\xi R} + 1 \tag{13.12}$$

where $\bar{N}_{RSU}$ is the average connectivity distance in the presence of the RSU and $\hat{N}$ is the average platoon size in a pure ad hoc network (stated in Equation 13.9).

As an open research problem, one can study the variability of connectivity metrics in the course of time. It should be noted that in the discussion above and almost all other works in the literature, a snapshot viewpoint of the network is considered. In other words, it is implicitly assumed that the communication is instantaneous. However, one may intend to study connectivity variability during a long-lasting communication. For this purpose it is necessary to obtain analytical expressions for connectivity duration, which is very beneficial in designing a VANET protocol. Another important problem of connectivity, as mentioned above, is connectivity in the presence of radio and channel effects such as fading and shadowing. Furthermore, in the case of hybrid VANETs, one may be interested in obtaining optimum number of RSUs and their distance distribution in order to provide an acceptable connectivity. The mentioned problem is very critical in the effective design of future VANETs.

13.10 Data Dissemination in VANETs

Data dissemination is referred to as the process of routing and forwarding information originated from a vehicle termed as the source to another vehicle termed as the destination. In the literature, terms "dissemination," "routing," and "forwarding" are employed interchangeably. Data dissemination mainly involves unicast forwarding in order to eliminate packet duplication due to overhead concerns. The main goal of message broadcasting is to enhance the achievable safety of vehicular traffic; however, dissemination mechanisms are aimed at facilitating future emerging traffic management solutions and comfort applications expected by different industrial and academic bodies. While safety applications mostly need local broadcast connectivity, it is expected that some emerging scenarios developed for ITSs would benefit from unicast communication over a multihop connectivity [64].

Dissemination techniques applicable in VANETs are classified into two broad categories based on their specific characteristics: topology-based and geographic (or position-based) routing. In the following, an overview of important features of these routing approaches is given and the most popular routing protocols proposed for VANETs by research bodies are introduced and discussed in more detail. It is important to note that some of the routing protocols mentioned hereafter are used only as a benchmark for comparison purposes and may not be properly applicable in vehicular networks.

13.10.1 Topology-Based Routing

Topology-based routing protocols work based on the concept of links and use links' information to forward data/warning packets. Depending on whether route discovery is involved in the routing process, these routing protocols can be classified into reactive or on-demand and proactive or table-driven. In proactive routing, control packets are periodically propagated in the network to acquire and maintain links existing between nodes in pair. As a part of the routing process, a table is constructed such that each entry in the table represents the next hop toward a potential destination. In table-driven routing schemes, the source node should maintain unused routes. Maintenance of unused routes utilizes a significant amount of bandwidth, which is undesired in vehicular networks due to high mobility and frequently changing network topology. A representative example of proactive routing protocols is fisheye state routing [65], which maintains a network map at each node and propagates link state updates to neighbor nodes.

In reactive routing protocols, upon receiving a packet from the application layer a node initiate route discovery to find a path to the intended destination. Path discovery is usually conducted by means of a query–reply procedure implemented in the form of route request (RREQ) and route reply (RREP) messages. When a path is discovered, it will be used by communication parties and will be maintained as long as it is used. If due to any reason a path failed, another route discovery procedure is triggered. Among various topology-based routing protocols, a few of them have been applied to vehicular networks. These protocols include Ad hoc On Demand Distance Vector (AODV) [66], Ad hoc On Demand Distance Vector - Preferred Group Broadcasting (AODV-PHB) [67], Dynamic Source Routing (DSR) [68], PRediction based AODV (PRAODV), and PRediction based AODV - Maximum predicted value (PRAODV-M) [69]. In AODV, upon receiving a routing query, an intermediate node records the address of a sender node and upon receiving in the destination a reply packet is sent back to the sender through the path that was taken by the query. During propagation of RREP, each intermediate node will identify and record the next hop, thereby creating a forwarding path. The main drawback of AODV if applied in VANETs is the delay associated with the route discovery procedure. Furthermore, as route discovery control packets are broadcasted in the network, packet collisions due to broadcast storm are inevitable. In addition, the routes created by AODV can break very frequently due to the dynamic nature of mobility involved. To suppress the effect of frequent route breakage and thus increasing the routing performance, some enhancement mechanisms have been proposed. In [67], a new broadcasting method referred to as preferred group broadcast (PGB) that aims at mitigating the broadcast overhead and route instability associated with AODV is introduced. According to the characteristics of the received signal, each node determines whether it can be a member of the preferred group. Among member vehicles, only one vehicle with the highest signal quality will finally forward the route query packet. In urban vehicular networks where interference due to the presence of obstacles is a common adverse phenomenon, the application of AODV-PGB outperforms AODV in that it suppresses interference effects to some extent [67]. However, in case preferred group is empty or because the vehicle chosen as the forwarder is not necessarily the closest node to the destination, delivery latency imposed by this protocol can be potentially high. PRAODV and PRAODV-M are prediction-based extensions to AODV. In these protocols, speed and location information of nodes is employed to predict the links' lifetimes. PRAODV constructs a new alternate route before the end of the estimated lifetime while AODV does it until route failure happens. PRAODV-M selects the maximum predicted lifetime path among multiple route options instead of selecting the shortest path in AODV and PRAODV. There are other modifications to AODV, which are worth mentioning here. In [70], AODV is modified to forward the RREQs only within the zone of relevance (ZOR). The basic idea is the same as the location-aided routing (LAR) [71]. The ZOR is usually specified as a rectangular or circular range, and it is determined by the particular application [72].

DSR is similar to AODV in that it forms a route on-demand when a transmitting computer requests one. However, it uses source routing instead of relying on the routing table at each intermediate device. Determining source routes requires accumulating the address of each device between the source and the destination during route discovery. The accumulated path information is cached by nodes processing the route discovery packets. The learned paths are used to route packets. To accomplish source routing, the routed packets contain the address of each node the packet will traverse. This may result in high overhead for long paths or large addresses, such as IPv6. To avoid using source routing, DSR optionally defines a flow id option that allows packets to be forwarded on a hop-by-hop basis. It is worth mentioning that some modifications of DSR with a focus on security issues have been proposed recently. Ariadne [73] extends DSR by security functions using symmetric cryptography. The authors suggest any of these three schemes: shared secrets between each pair of nodes, shared secrets between communication nodes combined with broadcast authentication, or digital signatures. Although the latter scheme does not use symmetric cryptography, it is also considered as an option due to its high reliability, but not as the preferred option because of high processing requirements.

The disadvantage of DSR is that the route maintenance mechanism does not locally repair a broken link. Stale route cache information could also result in inconsistencies during the route reconstruction phase. The connection setup delay is higher than in table-driven protocols. Even though the protocol performs well in static and low-mobility environments, the performance degrades rapidly with increasing mobility. Also, considerable routing overhead is involved due to the source-routing mechanism employed in DSR and this routing overhead is directly proportional to the path length.

13.10.2 Geographic (Position-Based) Routing

In position-based routing protocols, forwarding decision is mainly made according to the position of the destination as well as the position of nodes in the radio range of the source node. This implies that the position of the destination should be available to the source node prior to packet transmission. Furthermore, each 1-hop neighbor requires its position information at any given time instance in order to inform the immediate previous forwarder(s) of its position or use its own position to decide whether or not it is qualified to participate in the forwarding process. In geographic routing, it is assumed that nodes identify their current positions by means of GPS that is already available in existing transportation systems and is deemed to be ubiquitous in near future. Moreover, availability of navigation and location services [74] enables a source node to identify the position of the destination node.

Position-based routing techniques can be further classified based on specific strategies and indicators used by routing protocols [64].

■ DTN vs. non-DTN: A routing scheme is categorized as a DTN (delay tolerant network) routing protocol if the intermittent connectivity of a vehicular network is presumed in a routing protocol and accordingly forwarding mechanism is designed in a way that forwarding process continues in case of network fragmentation. On the contrary, in non-DTN routing protocols, a network is assumed to be connected by default. In case forwarding process encounters network fragmentation, a reactive recovery process is enabled or the packet is simply dropped.

■ Beacon-based vs. non-beacon-based: If a routing strategy uses beacon messages to gather information about 1-hop or 2-hop neighbors, it is referred to as beacon-based routing. Beacons usually play two roles in routing and forwarding process. They are used either to locate the closest neighbors to the destination or to identify anchor nodes (e.g., vehicles locating on junctions or on road turning points). The latter is usually performed via 2-hop beaconing.

■ Overlay vs. non-overlay: Some routing protocols rely on anchor nodes located in specific locations, such as junctions in urban environment, to make decision to which direction (or road segment) packets should be routed. Such routing strategies are classified as overlay routing. A key task in overlay routing is to locate and identify overlay nodes.

In the following, a number of well-known position-based routing schemes are investigated. This investigation will be concluded with a discussion on open issues and challenges of these routing protocols when employed in VANETs.

Vehicle assisted data delivery (VADD) [75] is a beacon-based and DTN routing protocol aiming at packet forwarding in sparsely connected vehicular networks. In VADD, it is assumed that vehicles find their neighbors through beacon messages. It is also assumed that vehicles are equipped with preloaded digital maps, which provide street-level map and traffic statistics such as traffic density and vehicle speed on roads at different times of the day. Digital maps are used in VADD to model delay associated with road segments. Parameters such as road density, average vehicle velocity, and road distance are employed to model road delay. This facilitates optimal path selection by nodes located in junctions. Forwarding decision is then made by choosing next forwarding node according to different strategies proposed in VADD. In location first probe (L-VADD) strategy, next forwarding node is the one closest to the selected forwarding path. Direction first probe (D-VADD) simply chooses a next forwarding node driving toward forwarding path. In hybrid strategy (H-VADD), a combination of two mentioned schemes is used for next forwarding node selection. After next forwarding hop is determined, it follows position-based strategy to further propagate the packet toward the destination. If network fragmentation is encountered, the current forwarding hop follows carry and forward strategy. According to [75], in light traffic conditions, VADD outperforms the Greedy Perimeter Stateless Routing (GPSR) [76] routing protocol in terms of delay and delivery ratio. However, performance of VADD in dense traffic conditions such as urban environment and in the presence of interference sources is not specified. Moreover, as mentioned, VADD relies on digital maps to model path delay. This imposes a delay overhead on the routing protocol as such maps should be downloaded and updated frequently through Internet access gateways. Alternative solutions with lower delay would be continuous path delay estimation by vehicles themselves or periodic dissemination of traffic conditions of roads by RSUs to nearby vehicles.

Geographical opportunistic routing (GeOpps) [77] is a DTN routing protocol that employs information provided by navigation systems installed onboard vehicles to opportunistically route data packets to a certain geographical location. GeOpps performs three subtasks to route a packet to the destination: (1) Neighbor vehicles that follow suggested routes to their driver's destination calculate the nearest point that they will get to the destination of the packet. (2) Afterward, they use the nearest point and their map in a utility function that expresses the minimum estimated time that this packet would need in order to reach its destination. (3) The vehicle that can deliver the packet quickly/closer to its destination becomes the next packet carrier. As GeOpps needs trajectory of neighbor vehicles to be available in the current forwarding vehicle, it may not be desirable from the security point of view.

GPSR [76] is a non-DTN, non-overlay, and beacon-based routing protocol. GPSR performs routing in two modes. Initially, it forwards packet in greedy fashion, meaning that among neighbors of a current forwarder the closest one is selected as the next hop to forward the packet. A node reaches a local maximum if it cannot find a neighbor closer than itself to the destination. In this case, GPSR triggers the recovery process and recovers from local maximum using a perimeter mode by means of the right-hand rule. If a node closer to the destination is not found in the perimeter mode, a face change (also called face routing) is performed. Upon finding a closer node to the destination, a greedy mode is enabled again until the packet is received in the destination. Operation of GPSR needs network graph to be planar. To fulfill this, GPSR uses some distributed algorithms to build planar graphs. Relative neighborhood graph [78] and Gabriel graph [79] are two planar graphs that are created in GPSR. The unit graph assumption is adopted in GPSR to create connected planar graphs. According to the unit graph assumption, two nodes are referred to as connected if their distance is less than a predetermined threshold; otherwise they are not connected. However, as in vehicular networks this assumption does not always hold due to the presence of obstacles and interference, applicability of GPSR is questionable [64]. Furthermore, the delay overhead imposed by GPSR in the greedy mode and the time needed to create planar graphs are not desirable in vehicular networks. Finally, in highly dynamic vehicular networks, positions of neighbors and destination change rapidly. This results in outdated position information in nodes currently deciding to forward the packet. In [67], an approach referred to as advanced greedy forwarding (AGF) is proposed as a solution to address this problem. Based on AGF, beacons are augmented with information reflecting dynamic behavior of vehicles. Information such as velocity vector (speed and direction) of each vehicle is included in beacons. Knowing velocity vector of neighbors, each vehicle can estimate their current locations. Moreover, each vehicle can estimate the current position of the destination. This strategy, although being preferable to GPSR, has some drawbacks. First, it does not work for vehicles changing their directions in junctions. Second, impacts of acceleration and deceleration are not considered in AGF. In [80], Schnaufer et al. proposed position-based routing with distance vector recovery as an alternative to the perimeter mode in GPSR when a transmitted packet reaches a local maximum. Upon receiving a packet from a vehicle currently at local maximum, the recipient vehicle checks if it is closer to the destination than the vehicle at local maximum. If it is not, then the packet is broadcasted again. If the recipient is closer to the destination, a reply packet is sent back to the vehicle at local maximum. Intermediate nodes will learn the previous nodes from which they received the reply packet. This enables the vehicle at local maximum to transmit through reply path. This solution eliminates the need for the creation of a planar graph as in GPSR. However, it inherits the drawbacks of AODV such as delay overhead as a result of the route discovery procedure. Greedy routing with abstract neighbor table (GRANT) [81] is another approach belonging to the GPSR family of protocols. The core idea of GRANT is to predict local maximum based on x-hop neighbor information. To avoid overhead imposed by *x*-hop beaconing, the plane is divided into areas and a single node is designated per area for beaconing purposes. The metric proposed in GRANT to select the next forwarding node is measured by the multiplication of distance between *x*-hop node and destination, distance between x-hop node and current forwarding node, and charge (or cost) per hop.

A slightly different category of geographic routing protocols adopts an overlay design. In these protocols, anchor points, i.e., nodes located in junctions, and turning points play critical roles in routing and forwarding process. Therefore, a key task in overlay protocols is to find anchor points and augment vehicles locating in anchor points with link state and traffic information of road segments connected to these points. Strategies used to identify anchor points fall into two categories. They use either a topological map or dynamic algorithms to determine anchor points. In what follows, a number of geographic overlay routing protocols are investigated and discussed in more detail.

One of the well-known geographic overlay protocols aiming at the elimination of the need for graph planarization is greedy perimeter coordinator routing (GPCR) [83]. In this scheme, junctions are identified by means of heuristic algorithms and are used as graph vertices. The resultant graph is naturally planar, and hence the planarization process is skipped. In case of encountering a local maximum, recovery will be performed by representative vehicles located on junctions. Packets are forwarded in greedy fashion along streets and stops whenever they reach a junction, where they are decided on which direction to proceed. GPCR differs from GPSR in that it does not need graph planarization. This may result in significant time saving in GPCR compared to GPSR. This is due to the fact that in GPSR, the network graph is formed by vehicles and not by junctions. Considering the fact that vehicles' mobility causes the network topology and hence the graph shape to change, a palanrization process must be invoked to create a new planar graph corresponding to the current network topology.

Geographical source routing (GSR) [83] uses a street map to identify junctions. Using such a map, GSR builds a graph with road segments indicating graph edges as well as junctions representing vertices. A sequence of junctions represents a path, and Dijkstra's shortest-path algorithm is employed to find shortest paths. When such a path is found, packets are forwarded in greedy fashion between junctions comprising the path. Unlike GSR, A-STAR [84] is a connectivity-aware routing approach that eliminates those anchor points in the shortest path that are not located on a connected path. Similar to GSR, A-STAR adopts the anchor-based routing approach with street awareness. The term "street awareness" is preferred over "spatial awareness" to describe more precisely the use of street map information in the routing scheme for anchor path computation [84], that is, using a street map to compute the sequence of junctions (anchors) through which a packet must pass to reach its destination. Unlike GSR, A-STAR computes the anchor paths with traffic awareness. "Traffic" herein refers to vehicular traffic, including cars, buses, and other roadway vehicles. A possible drawback of A-STAR is its dependence on RSUs that impose an overhead delay on routing protocols. In [85], an alternative solution referred to as street topology-based routing (STBR) is proposed to calculate connectivity of paths. In this approach, a representative node at each junction is responsible for tracking the connectivity status of road segments connected to the junction. This is performed by disseminating beacon messages by each representative vehicle on a junction to all other neighboring junctions' representatives. This strategy provides each representative vehicle in a junction with a 2-hop neighbor junctions' link information. Unlike GSR, Dijkstra's algorithm is not used to find the shortest path [64]. Instead, packets are geographically forwarded from street node (source) to junction, from junction to junction, and from junction to street node (destination). STBR has two major drawbacks: first, it does not specify any strategy when no junction makes progress toward the destination. Second, to build and maintain a neighbor table in each representative node, a significant number of beacons should be exchanged between junctions. This puts an excessive overhead on the routing protocol and wastes channel resources at the cost of safety applications.

To further enhance geographic routing protocols with connectivity awareness, a number of approaches have been proposed in the literature. In [86], Jerbi et al. proposed greedy traffic aware routing protocol (GyTAR). In GyTAR, it is assumed that each vehicle in the network knows its own position and current geographical position of the destination in order to make the routing decision. Moreover, it is assumed that each vehicle can determine the position of its neighboring junctions through preloaded digital maps, which provides a street-level map. It is also assumed that every vehicle is aware of the vehicular traffic (number of vehicles between two junctions). According to Jerbi et al., this information is provided by Infrastructure-Free Traffic Information. System (IFTIS): a decentralized mechanism for the estimation of traffic density in a road traffic network. In GyTAR, the different junctions the packet has to traverse in order to reach the destination are chosen dynamically and one by one, considering both vehicular traffic variation and distance to the destination: when selecting the next destination junction, a node (the sending vehicle or an intermediate vehicle in a junction) looks for the position of the neighboring junctions using the map. A score is given to each junction considering the traffic density and the geometric distance to the destination. The best destination junction (the junction with the highest score) is the one that is geographically closest to the destination vehicle and has the highest vehicular traffic. Landmark overlays for urban vehicular routing environment (LOUVRE) [87] is slightly different in the way it estimates connectivity. Instead of relying on RSUs, LOUVRE estimates traffic density in a peer-to-peer fashion by means of beacon messages exchanged by vehicles to their neighbors. Above a threshold density determined by road length and radio communication range, a road is said to be connected. The drawback associated with both GyTAR and LOUVRE is that vehicle density is not a suitable metric for connectivity measurement unless vehicles are distributed uniformly along the road under investigation. This assumption can be safely applied to highway scenario to some extent but in urban scenario such an assumption is not realistic. Connectivity-aware routing (CAR) [67] addresses connectivity from a different viewpoint. CAR employs AGF [80] to predict mobility of the destination and vehicles that were the neighbors of the current forwarding vehicle. This way, without too frequent beaconing and triggering location service, a vehicle can adaptively keep locations of its neighbors and the destination. This enables forwarding vehicle to have fresh information about its local connectivity. CAR is similar to AODV in that it uses a similar approach for route discovery. However, route discovery in CAR differs from that in AODV in two ways: first, a limited broadcast strategy is implemented by means of PGB [67]. Second, only anchor points (located on junctions) are reported in the reply packet. By receiving the information of anchor points within the reply packet, the source node employs AGF to forward packets explicitly to those anchor points and implicitly to the destination.

Contention based forwarding (CBF) [88] adopts a different strategy than those protocols mentioned before in that it does not rely on beaconing. Instead, CBF uses contention to implicitly select the next hop in the communication path. Each potential forwarder computes the time t it must wait before forwarding the packet depending on its suitability, i.e., its progress toward the destination defined as:

$$P(F, D, N) = \max\left\{0, \frac{\text{dist}(F, D) - \text{dist}(N, D)}{r_{\text{radio}}}\right\} \qquad (13.13)$$

$$t(P) = \begin{cases} \max\{0, \ T(1-P)\}, & P > 0 \\ \infty, & \text{otherwise} \end{cases} \qquad (13.14)$$

where P is the progress function depending on the positions of the last forwarder F, of the final destination node D of the packet, and of the receiving node N. The Euclidean distance between two positions is expressed as dist( ). r_{radio} denotes the maximum 1-hop distance toward the destination a message can travel and T defines the maximum contention time. $t(P)$ is the assigned waiting time to a node. Observe that CBF does not take into account the role of traffic density in the progress function P. This leads to actual forwarding progress to be less than the amount of progress estimated by (13.13). However, as CBF takes advantage of both geographic routing and opportunistic forwarding, it is categorized as a geo-opportunistic routing scheme. This family of protocols is more preferable in the presence of interference due to robustness of opportunistic forwarding against interference.

13.10.3 Open Issues

In addition to drawbacks identified for each routing protocol surveyed in previous sections, two other major challenges observable in all routing protocols are connectivity and interference awareness. In our terminology, interference is a by-product of at least two components: First, it is caused by static and dynamic physical objects such as buildings and vehicles in urban environments. High contention over channel access in high traffic densities such as queues built behind junctions or jams caused by accidents is considered as a second interference source. The latter is also caused by broadcast storm and hidden terminal effects and is intensified in dense traffic conditions. As a general fact, traffic density contributes to both interference types assuming that traffic density is proportional to packet transmission rate. This assumption is realistic in VANETs since, as specified in standard, all vehicles need to propagate beacon messages frequently.

In table-driven routing strategies due to high topology change in VANETs, the update rate of link states increases dramatically, which results in a high packet collision rate. Another source of collision is periodic link state exchange as specified by the protocol. In addition, as link state updates are exchanged only with neighbors and each node records only next-hop information toward a destination, a notion of end-to-end connectivity is not considered in table-driven protocols. When a packet fails to reach the destination, it is simply dropped and retransmitted, which in turn leads to delay overhead and wasteful bandwidth usage.

Delay associated with the route discovery procedure in reactive topology-based routing protocols hinders their applicability in VANETs. Furthermore, as route discovery control packets are broadcasted in the network, packet collisions due to broadcast storm are inevitable (AODV-PGB is an exception in that it limits the broadcast rate). In addition, as in VANETs the lifetime of connections is short, the routes created by reactive protocols are not stable. This causes the packets forwarded in predetermined routes to encounter route breakage due to network fragmentation or interference. Reactive protocols do not offer any strategy to predict route breakage dynamically and to handle packet forwarding when such incident occurs.

Geographic routing protocols deal with connectivity and interference in different ways. In DTN routing schemes such as epidemic, VADD, and GeOpps, a reactive strategy based on the store-carry-forward scheme is adopted when a forwarding node detects network fragmentation. These routing protocols ideally guarantee packet delivery at the cost of unbounded delay. Protocols in the GPSR family are similar to DTN protocols in that they react to network fragmentation only when it is detected during forwarding. However, instead of the store-carry-forward scheme, they rely on recovery schemes mainly implemented by perimeter traversal based on the right-hand rule. Based on whether or not a graph planarization procedure is required for a perimeter traversal purpose, protocols in the GPSR family adopt different strategy. As a general fact, graph planarization is not desired in VANETs due to its unit graph assumption. This invalid assumption makes protocols such as GPSR not robust against interference and less connectivity-aware owing to false link identification. On the contrary, overlay protocols such as GPCR perform routing on a natural planar graph composed of junctions and road segments. These protocols although being more promising in terms of robustness against environmental interference when they operate in the recovery mode, they do not propose an interference-aware forwarding strategy in greedy mode operation. A different class of geographic routing schemes is proactive in the way they deal with connectivity issues. GRANT, CAR, A-STAR, STBR, GyTAR, and LOUVRE are examples of this kind. Common to all of these protocols is rough estimation of traffic along the road segments and junctions and exploit it as a measure of connectivity. Traffic estimation mechanisms adopted in these protocols either are too simplistic or impose high overhead on the underlying network. A group of geographic routing protocols are neither reactive nor proactive and thus are not connectivity-aware. GSR and CBF are categorized in this group. An interesting advantage of CBF is its opportunistic forwarding strategy that makes it relatively robust against interference.

In conclusion, while routing protocols address interference and connectivity in VANETs to some extent, they sacrifice one to the advantage of another to achieve acceptable results for some proprietary traffic scenarios that by no means are extendable to various scenarios existing simultaneously in urban environments, for instance. In the best case the existing routing schemes predict connectivity prior to data forwarding, although the measurement strategies are not accurate or time-efficient. Thus, a fundamental analytical and experimental study is needed to investigate interference and intermittent connectivity of VANETs and the challenges caused by these issues especially in urban scenarios. A further step would be the application of the results of such a study to devise new data and message dissemination schemes.

13.11 Broadcast in VANETs

Unlike comfort applications, event-driven safety applications require that warning messages be broadcasted to a large number of vehicles instead of disseminating to a single vehicle or a fixed RSU. Message flooding is the most common scheme employed for message propagation in these applications. However, an immediate drawback of flooding approach is broadcast storm phenomenon that results in traffic overwhelming due to blind message propagation by vehicles acting as relays in multihop communications. The consequences of such a phenomenon are exacerbated in dense traffic conditions as the competition over channel access increases dramatically, which in turn leads to high packet collisions. An obvious result of packet collision is the low packet delivery ratio, implying for some vehicles not to receive the warning message at all. Considering that the delivery of warning messages is highly demanded in the case of emergency incidents, existing flooding-based approaches tend to reduce collisions by means of suppressing broadcast storm phenomenon. Broadcast approaches with flooding suppression can be categorized into four different classes [89] as described below.

13.11.1 Geographical-Limited Broadcasting

Geographical broadcasting (geocast) is a type of traditional multicasting. The difference is that geographical broadcasting is limited to vehicles spreading over a specific geographical area (i.e., geocast region). More specifically, in case of an emergency incident, only a subset of vehicles driving in the vicinity of the incident site are targeted as warning message recipients. Such vehicles are termed "geocast group." A geocast group is dynamic in that vehicles leaving the geocast region are removed from the group while vehicles driving into the geocast region at a given time become members of the group. In [90], Bachir and Benslimane proposed a novel geocast protocol for broadcasting warning messages based on their associated relevance to the current location of vehicles. This protocol aims at delivering messages to those vehicles approaching an accident site. Those irrelevant vehicles far from the accident location drop warning messages they receive, while relevant vehicles participate in rebroadcasting warning messages. Although the geocast approach reduces broadcasting to a limited region, still the problem of the broadcast storm in the geocast region remains unsolved.

13.11.2 Priority-Based Multihop Broadcast

Time critical warning messages are the most influenced by the broadcast storm phenomenon. One approach is to give higher priority to those nodes that need to propagate warning messages than those nodes that are transmitting periodic beacons or data packets. This approach has been introduced in [91], where an algorithm for classifying different nodes based on the type of message they intend to transmit is proposed. Then a scheduling algorithm is devised to transmit messages based on their holders' priorities. It is important to note that this approach does not address the broadcast storm problem; instead, the impacts of this problem are mitigated to the benefit of high-priority event-driven applications.

13.11.3 Distance-Based Multihop Broadcast

Distance-based approaches mainly aim at broadcast storm suppression by giving higher opportunity to those nodes farther from the current transmitter as the candidate next-hops participating in the rebroadcasting process. Ideas pursuing this approach fall into two main categories: (1) only the farthest node from the current sender rebroadcasts the warning message [92], (2) a group of vehicles situated in the broadcast range hearing the transmitted message participate in rebroadcasting according to a priority scheme. Priorities are given to vehicles based on their distance from the transmitter (as opposed to greedy geographic routing where the closest node to the destination is selected as the next hop). In [93,94], three different mechanisms are proposed based on the second approach: weighted p-persistence, slotted 1-persistence, and slotted p-persistence broadcasting.

Weighted p-persistence and slotted p-persistence are known as variations of gossip-based broadcasting approach, also referred to as probabilistic broadcasting [95], where vehicles receiving a transmitted packet rebroadcast it with some probability p. In weighted p-persistence, the broadcast probability of a vehicle driving in the broadcast range of the transmitter is determined according to the distance of this vehicle to the current transmitter. More specifically, upon reception of a packet in node j from node i, the forwarding probability p_{ij} is calculated as D_{ij}/R, where D_{ij} is the distance of node j from node i and R is the average transmission range. Before rebroadcasting, node j checks the packet ID and rebroadcasts it with the probability p_{ij} if it is the first time this packet is received by node j. The slotted p-persistence mechanism follows a slightly different approach. In this approach, the forwarding probability p is known *a priori* and is not calculated per-packet. Moreover, each recipient vehicle rebroadcasts with probability p at its assigned time slot if it is the first time it receives the packet and also has not received any duplicate packet before its assigned time slot. The longer the distance from the transmitter, the shorter is the time a recipient vehicle waits for rebroadcasting.

In the slotted 1-persistence mechanism, each recipient vehicle in the transmission range of the sender calculates its time slot in a way similar to that of the slotted p-persistence mechanism. However, it rebroadcasts with probability 1 if it is the first time it receives the packet and also has not received any duplicate packet before its assigned time slot. If packet delivery latency and penetration rate are of higher importance to safety application as compared with packet loss ratio, the slotted 1-persistence mechanism is preferred over other mechanisms. However, slotted p-persistence is more reliable from the packet loss point of view as long as the choice of p as a predetermined design factor is made carefully.

Another class of broadcasting schemes has been proposed in the literature based on the epidemic routing approach. These schemes differ from aforementioned broadcasting mechanisms in that messages are flooded network-wide to be hopefully delivered to a single destination instead of all vehicles or a group of them. Epidemic-based approaches can be categorized as unicast forwarding/routing techniques in which message communication occurs pairwise; i.e., messages are originated by a single sender and are targeted to a single destination. However, as already mentioned, the way they propagate messages is referred to as broadcasting. Epidemic-based message broadcasting is also classified as a type of delay-tolerant forwarding methods that work on the basis of the store-carry-forward scheme. The intention behind epidemic-based broadcasting is to cope with packet forwarding challenges arisen by frequent disconnection phenomenon in vehicular networks. To acquire a thorough insight into epidemic-based packet broadcasting, some well-known approaches and their derivations are investigated as follows.

13.11.4 Epidemic-Based Packet Routing Techniques

Epidemic routing [96] mimics the way an infectious disease spreads through direct contact in a population. If the disease is highly infectious and the individual contact frequencies are high, it is highly probable that the disease will spread through the entire population [89]. In a similar way, epidemic routing leads to all vehicles receiving a copy of the transmitted message with high probability; hence, any single destination targeted for the message will finally have the message with high probability. Basic epidemic routing is conducted in two phases: The first phase is initiated when two vehicles make their first contact. During this phase, vehicles exchange their message summary vectors containing the message IDs they have already received. In the second phase, contacting vehicles exchange new messages not being received previously. Each message has a time to live (TTL) field to restrict the number of contacts during which it can be exchanged. When a vehicle receives a message with TTL = 0, it is allowed to be exchanged only with the destination.

The apparent drawback of basic epidemic routing is flooding potentially the whole network In order to deliver a message to a single destination. In [97], the spray-and-wait approach is proposed to restrict the total number of broadcasted copies of the same message to a predetermined number L. In the spray phase, L copies of the message are forwarded to L distinct relays by the source vehicle and other receiving vehicles. In the wait phase, only direct transmission is allowed; i.e., relay nodes holding the message are allowed only to forward the message to the destination vehicle as soon as they contact. Restricting the number of message forwarding instances suppresses the negative impacts of the broadcast storm. Moreover, the spray-and-wait mechanism is scalable in the sense that with the increase in the number of vehicles (or equivalently traffic density), the number L of message copies is not needed to be increased in order to retain the same performance level achievable with a lower number of vehicles [89], i.e., when network is sparse. Despite positive aspects of the spray-and-wait approach, there is always a risk of unbounded delay as the message holder(s) may not ever have a contact with the destination vehicle during the wait phase.

To enhance epidemic routing further, Lindgren et al. [98] introduced PROPHET, a new epidemic-based forwarding mechanism targeted for delay-tolerant applications in vehicular environment. The key idea of PROPHET is delivery estimation made by each vehicle and represents the chance of message delivery from a typical vehicle to any other potential destination vehicle in the network. Delivery estimation is expressed in terms of delivery probability and calculated based on the contact history of a vehicle with other vehicles and updated upon any new contact of vehicles. According to Lindgren et al., the intuition behind PROPHET is that vehicle mobility in vehicular networks is not entirely random and vehicles tend to visit some locations more often or vehicles visiting one another in the past are more likely to have contacts in future. When nodes A and B contact, they initially exchange their message summary as well as delivery probability vectors and update their own delivery probability vectors according to the new information acquired during this contact. If vehicle A realizes that vehicle B has some messages for which it has higher delivery probability, then vehicle A chooses and transfer those messages for later delivery to the intended destination specified by vehicle B.

Another approach used for delivery estimation different from what is employed in PROPHET is based on virtual Euclidean mobility pattern space referred to as MobySpace [99]. According to MobySpace, messages should be forwarded to a next-hop node that shows a mobility pattern similar to that of the destination node. This approach has two drawbacks: First, vehicles must have a stable mobility pattern in order for MobySpace to be efficient [89]. Second, a similar mobility pattern in spatial dimension does not necessarily mean frequent contacts of vehicles. Mobility patterns should be similar in both spatial–temporal dimensions for rendering high message forwarding performance in MobySpace.

13.11.5 Open Issues

Broadcast storm as the major challenge of packet broadcasting in vehicular networks still stands without being fully solved. Approaches proposed to address this challenge either suppress it to a limited extent or lose some performance indicators such as delivery latency when they aim at resolving broadcast storm significantly. In geographic-oriented message broadcasting mechanism, although the impact of broadcast storm is reduced to a limited region rather than the whole network, still packet collisions and blind usage of transmission resources unavoidable in the geocast region targeted for message propagation. Priority-based broadcast techniques sacrifice data packets and beacon messages to the benefit of time critical messages. Besides, scalability of these techniques is questionable when the number of time critical messages increases in the network. Distance-based broadcast mechanism works well when traffic density is relatively low. In high-traffic-density conditions, there will be many nodes with identical or nearly equal forwarding probability or equally assigned time slot since these parameters are calculated according to the distance of vehicles to the message transmitter. Furthermore, in the slotted *p*-persistence broadcast mechanism, the choice of the parameter *p*, which plays a key role in the ultimate performance of broadcasting scheme, is assumed *a priori* for each vehicle. This imposes another challenge to the applicability of this mechanism. As a general fact, network connectivity is ignored in all broadcasting schemes as they assume that event-driven alarm messages are not needed to be delivered to the isolated regions and locations far from the incident location. Whether such assumption always holds is arguable as it is desirable to inform vehicles (perhaps far from the incident location) to proactively select alternative routes and avoid traffic jams caused by accidents.

Apart from basic epidemic routing, in most of the delay-tolerant broadcasting approaches the broadcast storm challenge is addressed remarkably and thereby significant improvement is achieved. However, in almost all of these mechanisms high latency is unavoidable and in some cases latency may be unbounded. For time critical alarm messages and real-time data packet dissemination, such mechanisms are not applicable. However, as a strong feature, epidemic-based forwarding and its associated variations are robust against intermittent connectivity in vehicular networks.

13.12 WiMAX in Vehicular Networks

Due to a recent surge of broadband wireless access, WiMAX has been attracting much attention from both industry and academic communities. The early proposal of WiMAX was based on IEEE 802.16-2004 (also called IEEE.802.16d), which supports both LOS (line of sight) and non-LOS for fixed nodes [100]. However, later in the framework of IEEE 802.16e (also called mobile WiMAX) [101], the support of mobility is added. The IEEE 802.16 standard defines a mesh operating mode along with a centralized mode called PMP (point to multipoint). In the former, data traffic occurs directly between subscriber station (SS) nodes; however, in the latter, data traffic should be handled by a centralized node called base station.

In general, WiMAX can be a potential candidate to be used in the vehicular network due its large transmission range and support of QoS. Actually the MAC layer of WiMAX is deterministic Time Division Multiple Access (TDMA)-like (as opposed to stochastic and contention-based nature of CSMA/CA of IEEE 802.11p). This is quite beneficial in avoiding MAC layer collision, in particular for beacon-based and alarm-based safety applications, which in turn leads to a higher safety level. Furthermore, WiMAX substantiates a class-based QoS support, which is very advantageous for comfort applications as well. The current QoS classes [100] include unsolicited grant service, real-time polling service, extended real-time polling service, non-real-time polling service, and best effort, which can be assigned for different kinds of applications, including safety and comfort applications.

It is expected that WiMAX in the PMP mode can be easily used for V2I scenarios. This issue has been studied in [102], and the preliminary results show that WiMAX offers larger coverage (a few kilometers), acceptable bandwidth, and even lower delay. Therefore, by using fewer RSUs the cost of network deployment is decreased noticeably. In this case the cost of base stations of WiMAX is a concern, which is much higher than that of IEEE 802.11p based RSUs. However, as Ge et al. [103] suggested, one can invoke IEEE 802.16j, which is aimed at supporting multi-hop relaying. In other words, by taking advantage of relay nodes the cost of implementation of WiMAX is reduced. Another middle-ground approach is to make use of integrated IEEE 802.11p and IEEE 802.16 standards for achieving both high data rate and high coverage area, as suggested in [104]. It should be noted that the coverage area of wireless technologies for highway and urban scenarios not only depends on the coverage of the corresponding technology but also on other factors including number of vehicles, environment obstacles, and their service demand. An example of such a computation is provided in [105]. Depending on the aforementioned conditions, the coverage range of WiMAX is in the range of a few kilometers and that of IEEE 802.11p is in the range of a few hundred meters.

When it comes to life safety application, the view of WiMAX is not that much clear. First, WiMAX should operate in the mesh (ad hoc) mode due to the need of V2V communication, which is not supported in the current mobile WiMAX standard. Second, WiMAX is a connection-oriented MAC layer; thus, for each message transmission, several steps should be followed until a connection is established. This is not in favor of safety applications due to the fact that safety messages (in particular alarm safety messages) are issued in response to an unexpected event for which there is no previous intention of message transmission. Therefore, the time needed to set up the connection may violate timeliness requirement of safety applications.

The final point here is that even if the above ideas are realized, a vehicle has to support both mesh mode (for safety applications) and PMP and relay modes (for comfort applications) of WiMAX. However, the possibility and efficiency of such a solution should be investigated due to the substantial difference between operation and protocols of these two modes.

13.13 Current Status and Future Trends

Vehicular networks have motivated joint cooperation between scientific and industrial bodies in the recent decade. During these years the view on VANETs has changed slightly as depicted by the feedbacks taken from prototyped industrial projects. Currently, the ambitious view of automatic driving is replaced by driver-assistant safety applications because it seems that with current DSRC standards and technology it takes a relatively long time until fully automatic driving is substantiated. The main reason is that automatic driving, by its nature, demands very high performance and QoS metrics to be satisfied. Another major motivation for VANETs research and deployment will be comfort applications. Due to their business profit and lighter QoS requirements, comfort applications can be potentially very promising in the future. For safety applications, V2V communication is primordial, but for comfort applications, both V2V and V2I communications are required. The current DSRC standard, which is based on the IEEE 802.11p MAC layer, may remain *de facto* standard for V2V communications. But it seems that the current trend is toward using the IEEE 802.16e standard (also called mobile WiMAX) or another long-ranged/high-bandwidth wireless standard for V2I communications as a complementary or replacing technology.

In this chapter, several challenging issues with VANETs have been introduced and related solutions are discussed briefly. This work can be considered as an updation and complementary to [106]. Table 13.2 summarizes the current status and future trends discussed in this chapter.

Table 13.2 Current Issues and Future Challenges in Vehicular Ad Hoc Networks

Vehicular Ad hoc Network: Challenging Issues	*Relevant Type of Application*	*Current Status*	*Future Trend*
Performance evaluation metrics (refer to Section 13.4)	Safety	• As ordinary networking scenarios (focused on average values)	• New metrics focused on safety requirements (e.g., worst-case vales)
	Comfort		• As ordinary networking scenarios
Simulation (refer to Section 13.5)	Safety and comfort	• Integrated off-line imulation	• Integrated co-simulation
Transport layer protocols (refer to Section 13.6)	Safety	• Using UDP	• UDP along with some application layer reliability support for unrepeatable safety messages
	Comfort	• Customized versions of TCP	• Customized versions of TCP • UDP along with FEC techniques
Vehicle to RSU communications (refer to Section 13.7)	Comfort	• Intra-RSU data scheduling • Single-class inter-RSU data dissemination	• Multiclass data distribution • Inter-RSU scheduling algorithms • Optimum placement of RSUs • Design of dissemination protocols
Beacon-based safety applications (refer to Section 13.8)	Safety	• Power control • Application level scheduling • Transmission interval control	• Adaptive power control • Adaptive application level scheduling • Dynamic transmission interval control
Connectivity (refer to Section 13.9)	Safety and comfort	• Study of steady-state connectivity in sparse traffic • Simplified analytical modeling (mostly hold for free-flow traffic phase)	• Study of connectivity duration • Effect of traffic parameter on connectivity • Connectivity of hybrid VANETs (both V2V and V2I) • Connectivity impairment due to channel and radio conditions • Study of intermittent connectivity

(Continued)

Table 13.2 (Continued) Current Issues and Future Challenges in Vehicular Ad Hoc Networks

Vehicular Ad hoc Network: Challenging Issues	Relevant Type of Application	Current Status	Future Trend
Data dissemination in VANETs (refer to Section 13.10)	Comfort	• Position-based greedy routing schemes • Selective forwarding • Routing recovery	• Hybrid DTN and non-DTN routing • Connectivity- and interference-aware routing and forwarding • Geographic-opportunistic routing schemes
Broadcast in VANETs (refer to Section 13.11)	Safety	• Message flooding • Geographical restricted broadcast • Node and packet prioritization	• Interference-ware broadcast • Broadcast adaptation based on safety application requirements
WiMAX in vehicular networks (refer to Section 13.12)	Safety	• At the starting points (the application is in doubt)	• Support of mesh and ad hoc mode of mobile WiMAX (IEEE 802.16e) (the current standard does not support)
	Comfort	• Preliminary research results show promises	• Invoking WiMAX for V2I applications • Implementation of joint DSRC/WiMAX infrastructure • Invoking WiMAX with relay nodes (IEEE 802.16j) to reduce the cost of implementation

For a more detailed discussion, please refer to related sections indicated in the table. It should be stressed that the challenges that should be addressed in VANETs directly depend on the application in mind (i.e., safety or comfort). This is because different types of applications have different QoS requirements.

Acknowledgment

The authors thank Maissam Asgari, Hamid Karimi, Hassan Nassiraei, and Masoud Rezayat (the MSc students of Urmia University) for their assistance in preparation of this chapter.

References

1. Fed. Commun. Comm., FCC 03-324. FCC Rep. Order, February 2004.
2. IEEE Standard 802.11p Draft Amendment. Wireless LAN Medium Access Control (MAC) and Physical Layer (PHY) Specifications: Wireless Access in Vehicular Environments (WAVE), IEEE, 2005.
3. National Center for Statistics and Analysis. Traffic Safety Facts 2003, Report DOT HS 809 767. Washington, DC: National Highway Traffic Safety Administration, U.S. Department of Transportation, 2004.
4. C.J. Adler. Information dissemination in vehicular ad hoc networks. Diploma Thesis, University of Munich, Germany, 2006. Available (in German) at http://www.destatis.de/ (Accessed on June 2011).
5. R.P. Roess, E.S. Prassas, and W.R. McShane. *Traffic Engineering*, 3rd ed. Upper Saddle River, NJ: Pearson Prentice Hall, 2004.
6. H. Morimoto, M. Koizumi, H. Inoue, and K. Nitadori. AHS road-to-vehicle communication system. In *Proceedings of the IEEE International Conference on Intelligent Transportation Systems, Tokyo, Japan*, 1999, pp. 327–334.
7. O. Andrisano, M. Nakagawa, and R. Verdone. Intelligent transportation systems: the role of third generation mobile radio networks. *IEEE Communications Magazine*, vol. 38, no. 9, 2000, pp. 144–151.
8. COMeSafety. Available at http://www.comesafety.org/ (Accessed on June 2011).
9. COM2REACT. Available at http://www.com2react-project.org/ (Accessed on June 2011).
10. COOPERS. Available at http://www.coopers-ip.eu/ (Accessed on June 2011).
11. CVIS. Available at http://www.cvisproject.org/ (Accessed on June 2011).
12. Intelligent Transport System. Cooperative intersection collision avoidance systems. http://www.its.dot.gov/cicas/ (Accessed on June 2011).
13. The NOW: Network on Wheels Project. Available at http://www.network-on-wheels.de (Accessed on June 2011).
14. U.S. Department of Transportation, Vehicle Safety Communications Project, 2006. Final Report, Public Document, Crash Avoidance Metrics Partnership.
15. H. Füßler, M. Transier, M. Torrent-Moreno, H. Hartenstein, and A. Festag. Thoughts on a protocol architecture for vehicular ad-hoc networks. In 2nd International Workshop on Intelligent Transportation (WIT 2005), Hamburg, Germany, 2005.
16. S. Yousefi and M. Fathy. Metrics for performance evaluation of safety applications in vehicular ad hoc networks. *Transport Vilnius: Technika*, vol. 23, no. 4, 2008, pp. 291–298.
17. J. Yin, T. ElBatt, G. Yeung, B. Ryu, S. Habermas, H. Krishnan, and T. Talty. Performance evaluation of safety applications over DSRC vehicular ad hoc networks. In *Proceedings of the 1st ACM International Workshop on Vehicular Ad Hoc Networks*, 2004, pp. 1–9.
18. T. Camp, J. Boleng, and V. Davies. A survey of mobility models for ad hoc network research. *Wireless Communications & Mobile Computing (WCMC)*, vol. 2, no. 5 (Special issue on Mobile Ad Hoc Networking: Research, Trends and Applications), 2002, pp. 483–502.
19. Quadstone Paramics, 2008. Available at http://www.paramicsonline.com (Accessed on June 2011).
20. *CORSIM User Manual*, Version 1.01. Washington, DC: U.S. Department Of Transportation, Federal Highway Administration, 1996.
21. D. Krajzewicz and C. Rossel. Simulation of urban mobility (SUMO). German Aerospace Centre, 2007. Available at http://sumo.sourceforge.net/index.shtml (Accessed on June 2011).
22. MOVE (MObility model generator for VEhicular networks): rapid generation of realistic simulation for VANET, 2007. Available at http://lens1.csie.ncku.edu.tw/MOVE/index.htm (Accessed on June 2011).
23. FreeSim, 2008. Available at http://www.freewaysimulator.com/ (Accessed on June 2011).
24. F.J. Martinez, J.C. Cano, C.T. Calafate, P. Manzoni. Citymob: a mobility model pattern generator for VANETs. In IEEE Vehicular Networks and Applications Workshop (Vehi-Mobi, held with ICC), Beijing, China, May 2008.
25. J. Haerri, M. Fiore, F. Fethi, and C. Bonnet. VanetMobiSim: generating realistic mobility patterns for VANETs. Institut Eurécom and Politecnico Di Torino, 2006. Available at http://vanet.eurecom.fr/ (Accessed on June 2011).

26. STRAW—STreet RAndom Waypoint—vehicular mobility model for network simulations (e.g., for networks). 2008. Available at http://www.aqualab.cs.northwestern.edu/projects/STRAW/index.php (Accessed on June 2011).

27. F.J. Martinez, C.K. Toh, J.C. Cano, C.T. Calafate, and P. Manzoni. A survey and comparative study of simulators for vehicular ad hoc networks (VANETs). *Wireless Communications and Mobile Computing*, 2009. Available at http://onlinelibrary.wiley.com/doi/10.1002/wcm.859/abstract (Accessed on June 2011).

28. OPNET Technologies, 2008. Available at http://www.opnet.com/ (Accessed on February 2011).

29. Scalable Network Technologies. QualNet. Scalable Network Technologies, Inc., 2006. Available at http://www.scalable-networks.com/products/qualnet/ (Accessed on June 2011).

30. The Network Simulator (ns-2). Available at http://www.isi.edu/nsnam/ns/ (Accessed on June 2011).

31. GloMoSim Network Simulator. Available at http://pcl.cs.ucla.edu/projects/glomosim/ (Accessed on June 2011).

32. OMNET++ Network Simulation Framework. Available at http://www.omnetpp.org (Accessed on June 2011).

33. JiST/SWANS: Java in Simulation Time/Scalable Wireless Ad hoc Network Simulator, 2004. Available at http://jist.ece.cornell.edu/ (Accessed on June 2011).

34. K. Walsh and E.G. Sirer. A staged network simulator (SNS). Computer Science Department, Cornell University, 2003. Available at http://www.cs.cornell.edu/people/egs/sns/ (Accessed on June 2011).

35. TraNS (Traffic and Network Simulation Environment). Available at http://trans.epfl.ch/ (Accessed on June 2011).

36. Veins (Vehicles in Network Simulation). Available at http://www7.informatik.uni-erlangen.de/~sommer/omnet/traci/ (Accessed on June 2011).

37. NCTUns 5.0, 2008. Available at http://nsl10.csie.nctu.edu.tw/ (Accessed on June 2011).

38. S. Schutz, L. Eggert, S. Schmid, and M. Brunner. Protocol enhancements for intermittently connected hosts. *ACM SIGCOMM Computer Communication Review*, vol. 35 no. 3, 2005, pp. 5–18.

39. T. Goff, J. Moronski, D.S. Phatak, and V. Gupta. Freeze-TCP: a true end-to-end TCP enhancement mechanism for mobile environments. In Proceedings of the 19th IEEE Conference on Computer Communications (INFOCOM), 2000.

40. A. Al Hanbali, E. Altman, and P. Nain. A survey of TCP over ad hoc networks. *IEEE Communications Surveys & Tutorials*, Vol. 7, no. 3, 2005, pp. 22–36.

41. M. Bechler, S. Jaap, and L. Wolf. An optimized TCP for Internet access of vehicular ad hoc networks. In *The Networking 2005*. LNCS No. 3462. Berlin: Springer, 2005.

42. S. Yousefi, T. Chahed, M. Moosavi, and K. Zayer. Comfort applications in vehicular ad hoc networks based on fountain coding. In IEEE WiVec 2010, Taipei, May 2010.

43. D.J. MacKay. *Information Theory, Inference, and Learning Algorithms*. Cambridge, UK: Cambridge University Press, 2003.

44. M.A. Shokrollahi. Raptor codes. In IEEE International Symposium on Information Theory, July 2003.

45. V. Bychkovsky, B. Hull, A. Miu, H. Balakrishnan, and S. Madden. A measurement study of vehicular Internet access using in situ Wi-Fi networks. In Proceedings of the 12th Annual International Conference on Mobile Computing and Networking (MobiCom 2006), 2006, pp. 50–61.

46. Y. Zhang, J. Zhao, and G. Cao. On scheduling vehicle-roadside data access.. In Proceedings of ACM VANET'07, in conjunction with Mobicom'07, Montreal, Canada, September 2007.

47. M. Shahverdy, M. Fathy, and S. Yousefi. Scheduling algorithm for vehicle to road-side data distribution. In *ICHCC-ICTMF 2009, China. Communications in Computer and Information Science* (CCIS Book Series), vol. 66. Berlin: Springer, 2010, pp. 30–38.

48. J. Mittag, F. Thomas, J. Härri, and H. Hartenstein. A comparison of single and multi-hop beaconing in VANETs. In Proceedings of the Sixth ACM International Workshop on VehiculAr InterNETworking (VANET'09), 2009.

49. S. Yousefi, M. Fathy, and A. Benslimane. Performance of beacon safety message dissemination in Vehicular Ad hoc NETworks (VANETs). *Journal of Zhejiang University Science A*, vol. 8, no. 12, 2007, pp. 1990–2004.

50. M. TorrentMoreno, I. Mittag, P. Santi, and H. Hartenstein. Vehicle-to-vehicle .communication: fair transmit power control for safety-critical information. *IEEE Transactions on Vehicular Technology*, vol. 58, no. 7, 2009, pp. 3684–3703.

51. M. Armaghan, M. Fathy, and S. Yousefi. Improving the performance of beacon safety message dissemination in vehicular networks using Kalman filter estimation. In *FGCN 2009, Korea. Communications in Computer and Information Science* (CCIS Book Series), vol. 56. Berlin: Springer, 2009, pp. 74–82.

52. V. Sadatpour, M. Fathy, S. Yousefi, A.M. Rahmani, E.-S. Cho, and M.-K. Choi. Scheduling algorithm for beacon safety message dissemination in vehicular ad-hoc networks. In *FGCN 2009, Korea. Communications in Computer and Information Science* (CCIS Book Series), vol. 56. Berlin: Springer, 2009, pp. 133–140.

53. S. Rezaei, R. Sengupta, H. Krishnan, X. Guan, and R. Bhatia. Tracking the position of neighboring vehicles using wireless communications. *Transportation Research Part C: Emerging Technologies*, vol. 18, no. 3, 2010, pp. 335–350.

54. G. Welch and G. Bishop An introduction to the Kalman filter. Technical Report. UMI Order Number TR95-041, University of North Carolina at Chapel Hill, 1995.

55. W.R. Chang, H.T. Lin, and B.X. Chen. TrafficGather: an efficient and scalable data collection protocol for vehicular ad hoc networks. In Proceedings of the IEEE CCNC, 2008.

56. M.M. Artimy, W. Robertson, and W.J. Phillips. Connectivity with static transmission range in vehicular ad hoc networks. In *Proceedings of the 3rd Annual Communication Networks and Services Research Conference*, 2005, pp. 237–242.

57. M.M. Artimy, W. Robertson, and W.J. Phillips. Assignment of dynamic transmission range based on estimation of vehicle density. In *Proceedings of the 2nd ACM International Workshop on Vehicular Ad hoc Networks (VANET)*, 2005, pp. 40–48.

58. S. Yousefi, E. Altman, R. El-Azouzi, and M. Fathy. Analytical model for connectivity in vehicular ad hoc networks. *IEEE Transactions on Vehicular Technology*, vol. 57, no. 6, 2008, pp. 3341–3356.

59. N. Wisitpongphan, F. Bai, P. Mudalige, V. Sadekar, and O. Tonguz. Routing in sparse vehicular ad hoc wireless networks. *IEEE Journal on Selected Areas in Communications*, vol. 25, no. 8, 2007, pp. 1538–1556.

60. D. Miorandi and E. Altman. Connectivity in one-dimensional ad hoc networks: a queuing theoretical approach. *Wireless Networks*, vol. 12, no. 6, 2006, pp. 573–587.

61. M. Kijima and M. Ohnishi. Stochastic orders and their applications in financial optimization. *Mathematical Methods of Operations Research*, vol. 50, no. 2, 1999, pp. 351–372.

62. M. Rudack, M. Meincke, and M. Lott. On the dynamics of ad hoc networks for inter-vehicles communications (IVC). In *Proceedings of ICWN*, 2002, pp. 40–48.

63. S. Yousefi, E. Altman, R. El-Azouzi, and M. Fathy. Improving connectivity in vehicular ad hoc networks: an analytical study. *Computer Communications*, vol. 31, no. 9, 2008, pp. 1653–1659.

64. K.C. Lee, U. Lee, and M. Gerla. Survey of routing protocols in vehicular ad hoc networks. In Advances in Vehicular Ad-Hoc Networks: Developments and Challenges, IGI Global, October 2009.

65. A. Iwata. Scalable routing strategies for ad-hoc wireless networks. In *IEEE JSAC*, August 1999, pp. 1369–1379.

66. C.E. Perkins and E.M Royer. Ad-hoc on-demand distance vector routing. In *Proceedings of the IEEE WMCSA'99, New Orleans, LA*, February 1999, pp. 90–100.

67. V. Naumov, R. Baumann, and T. Gross. An evaluation of inter-vehicle ad hoc networks based on realistic vehicular traces. In Proceedings of the ACM MobiHoc'06 Conference, May 2006.

68. D.B. Johnson and D.A Maltz. Dynamic source routing in ad hoc wireless networks. In T. Imielinski and H. Korth (eds.), *Mobile Computing*. Massachusetts: Kluwer Academic Publishers, 1996, Ch. 5, pp. 153–181.

69. V. Namboodiri, M. Agarwal, and L. Gao. A study on the feasibility of mobile gateways for vehicular ad-hoc networks. In *Proceedings of the First International Workshop on Vehicular Ad Hoc Networks*, 2004, pp. 66–75.

70. C.-C. Ooi and N. Fisal. Implementation of geocast-enhanced AODV-bis routing protocol in MANET. In *Proceedings of the IEEE Region 10 Conference*, vol. 2, 2004, pp. 660–663.

71. Y.-B. Ko and N.H. Vaidya. Location-aided routing (LAR) in mobile ad hoc networks. *Wireless Networks*, vol. 6, no. 4, 2000, pp. 307–321.

72. L. Briesemeister, L. Schäfers, and G. Hommel. Disseminating messages among highly mobile hosts based on inter-vehicle communication. In *Proceedings of the IEEE Intelligent Vehicles Symposium*, 2000, pp. 522–527.

73. Y.-C. Hu, A. Perrig, and D.B. Johnson. ARIADNE: a secure on-demand routing protocol for ad hoc networks. In *Proceedings of ACM International Conference on Mobile Computing and Networking (MOBICOM'02), Atlanta, GA,* September 2002, pp. 12–23.

74. T. Yu, G.H. Lu, and Z.L. Zhang. Enhancing location service scalability with high grade. In IEEE International Conference on Mobile Ad-hoc and Sensor Systems, October 25–27, 2004, pp. 164–173.

75. J. Zhao and G. Cao. VADD: vehicle-assisted data delivery in vehicular ad hoc networks. In *Proceedings of 25th IEEE International Conference on Computer Communications*, April 2006, pp. 1–12.

76. B. Karp and H.T. Kung. GPSR: greedy perimeter stateless routing for wireless networks. In *Proceedings of Mobile Computing and Networking*, 2000, pp. 243–254.

77. I. Leontiadis and C. Mascolo. GeOpps: geographical opportunistic routing for vehicular networks. In *IEEE International Symposium on a World of Wireless, Mobile and Multimedia Networks (WoWMoM)*, June 18–21, 2007, pp. 1–6.

78. G. Toussaint. The relative neighborhood graph of a finite planar set. *Pattern Recognition*, vol. 12, no. 12, 1980, pp. 231–268.

79. K.R. Gabriel and R. Sokal. A new statistical approach to geographic variation analysis. *Systematic Zoology*, vol. 18, 1969, pp. 231–268.

80. S. Schnaufer, H. Füßler, M. Transier, and W. Effelsberg. Unicast ad-hoc routing in vehicular city scenarios. TR-2007-012, University of Mannheim, Mannheim, Germany.

81. S. Schnaufer and W. Effelsberg. Position-based unicast routing for city scenarios. In *International Symposium on a World of Wireless, Mobile and Multimedia Networks*, June 23–26, 2008, pp. 1–8.

82. C. Lochert, M. Mauve, H. F¨ussler, and H. Hartenstein. Geographic routing in city scenarios. *SIGMOBILE Mobile Computing and Communications Review*, vol. 9, no. 1, 2005, pp. 69–72.

83. C. Lochert, H. Hartenstein, J. Tian, H. Fussler, D. Hermann, and M. Mauve. A routing strategy for vehicular ad hoc networks in city environments. In *Proceedings of IEEE Intelligent Vehicles Symposium*, June 9–11, 2003, pp. 156–161.

84. B.C. Seet, G. Liu, B.S. Lee, C.H. Foh, K.J. Wong, and K.K. Lee. A-STAR: a mobile ad hoc routing strategy for metropolis vehicular communications. In *Proceedings of the 3rd International IFIP-TC6 Networking Conference*, Athens, Greece, Lecture Notes in computer Science (LNCS), May 9–14, 2004, pp. 989–999.

85. D. Forderer. Street-topology based routing. Master's thesis, University of Mannheim, May 2005.

86. M. Jerbi, S.M. Senouci, R. Meraihi, and Y. Ghamri-Doudane. An improved vehicular ad hoc routing protocol for city environments. In *IEEE International Conference on Communications, 2007 (ICC'07)*, June 24–28, 2007, pp. 3972–3979.

87. K. Lee, M. Le, J. Haerri, and M. Gerla. Louvre: landmark overlays for urban vehicular routing environments. In Proceedings of IEEE WiVeC, 2008.

88. H. Füßler, H. Hannes, W. Jorg, M. Martin, and E. Wolfgang. Contention-based forwarding for street scenarios. In *Proceedings of the 1st International Workshop in Intelligent Transportation (WIT 2004), Hamburg, Germany,* March 2004, pages 155–160.

89. Y. Shao, C. Liu, and J. Wu. Delay tolerant networks in VANETs. In *Vehicular Networks: From Theory to Practice,* S. Olariu and M. C. Weigle (Eds.) Chapman and Hall/CRC, 2009.

90. A. Bachir and A. Benslimane. A multicast protocol in ad hoc networks inter-vehicle geocast. In Proceedings of IEEE VTC, 2003.

91. M. Torrent-Moreno, D. Jiang, and H. Hartenstein. Broadcast reception rates and effects of priority access in 802.11-based vehicular ad-hoc networks. In Proceedings of ACM VANET, 2004.

92. G. Korkmaz, E. Ekici, F. Ozguner, and U. Ozguner. Urban multi-hop broadcast protocol for inter-vehicle communication systems. In Proceedings of ACM VANET, October 2004.

93. N. Wisitpongphan, O.K. Tonguz, J.S. Parikh, P. Mudalige, F. Bai, and V. Sadekar. Broadcast storm mitigation techniques in vehicular ad hoc networks. *IEEE Wireless Communication*, vol. 14, 2007, pp. 84–89.

94. G. Korkmaz, E. Ekici, and F. Ozguner. An efficient fully ad-hoc multi-hop broadcast protocol for inter-vehicular communication systems. In Proceedings of IEEE ICC, 2006.

95. Z. Haas, J.Y. Halpern, and L. Li. Gossip-based ad hoc routing. In Proceedings of IEEE INFOCOM, 2002.

96. A. Vahdat and D. Becker. Epidemic routing for partially connected ad hoc networks, Technical Report (CS-2000-06), Department of Computer Science, Duke University, Durham, NC, April 2000.

97. T. Spyropoulos, K. Psounis, and C.S. Raghavendra. Spray and wait: an efficient routing scheme for intermittently connected mobile networks. In Proceedings of ACM SIGCOMM Workshop on Delay-Tolerant Networking, 2005.

98. A. Lindgren, A. Doria, and O. SchelSn. Probabilistic routing in intermittently connected networks. In Poster of ACM MOBIHOC, 2003.

99. J. Leguay, T. Friedman, and V. Conan. Evaluating mobility pattern space routing for DTNs. In Proceedings of IEEE INFOCOM, 2006.

100. IEEE 802.16-2004. Local and metropolitan networks—Part 16: air interface for fixed broadband wireless access systems, 2004.

101. IEEE 802.16e-2005. Local and metropolitan networks—Part 16: air interface for fixed and mobile broadband wireless access systems, amendment 2: physical and medium access control layers for combined fixed and mobile operation in licensed bands and corrigendum 1, 2006.

102. A. Costa, P. Pedreiras, J. Fonseca, H. Proença, Á. Gomes, and J.S. Gomes. Evaluating WiMAX for vehicular communication applications. In IEEE International Conference on Emerging Technologies and Factory Automation (ETFA 2008), 2008.

103. Y. Ge, S. Wen, Y.-H. Ang, and Y.-C. Liang. Optimal relay selection in IEEE 802.16j multihop relay vehicular networks. *IEEE Transactions on Vehicular Technology*, vol. 59, no. 5, 2010, pp. 2198–2206.

104. J. Mar, S.E. Chen, and Y.R. Lin. The effect of the MS speed on the traffic performance of an integrated mobile WiMAX and DSRC multimedia networks on the highway. In *Proceedings of the ACM IWCMC*, August 2007, pp. 43–48.

105. T. Tsuboi, J. Yamada, N. Yamauchi, and N. Yoshikawa. WAVE design for next DSRC applications. In IEEE Proceedings of the 9th Conference on Wireless Telecommunications Symposium, 2010.

106. S. Yousefi, M.S. Mousavi, and M. Fathy. Vehicular ad hoc networks (VANETs): challenges and perspectives. In *Proceedings of the 6th IEEE International Conference ITST, Chengdu, China*, 2006, pp. 761–766.

Chapter 14

Underwater Wireless Ad Hoc Networks: A Survey

Miguel García, Sandra Sendra,
Marcelo Atenas, and Jaime Lloret

Contents

Underwater ad hoc communication networks have become an important field of research for many research groups in the recent years. These types of networks are envisioned to enable applications for oceanographic data collection, ocean sampling, environmental and pollution monitoring, and others. This chapter presents an overview about the main issues in underwater wireless ad hoc networks. The technologies currently used have been analyzed; we can observe that the acoustic waves are best adapted to the aquatic environment. Moreover, the ad hoc routing protocols are going to be discussed. We will see that reactive protocols [such as dynamic source routing (DSR) and ad hoc on-demand distance vector (AODV)] are the most used in underwater networks. Finally, several applications have been described to see that there are projects based on ad hoc communications in underwater environment.

14.1 Introduction and Related Work

A wireless ad hoc network is a group of communication nodes that can set up and maintain a network among themselves without the support of a base station or a central controller (infrastructure). From the applications perspective, wireless ad hoc networks are useful for situations that require mobile or infrastructure-less local network deployment, such as crisis response, military applications, or recently for underwater acoustic networks in the detection of signals for oceanographic applications.

The global warming became a priority issue in the past years due to its direct involvement in the climate change. If the global warming keeps rising in the future, the polar ice sheets will increase their melting process, contributing to the sea level rise. So, it is important to deploy a system to control the change in the sea level, for example, with an autonomous underwater ad hoc network, capable of providing accurate and timely information to either governmental institution or private institution. For this reason the underwater acoustic communication networks have become an important research field for many research groups in the recent years.

Nowadays, there are several works where the authors explain several issues about underwater communications. For example, in [1], the authors discuss several fundamental key aspects of underwater acoustic communications. They describe the communication architecture of underwater sensor networks, the factors that influence underwater network design, and so on.

Akyildiz et al. in [2] analyze in detail the current solutions for MAC, network, and transport layer protocols. They show the open research issues. Moreover, they discuss architectures for two-dimensional (2D) and three-dimensional (3D) underwater sensor networks.

Another work is presented in [3]; in this case, the authors summarize a number of practical issues differentiating underwater acoustic networks from terrestrial radio-based sensor networks. They show the main features of underwater networks and their limitations.

There are other works published, which explain individually a protocol, an application, or something related to underwater communication. For example, Peleato and Stojanovic [4] propose a channel sharing protocol for ad hoc underwater networks that saves energy by avoiding collisions while maximizing the throughput. It is based on the minimization of the duration of a handshake by taking advantage of the receiver's tolerance to the interference when the two nodes are closer than the maximal transmission range.

Finally, another work is presented in [5]; in this case, the authors present a novel platform for underwater sensor networks to be used for long-term monitoring of coral reefs. The sensor network consists of static and mobile underwater sensor nodes. The nodes communicate point to point using a novel high-speed optical communication system, and they broadcast using an acoustic protocol.

There are several works that present a survey of several aspects related to underwater communications, but all are focused on sensor networks. The main objective of this chapter is to present a survey of ad hoc communications in underwater networks, with a more general purpose, which may include any type of ad hoc communication—whether sensors, vehicles, or any other device.

This chapter is organized as follows. Section 14.2 shows the main issues in this type of communication. The protocols used at the data link and network level are discussed in Section 14.3. In Section 14.4, we present different underwater applications. Finally, the conclusion is drawn in Section 14.5.

14.2 Underwater Communication Issues

Current systems for underwater communication essentially use three methods for transmitting information. These methods are based on sound waves, electromagnetic (EM) waves, and optical signals. Each one of these techniques has advantages and disadvantages mainly due to the physical constraints [6,7].

The signal propagation can be explained by the physicochemical properties of the water and the physical properties of the light.

The physicochemical factors that influence the light properties are the transparency, i.e., the amount of light transmitted in the seawater; the absorption; the amount of radiation retained; and turbidity, which reduces water clarity by the presence of suspended matter.

The physical properties of the light are reflection, refraction, and extinction, which measure the degree of the light that can penetrate into the marine environment. In addition, the two most important factors in seawater are the temperature and salinity, which determine the density of the water. In the ocean, the density tends to increase with depth, so the upper water layers always have higher density. All these factors are very important to know how to spread the light and sound in the ocean.

This section will show some of the main expressions of different types of waves in underwater environment and the factors that may cause difficulties in communication operation.

14.2.1 Underwater Communications Based on EM Waves

The use of EM waves to transmit signals in the water is characterized to be a fast and efficient communication between network nodes.

Moreover, the use of EM waves in the radio frequency band has several advantages over the acoustic waves: it is mainly faster and can be used in higher work frequencies (which results in a higher bandwidth). Furthermore, there are several factors that limit the use of EM waves in the water. EM waves are propagated in very different ways depending on the type of water where the system is implemented.

14.2.1.1 EM Waves in Freshwater

Freshwater is a medium that has low loss. The propagation speed c of the signals can be expressed by the following approximation [8]:

$$c \approx \frac{1}{\sqrt{\varepsilon \mu}} \tag{14.1}$$

where ε represents the dielectric permittivity and μ refers to the magnetic permeability of the material (in this case water). Its value has no significant changes for most nonmagnetic medium.

The dielectric permittivity of a material is usually expressed in relation to the permittivity of the vacuum, which is called *relative permittivity* (also called dielectric constant). The absolute permittivity is calculated by multiplying the relative permittivity by the permittivity of the vacuum:

$$\varepsilon = \varepsilon_r \varepsilon_0 = (1 + \chi_e)\varepsilon_0 \tag{14.2}$$

where χ_e is the electric susceptibility of the material.

On the contrary, the magnetic permeability of a substance is the product of the relative magnetic permeability and the magnetic permeability of the material in free space:

$$\mu = \mu_r \mu_0 \tag{14.3}$$

The magnetic permeability in free space is given by

$$\mu_0 = 4\pi \times 10^{-7} \text{ N/A}^2 \tag{14.4}$$

Although the dielectric permittivity of water is approximately 81, the propagation speed of signals is reduced by a factor 9 with respect to the speed of light in free space. However, this speed is much greater than the speed of sound waves (more than 4 orders of magnitude).

The absorption coefficient α for the propagation of EM in freshwater can be approximated by [8]:

$$\alpha \approx \frac{\sigma}{2}\sqrt{\frac{\mu}{\varepsilon}} \tag{14.5}$$

where σ is the electrical conductivity, ε is the material permittivity, and μ is the magnetic permittivity of the material. As can be seen in expressions 14.1 and 14.5, the speed propagation and absorption coefficient in freshwater are independent of the work frequency of the transmitted signals.

14.2.1.2 EM Waves in Seawater

Unlike freshwater, seawater is a medium that has large losses. The conductivity of seawater is mainly due to the concentration of total dissolved solids in water. The average salinity of seawater in the oceans is about 34 parts per thousand (ppt).

The electrical conductivity in seawater is about 2 orders of magnitude higher than in the freshwater. The electrical conductivity of seawater is a function of salt content (salinity) and temperature. At frequencies below 1 GHz, its value is given by

$$\sigma = 0.18C \times 0.93[1 + 0.02(T - 20)] \ (\text{S/m}) \tag{14.6}$$

where C is the salt content in ppt and T is the temperature in °C. At 20°C, seawater has an average value of 5 S/m, while freshwater features around 0.005–0.05 S/m.

The dielectric permittivity of seawater is also a function of salinity and temperature. Often a value of 80 is used for the relative permittivity of seawater at 20°C, although the real value at a low frequency is about 70. However, at frequencies below about 100 MHz, ε_r is much lower than 60.

In media with high conductivity used for the signals transmission, the speed of propagation and absorption losses of EM waves are directly proportional to the carrier frequency. The speed propagation of EM waves in seawater can be expressed by the approximation [8]:

$$C \approx \sqrt{\frac{4\pi f}{\mu\sigma}} \tag{14.7}$$

where μ is the magnetic permittivity of the material, f is the frequency of the carrier signal, and σ is the material conductivity. Moreover, the absorption losses in seawater can be approximated by [8]:

$$\alpha \approx \pi f \mu\sigma \tag{14.8}$$

where μ is the magnetic permittivity of the material, f is the frequency of the carrier signal, and σ is the electrical conductivity of the material.

Figures 14.1 and 14.2 show the evolution of speed propagation and absorption coefficient in the case of freshwater and seawater, depending on the frequency.

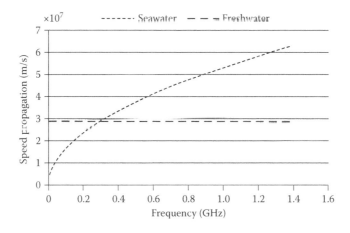

Figure 14.1 Speed propagation.

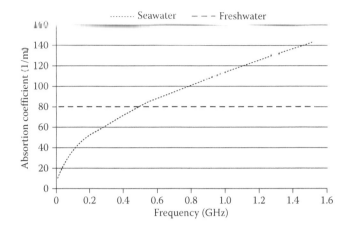

Figure 14.2   Absorption coefficient.

As we can see in these graphs, at high frequencies, there is a greater coefficient of absorption; therefore, it would be better to work at low frequencies while the higher frequencies show a faster rate. However, due to high absorption of seawater, an EM wave can hardly propagate in the seawater. This is the main motivation for working at a lower frequency in media of high conductivity. Seawater is a good example of this type of medium. Therefore, communication in seawater is impractical using classical approaches based on wave propagation.

14.2.2 Underwater Communications Based on Optical Signals

The light is a radiation mix of different frequencies. The optical signals used for wireless communications are generally limited to short distances because the water has a fairly high absorption factor in the optical waveband.

Not all frequencies within the spectrum of light are affected the same way. Each color is of a particular frequency radiation. For this reason, the red colors are the first to disappear and then the blue and green colors. The wavelengths of blue and green colors offer a good performance for broadband communication (10–150 Mbps) to moderate distances between 10 and 100 m. In theory, by using optical signals for communication underwater, speeds of 1 Gbps could be reached.

However, the optical signals have two main disadvantages: first, the suspended particles cause light dispersion, and second, due to the physical properties of water, the optical signals are absorbed quickly.

The light propagation depends on the medium traversed. For this reason, the light does not travel at the same speed in air than in water. When light propagates in an aqueous medium, its intensity decreases exponentially mainly due to the attenuation that is produced by two main causes:

■ Absorption: Absorption of light energy is converted into another type of energy, usually heat or chemical energy. This absorption is produced by
  – Algae, which use light as an energy source
  – Organic and inorganic particulate matter in suspension

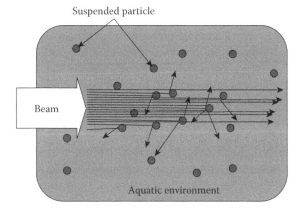

Figure 14.3 Scattering effect.

- - Inorganic compounds dissolved
- - The water itself
- ■ Scattering: This phenomenon is the result of the collision of the beam with particles in suspension, causing multiple reflections. More turbidity in water causes higher scattering effect, making the light penetration difficult. Figure 14.3 shows a diagram of the scattering effect in a beam when they incident on the particles in suspension.

As Figure 14.3 shows, the light in the ocean suffers the diffuse scattering effect. This means that the light does not always follow the shortest way to highlight an object, but many reflected rays arrive and divert from its original path. The amount of organic matter in suspension contained in seawater causes the light intensity to diminish in the propagation direction because it is absorbed by these particles; this parameter is called absorption coefficient or extinction factor of light and provides the value of transparency of the water. Many studies of the physical properties of the sea have been carried out using a white disc of 30 cm diameter called Secchi disc [9].

It is possible to calculate the total energy scattered by a particle through extremely complicated calculations [10] and using the expressions of the Mie scattering theory [11], which is valid for all possible relationships of particle diameter and wavelength of the signal. According to the Mie theory when the wavelength of light is similar to the diameter of the particle, light interacts with particles in a cross-sectional area larger than the geometric cross section of the particle. This allows us to express the total energy scattered by a particle in all directions [11]:

$$C_{SCA} = \frac{\int_0^{2\pi} \int_0^{2\pi} I_{SCA} r^2 \sin\phi \, d\phi \, d\theta}{I_0} \tag{14.9}$$

where I_{SCA} is the scattered light intensity, I_0 is the incident light intensity, r is the radius of the particle, and $\sin\phi \, d\phi \, d\theta$ represents the solid angle for the calculation of the entire surface area of the sphere. With this expression we would be able to calculate the amount of light beam energy lost due to beam collisions with particles suspended in the water. The larger the particles or turbidity in water, the more absorption of the light beam would be in the medium.

14.2.3 Underwater Communications Based on Acoustic Waves

The sound is produced by a vibration of the molecules in an elastic substance. Mechanical energy propagation of sound is absorbed in the medium by which it is spread. In water, the sound propagates faster and has lower energy losses than in the air. Sounds and ultrasonic sound waves are transmitted into the sea at a speed between 1400 and 1600 m/s, while in the air the speed propagation is 340 m/s. This is because seawater is not compressed; i.e., it cannot be reduced to a lower volume so that the absorption of sound waves is minimal, contrary to what happens in the atmosphere, where sounds are absorbed at very short distances.

There are several factors that influence the distance that sound can travel underwater. On the one hand, particles of seawater can reflect, scatter, and absorb certain frequencies of sound as well as certain particles in the atmosphere can reflect, scatter, and absorb certain wavelengths of light. Seawater absorbs 30 times the amount of sound absorbed by distilled water, attenuating certain frequencies of sound. Low-frequency sounds and passover tiny particles tend to travel farther without causing any loss by absorption or scattering. Studies have shown that factors such as salinity, temperature, and pressure changes cause the speed of sound propagation underwater.

In general, we can say that the ocean is divided into horizontal layers in which the speed of sound depends greatly on the temperature in the upper regions and the pressure in the lower regions. The top layer is heated by the sun whose temperature varies depending on the season. In middle latitudes, the water is perfectly mixed by the action of waves and currents. But there is also a transitional layer called the thermocline, where temperature drop continuously with depth. As the temperature drops, the speed of sound decreases. However, there is a point, ranging from 600 m to 1 km below the surface, where temperature changes are slight. In this case, the main factor influencing the speed of sound is increasing pressure, which causes the sound speed to increase.

Expression 14.10 relates the value of the sound speed propagation in water with the temperature (T) in °C, salinity (S) in ppt, and the depth (z) in m [12]

$$c(T,\ S,\ z) = a_1 + a_2 T + a_3 T^2 + a_4 T^3 + a_5 (S-35) + a_6 z + a_7 z^2 + a_8 T(S-35) + a_9 T z^3$$

$$(14.10)$$

where $a_1 \dots a_9$ represent constants whose values are given in Table 14.1.

If we take expression 14.10 and represent it graphically, we obtain Figure 14.4, which shows the values acquired by the speed of sound propagation, taking into account the effect of temperature and pressure variation according to the depth.

Underwater acoustic communications are mainly influenced by the physical properties of the medium's own sound wave and by path loss, noise, multipath, Doppler spread, and high and variable propagation delay.

Table 14.1 Values of the Constants

Constant	Value	Constant	Value	Constant	Value
a_1	1448.96	a_4	2.374×10^{-4}	a_7	1.675×10^{-7}
a_2	4.591	a_5	1.340	a_8	-1.025×10^{-2}
a_3	-5.304×10^{-2}	a_6	1.630×10^{-2}	a_9	-7.139×10^{-13}

- Propagation speed: The propagation speed of sound through water is much smaller than the propagation of electromagnetic signals. The speed of sound in water depends on the water properties as temperature, salinity, and pressure, as seen in Figure 14.4. A typical speed of sound in water near the ocean surface is about 1,525 m/s, which is about four times faster than the speed of sound in air, but 5 orders of magnitude slower than the speed of light. In approximate terms, the sound speed increases by 4.0 m/s for every degree Celcius rise in water temperature. When salinity increases by 1 practical salinity unit, the sound speed in water increases by 1.4 m/s .When the pressure of water (each additional 1 km deep), the speed of sound increases by about 17 m/s.
- Absorption: When there is swell at sea, the waves are converted into energy that is absorbed by the medium due to its intrinsic characteristics. When the acoustic signals propagate through the seawater, they suffer absorption as well, reducing its initial energy. The absorption phenomenon is more pronounced in some frequencies; high frequencies (above 100 kHz) are more sensitive, so we can ensure that it is frequency dependent.

We can express this behavior by expression 14.11, which has the contribution from the main salts in seawater [13]:

$$\alpha = 0.106 \frac{f_1 f^2}{f_1^2 + f^2} e^{(pH-8)/0.56} + 0.52\left(1+\frac{T}{43}\right)\left(\frac{S}{35}\right)\frac{f_2 f^2}{f_2^2 + f^2} e^{-z/6} + 0.00049 f^2 e^{-[(T/27)+(z/17)]} \tag{14.11}$$

where α is the attenuation in dB/km, f is the frequency in kHz, z is the depth in km, T is the temperature in °C, and S is the salinity in ppt. The first term is the contribution from boric acid, the second term is the contribution from magnesium sulfate, and the last term is the pure water contribution.

When a signal is transmitted through a medium that can generate losses, several factors must be taken into account. The main mechanisms for energy losses are the geometric spreading, absorption losses, and scattering losses.

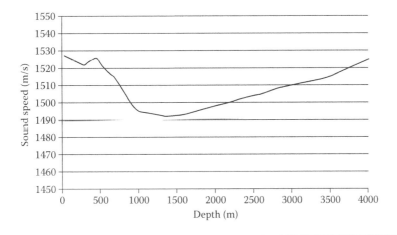

Figure 14.4 Sound speed as a function of depth.

■ Attenuation: It is mainly caused by absorption when the acoustic energy is converted into heat energy. The attenuation is caused by scattering and reverberation (a deep), refraction, and dispersion (surface). The attenuation also increases with distance and frequency. We have shown this phenomenon in Figures 14.1 and 14.2.

■ Geometric expansion: Geometric expansion is the loss of energy in the acoustic wave, which is spreading, due to conservation of energy. This refers to the diffusion of sound energy as a result of the expansion of the wave fronts; i.e., when a pulse of an acoustic wave propagates as the wave front moves away from the origin, sound energy covers a larger area. Therefore, the wave energy per unit area becomes smaller. This phenomenon is independent of the frequency. We can categorize geometric expansion into two classes:

– Spherical, where the source spreads out omnidirectional and characterizes the communications in deep water. The energy losses caused by the geometric expansion are proportional to the square of the distance.

– Cylindrical, which takes into account the horizontal radiation only and characterizes the communications in shallow water. The energy losses caused are proportional to the distance.

Other factors such as ambient noise, dispersion, or multipath effect may also hinder communication underwater.

■ Noise: Acoustic communication noise is defined as any unwanted sound that interferes with the communication between communication systems. Basically we can distinguish two types of noises:

– Generated noise: It is the noise caused by machinery such as engines and natural activity of the medium, as animal life.

– Ambient noise: It is related to the movements of water, tides, and currents; storm or water bubbles when it hits the coast; etc., which can generate losses up to 26 dB/m. Also seismic and biological phenomena can be considered. All of these phenomena are known as *hydrodynamic* [14].

■ Delay and delay variation: Due to the nature of the medium, the variation of signal delay can be very high and it may turn out to be very detrimental to the efficient design of a protocol and would complicate the estimation of the round trip time, an important parameter for many common communication protocols.

■ Doppler spread and scattering: Scattering is the phenomenon whereby a set of particles moving in one direction can bounce with the particles on the environment, and its effect can deviate from a predefined straight line movement [14]. Moreover, if the sea is calm, the signals near the surface are reflected almost perfectly, with the only strain that can generate phase difference. However, when the sea is rough, the waves can move the point of reflection of the acoustic wave and causes the dispersion of energy. These two phenomena can cause losses of signal strength due to the multipath effect.

■ Multipath: Multipath propagation can cause degradation of the acoustic communication signal mainly due to the generation of interference intersymbol and the destruction of information. The generation of multiple paths between the transmitter and the receiver depends on the link configuration. Communications where you create vertical channels do not suffer a very high scattering loss mainly because the signal does not found obstacles to reflect, while the horizontal channels may experience greater losses due to the surface effect. The diffusion level is a function of depth and distance between the transmitter and the receiver.

14.3 Communication Protocols

In this section, the data link layer and the network layer protocols will be explained as a part of integral communication architecture in underwater ad hoc networks. As we shall see below, many of the protocols used for underwater ad hoc networks are the same as those used in terrestrial ad hoc communications.

First we analyze the protocols used to access to the media and then we describe the network protocols.

14.3.1 Data Link Layer Protocols

The main solutions for data link layer protocols are mainly focused on carrier sensing multiple access (CSMA or CDMA), because frequency division multiple access is not suitable for underwater ad hoc networks, due to the narrow bandwidth in underwater channels and the vulnerability of limited band systems to fading and multipath. Moreover, time division multiple access (TDMA) shows limited bandwidth efficiency because of the long-time guards required in the underwater channel. There are several protocols based on CSMA or CDMA, but the most interesting are multiple access collision avoidance/MACA for wireless (MACA/MACAW), floor acquisition multiple access (FAMA), and underwater medium access control (UW-MAC).

14.3.1.1 MACA and MACAW

The MACA [15] protocol makes use of request to send (RTS), clear to send (CTS), data packet (DATA), and acknowledgement (ACK) sequences to operate, which have shown to be effective for underwater use, and specially in the Seaweb project [16].

MACA was the first modern protocol that used RTS/CTS exchange and underscored the benefit of it over the then existing protocols (which were largely carrier sense multiple access with collision avoidance (CSMA/CA) based). The main motivation was again the hidden terminal problem. In MACA, before a station sends the data it sends an RTS message to the receiver. On success the receiver responds with a CTS. The nearby stations are also listening to this exchange. If a station hears RTS it waits for the corresponding CTS. If it does not hear CTS, it means any transmission it has will not interfere with the receiver. The assumption here is if you cannot hear the receiver, the receiver cannot hear you too. This helps alleviate the exposed terminal problem.

Any station, other than the original RTS sender, on hearing CTS will defer its transmission. The time for which transmission is to be deferred depends on the packet length to be transmitted, which is contained in the CTS packet. This takes care of the hidden terminal problem. Binary exponential backoff was used in the case of collisions of RTS packets. MACA requires much simpler hardware because of the absence of carrier sense.

MACAW [17] is proposed as a series of improvements to the basic MACA algorithm. First, the authors suggest a less aggressive backoff algorithm. Second, they propose that receivers should send an ACK to the sender after successfully receiving a data message, and third, they propose two related techniques for allowing transmitters to avoid contention. Also, separate backoff parameters were introduced to avoid this copied parameter to spread widely even to areas with no congestion. It also proposed a multiple stream model for fairness among streams emerging from the same station.

MACAW acknowledged the importance of link layer acknowledgments and made the protocol from RTS–CTS–DATA to RTS–CTS–DATA–ACK. There are two ways of dealing with this: carrier sense or an explicit packet specifying the length of the transmission at the start of it.

MACAW takes the latter approach to keep the hardware simple and calls this packet DS (data sending). Another control packet's request for RTS was added to let the receiver commend for the sender to improve fairness in cases when there are two receivers in the vicinity of each other (thus only one can receive). By making the protocol significantly more complex, MACAW lost performance when the channel was lightly loaded but led to much better throughput and fairer allocation in the presence of high loads.

14.3.1.2 FAMA

The FAMA is another MACA-based scheme that requires every transmitting station to acquire control of the floor (i.e., the wireless channel) before it actually sends any data packet [18]. Unlike MACA or MACAW, FAMA requires that collision avoidance should be performed both at the sender and at the receiver.

In order to "acquire the floor," the sending node sends out an RTS using either nonpersistent packet sensing or nonpersistent carrier sensing. The receiver responds with a CTS packet, which contains the address of the sending node. Any station overhearing this CTS packet knows about the station that has acquired the floor. The CTS packets are repeated long enough for the benefit of any hidden sender that did not register another sending node's RTS. The authors recommend the nonpersistent carrier sensing variant for ad hoc networks since it addresses the hidden terminal problem effectively.

14.3.1.3 UW-MAC

In [19], Pompili et al. propose a distributed medium access control (MAC) protocol called UW-MAC for underwater acoustic sensor networks (UW-ASNs). UW-MAC is a transmitter based on the CDMA scheme that incorporates a novel closed-loop distributed algorithm to set the optimal transmit power and code length to minimize the near-far effect. It compensates for the effect of multipath by exploiting the time diversity in the underwater channel, thus achieving high channel reuse and low number of packet retransmissions, which results in decreased battery consumption and increased network throughput.

It is shown that UW-MAC manages to simultaneously meet the three objectives in deep water communications, which are not severely affected by multipath, while in shallow water communications, which are heavily affected by multipath, UW-MAC dynamically finds the optimal trade-off among high throughput and low access delay and energy consumption, according to the application requirements.

Main features of UW-MAC are as follows: (1) it provides a unique and flexible solution for different architectures such as static 2D deep water and 3D shallow water, and architectures with mobile autonomous underwater vehicles (AUVs); (2) it is fully distributed since code and transmit power are distributively selected by each sender without relying on a centralized entity; (3) it is intrinsically secure since it uses chaotic codes; (4) it efficiently supports multicast transmissions since spreading codes are decided at the transmitter side; (5) it is robust against inaccurate node position and interference information caused by mobility, traffic unpredictability, and packet loss due to channel impairment.

14.3.2 Network Layer Protocols

This subsection is divided into several subsections. This division has been made because in this way the reader can distinguish the protocols according to their performance. In ad hoc environments there are many routing protocols, but some of them cannot be used in underwater situations because they require other elements that cannot be transmitted in this underwater environment. For example, in terrestrial ad hoc networks there are protocols based on the situation of the nodes using the GPS (global positioning system) signal. These routing protocols cannot be used in underwater communication because the GPS signal cannot be transmitted underwater. For this reason, in this subsection we have included the most important protocols in underwater communications, but still there are many more that could be used.

14.3.2.1 Proactive Protocols

In networks utilizing a proactive routing protocol, every node maintains one or more tables representing the entire topology of the network. These tables are updated regularly in order to maintain up-to-date routing information from each node to every other node.

To maintain the up-to-date routing information, topology information needs to be exchanged between the nodes on a regular basis, leading to a relatively high overhead on the network. On the contrary, routes will always be available on request. The main disadvantages of such algorithms are

■ Respective amount of data needed for maintenance
■ Slow reaction on restructuring and failures

14.3.2.1.1 DSDV

The destination sequenced distance vector (DSDV) [20] protocol is a proactive routing protocol; it is based on the distance vector algorithms.

Each node in the network has a routing table for each destination, indicating how many hops are needed and what the next node is. The process of updating routing tables is produced by the exchange of information between nearby nodes and reapplying the shortest-path algorithms with lower costs. Each path is labeled with a sequence number, which gives a time indication on the validity of that path: higher sequence numbers imply more reliable paths. When two paths have the same sequence number, it is chosen which have the lowest cost (i.e., the least number of hops to arrive to destination).

If a node notices that a path to a destination does not work, it assigns a high value of jump number (meaning infinity) and to the sequence number an odd number. A sequence number identified with an odd number indicates that this path is unreachable, whereas an even number indicates that the destination itself is achievable.

14.3.2.1.2 OLSR

The optimized link state routing (OLSR) [21] protocol is a proactive routing protocol for mobile ad hoc networks (MANETs). The protocol inherits the stability of a link state algorithm and has the advantage of having routes immediately available when needed due to its proactive nature. OLSR is an optimization over the classical link state protocol tailored for MANETs.

OLSR minimizes the overhead from flooding of control traffic by using only selected nodes, called MPRs, to retransmit control messages. This technique significantly reduces the number of retransmissions required to flood a message to all nodes in the network. Then, OLSR requires only partial link state to be flooded in order to provide shortest-path routes. The minimal set of link state information required is that all nodes, selected as MPRs, *must* declare the links to their MPR selectors. Additional topological information, if present, *may* be utilized, for example, for redundancy purposes.

OLSR may optimize the reactivity to topological changes by reducing the maximum time interval for periodic control message transmission. Furthermore, as OLSR continuously maintains routes to all destinations in the network, the protocol is beneficial for traffic patterns where a large subset of nodes are communicating with another large subset of nodes, and where the (source, destination) pairs are changing over time. The protocol is particularly suited for large and dense networks, as the optimization done using MPRs works well in this context. The larger and more dense a network, the more optimization can be achieved as compared with the classic link state algorithm.

14.3.2.1.3 TBRPF

Topology dissemination based on reverse-path forwarding (TBRPF) is a proactive link state routing protocol [22] designed for MANETs, which provides hop-by-hop routing along shortest paths to each destination. Each node running TBRPF computes a source tree (providing shortest paths to all reachable nodes) based on partial topology information stored in its topology table, using a modification of Dijkstra's algorithm. To minimize overhead, each node reports only a part of its source tree to neighbors.

TBRPF uses a combination of periodic and differential updates to keep all neighbors informed of the reported part of its source tree. Each node also has the option to report additional topology information (up to the full topology) to provide improved robustness in highly mobile networks.

TBRPF performs neighbor discovery using "differential" Hello messages that report changes only in the status of neighbors. This results in Hello messages that are much smaller than those of other link state routing protocols such as open shortest path first (OSPF) [6].

TBRPF consists of two modules: the neighbor discovery module and the routing module (which performs topology discovery and route computation).

The TBRPF neighbor discovery protocol allows each node i to quickly detect the neighbor nodes j such that a bidirectional link (I, J) exists between an interface I of node i and an interface J of node j. The protocol also quickly detects when a bidirectional link breaks or becomes unidirectional.

Each node running TBRPF maintains a source tree, denoted T, which provides shortest paths to all reachable nodes. Each node computes and updates its source tree based on partial topology information stored in its topology table, using a modification of Dijkstra's algorithm. To minimize overhead, each node reports only part of its source tree to neighbors.

14.3.2.1.4 WRP

The wireless routing protocol (WRP) [23] is a proactive unicast routing protocol for MANETs. WRP uses an enhanced version of the distance vector routing protocol, which uses the Bellman–Ford algorithm to calculate paths.

WRP, similar to DSDV, inherits the properties of the distributed Bellman–Ford algorithm. In this protocol, each mobile host is a specialized router that periodically advertises its view of the interconnection topology with other mobile hosts within the network to maintain up-to-date information about the status of the network. Unfortunately, in DSDV a node has to wait until it receives the next updated message originated by the destination in order to update its distance-table entry for that destination.

This protocol relies on the exchange of short control packets forming a query–reply process. It also has the ability to maintain multiple paths to a given destination. This is a destination-oriented protocol in which separate versions of the algorithm run independently for each destination. Routing is source-initiated, which means that routes are maintained by those sources that actually desire routes. Even though this algorithm provides multiple paths to the destination because of the query-based synchronization approach to achieve loop-free paths, the communication complexity could be high.

It differs from DSDV in table maintenance and in the update procedures. While DSDV maintains only one topology table, WRP uses a set of tables to maintain more accurate information. The tables that are maintained by a node are the following: distance table, routing table, link cost table, and a message retransmission list.

14.3.2.2 Reactive Protocols

Unlike proactive routing protocols, reactive routing protocols do not make the nodes initiate a route discovery process until a route to a destination is required. This leads to higher latency than with proactive protocols, but lower overhead. The main disadvantages of such algorithms are as follows:

- High latency time in route finding
- Excessive flooding can lead to network clogging

14.3.2.2.1 AODV

The ad hoc on-demand distance vector (AODV) [24] algorithm enables dynamic, self-starting, multihop routing between participating mobile nodes wishing to establish and maintain an ad hoc network. AODV allows mobile nodes to obtain routes quickly for new destinations and does not require nodes to maintain routes to destinations that are not in active communication. AODV allows mobile nodes to respond to link breakages and changes in network topology in a timely manner. The operation of AODV is loop-free, and by avoiding the Bellman–Ford "counting to infinity" problem offers quick convergence when the ad hoc network topology changes (typically when a node moves in the network). When links break, AODV causes the affected set of nodes to be notified so that they are able to invalidate the routes using the lost link.

One distinguishing feature of AODV is its use of a destination sequence number for each route entry. The destination sequence number is created by the destination to be included along with any route information it sends to requesting nodes. Using destination sequence numbers ensures loop freedom and is simple to program. Given the choice between two routes to a destination, a requesting node is required to select the one with the greatest sequence number.

Route requests (RREQs), route replies (RREPs), and route errors (RERRs) are the message types defined by AODV. These message types are received via UDP, and normal IP header processing applies. So, for instance, the requesting node is expected to use its IP address as the originator IP address for the messages. For broadcast messages, the IP-limited broadcast address (255.255.255.255) is used.

As long as the endpoints of a communication connection have valid routes to each other, AODV does not play any role. When a route to a new destination is needed, the node broadcasts an RREQ to find a route to the destination. A route can be determined when the RREQ reaches either the destination itself or an intermediate node with a "fresh enough" route to the destination.

A fresh enough route is a valid route entry for the destination whose associated sequence number is at least as great as that contained in the RREQ. The route is made available by unicasting an RREP back to the origination of the RREQ. Each node receiving the request caches a route back to the originator of the request, so that the RREP can be unicast from the destination along a path to that originator or likewise from any intermediate node that is able to satisfy the request.

14.3.2.2.2 DSR

The dynamic source routing (DSR) [25] protocol is a simple and efficient routing protocol designed specifically for use in multihop wireless ad hoc networks of mobile nodes. Using DSR, the network is completely self-organizing and self-configuring, requiring no existing network infrastructure or administration.

Network nodes cooperate to forward packets for each other to allow communication over multiple "hops" between nodes not directly within wireless transmission range of one another. As nodes in the network move about or join or leave the network and as wireless transmission conditions such as sources of interference change, all routing is automatically determined and maintained by the DSR protocol. Since the number or sequence of intermediate hops needed to reach any destination may change at any time, the resulting network topology may be quite rich and rapidly changing.

In designing DSR, we sought to create a routing protocol that had very low overhead yet been able to react very quickly to changes in the network. The DSR protocol provides a highly reactive service in order to help ensure successful delivery of data packets in spite of node movement or other changes in network conditions.

The DSR protocol is composed of two main mechanisms that work together to allow the discovery and maintenance of source routes in the ad hoc network:

- Route discovery is the mechanism by which a node S wishing to send a packet to a destination node D obtains a source route to D. It is used only when S attempts to send a packet to D and does not already know a route to D.
- Route maintenance is the mechanism by which node S is able to detect, while using a source route to D, whether the network topology has changed such that it can no longer use its route to D because a link along the route no longer works. When route maintenance indicates that a source route is broken, S can attempt to use any other route it happens to know to D or it can invoke route discovery again to find a new route for subsequent packets to D. Route maintenance for this route is used only when S is actually sending packets to D.

In DSR, route discovery and route maintenance each operate entirely "on demand." In particular, unlike other protocols, DSR requires no periodic packets of any kind at any layer within the network. For example, DSR does not use any periodic routing advertisement, link status sensing, or neighbor detection packets and does not rely on these functions from any underlying protocols in the network. This entirely on-demand behavior and lack of periodic activity allow the number of overhead packets caused by DSR to scale all the way down to zero when all nodes are approximately stationary with respect to each other and all routes needed for current communication have already been discovered. As nodes begin to move more or as communication patterns change, the routing packet overhead of DSR automatically scales to only what is needed to track the routes currently in use. Network topology changes not affecting routes currently in use are ignored and do not cause reaction from the protocol.

14.3.2.2.3 DYMO

The dynamic MANET on-demand (DYMO) [26] routing protocol enables reactive, multihop unicast routing among participating DYMO routers. The basic operations of the DYMO protocol are route discovery and route maintenance.

During route discovery, the originator's DYMO router initiates dissemination of an RREQ throughout the network to find a route to the target's DYMO router. During this hop-by-hop dissemination process, each intermediate DYMO router records a route to the originator. When the target's DYMO router receives the RREQ, it responds with an RREP sent hop-by-hop toward the originator. Each intermediate DYMO router that receives the RREP creates a route to the target and then the RREP is unicast hop by hop toward the originator. When the originator's DYMO router receives the RREP, routes have then been established between the originating DYMO router and the target DYMO router in both directions.

Route maintenance consists of two operations. In order to preserve routes in use, DYMO routers extend route lifetimes upon successfully forwarding a packet. In order to react to changes in the network topology, DYMO routers monitor routes over which traffic is flowing.

When a data packet is received for forwarding and a route for the destination is not known or the route is broken, then the DYMO router of the source of the packet is notified. An RERR is sent toward the packet source to indicate that the route to that particular destination is invalid or missing. When the source's DYMO router receives the RERR, it deletes the route. If this source's DYMO router later receives a packet for forwarding to the same destination, it will need to perform route discovery again for that destination.

DYMO uses sequence numbers to ensure loop freedom. Sequence numbers enable DYMO routers to determine the temporal order of DYMO route discovery messages, thereby avoiding use of stale routing information.

14.3.2.2.4 FSDSR

The flow state in the dynamic source routing (FSDSR) [27] protocol is an on-demand routing protocol. It allows the routing of most packets without an explicit source route header in the packet, further reducing the overhead of the protocol while still preserving the fundamental properties of DSR's operation.

A source node sending packets to some destination node may use the DSR flow state extension described here to establish a route to that destination as a flow. A *flow* is a route from the source to the destination represented by hop-by-hop forwarding state within the nodes along the route. Each flow is uniquely identified by a combination of the source node address, the destination node address, and a flow identifier (flow ID) chosen by the source node. Each flow ID is a 16-bit unsigned integer.

A DSR flow state header in a packet identifies the flow ID to be followed in forwarding that packet. From a given source to some destination, any number of different flows may exist and be in use, for example, following different sequences of hops to reach the destination. One of these flows may be considered to be the "default" flow from that source to that destination. A node receiving a packet, with neither a DSR header [25] specifying the route to be taken (with a source route option in the DSR header) nor a DSR flow state header specifying the flow to be followed, is forwarded along the default flow for the source and destination addresses specified in the packet's IP header.

In establishing a new flow, the source node generates a nonzero 16-bit flow ID greater than any unexpired flow IDs for this (source, destination) pair. If the source wishes for this flow to become the default flow, the low bit of the flow ID must be set (the flow ID is an odd number); otherwise, the low bit must not be set (the flow ID is an even number).

The source node establishing the new flow then transmits a packet containing a DSR header with a source route option as defined in the base specification for DSR [25]; to establish the flow, the source node must also include in the packet a DSR flow state header, with the flow ID field set to the chosen flow ID for the new flow and must include a timeout option in the DSR header, giving the lifetime after which information about this flow is to expire.

The source node also records this flow in its flow table for future use, setting the time to live (TTL) in this flow table entry to be the value used in the TTL field in the packets' IP header and setting the lifetime in this entry to be the lifetime specified in the timeout option in the DSR header.

Any further packets sent with this flow ID before the timeout that also contain a DSR header with a source route option must use this same source route in the source route option.

14.3.2.3 Flow-Oriented Protocols

These types of protocols find a route on demand by following present flows. One option is to unicast consecutively when forwarding data while promoting a new link. The main disadvantages of such algorithms are as follows:

- Takes long time when exploring new routes without prior knowledge
- May refer to estimative existing traffic to compensate for missing knowledge on routes

14.3.2.3.1 MOR

The multipath on-demand routing (MOR) [28] protocol is a protocol to connect nodes in wireless sensor networks. It is an ad hoc routing protocol that is reactive or on-demand, meaning that it establishes routes as needed. The advantage of this approach is obvious if only a few routes are needed, since the routing overhead is less compared with the proactive approach of establishing routes whether or not are needed. The disadvantage of on-demand establishment of routes is that connections take more time if the route needs to be established.

MOR lessens the disadvantages of on-demand routing in wireless sensor networks by having the likely targets of communication perform an initial broadcast. This allows all recipients to have a route to these nodes.

The main characteristic distinguishing MOR from other ad hoc routing protocols is that it maintains multiple routes to each destination, when available, whereas most other such protocols only keep a single route. There are many advantages to having multiple routes when possible, including

- Increased reliability
- Potentially better load balancing
- More even energy consumption (a consequence of better load balancing)

Each node in MOR remembers all next-hop nodes that are closer to a given destination for which a route exists. It then sends successive packets to each such node in round-robin fashion. If a next-hop node fails to acknowledge a given packet, the retransmission is attempted to another node, again if possible. This allows automatic and graceful recovery from occasional localized congestion as well as longer-term reasons for node unavailability.

14.3.2.3.2 LUNAR

Lightweight underlay network ad hoc routing (LUNAR) [29] is a routing protocol based on address resolution protocol (ARP) forwarding. A source node that desires to send an IP packet to a destination in the same LUNAR network emits internally a standard ARP request. The LUNAR layer intercepts this ARP request and maps it to a LUNAR RREQ message that is broadcasted and forwarded inside the network. Once the destination node receives this RREQ message, it will send back an RREP and build two independently managed unicast routes (one for each direction) between the source and the destination nodes. When the source receives the RREP message, it maps it back to an ARP reply message, permitting the source's IP stack to send IP datagrams to the destination node. Thus, none of the involved IP stacks is aware of the potential multihop nature of the underlying LUNAR network.

LUNAR routes have a limited lifetime and need to be reestablished at regular intervals. The corresponding LUNAR system parameter has been set to 3 s; i.e., a data route is in use for 3 s before it is replaced by another newly constructed one.

The routing and forwarding strategy of LUNAR is simpler than other MANET routing protocols: LUNAR does not feature route repair, route caching, or packet salvation. Nevertheless, simulations revealed that LUNAR scales without problems to networks of 40 nodes and 10 hops. Despite these observations, this document conservatively recommends LUNAR operations for networks with less than 15 nodes and even imposes a network diameter of 3 hops. Factors such as TCP unfairness over IEEE 802.11 (WiFi) become so dominant at this range that the best thing the LUNAR protocol can do is to impose an artificial 3-hop zone in order to let users operate with parameters yielding acceptable network performance.

LUNAR has four building blocks:

- The IP adaption and steering layer is responsible for translating events at the border to the IP stack into LUNAR-specific actions and vice versa. Among others, this includes the intercepting of ARP and DHCP requests and mapping them to the corresponding LUNAR primitives. This layer also takes care of the forced rediscovery of routes every 3 s.
- The LUNAR protocol engine implements the core routing algorithm. It handles the generation of LUNAR RREQ and RREP messages and their processing at intermediate and destination nodes. As a result, data routes will be created.
- The extensible resolution protocol is the request/reply protocol and data format through which the LUNAR protocol engines interact with each other.
- The selector network is a forwarding layer that supports the creation of data routes composed of network pointers. It is independent of IP addresses and the IP routing layer and works directly on top of the link layer.

14.3.2.3.3 SrcRR

The basic operation of source route (SrcRR) is similar to DSR with link caches: SrcRR is a reactive routing protocol with source-routed data traffic [30].

Every node running SrcRR maintains a link cache, which tracks the expected transmission count (ETX) metric values for links it has heard about recently. Whenever a change is made to the link cache, the node locally runs Dijkstra's weighted shortest-path algorithm on this database to find the current, minimum-metric routes to all other nodes in the network. To ensure that only fresh information is used for routing, if a link metric has not been updated within 30 s it is dropped from the link cache.

When a node wants to send data to a node to which it does not have a route, it floods an RREQ. When a node receives an RREQ, it appends its own node ID, as well as the current ETX metric from the node from which it received the request, and rebroadcasts it. A node will always forward a given RREQ the first time it receives it. If it receives the same RREQ again over a different route, it will forward it again if the accumulated route metric is better than the best metric it has forwarded so far. This ensures that the target of the RREQ will receive the best routes.

When a node receives an RREQ for which it is the target, it reverses the accumulated route and uses this as the source route for an RREP. When the original source node receives this reply, it adds each of the links to its link cache and then source-routes data over the minimum-metric path to the destination.

When an SrcRR node forwards a source-routed data packet, it updates its entry in the source route to contain the latest ETX metric for the link on which it received the packet. This allows the source and destination to maintain update link caches and discover when a route's quality has declined enough that an alternate route would be better. This allows the source and destination to learn of the existence and metric of some alternate links. As with all changes to the link cache, this prompts recomputation of all the best routes by using Dijkstra's algorithm.

14.3.2.4 Hybrid Protocols

These types of protocols combine the advantages of proactive and reactive routing. The routing is initially established with some proactively prospected routes and then serves the demand from additionally activated nodes through reactive flooding. The choice for one or the other method requires predetermination for typical cases. The main disadvantages of such algorithms are as follows:

- Advantage depends on the amount of nodes activated
- Reaction to traffic demand depends on the gradient of traffic volume

14.3.2.4.1 TORA

The temporally ordered routing algorithm (TORA) [31] is an adaptive routing protocol for multihop networks that possess the following attributes:

- Distributed execution
- Loop-free routing
- Multipath routing
- Reactive or proactive route establishment and maintenance
- Minimization of communication overhead via localization of algorithmic reaction to topological changes

TORA is distributed, in that routers need only maintain information about adjacent routers (i.e., one-hop knowledge). Like a distance vector routing approach, TORA maintains state on a per-destination basis. However, TORA does not continuously execute a shortest-path computation and thus the metric used to establish the routing structure does not represent a distance. The destination-oriented nature of the routing structure in TORA supports a mix of reactive and proactive routing on a per-destination basis. During reactive operation, sources initiate the establishment of routes to a given destination on-demand. This mode of operation may be advantageous in dynamic networks with relatively sparse traffic patterns, since it may not be necessary (nor desirable) to maintain routes between every (source, destination) pair at all times. At the same time, selected destinations can initiate proactive operation, resembling traditional table-driven routing approaches. This allows routes to be proactively maintained to destinations for which routing is consistently or frequently required (i.e., servers or gateways to hardwired infrastructure).

TORA has been designed to work on top of lower layer mechanisms or protocols that provide the following basic services between neighboring routers:

- Link status sensing and neighbor discovery
- Reliable, in-order control packet delivery
- Link and network layer address resolution and mapping
- Security authentication

Events such as the reception of control messages and changes in connectivity with neighboring routers trigger TORA's algorithmic reactions.

A logically separate version of TORA is run for each "destination" to which routing is required. The following discussion focuses on a single version of TORA running for a given destination. The term *destination* is used herein to refer to a traditional IP routing destination, which is identified by an IP address and mask (or prefix). Thus, the route to a destination may correspond to the individual address of an interface on a specific machine (i.e., a host route) or an aggregation of addresses (i.e., a network route).

TORA assigns directions to the links between routers to form a routing structure that is used to forward datagrams to the destination. A router assigns a direction ("upstream" or "downstream") to the link with a neighboring router based on the relative values of a metric associated with each router. The metric maintained by a router can conceptually be thought of as the router's "height" (i.e., links are directed from the higher router to the lower router). The significance of the heights and the link directional assignments is that a router may only forward datagrams downstream. Links from a router to any neighboring routers with an unknown or undefined height are considered undirected and cannot be used for forwarding.

14.3.2.4.2 ZRP

The zone routing protocol (ZRP) [32] is a protocol that divides the topology into zones and seeks to utilize different routing protocols within and between the zones based on the weaknesses and strengths of these protocols. The ZRP is totally modular, meaning that any routing protocol can be used within and between zones. The size of the zones is defined by a parameter *r* describing the radius in hops. Figure 14.5 illustrates a ZRP scenario with *r* set to 1. Intrazone routing is done by a proactive protocol since these protocols keep an up-to-date view of the zone topology, which results in no initial delay when communicating with nodes within the zone. Interzone routing is done by a reactive protocol. This eliminates the need for nodes to keep a proactive fresh state of the entire network.

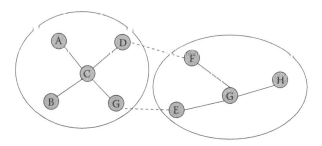

Figure 14.5   ZRP scenario with *r* set to 1.

The ZRP defines a technique called the bordercast resolution protocol (BRP) to control traffic between zones. If a node has no route to a destination provided by the proactive interzone routing, BRP is used to spread the reactive RREQ.

14.3.2.5 Hierarchical Protocols

With these types of protocols, the choice of proactive and reactive routing depends on the hierarchical level where a node resides. The routing is initially established with some proactively prospected routes and then serves the demand from additionally activated nodes through reactive flooding on the lower levels. The choice for one or the other method requires proper attribution for respective levels. The main disadvantages of such algorithms are as follows:

■ Advantage depends on depth of nesting and addressing scheme
■ Reaction to traffic demand depends on meshing parameter

14.3.2.5.1 CBRP

The cluster based routing protocol (CBRP) [33] is a routing protocol designed for use in MANETs. The protocol divides the nodes of the ad hoc network into a number of overlapping or disjoint 2-hop-diameter clusters in a distributed manner. A cluster head is elected for each cluster to maintain cluster membership information. Intercluster routes are discovered dynamically using the cluster membership information kept at each cluster head. By clustering nodes into groups, the protocol efficiently minimizes the flooding traffic during route discovery and speeds up this process as well.

Furthermore, the protocol takes into consideration the existence of unidirectional links and uses these links for both intracluster and intercluster routing.

14.3.2.5.2 FSR

The fisheye state routing (FSR) [34] protocol is a table-driven or proactive routing protocol. It is based on the link state protocol and has the ability of immediately providing route information when needed. The fisheye scope technique allows exchanging link state messages at different intervals for nodes within different fisheye scope distances, which reduces the link state message size. Further optimization allows FSR to broadcast only topology message to neighbors [4] in order to reduce the flood overhead. With these optimizations, FSR significantly reduces the topology exchange overhead and scales well to large network size.

FSR routes each data packet according to locally computed routing table. The routing table uses most recent topology information. The fisheye scope message updating scheme will not lose routing accuracy for inner scope nodes. For outer scope nodes, information in routing entries may blur due to longer exchange interval, but the extra work of "finding" the destination (as in on-demand routing) is not necessary. Thus, low single packet transmission latency can be maintained. In a mobile environment, this inaccuracy for remote nodes will increase. However, when a packet approaches its destination, it finds increasingly accurate routing instructions as it enters sectors with a higher refresh rate.

FSR does not trigger any control messages when a link failure is reported. Thus it is suitable for high topology change environment. The broken link will not be included in the next fisheye scope link state message exchange. Sequence number and table refreshment enables the FSR to maintain the latest link state information and loop-free in an unreliable propagation media and highly mobile network.

The protocol works independently with the IP format of packets and is a distributed protocol. It can be implemented either in network layer or in application layer. It only deals with routing table management of the network system.

14.3.2.5.3 HSR

The characteristic features of hierarchical state routing (HSR) [35] are multilevel clustering and logical partitioning of mobile network nodes. The physical network nodes are partitioned into clusters and cluster heads are elected as in a cluster-based algorithm.

In addition, cluster heads at a low level become members of the next higher level. These new virtual cluster members organize themselves again into clusters and so on; this process leads to a hierarchical topology. The ID's at level 0 are physical addresses (similar to MAC addresses) and thus unique. And, for example, level 1 and level 2 clusters are generated by the recursive selection of cluster heads. Thus, those upper level clusters are only virtual with the so-called virtual links between nodes.

Nodes belonging to a physical cluster broadcast their link information to each other. Each cluster head summarizes all information about its cluster and sends it to neighboring cluster heads via gateway. This knowledge of neighbor cluster heads leads to the formation of next level clusters.

As already mentioned, cluster heads are members of a virtual cluster on a next higher level and they exchange their own link information as well as the summarized lower-level information among each other.

A node in a virtual cluster level floods the information that it obtains to its lower level. So the lower level will know about the hierarchical topology, meaning that each node has a hierarchical address called HID (hierarchical ID).

A hierarchical address is assigned by using the cluster head (or node) IDs on the way from the root (top-level) down to the node (physical level). An HID can be considered as a series of MAC addresses.

As a gateway can be reached from the top level via more than one path, it can have more than one HID. A hierarchical address is enough to ensure delivery from anywhere in the network to a specific host. Each node will dynamically keep its HID up-to-date upon receiving updates from higher-level nodes.

14.3.2.5.4 LANMAR

The landmark (LANMAR) [36] routing protocol utilizes the concept of landmark for scalable routing in large MANETs.

It relies on the notion of group mobility; i.e., a logical group (e.g., a team of coworkers at a convention) moves in a coordinated fashion. The existence of such a logical group can be efficiently reflected in the addressing scheme. It assumes that an IP like address is used consisting of a group ID (or subnet ID) and a host ID, i.e., <Group ID, Host ID>. A landmark is dynamically elected in each group. The route to a landmark is propagated throughout the network using a distance vector mechanism.

Separately, each node in the network uses a scoped routing algorithm (i.e., FSR) to learn about routes within a given (maximum number of hops) scope. To route a packet to a destination outside its scope, a node will direct the packet to the landmark corresponding to the group ID of such a destination. Once the packet approaches the landmark, it will typically be routed directly to the destination. A solution to nodes outside of the scope of their landmark (i.e., drifters) is also addressed in the draft. Thus, by summarizing in the corresponding landmarks the routing information of remote groups of nodes and by using the truncated local routing table, LANMAR dramatically reduces routing table size and routing update overhead in large networks. The dynamic election of landmarks enables LANMAR to cope with mobile environments.

LANMAR is well suited to provide an efficient and scalable routing solution in large, mobile, ad hoc environments in which group behavior applies and high mobility renders traditional routing schemes inefficient.

14.4 Existing Communication Systems

Several existing communication systems can be seen in this section. This section is divided into (a) applications of underwater acoustic systems, (b) underwater acoustic ad hoc networks, (c) communication architecture, and (d) sensor networks with AUVs. In this section we attempt to show the several underwater systems, focusing on the ad hoc networks used in the underwater environment.

14.4.1 Applications of Underwater Acoustic Systems

14.4.1.1 Underwater Acoustics for Military Use

Currently there are many applications of underwater acoustics. The first steps into practical underwater acoustic applications were developed for military use. At the beginning of World War II, the progress in electronics and in the radio industry was advanced enough to build sonar, which played an important role in detecting the threat of German submarines that were destroying the allies' ships.

Passive military sonar is designed for detection, tracking, and identification of submarines. They work at very low frequencies, between a few tens of hertz and a few kilohertz. The detection ranges are also longest at low frequencies due to the smaller absorption losses. Modern passive sonar is characterized by the deployment of towed linear arrays, very long, able to efficiently detect and locate low-frequency noise sources.

14.4.1.2 Underwater Acoustics for Civilian Use

While the military industry developed underwater acoustic systems, the private industry was able to profit from the development of underwater acoustics. Acoustic sounders quickly replaced the traditional lead line to measure the water depth below ship or to detect obstacles. These systems are indispensable tools for sea fishing, navigation, and scientific monitoring of biomass. Other useful underwater system is sidescan sonar [37], which is employed to obtain acoustic images from the seabed used for geology in the construction of seafloor maps, which are of relevant importance in the offshore oil industry.

In the recent years, techniques of acoustic monitoring have been deployed to monitor the evolution of the average temperature of large ocean basins on a permanent basis as a part of global climate studies.

Lurton [38] states that underwater acoustic systems use a restricted variety of signals, chosen for their capacity to carry the information sought by the end user in specific applications. He also explains that there are two main aspects to the good functioning of an underwater acoustic system:

- The definition and use of the signal well suited to the objective and to the environmental conditions known *a priori*.
- The use in the reception chain of processing techniques combining the best performance achievable and the level of complexity and cost compatible with the objectives of the system.

14.4.1.3 Underwater Acoustic Data Transmissions

Underwater acoustic data transmissions have a wide range of uses and applications. As mentioned before, the military, naval, and industrial domains use them mainly for the following:

- For control of an ROV (remotely operated vehicle) or an AUV
- For communication between a submarine and a node anchored to the seafloor for transmission of oceanographic data
- For video surveillance of underwater structures
- For underwater backup safety systems
- For collecting data from seafloor nodes after a geological study or an earthquake
- For audio communication between divers and an underwater support vehicle
- For underwater pollution monitoring

14.4.2 Underwater Acoustic Ad Hoc Networks

The actual work in underwater acoustic communication and networking is generating a huge amount of different systems with specific and custom protocols of communications for every new application. However, currently, only a couple of standards exist for underwater communications. They were created by different manufacturers for their proprietary systems, but in general, these standards create incompatibility and interoperability when one tries to build a homogeneous system of underwater communication. So standardization became a relevant issue for underwater acoustic communication systems.

Otnes et al. [39] propose a possible roadmap to standardization, which is a prerequisite to reaching a state of ubiquitous underwater acoustic communications and networking. They also state that research in radio communications is mostly based on simulations with the use of established models, but research in underwater acoustic communications is mostly based on sea trials, which are expensive and time-consuming.

Underwater ad hoc networks are envisioned to enable applications for oceanographic data collection, ocean sampling, environmental and pollution monitoring, offshore exploration, disaster prevention, tsunami or seaquake early warning, assisted navigation, and video surveillance systems for underwater vehicles and structures. To make these applications viable, it is necessary to create an underwater communication platform among underwater ad hoc devices. The underwater communication platform must be created using fixed and mobile nodes with self-configuration capabilities in order to have some flexibility in cases of lost of remote supervision or in a failure event. The nodes must be able to coordinate their operation with the nearest hop and exchanging configuration, operational status, location, and movement information. Also they must relay monitored data when required by the onshore station or by another system. So there is a significant interest in getting the information on aquatic environments for scientific, commercial, governmental, environmental, and military reasons and on where current methods of aquatic communications cannot satisfy the actual needs.

Underwater acoustic sensor networks (UWASNs) [1] consist of sensors/nodes that are deployed to perform collaborative monitoring tasks over a given region. UWASN communication links are mainly based on acoustic wireless technology, which poses unique challenges due to the harsh underwater environment, such as limited bandwidth capacity [40], temporary losses of connectivity caused by multipath and fading phenomena [41], high and variable propagation delays [9], and high bit error rates.

14.4.3 Communication Architecture

A reference architecture for UWASN has been studied [1,2,42,43] for the deployment of underwater communications in the recent years, but still the underwater ad hoc network topology is an open research issue in itself that needs further analytical and simulative investigation from the research community.

The network topology is in general a crucial factor in determining the energy consumption and the capacity and the reliability of a network. Hence, the network topology should be carefully engineered and postdeployment topology optimization should be performed, when possible.

The network must be highly reliable in order to avoid failure of monitoring. So, the network design is a crucial and important factor, where a single point of failure should be avoided in the network topology. Also, Akyildiz and coworkers [2,42,43] introduce reference architectures for 2D and 3D underwater acoustic networks and present several types of fixed and mobile nodes that can enhance capability of underwater ad hoc networks.

14.4.3.1 2D Underwater Ad Hoc Network

A 2D underwater ad hoc network [42,43] is a set of underwater nodes that are anchored to the seafloor, organized in a cluster-based architecture, but able to move due to anchor drift or disturbance from external effects.

Underwater nodes are interconnected to one or more underwater gateways (UWGs) by means of acoustic links. Since data are not stored in the underwater node, data loss is prevented as long as isolated node failures can be circumvented by reconfiguring the network.

The UWGs are equipped with two acoustic transceivers. The first one is a long-range vertical transceiver (of up to 10 km) used by the UWG to relay collected monitored data from the seafloor network to a surface station, and the second one is a horizontal transceiver used by the UWG to establish communication with nodes in order to send commands and configurations to them and to collect monitored data from the seafloor network.

The surface station is able to handle multiple parallel communications with the deployed UWGs using a multiconnection acoustic transceiver and with a long-range radio transmitter and/or satellite transmitter, which is needed to communicate with an onshore station and/or offshore ship.

The underwater nodes are able to connect to UWG via direct links or through multihop paths. Although direct link connection is the simplest way to create an underwater wireless ad hoc network, it may not be the most energy-efficient solution [40,44]. Like in terrestrial ad hoc networks [45], where information of a node is relayed by intermediate nodes until it reaches its final destination. A 2D underwater ad hoc network architecture may be implemented using the same above principle by creating intracluster communication or extended to intercluster communication as a measure to support a failed UWG or an overloaded cluster/node. However, a custom routing protocol may be required to handle this functionality.

Figure 14.6 illustrates a 2D underwater ad hoc network architecture, which can support a mobile underwater ad hoc device, such as ROV or AUV.

14.4.3.2 3D Underwater Ad Hoc Network

A 3D underwater network [42,43] is a set of fixed and mobile underwater nodes that float at different depths to perform cooperative tasks. A wide range of tasks can be performed by this depth network; it may be used for surveillance applications or monitoring and sampling of the 3D ocean environment in order to detect and observe unusual ocean phenomena, such as biogeochemical processes, water streams, and pollution that cannot be adequately detected or observed by means of a seafloor network.

An easy and quick deployment of a 3D underwater network is attaching each node to a surface buoy, by means of retractile cables, in order to adjust the depth of each node [46]. However,

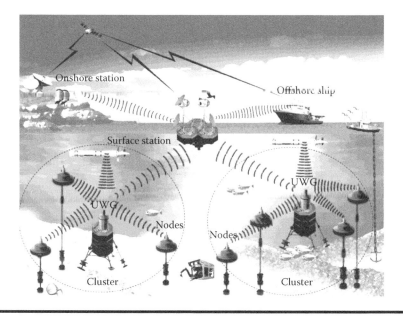

Figure 14.6   Architecture for 2D underwater ad hoc network.

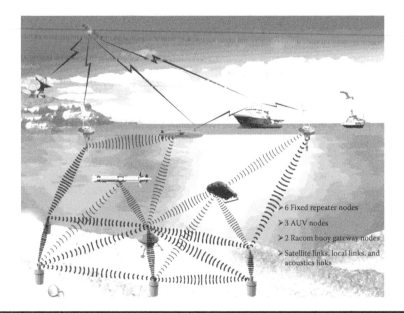

6 Fixed repeater nodes
3 AUV nodes
2 Racom buoy gateway nodes
Satellite links, local links, and acoustics links

Figure 14.7   Architecture for 3D underwater ad hoc network.

having multiple floating buoys may obstruct commercial fishing, ships navigating water sports, or any other activity that takes place on the surface. Furthermore, floating buoys are vulnerable to weather and tampering or pilfering.

One approach to resolve the above issue is to anchor each node device to the seafloor and equip it with a floating buoy that can be inflated by a pump, which pushes the node toward the sea surface. The depth of the node can then be regulated by adjusting the length of the retractile cable that connects the node to the anchor by means of an electronically controlled engine that resides inside of the node. Figure 14.7 illustrates this kind of architecture, which has the same communication hardware of 2D underwater ad hoc network architecture.

There are a couple of considerations concerning a 3D underwater network deployment:

■ Node coverage: Nodes should collaboratively regulate their depth in order to achieve 3D coverage of the ocean column.
■ Communication coverage: In 3D underwater networks, nodes should be able to relay information to the surface station via multihop paths using for this the fixed nodes or mobiles nodes such as ROV or AUV. So, network devices should coordinate their depths in such a way that the network layer is always active and running.

14.4.4 Sensor Networks with AUVs

14.4.4.1 Actual Systems

14.4.4.1.1 AquaNodes

Currently there are a few systems for underwater ad hoc acoustic networks. One of them is AquaNodes [47]. It is a group of underwater nodes with dual communication and support for sensing and mobility, where they can be deployed in lakes, rivers, and even the oceans.

The systems consist of fixed and mobile robots that are dually networked, optically with an optical modem implemented using green light, for a point-to-point transmission at 330 kb/s and acoustically for an acoustic modem for broadcast communication over ranges of hundreds of meters at 300 b/s, using TDMA as a communication protocol.

Each node has a built-in camera and sensors for temperature, pressure, and inputs for water chemistry sensors.

14.4.4.1.2 Seaweb

The development of underwater wireless networks by US Navy follows a concept of operations called Seaweb [48,49]. The Seaweb system uses underwater acoustic modems to connect underwater ad hoc nodes to a platform located at the sea surface that acts like a gateway. The gateway makes a bidirectional radio communication between the underwater Seaweb domain and the onshore station to send and receive data.

Seaweb networking (see Figure 14.8) provides acoustic ranging, localization, and navigation functionality and thereby supports the participation of mobile nodes, including submarines and collaborative swarms of AUVs.

The system has proven to be effective in shallow waters such as the intracoastal waterway and in waters up to 300 m deep off the coasts of Nova Scotia (Canada), San Diego, Long Island, and Florida (USA). It has been demonstrated in the Pacific and Atlantic oceans, in the Mediterranean and Baltic seas, in Norwegian fjords, and under the Arctic ice shelf.

14.4.4.1.3 AOSN

The autonomous ocean sampling network (AOSN) is a US project from the Monterey Bay Aquarium Research Institute [50]. This system is very similar to Seaweb, which uses a new robotic

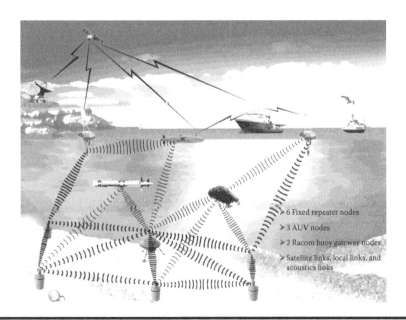

Figure 14.8 Seaweb underwater networks.

type of AUVs with advanced ocean models to improve the ability of humans to observe and predict the ocean. The AO3N system includes data collection by smart and adaptive devices and sensors that relay information to an onshore station in near real time, where it is assimilated into numerical models, which create 4D fields and predict future conditions of the ocean Also, the system has the ability to predict physical properties of the ocean, such as temperature, salinity, and current, as well as biological and chemical agents.

14.4.4.2 Cluster for Underwater Ad Hoc Vehicles

A multicluster communication system for ad hoc mobile underwater acoustic networks [51] is a system that consists of a varying number of vehicles that are required to perform collaborative tasks over a given area. To do so, vehicles must be able, at a minimum, to coordinate their operation by exchanging location and movement information by using a multiple access scheme based on clustering that provides efficient scalability by the spatial reuse of channel resources.

The performance of the system is evaluated in terms of measures of connectivity, successful transmission rate, average delay, and energy consumption. An important issue discussed in this system is the cluster size, which is determined by the maximal number of nodes per cluster for which the network connectivity is maximized.

14.5 Conclusions

Underwater communication networks have became an important field of investigation for many research groups in the recent years. In this chapter the main concepts in underwater ad hoc communications have been analyzed. We have presented an overview of the state of the art of underwater ad hoc communication networks and the potential applications of these new types of systems mainly used by military, naval, and industrial domains.

In this chapter, first, the main problems related to transmission have been shown. We have described the most used technologies in underwater communications. Acoustic communication is the most versatile and widely used method in underwater environments due to the low attenuation of sound in water. However, the use of acoustic waves in shallow waters may be adversely affected by temperature gradients, ambient background noise, and multipath propagation due to reflection and refraction. The much slower speed of acoustic propagation in water, about 1,500 m/s compared with electromagnetic and optics waves, is another limiting factor for effective communication and networking. Still, the best technology for underwater communications is acoustic.

Then, the ad hoc network protocols have been described. Since there are several ad hoc network protocols, they have been divided into groups. Each group uses a way to create and maintain their routing tables besides having an architecture and/or topology. According to the references the reactive protocols are most commonly used. Within this group we could highlight the DSR and AODV protocols. The underwater communications use this type of protocols, because they do not need to save the routes for a long time. They create their routes when they need to transmit some data to another node. In the case of using a hierarchical architecture, the cluster-oriented routing protocols are mostly used.

Finally, the current applications have been exposed. At present, the work in underwater acoustic communication and networking is generating a huge amount of different systems, sometimes incompatible to operate between different manufacturers. So standardization has become a relevant issue for these systems. Furthermore, there are other crucial factors in underwater acoustic network to be considered; these are energy consumption, capacity, and reliability of the network, which need to be studied in depth to deliver a proper quality of service.

Underwater ad hoc networks are envisioned to enable applications for oceanographic data collection, ocean sampling, environmental and pollution monitoring, and others. To make these applications viable, communication among underwater fixed and mobile nodes is required; by creating a 2D or 3D underwater ad hoc network we are able to relay monitored data to a surface or onshore station.

References

1. I.F. Akyildiz, D. Pompili, and T. Melodia. Underwater acoustic sensor networks: research challenges. *Ad Hoc Networks*, vol. 3, no. 3, 2005, pp. 257–279.
2. I.F. Akyildiz, D. Pompili, and T. Melodia. State-of-the-art in protocol research for underwater acoustic sensor networks. In *Proceedings of the 1st ACM International Workshop on Underwater Networks (WUWNet'06), Los Angeles, CA*, September 25, 2006, pp. 7–16.
3. J. Partan, J. Kurose, and B. Levine. A survey of practical issues in underwater networks. *SIGMOBILE Mobile Computing and Communications Review*, vol. 11, no. 4, 2007, pp. 23–33.
4. B. Peleato and M. Stojanovic. A MAC protocol for ad-hoc underwater acoustic sensor networks. In *Proceedings of the 1st ACM International Workshop on Underwater Networks (WUWNet'06), Los Angeles, CA*, September 25, 2006, pp. 113–115.
5. I. Vasilescu, K. Kotay, D. Rus, M. Dunbabin, and P. Corke. Data collection, storage, and retrieval with an underwater sensor network. In *Proceedings of the 3rd International Conference on Embedded Networked Sensor Systems (SenSys'05)*, San Diego, CA, November 2–4, 2005.
6. L. Liu, S. Zhou, and J.-H. Cui. Prospects and problems of wireless communications for underwater sensor networks. *Wireless Communications and Mobile Computing*, vol. 8, no. 8 (Special Issue on Underwater Sensor Networks), 2008, pp. 977–994.
7. M. Stojanovic. Underwater acoustic communication. In J.G. Webster (ed.), *Encyclopedia of Electrical and Electronics Engineering*. John Wiley & Sons, 1999, vol. 22, pp. 688–698.
8. C.A. Balanis. *Advanced Engineering Electromagnetics*. New York: John Wiley & Sons, 1989.
9. D.G. Borkman and T.J. Smayda. Long-term trends in water clarity revealed by Secchi-disk measurements in lower Narragansett Bay. *ICES Journal of Marine Science*, vol. 55, 1998, pp. 668–679.
10. M.I. Mishchenko, L.D. Travis, and A.A. Lacis. *Scattering, Absorption, and Emission of Light by Small Particles*. New York: Cambridge University, 2002.
11. R.G. Newton. *Scattering Theory of Waves and Particles*, 2nd ed. New York: Springer-Verlag, 2002.
12. M.J. Buckingham. Ocean-acoustic propagation models. *Journal d'Acoustique*, vol. 3, 1992, pp. 223–287.
13. M.A. Ainslie and J.G. McColm. A simplified formula for viscous and chemical absorption in sea water. *Journal of the Acoustical Society of America*, vol. 103, no. 3, 1998, pp. 1671–1672.
14. J. Preisig. Acoustic propagation considerations for underwater acoustic communications network development. In *Proceedings of First ACM International Workshop on Underwater Networks (WUWNet)*, Los Angeles, CA, September 2006.
15. P. Karn. MACA — a new channel access method for packet radio. In *Proceedings of the 9th Computer Networking Conference ARRL/CRRL Amateur Radio, London, Ontario, Canada*, September 22, 1990, pp. 134–140.
16. J. Rice, B. Creber, C. Fletcher, P. Baxley, K. Rogers, K. McDonald, D. Rees, et al. Evolution of Seaweb underwater acoustic networking. In *OCEANS 2000 MTS/IEEE Conference and Exhibition*, Providence, RI, September 11–14, 2000, vol. 3, pp. 2007–2017.
17. V. Bhargavan, A. Demers, S. Shenker, and L. Zhang. MACAW: a media access protocol for wireless LAN's. *ACM SIGCOMM Computer Communication Review*, vol. 24, no. 4, 1994, pp. 212–225.
18. C.L. Fullmer and J.J. Garcia-Luna-Aceves. Floor acquisition multiple access (FAMA) for packet-radio networks. In *Proceedings of the ACM SIGCOMM*, Cambridge, MA, August 28–September 1, 1995.

19. D. Pompili, T. Melodia, and I.F. Akyildiz. A CDMA-based medium access control protocol for underwater acoustic sensor networks. *IEEE Transactions on Wireless Communications*, vol. 8, no. 4, 2009, pp. 1899–1909.

20. C.E. Perkins and P. Bhagwat. Highly dynamic destination-sequenced distance-vector routing (DSDV) for mobile computers. *SIGCOMM Computer Communication Review*, vol. 24, no. 4, 1994, pp. 234–244.

21. T. Clausen and P. Jacquet. Optimized link state routing protocol (OLSR), RFC 3626, 2003. Available at: http://www.ietf.org/rfc/rfc3626.txt (Accessed on June 25 2011).

22. R. Ogier, F. Templin, and M. Lewis. Topology dissemination based on reverse-path forwarding (TBRPF), RFC 3684, 2004. Available at: http://www.rfc-editor.org/rfc/rfc3684.txt (Accessed on June 25 2011).

23. S. Murthy and J.J. Garcia-Luna-Aveces. An efficient routing protocol for wireless networks. *AACM/ Baltzer Journal on Mobile Networks and Applications* vol. 1, no. 2 (Special Issue on Routing in Mobile Communication Networks), 1996, pp. 183–197.

24. C. Perkins, E. Belding-Royer, and S. Das. Ad-hoc on-demand distance vector (AODV) routing, RFC 3561, 2003. Available at: http://www.ietf.org/rfc/rfc3561.txt (Accessed on June 25 2011).

25. D. Johnson, Y. Hu, and D. Maltz. The dynamic source routing protocol (DSR) for mobile ad-hoc networks for IPv4. RFC 4728, 2007. Available at: http://www.ietf.org/rfc/rfc4728.txt (Accessed on June 25 2011).

26. I. Chakeresand and C. Perkins. Dynamic MANET on-demand routing protocol (DYMO). Internet Draft, July 2010. Available at http://tools.ietf.org/html/draft-ietf-manet-dymo (Accessed on June 25 2011).

27. Y. Hu, D. Johnson, and D. Maltz. Flow state in the dynamic source routing protocol. Internet Draft, June 2001. Available at http://tools.ietf.org/html/draft-ietf-manet-dsrflow (Accessed on June 25 2011).

28. E. Biagioni and S.H. Chen. A reliability layer for ad-hoc wireless sensor network routing. In *Proceedings of the 37th Annual Hawaii International Conference on System Sciences (Hicss'04), Big Island, Hawaii*. Track 9, vol. 9, January 5–8, 2004, p. 90300.

29. C. Tschudin, R. Gold, O. Rensfelt, and O. Wibling. LUNAR: a lightweight underlay network ad-hoc routing protocol and implementation. In Proceedings of the Next Generation Teletraffic and Wired/ Wireless Advanced Networking (NEW2AN'04), St. Petersburg, February 2004.

30. D. Aguayo, J. Bicket, and R. Morris. SrcRR: a high-throughput routing protocol for 802.11 mesh networks (DRAFT). Available at http://pdos.csail.mit.edu/~rtm/srcrr-draft.pdf (Accessed on June 25 2011).

31. V. Park and S. Corson. Temporally-ordered routing algorithm (TORA) version 1, functional specification. Internet Draft, IETF MANET Working Group, June 2001. Available at http://tools.ietf.org/html/draft-ietf-manet-tora-spec-04 (Accessed on June 25 2011).

32. Z. Haas, M. Pearlman, and P. Samar. The zone routing protocol (ZRP) for ad-hoc networks. Internet Draft, July 2002. Available at http://tools.ietf.org/html/draft-ietf-manet-zone-zrp (Accessed on June 25 2011).

33. M. Jiang, J. Li, and Y.C. Tay. Cluster based routing protocol (CBRP). Functional Specification Internet Draft, June 1999. Available at http://tools.ietf.org/html/draft-ietf-manet cbrp spec (Accessed on June 25 2011).

34. M. Gerla, G. Pei, X. Hong, and T. Chen. Fisheye state routing protocol (FSR) for ad-hoc networks. Internet Draft, June 2001. Available at http://tools.ietf.org/html/draft-ietf-manet-fsr (Accessed on June 25 2011).

35. A. Iwata, C.-C. Chiang, G. Pei, M. Gerla, and T.-W. Chen. Scalable routing strategies for ad-hoc wireless networks. *IEEE Journal on Selected Areas in Communications*, vol. 17, 1999, pp. 1369–1379.

36. G. Pei, M. Gerlaand, and X. Hong. LANMAR: landmark routing for large scale wireless ad-hoc networks with group mobility. In *Proceedings of the 1st ACM International Symposium on Mobile Ad-Hoc Networking and Computing, Boston, MA*, August 2000, pp. 11–18.

37. T. Reed IV and D. Hussong. Digital image processing techniques for enhancement and classification of SeaMARC II sidescan sonar imagery. *Journal of Geophysical Research*, vol. 94, no. B6, 1989, pp. 7469–7490.

38. X. Lurton. *An Introduction to Underwater Acoustics: Principles and Applications*. London: Springer-Verlag, 2010.

39. R. Otnes, T. Jenserud, J.E. Voldhaug, and C. Soldberg. A roadmap to ubiquitous underwater acoustic communications and networking. In Proceedings Underwater Acoustic Measurement: Technologies and Results, Nafplion, Greece, June 21–26, 2009.

40. E. Sozer, M. Stojanovic, and J. Proakis. Underwater acoustic networks. *IEEE Journal of Oceanic Engineering*, vol. 25, no. 1, 2000, pp. 72–83.

41. M. Stojanovic. Acoustic (underwater) communications. In J.G. Proakis (ed.), *Encyclopedia of Telecommunications*. John Wiley & Sons, 2003.

42. D. Pompili, T. Melodia, and I.F. Akyildiz. Deployment analysis in underwater acoustic wireless sensor networks. In The First ACM International Workshop on UnderWater Networks (WUWNet06), Los Angeles, CA, September 25, 2006.

43. I.F. Akyildiz, D. Pompili, and T. Melodia. Challenges for efficient communication in underwater acoustic sensor networks. *ACM SIGBED Review*, vol. 1, no. 2, 2004, pp. 3–8.

44. J.G. Proakis, E.M. Sozer, J.A. Rice, and M. Stojanovic. Shallow water acoustic networks. *IEEE Personal Communications*, vol. 39, 2001, pp. 114–119.

45. I.F. Akyildiz, W. Su, Y. Sankarasubramaniam, and E. Cayirci. Wireless sensor networks: a survey. *Computer Networks*, vol. 38, no. 4, 2002, pp. 393–422.

46. E. Cayirci, H. Tezcan, Y. Dogan, and V. Coskun. Wireless sensor networks for underwater surveillance systems. *Ad-hoc Networks*, vol. 4, No. 4, 2006, pp. 431–446.

47. I. Vasilescu, C. Detweiler, and D. Rus. Aquanodes: an underwater sensor network. In The Second ACM International Workshop on UnderWater Networks (WUWNet07), Montreal, Quebec, Canada, September 14, 2007.

48. J. Rice, R. Creber, C. Fletcher, P. Baxley, D. Davison, and K. Rogers. Seaweb undersea acoustic nets. In *Biennial Review 2001*, SSC San Diego Technical Document TD 3117, August 2001, pp. 234–250.

49. J. Rice. SeaWeb acoustic communication and navigation networks. In Proceedings Underwater Acoustic Measurement: Technologies and Results, Heraklion, Crete, Greece. June 28–July 1, 2005.

50. Monterey Bay Aquarium Research Institute. Available at http://www.mbari.org/Last. Accessed November 2010.

51. F. Salva-Garau and M. Stojanovic. Multi-cluster protocol for ad-hoc mobile underwater acoustic networks. In Proceedings of the IEEE OCEANS'03 Conference, San Diego, CA, September 2003.

Chapter 15

Underwater Sensor Networks

Luiz Filipe M. Vieira

Contents

Underwater sensor networks (UWSNs) are mobile ad hoc networks (MANETs) that present the opportunity for many applications, such as coast surveillance, four-dimensional (4D) monitoring (space and time) such as for ocean and biology studies, and oil and gas field monitoring. We present the current state of the art of UWSNs. It includes an overview of the research in this area, as well as a description of some specific proposals that show current trends and challenges. The UWSN is a recent research topic that has many applications. Examples include studies in oceanography, marine biology, interaction between ocean and atmosphere, deep-sea archaeology, seismic predictions, pollution detection, oil and gas field monitoring, lost treasure discovery, lost item recovery, hurricane disaster recovery, oil and chemical spill monitoring, anti-submarine missions, and surveillance. Many challenges are presented in developing UWSNs. Electromagnetic waves are rapidly absorbed by water, and therefore they do not propagate for long distances inside water. Thus, the current research uses acoustic communication that suffers from high latency (speed of sound in water is approximately 1,500 m/s, 5 orders of magnitude slower than the speed of light in the vacuum), small bandwidth, high bit error rate among many other characteristics that make communication difficult and demand smart solutions. Furthermore, sensor nodes can move with water currents, which allows a 4D environment monitoring (space and time). However, this mobility increases the difficulty in the development of UWSNs.

15.1 Introduction

Why should we study problems related to aquatic environment? The earth is mainly composed of water. About two thirds of the earth's surface is covered by oceans, which is mainly unexplored. Furthermore, there is a large amount of natural resources that can be discovered and explored. In addition, we know that oceans have a great impact on weather. Studying water temperature of the oceans can help us in answering questions on global warming. In some situations, such as in warfare, oceans can be a good place for defense and attack. Finally, there are many potential applications for mobile underwater ad hoc networks, such as oil and gas field monitoring [1,2].

Mobile underwater ad hoc networks are composed of sensor nodes that possess communication capabilities. For long-term monitoring, sensor nodes can be tied to the ocean bottom or to buoys. For short-term investigations, sensor nodes can move with water currents at various different levels; one way to implement this scenario is to throw sensor nodes by planes. The mobility of the sensor nodes allows a 4D environment monitoring (space and time) and a dynamic coverage.

Figure 15.1 illustrates the prototype of a UWSN. Different from terrestrial sensor networks, UWSNs need to support the water pressure and, therefore, have a more resistant encapsulation.

Figure 15.1   Underwater sensor networks composed of sensor nodes. Courtesy of Dr. Schurgers and Dr. Jaffe http://jaffeweb.ucsd.edu/node/81

15.1.1 Challenges

Compared with ground-based sensor networks, mobile underwater ad hoc networks present a number of novel challenges:

- Acoustic communication: Electromagnetic waves cannot propagate over long distances in water. Therefore, a sea-swarm network must rely on acoustic signals that are affected by large propagation delay, narrow communication bandwidth, multipath channel fading, and high bit error rate. The bandwidth × range product is about 40 kbps × km, which is very low compared to that of the radio channel (ratio of 1:100). The acoustic signal propagation is orders of magnitude larger, approximately 1.5×10^3 vs. 3×10^8 m/s (ratio of 1:10,000).
- Sensor node mobility: Empirical observations suggest that sensor nodes will move at a speed of 3–6 km/h. A sea-swarm network must identify nodes location (localization), with the majority of the network components traveling within the flow.
- Limited energy: Battery power is limited and usually batteries cannot be recharged.

Recent works have addressed some of the challenges presented by underwater sensors [1,3,4]. Since UWSN is an emerging topic, up to now most of the researches have been mainly focused on fundamental sensor networking problems, such as data gathering [5], synchronization [6], routing protocols [7,8], energy minimization, and medium access [9,10] issues.

15.1.2 Differences between Underwater and Terrestrial MANETs

The main differences between underwater and terrestrial MANETs are as follows:

- Cost: Underwater sensors are expensive devices, mainly because of two reasons: first, because of hardware protection needed in the extreme underwater environment; second, the small-scale production of underwater sensor nodes.
- Density: Underwater deployments are generally sparser than terrestrial ones.
- Power: Acoustic underwater communications requires more power than terrestrial radio communications. This is due to longer distances and impairments of the channel.

15.1.3 Current Status and Future Trends

The underwater acoustic channel can be characterized as follows [11]:

- Narrow bandwidth: Just a few hundred kilohertz, limited by absorption [1]
- Multipath fading
- High attenuation
- Bandwidth dependent on frequency and range: According to acoustic telemetry studies [12], the bandwidth × range product is limited, approximately 40 kbps × km—a low value as compared to that of the terrestrial radio. For example, in the IEEE 802.11b/a/g standard, it is 5 Mbps × km (ratio of 1:100).
- High latency: The speed of sound in water is approximately 1.5×10^3 m/s. The speed of light in vacuum, used in terrestrial communication, is almost 3×10^8 m/s (ratio of 1:10,000, i.e., a difference of 5 orders of magnitude).

The speed v of sound in water is modeled by the equation [13]

$$v = 1449.05 + 45.7t - 5.21t^2 + 0.23t^3 + (1.333 - 0.126t + 0.009t^2)(S - 35) + 16.3z + 0.18z^2 \quad (15.1)$$

where t is the water temperature in Celsius, z is the depth in meters, and S is water salinity degree in parts per thousand.

As noted above, underwater communication is very restricted and this uniquely characterizes mobile underwater ad hoc networks. In addition, we have mobility due to water currents, around 1–1.5 m/s [1].

The current state-of-the-art researches use the instrument called RAFOS [14] (see Figure 15.2), which has sensors to estimate temperature, salinity, and depth and is also capable of moving vertically by changing its density. Once at the surface, these sensors communicate via satellites. However, while underwater, they do not communicate. It is desirable that they could communicate while underwater. The current technology also presents acoustic modems [15] that allow point-to-point communication. It is desirable to have communication with the current

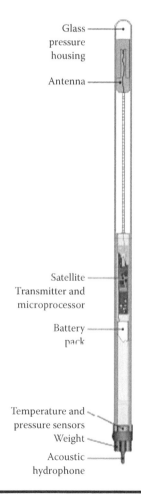

Glass pressure housing

Antenna

Satellite Transmitter and microprocessor

Battery pack

Temperature and pressure sensors

Weight

Acoustic hydrophone

Figure 15.2 Current technology: RAFOS. Courtesy from http://www.whoi.edu/instruments/ viewInstrument.do?id=1061

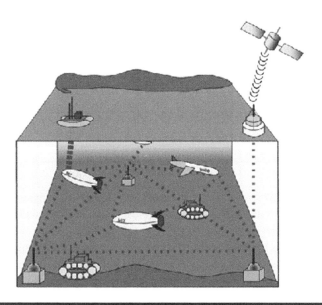

Figure 15.3   Ad hoc networks and underwater sensor networks.

underwater instruments, which would enable mobile underwater sensing networking and real-time monitoring.

Future trends point to autonomous networks for monitoring oceans, seas, and dams. Those networks could be composed of ad hoc sensors and autonomous underwater vehicles, as shown in Figure 15.3.

15.1.4 Text Organization

This chapter is organized as follows. Section 15.2 describes physical layer models for signal propagation and underwater packet delivery estimation. Section 15.3 discusses routing in mobile underwater ad hoc networks. Localization is a challenging task in underwater environment. It is required to tag sensed data. It is also essential for position-based routing algorithms. Section 15.4 addresses current energy- and message-efficient localization schemes.

15.2 Physical Layer

15.2.1 Signal Propagation

In terrestrial 3D MANETs, the signal propagation is described by the power received (PL) in decibels at a distance d, which is given by

$$\mathrm{PL}(d) = \mathrm{PL}(d0) + 10\alpha \log(d/d0) \tag{15.2}$$

where $\mathrm{PL}(d)$ is the signal power at a distance d and α is the loss exponent. $\mathrm{PL}(d0)$ is the average loss in decibels over a distance $d0$.

Unlike terrestrial networks, in UWSNs, the radio signal is not used due to the high attenuation. Instead, acoustic channels have been proposed by the current research.

Table 15.1 Bandwidth and Communication Range

Communication Range (km)	Bandwidth (kHz)
1,000	<1
10–100	2–5
1–10	10
0.1–1	20–50
<0.1	>100

Source: From Partan J., Kurose J., and Levine B.N., *WUWNet'06, Los Angeles, CA*, pp. 17–24, 2006.

An acoustic channel has low bandwidth [11] and large propagation latency. The speed of sound in water is about 5 orders of magnitude lower than the speed of light. An acoustic data transmission consumes more energy than does a terrestrial microwave data transmission. Moreover, high latency makes the whole network vulnerable to congestion due to packet collisions.

In deep water, the attenuation and loss of a signal can be expressed as [11]

$$A(d,f) = \left(\frac{d}{d_r}\right)^k a(f)^{d-d_r} \tag{15.3}$$

where f is the signal frequency and d is the transmission distance, taken in reference to some d_r. The path loss exponent k models the spreading loss.

Table 15.1 shows the communication range and bandwidth.

15.2.2 Underwater Packet Delivery Estimation

We use the following underwater acoustic channel models to estimate delivery probability [11,16].

The path loss over a distance d for a signal of frequency f due to large-scale fading is given as

$$A(d,f) = d^k a(f)^d \tag{15.4}$$

where k is the spreading factor and $a(f)$ is the absorption coefficient.

The geometry of propagation is described using the spreading factor ($1 \leq k \leq 2$); for a practical scenario, k is given to be 1.5. The absorption coefficient $a(f)$ is described by Thorp's formula [16].

The average signal-to-noise ratio (SNR) over a distance d is thus given by

$$\Gamma(d) = \frac{E_b / A(d,f)}{N_0} = \frac{E_b}{N_0 d^k a(f)^d} \tag{15.5}$$

where E_b and N_0 are the constants that represent the average transmission energy per bit and noise power density in a nonfading additive white Gaussian noise channel. As in [17] and [18], we use Rayleigh fading to model small-scale fading, where SNR has the following probability distribution:

$$p_{\mathrm{d}}(X) = \frac{1}{\Gamma(d)} e^{-X/\Gamma(d)} \tag{15.6}$$

The probability of error can be evaluated as

$$p_{\mathrm{e}}(d) = \int_0^\infty p_{\mathrm{e}}(X) p_{\mathrm{d}}(X)\, \mathrm{d}X \tag{15.7}$$

where $p_{\mathrm{e}}(X)$ is the probability of error for an arbitrary modulation at a specific value of SNR X. Here we use BPSK (binary phase shift keying) modulation that is widely used in the state-of-the-art acoustic modems [15].

In BPSK, each symbol carries a bit. In [19], the probability of bit error over a distance d is given by

$$p_{\mathrm{e}}(d) = \frac{1}{2}\left(1 - \sqrt{\frac{\Gamma(d)}{1+\Gamma(d)}}\right) \tag{15.8}$$

Thus, for any pair of nodes with distance d, the delivery probability of a packet of size m bits is given by

$$p(d, m) = [1 - p_{\mathrm{e}}(d)]^m \tag{15.9}$$

15.3 Routing

Routing protocols designed for terrestrial MANETs do not perform well underwater. The existing routing protocols can be divided into three categories: reactive (e.g., ad hoc on demand distance vector [20]), proactive (e.g., optimized link state routing [19]), and geographical. Proactive and reactive protocols require route discovery (through flooding) and/or route maintenance; therefore, they are not appropriate for limited bandwidth communications. In addition, there are message collisions and high-energy consumption; thus, in such cases, geographic routing [21–23] is preferable; nonetheless, it requires a location service to provide the destination position to the source.

Reactive protocols (e.g., ad hoc on demand distance vector [20] and dynamic source routing [24]) are appropriate for dynamic environments but possess high latency and still require source initiated flooding to establish routing paths. Reactive protocols also cause a high latency in the establishment of paths. This is amplified underwater due to the slow propagation of acoustic signals.

Proactive protocols (e.g., destination-sequenced distance-vector routing [24] and optimized link state routing [19]) have a large signaling overhead to establish routes for the first time. Moreover, each time the network topology is modified it causes signaling overhead because of mobility or node failures. In this way, each device is able to establish a path to any other node in the network.

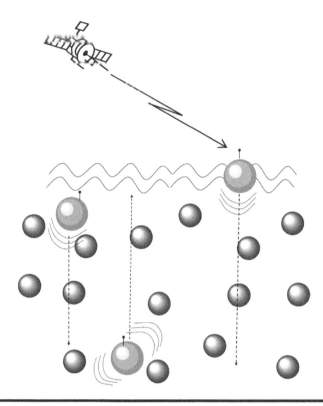

Figure 15.4 Localization by using dive and rise technique.

Geographical routing protocols (e.g., greedy-face-greedy (GFG) [25] and greedy perimeter stateless routing [26]) are very promising for their scalability feature. However, GPS (global positioning system) radio receivers do not work underwater. GPS uses waves in the 1.5-GHz band and those waves do not propagate underwater.

Recently, a routing protocol was proposed for underwater networks based on pressure [27]. The application scenario is specialized. A set of buoys at the surface receive data, which allows the problem to be solved by anycast geographic routing. It is sufficient to send the packet upward till it reaches the surface, as shown in Figure 15.4.

Given that a pressure sensor can estimate depth with a reasonable accuracy (mean error <1 m [28]), the depth information is used for anycast geographic routing, as discussed in [29].

Forwarding decisions are made locally. Packets are forward to neighbors with smaller pressure by using a greedy algorithm.

15.4 Localization

Localization is a challenging task in underwater environment. It is required to tag sensed data. It is also essential for position-based routing algorithms. The use of GPS is restricted to surface nodes, because the GPS signal does not propagate through water. In addition, acoustic medium enforces minimum message exchange. Here we describe current energy- and message-efficient localization schemes.

15.4.1 Localization with Autonomous Underwater Vehicles

Erol et al. [30] proposed the use of autonomous underwater vehicles (AUVs) to aid in localization in mobile underwater networks. In some applications, such as emergency ones, mobile nodes can be thrown inside water which stay there for a few days. Since the nodes are not fixed, they move with water currents. The AUV can help to establish a periodic localization process. The AUV announces its presence by sending a wakeup signal to nodes. The device that receives the signal will start the localization process by sending a request packet. The AUV replies to the packet with a packet that includes its coordinates. The propagation delay between the two packets is used to measure the distance (assuming uniform sound speed under water).

Localization with AUVs uses three messages: wakeup, request, and reply. The AUV sends the wakeup message to notify its presence to the nodes that are inside its communication range. In this case, it is assumed that the nodes are not synchronized and, therefore, a request and a reply message are used to compute the round trip propagation delay (RTPD). Assuming a uniform speed of sound underwater, it is possible to estimate the distance via speed of sound × RTPD/2. The reply message includes the AUV coordinates.

15.4.2 Localization by Using Dive and Rise technique

In long-term applications, the usual localization approach has been the Long Base-Line (LBL) technique. In these scenarios, equipments are placed kilometers apart and they collect data individually. They also transfer collected data to a central station via satellite links. However, they do not communicate with each other; i.e., they do not form a network. Current underwater applications, such as oceanography, demand networking capabilities. In this case, to achieve higher data rates, the range between nodes need to decrease. For localization in such underwater networks, the long-range pingers should be replaced with short-range alternatives. Location information needs to be forwarded iteratively to nodes that are not in the transmission range of the surface buoys. An alternative is to have some mobile nodes deliver the GPS-driven coordinates by moving in the vicinity of underwater nodes. To extend the global location information of the GPS service to the underwater environment, Erol et al. [32] proposed the dive and rise (DNR) technique shown in Figure 15.4.

DNR uses mobile beacons to distribute the GPS-driven coordinates in the underwater network. DNR beacons learn their coordinates at the surface of the ocean. Then, they periodically descend to the deepest level of the network and ascend to the surface. While diving and rising, DNR beacons broadcast localization messages. The underwater nodes passively listen to these messages and compute their localization. A DNR message includes a time-stamp field and the DNR beacon coordinates. The time-stamp field is used to compute the distance between the beacon and the node by using the time of arrival technique.

In DNR, when the underwater node receives messages from three or more beacons it itself calculates the coordinates by using lateration. If there are $n + 1$ or more beacon messages, lateration can be used to estimate n coordinates. Basically, the estimated coordinates should satisfy a set of equations

$$(x - x_i)^2 + (y - y_i)^2 + (z - z_i)^2 = d_i^2 \qquad (15.10)$$

where i denotes the beacon identifier, (x_i, y_i, z_i) are the beacon's coordinates, and d_i is the measured distance between the beacon and the node.

Note that three independent equations are sufficient for solving this nonlinear equation system for (x, y). Nodes have pressure sensors on board; therefore, the depth (the z coordinate) is known. The equation system is linearized by subtracting the $(n + 1)$th equation from the first n equations.

The coordinates are estimated with a least-squares estimator, by solving $AX - b$, where

$$A = \begin{bmatrix} 2(x_1 - x_n) & 2(y_1 - y_n) \\ \cdot & \cdot \\ \cdot & \cdot \\ \cdot & \cdot \\ 2(x_{n-1} - x_n) & 2(y_{n-1} - x_n) \end{bmatrix}$$

$$b = \begin{bmatrix} x_1^2 - x_n^2 + y_1^2 - y_n^2 + z_1^2 - z_n^2 - 2z(z_1 - z_n) + d_n^2 - d_1^2 \\ \cdot \\ \cdot \\ \cdot \\ x_{n-1}^2 - x_n^2 + y_{n-1}^2 - y_n^2 + z_{n-1}^2 - z_n^2 - 2z(z_{n-1} - z_n) + d_n^2 - d_{n-1}^2 \end{bmatrix}$$

The coordinates $\hat{\phi} = [\hat{x}\,\hat{y}]^\mathsf{T}$ are computed via the least-squares method: $\hat{\phi} = (A^\mathsf{T} A)^{-1} A^\mathsf{T} b$.

In DNR, a node is considered localized if the estimation error is less than the communication range R.

The error X is defined as the difference between the estimated distance and the measured distance [31]. The *estimated distance* is defined as the distance between the estimated node coordinates and the beacon coordinates. The measured distance is calculated using time of arrival.

$$\in = \frac{1}{n} \sum_{i=1}^{n} \sqrt{(x_i - \hat{x})^2 + (y_i - \hat{y})^2 + (z_i - z)^2} - d_i \tag{15.11}$$

If $\in > R$, then the node is defined as nonlocalized. Observe that the localization is done periodically; a nonlocalized node may become localized later and localized nodes may refine their localization estimates.

15.5 Future Research Directions

There are many research topics that need to be investigated in order to facilitate the development of UWSNs. Some of them are as follows:

■ Physical sensors: Many sensors exist for underwater measurements, such as analytical chemistry sensors used for estimating nitrate, phosphorous, pH, and dissolved oxygen. It is a challenge to design sensors that can last for long time. In addition, the sensors should be of low power owing to energy constraints.

- Software architectures: Current software architectures are application-specific. They should be generic enough to be easily used by the scientific community.
- Protocols: Communication protocols are an active research field. Medium access layer, routing, and transport are all challenging areas that demand efficient solutions and deals with the energy and bandwidth constraints. Protocols should be energy-aware.
- Localization: Localization is essential for monitoring applications, and new techniques should be investigated.

In the coming years, the development in technology will allow the constructions of mobile underwater ad hoc networks.

15.6 Conclusion

The UWSN is an important research area. In the coming years, there will be a growing interest in this field and in the utilization of UWSN, because of the many potential applications: oceanography, marine biology, interaction between ocean and atmosphere, deep-sea archaeology, seismic predictions, pollution detection, oil and gas field monitoring, lost treasure discovery, lost item recovery, hurricane disaster recovery, oil and chemical spill monitoring, anti-submarine missions, surveillance, and so on.

The most important difference between terrestrial and underwater MANETs is the communication medium. This aspect influences the application and protocol development. In this chapter, we described the physical layer, with the details on the signal propagation and packet delivery. We also discussed proactive, reactive, and geographical routing protocols. Finally, the chapter presented localization methods.

References

1. J. Kong, J. Cui, D. Wu, and M. Gerla. Building underwater ad-hoc networks and sensor networks for large scale real-time aquatic applications. In IEEE MILCOM, Atlantic City, NJ, 2005.
2. L.F.M. Vieira. Underwater SEA swarm. PhD thesis, University of California, Los Angeles, 2009.
3. I.F. Akyildiz, D. Pompili, and T. Melodia. Underwater acoustic sensor networks: research challenges. *Ad Hoc Networks*, vol. 2, no. 3, 2005, pp. 257–279.
4. J. Partan, J. Kurose, and B.N. Levine. A survey of practical issues in underwater networks. In *WUWNet'06, Los Angeles, CA,* 2006, pp. 17–24.
5. I. Vasilescu, K. Kotay, D. Rus, M. Dunbabin, and P. Corke. Data collection, storage, and retrieval with an underwater sensor network. In *SenSys'05: Proceedings of the 3rd International Conference on Embedded Networked Sensor Systems, San Diego, CA,* 2005, pp. 154–165.
6. A. Syed and J. Heidemann. Time synchronization for high latency acoustic networks. In Proceedings of the IEEE Infocom, Barcelona, Spain, April 2006.
7. D. Pompili and T. Melodia. Three-dimensional routing in underwater acoustic sensor networks. In *PE-WASUN'05: Proceedings of the 2nd ACM International Workshop on Performance Evaluation of Wireless Ad Hoc, Sensor, and Ubiquitous Networks, Montreal, Quebec, Canada,* 2005, pp. 214–221.
8. P. Xie, J.H. Cui, and L. Lao. VBF: Vector-based forwarding protocol for underwater sensor networks. In Proceedings of IFIP Networking'06, Portugal, 2006.
9. N. Chirdchoo, W.-S. Soh, and K.C. Chua. ALOHA-based MAC protocols with collision avoidance for underwater acoustic networks. In *INFOCOM 2007, Anchorage, AL,* May 2007, pp. 2271–2275.

10. D. Makhija, P. Kumaraswamy, and R. Roy. Challenges and design of MAC protocol for underwater acoustic sensor networks. In *4th International Symposium on Modeling and Optimization in Mobile, Ad Hoc and Wireless Networks, Boston, MA*, 3–6 April 2006, pp. 1–6.

11. M. Stojanovic. On the relationship between capacity and distance in an underwater acoustic communication channel. In *WUWNet'06: Proceedings of the 1st ACM International Workshop on Underwater Networks*. New York: ACM, 2006, pp. 41–47.

12. D.B. Kilfoyle and A.B. Baggeroer. The state of the art in underwater acoustic telemetry. *IEEE Journal of Oceanic Enginnering*, vol. 25, no. 1, 2000, pp. 4–27.

13. Robert J. Urick. Principles of Underwater Sound, 3rd Edition, 1996.

14. T. Rossby, D. Dorson, and J. Fontaine. The RAFOS system. *Journal of Atmospheric Oceanic Technology*, vol. 3, no. 4, 1986, pp. 672–679.

15. L. Freitag, M. Grund, S. Singh, J. Partan, P. Koski, and K. Ball. The WHOI micro-modem: an acoustic communications and navigation system for multiple platforms. In *Proc. IEEE OCEANS'05*, 2005, pp. 1086–1092.

16. L.M. Brekhovskikh and Y. Lysanov. *Fundamentals of Ocean Acoustics*. New York: Springer Verlag, 2003.

17. C. Carbonelli and U. Mitra. Cooperative multihop communication for underwater acoustic networks. In *WUWNet'06: Proceedings of the 1st ACM International Workshop on Underwater Networks*. New York: ACM Press, 2006, pp. 97–100.

18. M. Stojanovic. Recent advances in high-speed underwater acoustic communications. *IEEE Journal of Oceanic Engineering*, vol. 21, 1996, pp. 125–136.

19. T. Clausen and P. Jacquet. Optimized link state routing protocol (OLSR). IETF RFC 3626, October 2003.

20. C. Perkins. Ad hoc on demand distance vector (AODV) routing, 1997.

21. R. Flury and R. Wattenhofer. Randomized 3D geographic routing. In INFOCOM'08, Phoenix, AZ, April 2008.

22. H. Füßler, M. Käsemann, M. Mauve, H. Hartenstein, and J. Widmer. Contention-based forwarding for mobile ad-hoc networks. *Elsevier Ad Hoc Networks*, vol. 1, no. 4, 2003, pp. 351–369.

23. H. Kalosha, A. Nayak, S. Ruhrup, and I. Stojmenovi. Select-and-protest-based beaconless georouting with guaranteed delivery in wireless sensor networks. In INFOCOM'08, Phoenix, AZ, April 2008.

24. C.E. Perkins and P. Bhagwat. Highly dynamic destination-sequenced distance-vector routing (DSDV) for mobile computers. *ACM SIGCOMM - Computer Communication Review*, vol. 24, no. 4, 1994, pp. 234–244.

25. P. Bose, P. Morin, I. Stojmenovíc, and J. Urrutia. Routing with guaranteed delivery in ad hoc wireless networks. In *DIALM'99: Proceedings of the 3rd International Workshop on Discrete Algorithms and Methods for Mobile Computing and Communications*. New York: ACM, 1999, pp. 48–55.

26. B. Karp and H.T. Kung. GPSR: greedy perimeter stateless routing for wireless networks. In *MobiCom'00, Boston, MA*, 2000, pp. 243–254.

27. U. Lee, P. Wang, Y. Noh, L.F.M. Vieira, M. Gerla, and J.-H. Cui. Pressure routing for underwater sensor networks. In INFOCOM, San Diego, CA, 2010.

28. B. Jalving. Depth accuracy in seabed mapping with underwater vehicles, IEEE OCEANS'99. September 1999.

29. H. Yan, Z.J. Shi, and J.-H. Cui. DBR: Depth-based routing for underwater sensor networks. In *Networking*. Lecture Notes in Computer Science. Vol. 4982, Springer, 2008, pp. 72–86.

30. M. Erol, L.F.M. Vieira, and M. Gerla. AUV-aided localization for underwater sensor networks. In WASA'07, Chicago, IL, August 2007.

31. K. Langendoen and N. Reijers. Distributed localization in wireless sensor networks: a quantitative comparison. *Computer Networks*, vol. 43, no. 4, 2003, pp. 499–518.

32. M. Erol, L.F.M. Vieira, and M. Gerla. Localization with Dive'N'Rise (DNR) beacons for underwater acoustic sensor networks. In WUWNet'07, Montreal, Quebec, Canada, September 2007.

33. http://jaffeweb.ucsd.edu/node/81

34. http://www.whoi.edu/instruments/viewInstrument.do?id=1061

Chapter 16

Wireless Mesh Network: Architecture and Protocols

Christos K. Zachos, Jonathan Loo and Shafiullah Khan

Contents

16.1 Introduction

As Internet access becomes a need and the number of consumer end devices grows, new wireless access solutions toward the realization of anywhere, anytime connectivity emerge. Wireless mesh networks (WMNs) are promising solutions.

WMNs are relatively new concepts of the multi-hop wireless networks and a domain of active research. WMNs are not a single technology, topology, or architecture but rather a concept to move backhaul networks from "wired" connections to "wireless."

To the best of our knowledge, one of the earliest references to such a concept was in 1997 when Garcia-Luna-Aceves et al. published a paper titled "Wireless Internet Gateways (WINGS)" [1] as part of the Defense Advanced Research Projects Agency (DARPA) GloMo program. This paper describes a node model similar to a wireless mesh router and, most importantly, differentiates its applications from the military-focused Mobile Ad Hoc Networks (MANETs).

Later, in the Junes 2002 issue of *IEEE Spectrum* magazine, Schrick and Riezenman published an article about wireless broadband access [2]. Part of this article described a number of proprietary technologies under the subtitle "What a Mesh." Among these were Nokia's Wireless Routing Group and SkyPilot Network. The general architecture illustrated can be considered as WMNs interconnected through a central base station.

In 2004, Akyildiz et al. published an article titled "Wireless Mesh Networks: A Survey" [3] that summarized all the research results considering WMNs (directly or not). The article is cited in more than 1500 papers so far, and it is considered by many as the starting point for WMNs as an independent technology.

Since 2004, research efforts have produced a considerable amount of papers on all aspects of WMNs, and various vendors have presented commercial products based on proprietary technology. Although an Internet search about wireless mesh networks will return more than 750,000 results, the bibliography on the subject is limited to less than a dozen citations, the majority of which were published in 2009.

Despite these efforts, the operational framework of WMN is neither firmly set nor well defined; even the boundaries between WMNs and other ad hoc, multi-hop networks—such as MANETs—are blurred. However, both researchers and vendors tend to envisage WMN as a backhaul technology for wireless metropolitan area networks with a multi-tier hierarchical architecture using multi-radio 802.11 relay nodes.

16.2 Applications

The motivation behind the great effort in research and development of WMNs is the wide variety of their applications. WMNs are promising markets as many of their applications cannot be supported by other networks—wired or wireless—or, if they can, then the price is too high.

The "killer app" for WMNs lies in metropolitan area networks (MANs) in the form of municipal, community, and urban commercial networks. Over the last few years, there have been a continuously growing number of municipalities interested in providing Internet access to their citizens. The main reasons for the investment may be the promotion of economic development, e-government, and the bridging of the digital divide—as many citizens still cannot afford a broadband connection. In most cases, municipalities build a network of uninterconnected hot-spots—such as in the case of the municipalities of Trikala [4] and Rodos in Greece. WMNs are already important players in this field, especially in the United States. Projects such as Wireless Philadelphia [5], the city of Austin's WMN [6], and San Francisco's Municipal Network—just to name a few—are all based on proprietary WMN technologies. Open standards and the resulting lower cost may further boost the market. Surveillance and public safety applications can either add value to municipal networks or be the motivation to build special purpose WMNs.

WMNs can also serve the needs of telecommunication companies for fast deployment, particularly in cities with small coverage of optical or other wired networks. For example, the Russia's Golden Telecom [7] has built the world's largest urban wireless network in Moscow based on Nortel's proprietary mesh technology [8]. Golden Telecom provides indoor and outdoor broadband services using more than 5000 802.11 access points to more than 3.9 million households (Figure 16.1). The monthly fee for the always-on service is as low as Euro 12.5 and the only equipment required is an 802.11-enabled end device. The initial interest of telecommunication companies for 802.11 as local-loop faded due to the large operational costs of the wired backhaul. WMNs can face this shortcoming and make 802.11 a viable business [9].

Community networks such as Wi-Fi and MANET, on the other hand, cannot be seen as a market but they are steadily gaining dimensions of a movement. Only in Thessaloniki, with a resident population that barely exceeds one million, are there three such communities. The largest of these three already has already more than 10,000 members and more than 1600 relay nodes [10]. Generally, most community networks use 802.11 point-to-point links and traditional routing protocols like open shortest path first (OSPF) and Border Gateway Protocol (BGP) to form a backhaul. As central administration is not feasible, the distributed nature of WMN may fit them

Figure 16.1   Golden telecom's Wi-Fi coverage.

well, A lot of wireless communities not only have adopted WMN technologies but they actively participate in the development of new architectures and protocols as well.

Due to low implementation time, WMNs can help in disaster recovery situations. Probably, during the search and rescue phase, other wireless multi-hop networks such as MANETs can enable intra-rescue-team communications much more efficiently. However, after the first few days, WMNs can assist the community by providing campus communications and Internet access. One recent example is New Orleans. After Hurricane Katrina hit, authorities approved a WMN based on Tropos MetroMesh [11]. Even one year after, the only phone service available in many parts of the city was voice over Internet protocol (VoIP) via the WMN.

Basic virtues such as low cost, fast implementation, and large coverage make WMNs ideal candidates for helping developing countries and underserved areas close the digital divide. Institutions such as Meraca [12] have already chosen WMN as the platform for their vision for Wireless Africa. One of their first attempts was the WMN of Pretoria, South Africa, based of the Freifunk project (see Section 16.9).

Finally, other possible applications can emerge. Such applications may include the range expansion of indoor networks—such as broadband home and enterprise networking—as an inexpensive alternative to wires or where wired infrastructure is prohibited (e.g., archaeological sites and churches). Moreover, WMNs can provide the interconnection among distant sensor and personal area networks and vehicular ad hoc networks (VANETs).

16.3 WMN Architecture

So far, two categorizations of WMN architectures were proposed by Wang et al. (2005) and Manoj et al. in [3] and [13], respectively, with the former appearing more popular among researchers. In [3], WMN architectures are grouped into three categories: infrastructure/backbone, client, and hybrid.

Infrastructure/backbone WMNs (Figure 16.2) consist of interconnected mesh routers that provide access to clients. In this architecture, wireless clients are associated with one of the mesh routers and they are not forwarding traffic. The architecture can be enhanced with wired nodes, too. Moreover, the mesh routers can act as gateways to other networks, e.g., the Internet, sensor networks, etc. Although the dominating 802.11 technology might be the obvious choice, any wireless technology can be used.

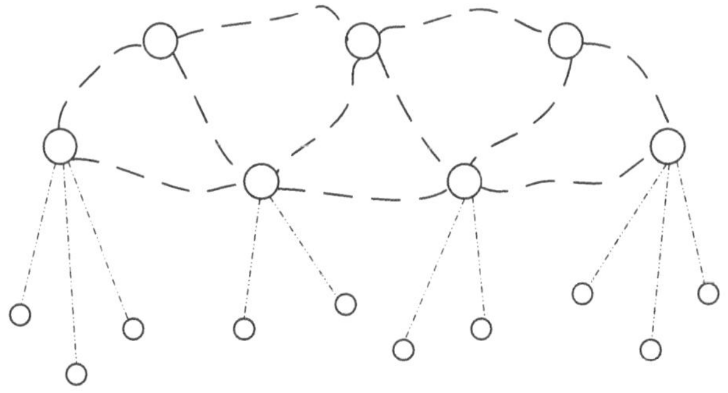

Figure 16.2   Infrastructure/backbone WMN.

In client WMNs (Figure 16.3), no mesh router is involved and the mesh clients form a peer-to-peer network, communicating directly with each other. Client nodes not only provide end applications but also have routing capabilities to forward each other's traffic. However, in this case, there is no possible interface to other networks, so all of the produced traffic remains local.

Hybrid WMN architecture is the combination of the previous two. Mesh routers form a backbone network and mesh clients may be connected either directly to a mesh router or through another mesh client that should forward its traffic to the backbone network.

The second classification, in [13] also groups WMN in three architectures: flat, hierarchical, and hybrid. The flat architecture is proportional to the client WMN. In hierarchical WMNs—which are similar to infrastructure WMNs, a backbone is formed by mesh routers that only use 802.11 links. If other technology, such as Worldwide Interoperability for Microwave Access (WiMAX), satellite, or cellular networks, is utilized in the hierarchy of the WMN, then this special case of hierarchical WMN is called *hybrid architecture.*

Once again, due to lack of standardization, WMN architecture is an open issue. The definition of client/flat network architecture is in its essence the definition of MANETs. The hybrid architecture in Figure 16.4 expands the WMN coverage but it may introduce difficulties if both static mesh routers and MANET clients should route under the same protocol, as their characteristics of mobility, power consumption, etc., can vary greatly.

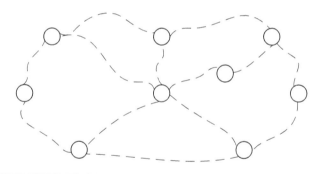

Figure 16.3   Client WMN.

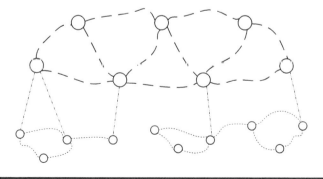

Figure 16.4   Hybrid WMN.

16.4 WMN Components

There are four distinct components that can be considered as parts of WMN architecture: mesh clients, mesh routers, mesh gateways, and MANET stations.

Mesh clients (MC) can, and should, be any devices that support wireless connection to today's single-hop wireless networks. It is crucial for the ease of deployment and the acceptance of WMN that network clouds be transparent to the end devices, and that their participation should be off-the-shelf without further modification.

Mesh routers (MR) do not originate or terminate any data flows. They are relay points that forward MC's traffic toward its destination. MR usually have multiple wireless interfaces and they can act as access points (AP) for MCs. Their main service is to discover and maintain constant paths in the mesh cloud.

Mesh gateways (MG) are MR with additional functionality that enables them to connect the mesh network with other networks—such as wired Ethernet, sensor networks, cellular networks, and ultimately to the Internet. Depending on the application, MG can have multiple interfaces of various technologies.

MANETs can coexist and supplement WMN. MANET stations—which can be considered as MC with routing capabilities—can interact with MR, inject routes, and further extend the WMN coverage. However, these additional capabilities presume a level of engagement from the WMN administration authority.

16.5 PHY and MAC Layers

16.5.1 802.11 and PHY Extensions

Physical (PHY) and media access control (MAC) layers encompass the concept of reliable transmission of upper layer data from one node in the path to its successive node. Given the total domination of the 802.11 standard in the wireless local and metropolitan area networks, all research efforts have this as their starting point.

Table 16.1 briefly summarizes the 802.11 standard. The initial 802.11 standard was introduced in June 1997 and included the PHY and MAC specifications for wireless data transmission over the industrial, scientific, and medical (ISM, 2.4 GHz) band. The PHY specification set two modes of operation: frequency hopping spread spectrum (FHSS) and direct sequence spread spectrum (DSSS). Both modes had data rates up to 2 Mbps and used a bandwidth of 20 MHz. The original standard is now considered legacy but the MAC was used with no modifications along with PHY enhancements up until 1997.

Table 16.1 Outline of 802.11 Standard Family

Standard	Release	BW (MHz)	Mode	PHY Data Rates (Mbps)	Outdoor Range
802.11	1997	20	DSSS	1,2	100 m
802.11a	1999	20	OFDM	6,9,12,18,24,36,48,54	120 m
802.11b	1999	20	DSSS	1,2,5.5,11	140 m
802.11g	2003	20	OFDM	1,2,6,9,12,18,24,36,48,54	140 m

Two years later, in September 1999, two enhancements on the 802.11 PHY were published. The first one, 802.11a, uses a completely new orthogonal frequency division multiplexing (OFDM) air interface over the unlicensed national information infrastructure (UNII, 5 GHz) band; it is capable of data rates up to 54 Mbps. The second, 802.11b, is an extension of the original 802.11 DSSS PHY specification. It achieves data rates up to 11 Mbps and it operates in ISM band. Although 802.11b had considerably lower data rates, it became the technology that established 802.11 as the de facto standard in wireless data communications, mainly because of the low price of 802.11b products and the fact that the 5 GHz band was not licensed worldwide, particularly in Europe, by that time.

The 802.11a technology in the ISM band was introduced in June 2003 by the 802.11g standard. However, 802.11 g is backward when compared with 802.11b. Mainly boosted by the concurrent raise of broadband Internet connections, 802.11g is now the most applied wireless local area technology, integrated in almost every end device in sales today. Given the widespread deployment, its use as a WMN's access interface will give the overall infrastructure instant acceptance.

Figure 16.5 shows the trade of PHY data rates of 802.11g as the distance between transmitter and receiver grows. As the 802.11 standard's main aim focuses on indoor local area networks, the deployment service area rarely exceeds 30 m in radius.

However, by using high gain antennas, 802.11's range can be expanded to cover metropolitan areas within the legal power limitations. Figure 16.6 illustrates the ISM and UNII frequency bands and the respective equivalent isotropically radiated power (EIRP) levels based on the current European Union (EU) regulations.

Although in the EU the ISM band includes two more channels (12 and 13) than in the Americas, the useful non-overlapping channels remain three (1, 6, and 11). Each channel width is 22 MHz to accommodate DSSS (802.11b) but only 20 MHz are used in OFDM (802.11g). The maximum EIRP is limited to 100 mW, enough to create access cells of an outdoor radius of around 150 m.

In the UNII band, only the non-overlapping channels are illustrated. The lower UNII band consists of UNII-1 (channels 36 to 48) and UNII-2 (channels 52 to 64) and the allowed EIRP is 200 mW. More appropriate for long-range links is the additional UNII band (channels 100 to 140) where the maximum EIRP can reach 1 W. However, as additional UNII is not used in the United States, equipment is hard to find. Most proprietary WMN solutions [6,8,11] use ISM band for access and UNII band to form the backhaul.

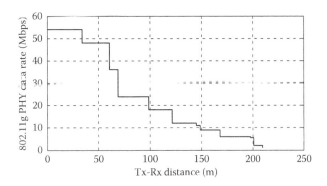

Figure 16.5 Outdoor 802.11g range—30 mW with 2.2 dBi gain patch antenna.

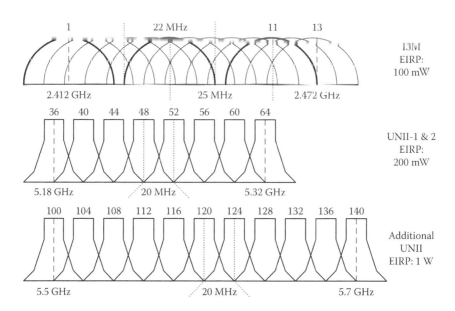

Figure 16.6   ISM and UNII channels and EIRP for EU.

16.5.2 802.11 Architecture and Components

Figure 16.7 illustrators the service sets (SS) of 802.11 infrastructure. The main component of an 802.11 network is the station (STA). STA is defined as any device that contains 802.11-conformant MAC and PHY interface to the wireless medium. STAs can form a self-contained ad hoc network or in IEEE terminology, an independent basic service set (IBSS). Membership in an IBSS does not imply that wireless communication with all other members of the BSS is possible. In ad hoc mode, an STA can communicate only with other members of the service set (SS) within range. Although IBSSs are used extensively with routing protocols in the third layer to form MANETs, no routing is defined or assumed in the standard.

With no central coordination and no entity to provide connection to outside wired or wireless networks, IBSSs are not formed to last. An access point (AP) is defined as the device that has STA functionality and provides access to the distribution services via the wireless medium for associated STAs. In other words, APs bridge the traffic from the wireless networks to other networks. When an AP is present, every member of the SS needs to be associated with it to form an infrastructure basic service set (BSS). When a BSS is formed, the communication among STAs is not direct, but it always flows through the AP.

A number of BSSs can be grouped to extend a wireless network's coverage and capacity. The resulting architecture defines an extended service set (ESS) which appears as a single BSS to the logical link control (LLC) layer at any STA associated with one of those BSSs.

The interconnection of the BSSs for the formation of an ESS is the function of the third 802.11 component, the distribution system (DS). In most cases, switched Ethernet plays the role of DS, but there is no limitation set by the standard. DS can be any system that conforms to the standard's services and this includes WMNs as well.

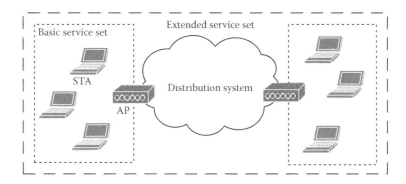

Figure 16.7   802.11 infrastructure service sets.

Table 16.2   Medium Access Timing

	SIFS (µS)	*Slot Time* (µS)	*DIFS* (µS)
802.11a	16	9	34
802.11b	10	20	50
802.11g	10	9	28

16.5.3 Basic Medium Access Mechanism

The 802.11 standard specifies two medium access mechanisms, the distributed coordination function (DCF) and the point coordination function (PCF). PCF is a deterministic access method where anyone who wants access should request it from a central control entity. A centralized entity is not feasible in the naturally distributed environment of a WMN and, anyway, PCF was never implemented in any commercial product. The basic access mechanism of 802.11 is DCF.

The DCF is based on carrier sense multiple access with collision avoidance (CSMA/CA) with binary exponential back-off and is common to all PHY technologies up to 802.11g. Similar to the access mechanism of Ethernet, an STA will listen to the medium before transmission, and the transmission will not start until the medium becomes idle. When the medium is sensed free for a specified period of time (DCF inter-frame space, DIFS) then the STA can start its transmission.

Unlike Ethernet, when a wireless single-radio STA transmits, it is not possible to detect a collision. Given the high bit error rate of the environment, an STA cannot assume the correct reception of the transmitted data. DCF's collision avoidance is based on a mechanism of positive acknowledgments (ACKs). Upon the correct reception of a frame, the receiver returns an ACK after another fixed period of time (short inter-frame space, SIFS). If no ACK is received, the transmitter assumes that a collision has occurred and it doubles the size of its contention window (CW). Then, it generates a random back-off number from 0 to CW. Now the sender is allowed to re-transmit only if the medium is free for DIFS plus a time equal to back-off times the PHY's time slot. DIFS, SIFS, and time slot duration for the various PHYs are summarized in Table 16.2.

16.5.4 802.11 Limitations in Multi-Hop Environments

IEEE 802.11 was not initially designed for use in multi-hop environments. Limited capacity is not usually an issue in small local area networks but it grows as the number of users and the distance among them grows. Despite the advantages in the physical layer, the advertised 54Mbps bandwidth is the peak link level data rate. As it is calculated in Section 3.4, when all the overheads are subtracted, the actual throughput available to applications is almost halved. Another issue is the inherent half-duplex operation of the current single-radio implementations.

The limited capacity issue is even more severe for multi-hop networks where all nodes operate over the same radio channel in order to keep the network connected. The result is interference between transmissions from neighbor nodes in the same path (interflow), as well as nodes in close paths (intraflow) which reduces the end-to-end capacity of the network [14,15,16].

Figure 16.8 illustrates an example of interflow and intraflow interference. Nodes T1 and T2 transmit to nodes D1 and D2, respectively, through two neighboring paths. Nodes R2, R4, and R7 are in the range of R3. When node R3 is active, literally no other node can transmit in the same or neighboring path, as all intermediate nodes are stacked in the listening state.

Moreover, inherent problems of DCF such as the hidden terminal problem further burden the performance. For example, in Figure 16.8, if T1 and D1 start transmitting toward each other, the packets arriving from R2 will eventually collide over R3 with the packets arriving from R4. The problem can be partially solved by the request to send, clear to send (RTS/CTS) function of 802.11 but the overhead is high. In such cases, the sending nodes, R2 and R4, should apply to R3 by sending an RTS packet to reserve the medium. R3 will reply with a CTS packet that allocates the medium to one of them for the applied duration. However, the RTS/CTS mechanism has some flaws. For example, as there is no way to predict a hidden terminal situation, the RTS/CTS mechanism is triggered by the packet length. If the packet threshold is high, the problem may not be encountered sufficiently. If the threshold is low, for the mechanism to be triggered—in most cases—the overhead is too high.

16.5.5 MIMO System and 802.11n

Major advances in 802.11's throughput and coverage could be achieved using multiple-input multiple-output (MIMO) systems. MIMO describes a system that uses a single transmitter with multiple antennas to transmit to a single receiver with multiple receiving antennas. Such systems are already used in 802.11a and 802.11g systems to add diversity and extend coverage. When a transmitter transmits a single stream of data through multiple antennas, the received signal is the composite signal of the multiple, usually two, antennas. On the receiver side, diversity is added

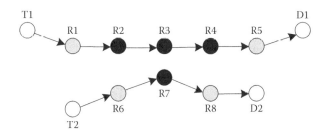

Figure 16.8 Interflow and intraflow interference.

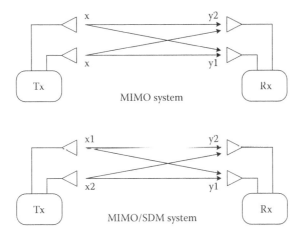

Figure 16.9   Multiple antenna systems.

by optimal combination of the received signals (Figure 16.9). Although the quality of the signal is improved, and the range of a given data rate can be extended, the system's data rate cannot be increased.

True throughput enhancement can be achieved when a MIMO system is used to transmit spatial streams on different antennas using spatial division multiplexing (SDM, Figure 16.9). In this case, PHY's data rate increases as a function of the independent data streams. MIMO/SDM is the main enhancement to 802.11's PHY by the IEEE802.11n-2009 standard.

IEEE 802.11n uses shorter guard intervals (400 ns instead of 800 ns), binding of two consecutive channels to form 40 MHz channels, a forward error correction (FEC) ratio of 5/6, and 64-QAM to achieve a raw data rate of 150 Mbps per stream [17]. An 802.11n system can use up to four streams using four antennas per transceiver; that leads to a total of 600 Mbps. Even if used with the traditional MAC, such systems can produce an average throughput of around 350 Mbps, a vast improvement compared to the less than 30 Mbps of 802.11a/g.

In contrast with previous 802.11 PHY amendments, 802.11n includes the concept of frame aggregation further improving throughput. Frame aggregation was introduced in 802.11e [18] but was never implemented as part of Wi-Fi Multimedia (WMM). When frame aggregation is used, per frame ACKs are suppressed. Instead, an extended ACK is used to acknowledge the reception of multiple frames. The block acknowledgment mechanism is mandatory to all 802.11n STAs.

16.5.6 Multichannel Systems

Interflow and intraflow interference is the result of using one channel throughout the network. However, as already described earlier in this chapter, there are a number of non-overlapping channels in the same frequency band. If multiple channels could be used for simultaneous transmissions, the networks capacity could be considerably increased. Every 802.11 interface can operate in all the channels in a band, so there is no challenge in PHY. The challenges lie to the upper layer protocols where MAC should determine the best channel available and routing should take the channel assignment decisions. Another approach to multichannel systems may be the use of multiple radios per node. If multiple interfaces are used, channel switching is not an issue as the channel assignment can be static and determined in advance.

16.6 Network Layers

While PHY and MAC layers provide hop-by-hop connectivity, the network layer provides the functional and procedural means for consistent end-to-end packet delivery. There are two types of protocols involved in the process: *routing* protocols discover the optimal, loop-free path from source to destination through the mesh, based on the addressing scheme organized by protocols that can be *routed*.

16.6.1 Routed Protocols

If there is a single point on which everyone agrees, it is which routed protocols to use. Internet Protocol in its two current versions, IPv4 [19] and IPv6 [20], is so catholically accepted that it is rarely discussed. However, the comparison between the two versions shows IPv6 as the most appropriate for WMNs:

- **Exhaustion of IPv4 addressing space.** It is a fact that IPv6 should have already been implemented in most current systems but the cost of upgrading systems and infrastructure may be a factor. WMNs as a new technology have no such problems.
- **Autoconfiguration.** IPv6 provides stateless autoconfiguration without the need of external protocols such as Dynamic Host Configuration Protocol (DHCP). Mesh users could just plug-and-play without prior contact with the administration.
- **Mobility.** In most applications, end users are free to roam while associated with the WMN. Mobile IPv6 [21] handles mobile users in a much more improved way than its predecessor, enabling them to maintain a permanent IP address while roaming among different mesh routers.
- **Security.** Although IP security protocol (IPsec), the protocol that encrypts an IP packet's payload, can be used with both IPv4 and IPv6, in the latter its use is mandatory.

One obstacle to IPv6 implementations is that currently only one routing protocol (OLSR [22]) supports it inherently. However, many experimental implementations are under development.

16.6.2 Routing Protocols

In essence, routing protocols for wireless multi-hop networks are not that different from the wired networks ones. However, although the goal is the same, the assumptions made as far as the link stability and network consistency are concerned are far from true. Even if the nodes are static, the wireless environment is highly dynamic, leading the traditional routing protocols to perform poorly. As a result, modifications and enhancements are crucial for the stability, scalability, and performance of wireless multi-hop networks.

The routing protocols used in WMN test beds evolved, or simply adapted, from the MANET ones. MANET is an area of multi-hop wireless networks, older and far more explored than WMNs. Although in many cases the boundaries between the two areas are blurred and in many cases overlapping, the main constraints of mobility and power efficiency are not present in WMNs—where scalability and high throughput are dominant.

16.6.2.1 Classification of Wireless Multi-hop Routing Protocols

The general taxonomy of MANET routing protocols [23,24,25], according to the scheduling of the path selection mechanism, is maintained in WMNs as well [3,26]. Moreover, further criteria that can be considered in the classification are:

- **State Information.** Protocols can be distinguished between topology-based—where every node in the network maintains constant information about every link (as is the case of link state protocols) and destination-based—where only information about neighboring links is maintained (as in the case of distant vector protocols).
- **Structure.** In a uniform, flat protocol all nodes have the same behavior toward routing. Non-uniform protocols can group or promote some of the nodes with the main objective to reduce routing traffic. For example, in cluster-based protocols the head-of-the-cluster node is responsible for flooding routing information and acts as gateway to other clusters and networks.
- **Communication Model.** Some of the proposed protocols aim for multi-channel communications, and in most cases a Time Division Multiple Access (TDMA) MAC scheme is proposed as well. Such protocols combine routing with channel allocation functionality. On the other hand, solutions which use CSMA/CA MAC (as in 802.11) aim for single-channel communications. However, at least one protocol [27] based on multi-channel multi-radio 802.11 interfaces [28] is proposed; it may be a viable and inexpensive solution for WMNs.

16.6.2.2 Proactive Routing Protocols

Proactive or table-driven ad hoc routing protocols maintain consistent routing information from each node to every other node in the network. Each node maintains one or more tables to store all possible destinations, along with additional information such as the next hop, path metrics, etc. In the event of a change in topology or any other aspect that routing algorithm is taking as input, nodes respond by flooding the network with routing updates in order for everyone to maintain a consistent view.

All proactive protocols, as do their wired counterparts, use Bellman-Ford [29] or Dijkstra [30] algorithms to calculate the optimal path to all destinations. In fact, some of the earliest ad hoc routing protocols—such as highly dynamic destination acquenced distance vector routing (DSDV) [31] and optimized link state routing (OLSR) [32]—were proactive and mere adaptations from interior gateway routing protocols such as routing information protocol (RIP) [33] and OSPF) [34]. In WMNs, OLSR and a modified version of DSDV (BABEL [35]) are widely deployed.

16.6.2.3 Reactive Routing Protocols

Reactive or on-demand routing protocols create routes only under the request of the source node. The main idea behind reactive protocols is that the environment is so dynamic that the routes can change many times even before a node may actually use them. When a node needs a route to a destination, it initiates the route discovery process—usually by broadcasting its request to its neighboring nodes. The request floods the network until it reaches its destination or an intermediate node that already has a path to it. Once the route is established, a second phase of route maintenance starts. Depending on the protocol's details, the route remains active until it is no longer needed or for a specific amount of time.

To the best of our knowledge, reactive routing protocols can use two approaches in route maintenance and forwarding. Protocols based on the Bellman Ford algorithm, such as ad hoc on demand distant vector (AODV) [36,37], use tables to store the results of the route discovery process and make forwarding decisions. Other protocols, such as dynamic source routing (DSR) [38], use source routing where the complete path for every given destination is cached in the source node's memory. Intermediate nodes make no forwarding decisions because the source node includes all the addresses in the path as part of the packet's header. Both protocols, AODV and DSR, are used as-is or modified in WMNs.

16.6.2.4 Hybrid Routing Protocols

Hybrid routing protocols try to combine the virtues of the aforementioned two categories. Common practice in hybrid protocols is the formation of groups or clusters. Routing overhead is minimized by running a proactive routing protocol within the group. When a destination belongs to another group, the path is established reactively. For example, zone routing protocol (ZRP) [39], one of the earliest protocols of its kind, groups the participants based on their distance in hops. The "proactive" zone of every node consists of its *n*-hop neighbors. If the destination exceeds this limit, the source initiates a path discovery by sending route requests to its border neighbors. If the destination belongs to their local routing zone, they send a route reply on the reverse path; otherwise, they will forward the request to their border neighbors.

As for the WMN implementation, the two most deployed and representative routing protocols are AODV and OLSR.

16.7 Routing Metrics for Wireless Multi-Hop Networks

The simplest and the most deployed routing metric for WMNs is the hop-count metric. Hop-count leads to paths with the smallest number of intermediate nodes. However, in wireless networks, it means that the protocol chooses the longest links which are usually prone to errors. Moreover, traditional metrics, such as static assignment of link bandwidth, are not appropriate for the wireless, highly dynamic environment. Dynamic parameters of WMN's links can only be measured by testing and so, the majority of the proposed metrics are based on active probing.

16.7.1 Per-Hop Round Trip Time

Per-hop round trip time (RTT) metric was proposed as part of the general link layer protocol multi-radio unification protocol (MUP) [28] and is a measure of the bidirectional delay between two neighbor nodes. MUP includes a channel allocation mechanism based on a two-way hand-shake which incorporates the messages channel select (CS) and channel select acknowledgement (CS-ACK). CS messages carry timestamps that can be used as probes.

To measure the channel quality, every node sends a CS unicast in a timely manner. The time interval is set to 30 seconds in MUP. The receiving node must respond immediately using a CS-ACK. When the sending node receives the ACK, it calculates the RTT in a weighted average called *smoothed RTT* (SRTT):

$$\text{SRTT}_{new} = a \cdot \text{RTT}_{new} + (1 - a) \cdot \text{SRTT}_{current} \tag{16.1}$$

where *a* is weighting.

If either CS or CS-ACK is lost or out-of-order, the handshake is repeated for a time equal to three times the current SRTT; if it fails, then the SRTT is set to three times the current SRTT. The path metric can be cost calculated as the summation of every hop SRTT or, as in the case of path vector routing protocols, the addition of the measured last-hop SRTT in the advertised path metric.

Per-hop RTT metric can be used outside the MUP framework with minor modifications of the probing mechanism. Although the probing consumes some of the network capacity, it is expected to perform well in WMNs as it comprises many crucial factors, such as queuing delays, link layer retransmission delays, and loads.

16.7.2 Per-Hop Packet Pair Delay

Per-hop packet pair delay (PktPair) metric was initially designed for wired networks. It was proposed in [40], and it was investigated for wireless use in [41] as part of a modified version of DSR.

Packet pair technique involves the periodic transmission of two probe packets, one small and one large, back-to-back to every neighbor. The receiver measures the delay between the receptions, and it reports the value back to the sender. Finally, the sender uses this value as a metric of the link quality. As in the case of per-hop RTT, the path metric derives from the summation of all PktPair metrics of the hops in the path.

The rationale behind the PktPair metric is that it maintains the advantages of active probing while it excludes queuing delays from the overall delay calculation. However, the packet pair may increase the routing traffic significantly due to the dual packet mechanism.

16.7.3 Expected Transmission Count and Variants

Expected transmission count (ETX) is one of the first active probing routing metrics proposed [42] to defy the poor performance of hop-count over wireless multi-hop networks. ETX is the expected number of retransmissions before a packet is delivered successfully over a link; and, in cases such as 802.11 where the successful delivery of a frame involves acknowledgment, it is calculated as

$$ETX = \frac{1}{d_f \cdot d_r} \tag{16.2}$$

where d_f is the probability of a successful frame transmission and d_r is the probability of a successful acknowledgment transmission.

To calculate the forward and reverse path probabilities, every node broadcasts probes to its neighbors at (almost) fixed interval τ. τ can vary up to 10% to avoid synchronization and collisions. The receiving nodes keep track of the successfully received probes in a time window w. The d_f of the sending node can be calculated as

$$d_{f,Tx} = \frac{count(t - w, t)}{w/\tau} \tag{16.3}$$

where $count(t - w, t)$ is the number of successful probes during w and w/τ is the actual number of the probes sent during w; w/τ is known to the receiving part included in the probes. As in the previous cases, the ETX of a path is the sum of the link metrics.

As compared to the per-hop RTT and PktPair, ETX introduces less overhead to the network as the probes are broadcasts rather than unicasts. Moreover, as broadcasts, the probes do not require acknowledgment.

16.7.4 Expected Transmission Time and Variants

Expected transmission time (ETT) and its path variant weighted cumulative ETT (WCETT) are an extension of ETX; they were proposed in [27] as part of the multi-radio link-quality source routing (MR-LQSR) protocol. ETT enhances ETX by taking packet length into consideration.

ETT is calculated as a "bandwidth-adjusted ETX" and it incorporates the overall ETX mechanism to calculate forward and reverse path probabilities. For a fixed packet size S and raw link bandwidth B:

$$\text{ETT} = \text{ETX} \cdot \frac{S}{B} \tag{16.4}$$

While the packet size is fixed, the link bandwidth must somehow be calculated or presumed. One way is to presume that it is fixed and constant [42] or to use this information from an 802.11 interface autorate function. The other, used in [27] is based on the packet pair mechanism of [40] as described in Section 16.7.2.

In a single-radio, single-channel environment, the ETT on a path of n-hops may be calculated as the sum of link ETTs

$$\text{ETT}_{path} = \sum_{i=1}^{n} \text{ETT}_i \tag{16.5}$$

In multi-channel environments, if two hops in a path are on the same channel, they will probably interfere with each other, resulting in poor performance. As a result, the path throughput is dominated by the bottleneck channel, which has the maximum ETT. If the system uses k channels, ETT per channel X_j would be

$$X_j = \sum_{i \, on \, j} \text{ETT}_i, \quad 1 \le j \le k \tag{16.6}$$

and the path ETT would be

$$\text{ETT}_{path} = \max_{1 \le j \le k} X_j \tag{16.7}$$

The WCETT path metric combines the previously calculated path metrics by taking their weighted average:

$$\text{WCETT} = (1 - \beta) \sum_{i=1}^{n} \text{ETT}_i + \beta \cdot \max_{1 \le j \le k} X_j \tag{16.8}$$

16.7.5 Metric of Interference and Channel Switching

Metric of interference and channel switching (MIC) is a path-weight function consisting of two routing metrics, the interference-aware resource usage (IRU) and the channel switch cost (CSC). MIC was introduced in [43] as part of the proactive routing protocol load and interference balanced routing algorithm (LIBRA). MIC improves the WCETT metric mainly by taking into account interflow interference. LIBRA is one of the few pure WMN proposals.

IRU is based on ETT measurements; it is an index of interflow interference, packet loss ratio, and transmission rate in the wireless links. If $N_i(c)$ is the set of neighbors which interfere with node i when it utilizes channel c, and $N_j(c)$ is the set of neighbors which interfere with node j in the same channel, then the total number of interfering nodes in a transaction between i and j would be $N_i(c) \cup N_j(c)$. The factor $N_i(c) \cup N_j(c)$ is used in the metric as an indication of interflow interference. IRU on the (i,j) link in channel c is then calculated as

$$\text{IRU}_{ij}(c) = \text{ETT}_{ij}(c).\left|N_i(c) \bigcup N_j(c)\right| \tag{16.9}$$

CSC metric aims only to intraflow interference between two successive hops. The concept is simple: if the two successive hops utilize different non-overlapping channels, then a weight w_1 is assigned to the path. If the two successive hops utilize the same channel, there would be intra-flow interference; then a different weight w_2 greater than the previous is assigned to the path. As a result, paths with less cumulative intraflow interference, lower weight, are preferable:

$$\text{CSC}_i = \begin{cases} w_1 & if\ \text{Ch}(prev(i)) \neq \text{Ch}(i) \\ w_2 & if\ \text{Ch}(prev(i)) = \text{Ch}(i) \end{cases} \tag{16.10}$$

MIC combines the two metrics to include intra- and interflow interference as well as the loss ratio and the transmission rate:

$$\text{MIC}(p) = \alpha \sum_{l \in P} \text{IRU}_i + \sum_{i \in p} \text{CSC}_i \tag{16.11}$$

where p is the path, i is a node in the path, and l is the link $(i-1, i)$. The positive factor α is a weight of how important for the network designer is the per-flow importance as compared with the load-balancing ability which is the main aim in [43] and in their test bed:

$$\alpha = \frac{1}{N \cdot min(\text{ETT})} \tag{16.12}$$

where N is the total number of nodes in the network. However, IRU could be incorporated as a metric itself in single-interface, single-channel designs.

16.7.6 Airtime Cost

Airtime cost (c_a) [44,45] is the default link metric of the forthcoming IEEE 802.11s standard, and it is an estimation of the consumed channel resources when a frame is transmitted over an 802.11 link; c_a can be considered a cross-layer metric as MAC layer statistics, along with active probing, are used in the calculation.

Similar to other active probing metrics:

$$c_a = \left[0_{ca} + 0_p + \frac{B_i}{r} \right] \frac{1}{1 - e_{fr}} \tag{16.13}$$

16.8 Cross-Layer Approaches

Protocols with cross-layer characteristics are both a need and a trend in WMNs. As described earlier in this chapter, much of the misfortunes of multi-hop wireless networks derive from intra-channel and inter-channel interference and other characteristics of the link layer. However, these misfortunes are transparent to the protocols that are responsible for the best path determination and that are placed traditionally in the network layer.

There are two approaches on the design of cross-layer routing protocols. The first and most deployed is the use of information gathered in the PHY and MAC layers as part of the composite metrics. Some of these metrics have already been described in the previous section, and they can work along with nearly every routing protocol with little modification. In the second approach, not only the metrics but also the routing protocols lie entirely in the second layer. These are known as layer 2.5 [46] protocols and their main advantage is that no modification is needed in the upper layers. However, their use and their integration with other networks are still under question.

Good examples of cross-layer routing design are protocols such as link quality source routing (LQSR), extremely opportunistic routing (ExOR), AODV-Spanning Tree (AODV-ST), and the proposed hybrid wireless mesh protocol (HWMP) for the forthcoming 802.11s standard.

LQSR [41] is a cross-layer adaptation of DSR. The overall functionality of DSR is maintained but the LQSR is implemented as layer 2.5 protocol; thus, the addressing scheme is based on MAC addresses rather than IP ones. Moreover, the metric can be either hop count or ETX, with the former performing better as the mobility of the nodes increases.

ExOR [47] is an integrated MAC and routing technique based on cooperative broadcasting of packets without explicitly setting up a routing path. Before a node starts sending packets, it reorganizes them in "batches" and it broadcasts them. Some subset of the available nodes receives the batches, and the one nearer to the destination rebroadcasts them. Which node is "nearer to the destination" can be calculated as the one with the maximum delivery probability calculated by ETX. The rebroadcasts continue until 90% of the batches reach their destination. The remaining 10% uses traditional minimum hop-count routing. In [47], the authors say that the protocol boosts end-to-end throughput by a factor of two.

AODV-ST [48] is a hybrid routing protocol, designed to support multi-radio infrastructure WMNs. AODV is used inside the mesh network; a spanning-tree based routing is used to bridge traffic to and from the gateways. Spanning-trees are built from every gateway by disabling some of the wireless interfaces to avoid loops. Furthermore, one of the enhancements over AODV is the use of the ETT metric.

16.9 Implementations and Test Beds

All experimental mesh networks have three common starting points: They are all based in existing 802.11 technologies; they all use Linux-based operating systems in the mesh routers; and they all use common code implementations of AODV (e.g., AODV-UU [49]), OLSR (e.g., OLSRd [50]),

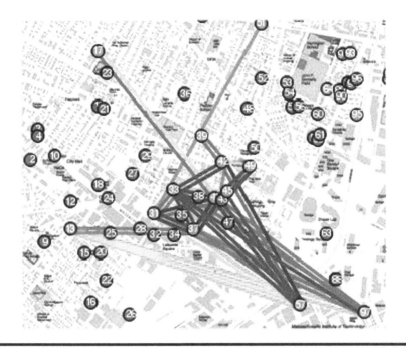

Figure 16.10 MIT Roofnet, active nodes, and links.

and DSR (e.g., DSR-UU [51]) for routing. From this point onwards, there is a solid foundation where protocols can be tested, modified, and ultimately, invented.

As expected, the majority of implementations and test beds have been developed as parts of academic research programs. However, for the first time, community networks have taken an active part in the development, testing, and popularization of new protocols. The number of WMN test beds is great and continuously growing. Two of the earliest and most representative examples are described in the following paragraphs.

Probably the most prominent experimental mesh network is Massachusetts Institute of Technology (MIT) Roofnet [52]. Roofnet provides broadband Internet access to users in Cambridge, and there are currently 20 active nodes (mesh gateways, Figure 16.10). The implementation is based on 802.11b/g radios and modified commercial products such as Netgear's WGT634U. The router's firmware is based on Linux with modified network drivers. As reported, link capacity varies from 1Kbps (orange) to 600Kbps (deep blue). Initially, Roofnet's routing protocol was a modified version of DSR. Gradually, two routing protocols emerged from this project, ExOR and ScrRR [53]. The latter is based on source routing and uses ETX as metric. ExOR has been proprietary since 2006, when two of its inventors spun off the Roofnet project to form Meraki, Inc. [54].

Freifunk [55] is one of the largest wireless communities and it supports networks in more than 20 major cities in Germany, Austria, and Switzerland. Freifunk is self-described as a system and movement for creating wireless mesh networks. The base of their system is a modified version of the Linux-based OpenWRT [56] distribution for embedded systems. OpenWRT can replace the original firmware of various low-cost 802.11-based routers and equip them with a number of additional capabilities such as wireless multi-hop routing protocols. Although OLSR is still running in some regions, Freifunk efforts resulted in the development of the Better Approach To Mobile Ad hoc Networking (B.A.T.M.A.N) routing protocol.

16.10 Standardization

Today, WMN implementations are in the form of academic test beds and commercial proprietary products. As is the case with every other networking technology, the establishment of WMNs will be the result of standardization. So far, the standardization process has poor results, all in the form of drafts and experimental Request for Comments (RFCs).

IEEE has formed the 802.11s task group since July 2004 to standardize the ESS [44]. Task groups (TGs) aim to standardize PHY and MAC layers for WMNs that extend coverage of 802.11 networks. Furthermore, respective task groups are formed in other fields, such as in 802.16j metropolitan area networks and 802.15 personal and sensor networks. Especially, 802.15.3a standard on ultra wide band (UWB) personal communications is expected to be used primarily in the form of mesh network.

The MAC layer of 802.11s is expected to follow the extended hybrid coordination function (HCF) as described in the 802.11e standard. Moreover, the routing protocols are located purely at layer 2 and therefore, they use MAC addresses. The default routing protocol is HWMP and it is an adaptation of AODV in the layer 2 which uses radio aware metrics. Optionally, 802.11s supports the proactive protocol Radio Aware–OLSR (RA-OLSR). RA-OLSR follows the specifications of OLSR but as in the previous case, instead of IP addresses it routes based on MAC addresses, and can work with radio-aware metrics such as Airtime.

16.11 Summary

WMNs can provide the infrastructure for the intelligent and integrated future metropolitan area networks: Free, best-effort Internet access for citizens; return of investment through high quality services to enterprises; e-government and interconnection for municipal sites; backhaul for surveillance, public safety, traffic control, and sensor networks, and many more applications.

Probably 802.11 is the most adequate option for hop-by-hop connectivity, mainly due to its vast deployment. However, 802.11 faces many problems in multi-hop environments as it was never designed for such networks. Improvements of MAC layer are under research, and special MAC standard for WMNs is under development. Furthermore, other wireless technologies, such as WiMax, may be used to extend coverage and capacity.

Routing protocols incorporated in WMNs were initially designed for the highly mobile environment of MANETs. New routing protocols, designed specifically for the WMNs' constraints of high throughput and scalability, are under active research. Enhancements, such as cross-layer metrics and active-probing metrics, can greatly improve the performance of the existing routing protocols. Hardware test beds developed by academia and communities provide the ideal testing and development platform for architectures and protocols.

References

1. Frivold Garcia-Luna-Aceves Beyer. 1997. Wireless Internet Gateways (WINGS). *IEEE MILCOM '97*, pp. 1271–1276.
2. Riezenman Schrick. Wireless broadband in a box. June 2002. *IEEE Spectrum* e, pp. 38–43.
3. Xudong Wang, Weilin Wang, and Ian F. Akyildiz. 2005. Wireless mesh networks: a survey. *Elsevier Computer Networks*, no. 47, pp. 445–487.
4. Wireless Municipal Network of Trikala. 2010. Available at http://www.e-trikala.gr (Accessed on August 2011).

5. Sharon E. Gillett. 2005. Municipal wireless broadband: hype or harbinger? Symposium on "Wireless Broadband: Is the U.S. Lagging?" Washington, DC, pp. 561–593.
6. Cisco Systems, Inc. 2006. Austin's Wireless Mesh Provides Free Access and Test Environment. Available at http://www.cisco.com/en/US/prod/collateral/wireless/ps5679/ps6548/prod_case_study 0900aecd80563c29.pdf (Accessed on August 2011).
7. Golden Telecom Wi-Fi. 2010. Available at http://www.goldenwifi.ru/en/ (Accessed on August 2011).
8. Nortel Networks. 2010. Wireless mesh network solution. Available at http://www2.nortel.com/go/ solution_content.jsp?segId = 0&catId = 0&parId = 0&prod_id = 47160&locale = en-US (Accessed on August 2011).
9. J. Tanner. December 2005. Wireless mesh still lives. Available at http://findarticles.com/p/articles/ mi_m0FGI/is_12_16/ai_n27869658/ (Accessed on August 2011).
10. Thessaloniki Wireless Metropolitan Network (TWMN). 2010. *Node Database.* Available at http:// wind.twmn.net/?page=nodes (Accessed on August 2011).
11. Tropos Networks, Inc. June 2005. Saving lives with Tropos MetroMesh, city of New Orleans, Louisiana. Available at http://www.tropos.com/pdf/case_studies/tropos_casestudy_new_orleans.pdf (Accessed on August 2011).
12. Meraka Institute—African Advanced Institute for Information and Communication Technology. 2010. *Wireless Africa.* Available at http://www.meraka.org.za/wireless.htm (Accessed on August 2011).
13. Yan Zhang , Jijun Luo, and Honglin Hu. 2007. *Wireless Mesh Networking.* Auerbach Publications, 2007.
14. Hong Fei and Bai Yu. 2007. Performance evaluation of wireless mesh networks with self-similar traffic. In International Conference on Wireless Communications, Networking and Mobile Computing (WiCom), pp. 1697–1700.
15. Kumar Gupta. March 2000. The capacity of wireless networks. *IEEE Transactions on Information Theory*, vol. 46, no. 2, pp. 388–404.
16. Sichitiu Jun. 2003. The nominal capacity of wireless mesh networks. *Wireless Communications*, vol. 10, no. 5, pp. 8–14.
17. Eldad Perahia and Robert Stacey. *Next Generation Wireless LANs: Throughput, Robustness and Reliability in 802.11n.* Cambridge University Press, 2008.
18. IEEE. 2005. "Amendment 8: Medium Access Control (MAC) Quality of Service Enhancements." IEEE Standard 802.11e-2005.
19. DARPA Internet Program. 1981. *Internet protocol.* IETF, RFC 791.
20. Hinden Deering. 1998. Internet protocol, version 6 (IPv6). IETF, RFC 2460.
21. Arkko Johnson Perkins. 2004. Mobility support in IPv6. IETF, RFC 3775.
22. P. Jacquet and T. Clausen. 2003. Optimized link state routing protocol (OLSR). IETF, Experimental RFC 3626.
23. E. M. Royer and C. K. Toh. April 1999. A review of current routing protocols for ad-hoc mobile wireless networks. *IEEE Personal Communications*, pp. 46–55.
24. L. M. Feeney. 1999. A taxonomy for routing protocols in mobile ad hoc networks. *SICS Technical Report T99/07.*
25. Petteri Kuosmanen. Classification of ad hoc routing protocols.1999. Finnish Defence Forces. Available at www.netlab.tkk.fi/opetus/s38030/k02/Papers/12-Petteri.pdf (Accessed on August 2011).
26. J. Wang, X. Jia, and M. Bahr. "Routing in wireless mesh networks," in J. Luo, H. Hu, and Y. Zhang, eds. *Wireless Mesh Networking.* Auerbach Publications, 2007, pp. 113–146.
27. Richard Draves, Jitendra Padhye, and Brian Zill. 2004. "Routing in Multi-Radio, Multi-Hop Wireless Mesh Networks," in Proceedings of the 10th annual international conference on Mobile computing and networking (MobiCom'04), pp. 114–128.
28. Atul Adya, Paramvir Bahl, Jitendra Padhye, Alec Wolman, and Lidong Zhou. 2004. "A Multi-Radio Unification Protocol for IEEE 802.11 Wireless Networks," in *Proceedings of the First International Conference on Broadband Networks (BroadNets),* pp. 344–354.
29. Richard Bellman. 1958. On a routing problem. *Quarterly of Applied Mathematics*, vol. 16, no. 1, pp. 87–90.
30. Edsger W. Dijkstra. 1959. A note on two problems in connexion with graphs. *Numerische Mathematik*, no. 1, pp. 269–271.

31. Charles E. Perkins and Pravin Bhagwat. 1994. Highly dynamic destination sequenced distance vector routing (DSDV) for mobile computers. In ACM Special Interest Group on Data Communication (SIGCOMM), pp. 234–244.

32. P. Jacquet, P. Muhlethaler, T. Clausen, A. Laouiti, and A. Qayyum, L. Viennot. 2001. Optimized link state routing protocol for ad-hoc networks. In IEEE International Multi Topic Conference (INMIC), pp. 62–68.

33. C. Hedrick. 1988. Routing information protocol. IETF, RFC 1058. Available at http://www.ietf.org/rfc/rfc1058.txt (Accessed on August 2011).

34. J. Moy. 1998. Open Shortest Path First Version 2. IETF, RFC2328. Available at http://www.ietf.org/rfc/rfc2328.txt (Accessed on August 2011).

35. J. Chroboczek. 2011. The Babel routing protocol. IETF, draft-chroboczek-babel-routing-protocol-05. (Accessed on August 2011).

36. Charles E. Perkins and Elizabeth M. Royer. 1999. Ad-hoc on-demand distance vector routing. *Proceedings of the 2nd IEEE Workshop on Mobile Computing Systems and Applications,* New Orleans, LA, pp. 90–100.

37. C. Perkins, E. Belding-Royer, and S. Das. 2003. Ad hoc on-demand distance vector (AODV) routing. IETF, Experimental RFC 3561. Available at http://www.ietf.org/rfc/rfc3561.txt

38. D. Johnson, Y. Hu, and D. Maltz. 2007. The dynamic source routing protocol (DSR) for mobile ad hoc networks for IPv4. IETF, Experimental RFC 4728. Available at http://www.rfc-editor.org/rfc/rfc4728.txt (Accessed on August 2011).

39. Zygmunt J. Haas, Marc R. Pearlman, and Prince Samar. 2002. The zone routing protocol (ZRP) for ad hoc networks. IETF, INTERNET-DRAFT draft-ietf-manet-zone-zrp-04.txt. Available at http://tools.ietf.org/id/draft-ietf-manet-zone-zrp-04.txt (Accessed on August 2011).

40. Srinivasan Keshav. A control-theoretic approach to flow control. 1991. SIGCOMM '91 Proceedings of the conference on Communications architecture & protocols.

41. Jitendra Padhye, Brian Zill, and Richard Draves. 2004. Comparison of routing metrics for static multi-hop wireless networks. SIGCOMM '04, 2004, pp. 133–144.

42. Daniel Aguayo, John Bicket, Robert Morris, and Douglas S. F. De Couto. 2005. A high-throughput path metric for multi-hop wireless routing. *Wireless Networks*, no. 11, pp. 419–434.

43. Jun Wang, Robin Kravets, and Yaling Yang. 2005. Interference-aware load balancing for multihop wireless networks. UIUCDCS-R-2005-2526. Available at http://www.mendeley.com/research/interferenceaware-load-balancing-multihop-wireless-networks/

44. IEEE P802.11s Draft 1.00 2006. Draft amendment: ESS mesh networking. IEEE.

45. Michael Bahr. Proposed routing for IEEE 802.11s WLAN mesh networks. In *The 2nd Annual International Wireless Internet Conference*, Boston, MA, 2006.

46. Alan Carlton. Defining layer 2.5. Available at http://www.google.co.uk/url?sa=t&source=web&cd=1&sqi=2&ved=0CCAQFjAA&url=http%3A%2F%2Fwww.ieee802.org%2F21%2Fdoctree%2F2004_Meeting_Docs%2F2004-05_meeting_docs%2F21-04-0035-00-0000-Layer2_5_concept.ppt&rct=j&q=alan%20carlton%20Defining%20layer%202.5&ei=9KdTTu-uEc7srQfE7_jNDg&usg=AFQjCNGxD5OdK5s6yxO41hY0Bj0CxX1d1A&cad=rja (Accessed on August 2011).

47. Sanjit Biswas and Robert Morris. ExOR: Opportunistic multi-hop routing for wireless networks. In SIGCOMM '05, Philadelphia, PA, 2005.

48. K. N. Ramachandran, M. M. Buddhikot, and G. Chandranmenon. On the design and implementation of infrastructure mesh networks. In IEEE WIMESH, 2005.

49. E. Nordström, AODV-UU, Ad-hoc On-demand Distance Vector Routing, Uppsala University. 2007. Available at http://core.it.uu.se/core/index.php/AODV-UU (Accessed on August 2011).

50. OLSRd An adhoc wireless mesh routing deamon. Available at http://www.olsr.org/ (Accessed on August 2011).

51. DSR-UU Implementation Available at http://core.it.uu.se/core/index.php/DSR-UU.

52. MIT Roofnet. Available at http://pdos.csail.mit.edu/roofnet/doku.php#about (Accessed on August 2011).

53. Daniel Aguayo, John Bicket, and Robert Morris. SrcRR: A high throughput routing protocol for 802.11 mesh networks. http://www.google.co.uk/url?sa=t&source=web&cd=1&ved=0CB0QFjAA&url=http%3A%2F%2Fciteseerx.ist.psu.edu%2Fviewdoc%2Fdownload%3Fdoi%3D10.1.1.71.2878%26rep%3Drep1%26type%3Dpdf&rct=j&q=A%20high%20throughput%20routing%20proto-col%20for%20802.11%20mesh%20networks&ei=AaRTTvvmO8bYrQefz4TVDg&usg=AFQjCNFVdR2zAoa3bsFI0yY2hT5OVv2IGQ&cad=rja (Accessed on August 2011).

54. Meraki, Inc. Meraki: wireless networks that simply work. Available at http://meraki.com/index.php (Accessed on August 2011).

55. Freifunk international Project for free wireless networks and frequencies (Open Spectrum). Available at http://wiki.freifunk.net/Kategorie:English (Accessed on August 2011).

56. OpenWRT Wireless Freedom. Available at http://openwrt.org/ (Accessed on August 2011).

Chapter 17

Wireless Mesh Network: Design, Modeling, Simulation, and Analysis

Christos K. Zachos and Jonathan Loo

Contents

In the previous chapter, the existing wireless technologies, architectures, and protocols, their strengths and weaknesses, and how they can be improved to accommodate the special needs of wireless mesh networks (WMNs) have been thoroughly discussed. This chapter aims to design a realistic form of a WMN that is functional, scalable, and low-cost as a municipality area network. As WMNs are not an established and well-defined technology, to set the realistic environment, their applications and their needs should be well-defined. A simulation model based on OPNET is created so the design's efficiency and theoretical performance can be analyzed and evaluated.

17.1 Network Design Goals and Constraints

For this study, the aim of the network design is to produce a functional, scalable, and low-cost municipality area network that should also be easy to implement and support.

First of all, the WMN is envisaged to provide Internet connectivity and basic communication services to citizens in a municipal area. The network resources should be sufficient to serve the population. The WMN network design should be scalable as well such that it is able to maintain linearity between the deployment cost and coverage.

The cost may be kept low by using existing, advanced—yet inexpensive—technology. That also means the end devices should be able to participate in the network without any modification or additional software, just like in an ordinary Wi-Fi hotspot.

The ease of implementation may be the one single design constraint critical enough to judge the feasibility of the project as it includes installation cost and time-to-market. As in such networks, municipality probably bounds all implementation and maintenance cost. Therefore, wireless mesh routers should be placed on sites owned by the municipality that has access to electric supplies such as lampposts. Furthermore, mesh gateways should be placed in public buildings with access to Internet service providers (ISPs) to avoid cabling costs.

Based on the aforementioned goals and constraints, an OPNET simulation model of a WMN in a metropolitan environment is created and its performances are analyzed and discussed later in this chapter.

17.2 Physical Topology

Thessaloniki's historical and economical center in Greece was selected as the reference metropolitan area for simulations. The service area is encompassed by the roads Kasandrou and Agiou Dimitriou to the north, Nikis avenue to the south, El. Venizelou to the west, and parts of Aristotle University and Thessaloniki International Fair to the east (Figure 17.1). The total service area

Figure 17.1   Service area and gateway placement.

is about 1.5 km² and the resident population is around 70,000 according to the latest census. Thessaloniki's city center represents a good example of a dense metropolitan area that already includes more than 5500 wireless LANs and, as a result, a sutured 2.4 GHz band.

In Figure 17.1, there are 10 feasible sites spotted that fulfill the requirements for the placement of mesh gateways. These sites include public buildings under direct control of the municipality, such as the city hall, museums, churches, university premises, etc.

The overall number of access interfaces is dictated by the criterion of 1% of the general population. According to vendors, the recommended number of wireless clients for sufficient throughput per 802.11 access point is 10 to 20. As the aim is to provide basic connectivity to up to 700 clients, the number of access interfaces would be 70. One assumption made is that the clients would be spread equally throughout the service area.

Mesh gateways would be placed on the positions mentioned in the figure. The positioning of the 60 mesh routers should meet three criteria:

- **Coverage.** The service area should have 100% coverage.
- **Distance from mesh gateway.** Every mesh router should be in range of at least one mesh gateway to minimize the distance, in hops, from the main service, the Internet. The effect of the distance in a multi-hop wireless environment is examined in detail later on in this chapter.
- **Number of 1-hop neighbors.** Kleinrock et al. in [1] proved that, for a static packet radio network, the optimum transmission power level is reached when each node has six one-hop neighboring nodes.

One of the possible combinations which meet the criteria is presented in Figure 17.2. In the simulation environment, there is 100% coverage of the service area using 802.11g interfaces

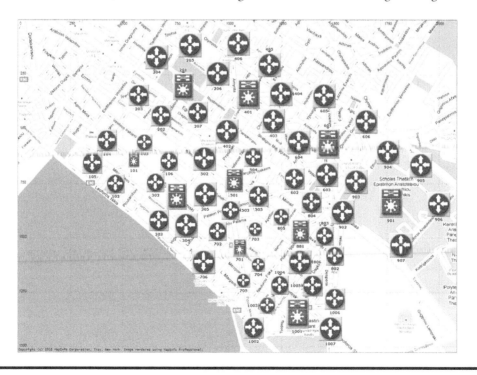

Figure 17.2   Physical topology of mesh routers and gateways.

and 5 mW transmit power. Moreover, every mesh router is within reach of at least one wireless gateway. The third criterion was the hardest to meet, and it is fully applied only in the case of mesh gateways.

17.3 Traffic Demand Model

Accurate design of a traffic demand model was one of the most difficult parts of the network design. As no user has the same behavior toward the Internet, standard application usage in public networks and Wi-Fi hotspots are considered such as:

- Web browsing.
- Streaming media. This category includes low quality video conferencing and streaming media, e.g., YouTube.
- Voice calls and instant messaging. Mainly using Skype and MSN messenger.
- File transfer. Rarely and only for small files, e.g., photos from cell phones uploaded to Facebook.
- E-mail. Mainly via webmail. As traffic, it can fall into the web browsing and file transfer categories.

The detailed user traffic profile and traffic characteristics are described in the simulation setup in Section 17.5.

17.4 Radio and Wireless Media Access

The existing 802.11-based wireless access technologies would be the best option for use in the implementation of a low-cost and publicly accessible WMN in the municipal area. As the majority of wireless end devices are equipped with 2.4 GHz radios, 802.11g radio should be used as the main wireless access interface.

Due to channel saturation and shortage in 2.4 GHz band, router-to-router communication should use the 5 GHz band or another technology such as 802.16. The 5 GHz band has three main advantages: a large number of non-overlapping channels, the regulations allow higher power levels than in 2.4 GHz band [2], and it is far less populated.

Taking into consideration the poor implementation of 802.16 in the OPNET modeler and the limitations of 802.11 to 802.16 bridging, the use of 802.11a radios for the router-to-router communication seems appropriate.

Both 802.11a and 802.11g use orthogonal frequency-division multiplexing (OFDM) with 52 sub-carries to transmit data at rates up to 54 Mbps. The general physical layer packet format used in 802.11a and in 802.11g with minor setting for compatibility is illustrated in Figure 17.3.

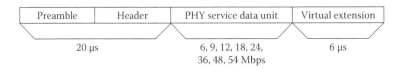

Preamble	Header	PHY service data unit	Virtual extension
20 µs		6, 9, 12, 18, 24, 36, 48, 54 Mbps	6 µs

Figure 17.3   OFDM packet format.

There are two methods of media access control (MAC) in 802.11: distributed coordination function (DCF) and point coordination function (PCF). The PCF was never implemented in a commercial product, and it is not part of the mandatory programs of WiFi alliance [3].

The physical layer service data unit (PSDU) consists of the applications' data bits and the lower layers' encapsulation. In a typical case where a packet is transmitted using standard 802.11 encapsulation, IPv4, and transmission control protocol (TCP) in the transport layer, the additional overhead would be 92 bytes:

■ Transport Layer: 20 bytes
■ Network Layer: 20 bytes
■ Data Link Layer: 52 bytes

17.5 802.2 Logical link control (LLC) and subnetwork access protocol (SNAP) headers: 8 bytes
17.6 802.11 MAC data header and frame check sequence (FCS): 44 bytes

From the 52 sub-carriers in each OFDM symbol, only 48 carry data. To achieve the 54 Mbps data rate, 802.11a physical layer uses 64-Quadrature Amplitude Modulation (QAM) modulation and forward error correction (FEC) ratio of ¾. The number of user data bits transmitted in every OFDM symbol would be:

$$64 - QAM \rightarrow 6 \text{ bits per OFDM sub} - \text{carrier}$$

$$6 \text{ bits per sub} - \text{carrier} \times 48 \text{ sub} - \text{carrier} = 288 \text{ bits per OFDM symbol}$$

$$288 \text{ bits per OFDM symbol} \times FEC = 288 \times 3/4 = 216 \text{ data bits per OFDM symbol} \quad (17.1)$$

The maximum length of unfragmented application data that can be transmitted over the network would be the MAC layer service data unit (MSDU) which is 2304 octets [4] minus the upper layer headers which is 48 octets as described earlier. However, most modern operating systems limit the Maximum Transmission Unit (MTU) at 1500 octets [5]; it is a more pragmatic approach as the same packet can be bridged to the wired part of the network—which in most cases is Ethernet—without fragmentation. In this case, the maximum length of application data would be 1452 octets.

The 1452 bytes of application data will grow to 1544 bytes when they reach physical layer, and they will be transmitted as 58 OFDM symbols, 216 bits each. At the maximum data rate of 54 Mbps, the symbol duration is 4 µs, so the total transmission duration for the application data is 232 µs. Moreover, OFDM adds 26 µs of preamble, header, and virtual extension that leads to a total duration of 258 µs for every 1452 bytes of application data.

Before the actual transmission, the transmitter should wait for at least Distributed Inter-Frame Space (DIFS), which is 34 µs, and after the transmission the received frame should be acknowledged. Before the acknowledgment, the receiver should wait for SIFS, which is 16 µs. The acknowledgment lasts one OFDM symbol plus the OFDM overhead for a total of 30 µs.

So, the total transaction at the physical layer lasts for 338 µs. However, as TCP is used in the transport layer, the reception of the application data should be acknowledged by the receiver and this transaction should be repeated the other way around.

The complete transaction over time is illustrated in Figure 17.4. The total duration is at least 456 µs which leads to a maximum throughput of 25.47 Mbps over one hop 802.11a link. Following the same methodology, the maximum theoretical throughput using UDP is 34.37 Mbps. Given the half-duplex nature of the communication, as a single radio is used for both uplink and downlink, the theoretical throughput is shared equally, in long term, among the users.

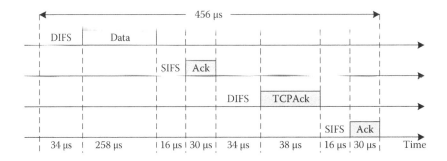

Figure 17.4 Sample transaction over time.

For this performance analysis, a number of assumptions were made to simplify the calculations. For example, the contention window (CW) is never 0 and so the sender must wait for a time longer than DIFS. Moreover, TCP congestion window size was set to one. Future packets may not need to be acknowledged in a one-to-one fashion, and so the throughput would grow. Furthermore, throughput is shared among the users but the network throughput itself is decreased as the number of users is increased and so is the possibility of collisions. However, the theoretical throughput is not deviating much from experimental results [6]:

Moving to the multi-hop domain, wireless routers forward packets in a store-and-forward manner which means that the whole 802.11 frame should be received and acknowledged before moving to the next hop. Figure 17.5 summarizes the transaction of 1452 bytes over a two-hop wireless network. Again, the assumptions are the same and no queuing or processing delay is taken into consideration.

17.5 Logical Topologies

WMN can be organized into two possible logical topologies—namely flat and hierarchical infrastructure mesh, and they will be put to test in this section.

Figure 17.6 illustrates the flat infrastructure mesh. Clients are ordinary end devices equipped with 802.11g interfaces. No routing is involved in the access layer. Clients are associated with the mesh routers that play the part of access points. Wireless mesh routers are equipped with two wireless interfaces, one 802.11g—to serve the clients, and one 802.11a—to form the WMN. Mesh gateways connect the WMN to the outside world via an 802.3 interface.

Logical addressing is planned accordingly. All members of the ad hoc network belong to the same service set, and they are addressed by the pool of 192.0.1.0/24 subnet. The clients are separated into different logical domains according to the associated access point. Each of the wireless local area networks formed in the access layer are addressed by a different class C network.

Dynamic routing protocols are running in all wireless and wired interfaces of the mesh routers and gateways to form a single IP routing table. However, for the wired routes to be redistributed to the WMN and vice versa, the 802.3 interfaces should run both routing protocols served in the wired and wireless part of the network. For example, if AODV is running on the 802.11 interfaces and RIP is running on the wired infrastructure, 802.3 interfaces should support both.

Figure 17.7 illustrates hierarchical infrastructure mesh. The wireless mesh routers are grouped into clusters at the "distribution" level and mesh gateways act as cluster heads at "core" level.

In the access level, clients are associated with the 802.11g interfaces of wireless routers and gateways to form the wireless local area networks. Again, clients are not involved in the path determination procedure and the addressing scheme remains the same.

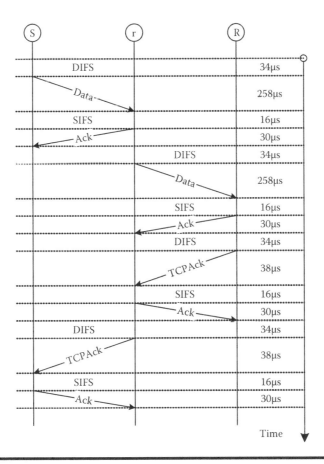

Figure 17.5   TCP transaction over two wireless hops.

Every router and gateway participates in the distribution level using one 802.11a interface, as in the case of the flat infrastructure. Here, a number of wireless routers form a cluster with a mesh gateway as the cluster head. This results in a formation of a number of independent ad hoc networks. Each one of them is a different service set and has its own addressing scheme. Links in the distribution layer use the lower UNII (unlicensed national information infrastructure) radio band and the channel assignment is so as to avoid interference among adjacent clusters.

Finally, in the core layer, cluster heads form an ad hoc network to interconnect the cluster members. To do so, mesh gateways are now equipped with a third 802.11a wireless interface which operates in the additional UNII band. The use of the additional UNII band permits high-power, long-distance links in a relatively interference-free environment.

The hierarchical approach is expected to have certain advantages over the flat infrastructure mesh architecture. The most prominent one is the route aggregation in the cluster level. If the addressing scheme follows the hierarchy of the cluster, every cluster head can advertise one route to the core mesh network. For example, in Figure 17.8, every collision domain—in other words, every subnet—is assigned to a class C 192.168.x.0/24 address. The whole network behind the cluster head can be represented with the classless [/] address 192.168.0.0/21. Even in this oversimplified example, if the routes were not aggregated, the number of advertised routes would grow to five.

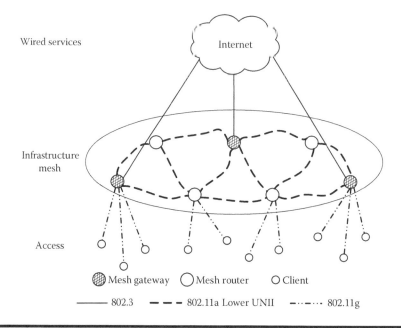

Figure 17.6 Flat infrastructure mesh.

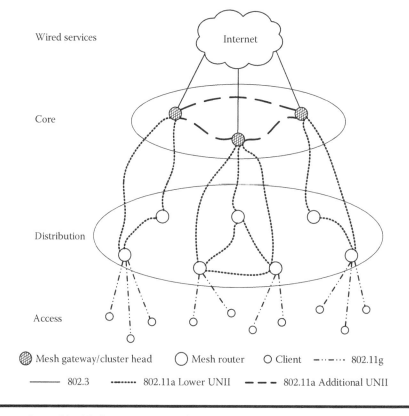

Figure 17.7 Hierarchical infrastructure mesh.

As a result, route aggregation has the advantages in three main areas:

- **Cost.** Route aggregation leads to far smaller routing tables, meaning that the routing process is less intensive for the CPU and less memory is needed.
- **Performance.** Less routes to advertise means less and smaller advertisements and so more throughput to the users. Moreover, a smaller routing table means quicker look-up and so less processing and queuing delays.
- **Network Stability.** Especially if a proactive routing protocol is in use. If, for example, the link between the mesh router of subnet 192.168.1.0/24 and the cluster head fails (Figure 17.8), the route discovery process will be activated within the cluster but the instability will not spread throughout the rest of the municipal network as the aggregated route is still active.

In the flat infrastructure architecture, every mesh router introduces two routing entries to the network, one to the meshed interface, e.g., 10.200.150.11/32, and one to the access network it serves, e.g., 192.168.22.0/24. So, the number of the route entries in the routing table of every mesh router in the network can be calculated as the routes to every served network (Ri), plus the routes to all meshed interfaces (Ri), minus its own meshed interface: $2 \cdot Ri - 1$.

In the hierarchical infrastructure mesh, every mesh router needs the following information to have complete knowledge of the network: the aggregated routes to all the clusters except the one in which it participates—that is equal to the number of cluster heads (Rg) minus one; a route to every cluster head (Rg); a route of every other member of the cluster which, in this case, is 6; and routes to every access subnet of the cluster which is equal to the number of the cluster member. In this case it is 7 minus the cluster head whose route was taken into account before. The total number of routing entries would be $2 \cdot Rg + 11$.

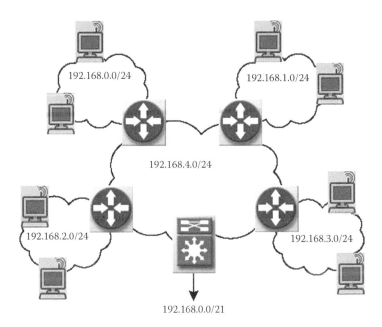

Figure 17.8   Route aggregation in cluster level.

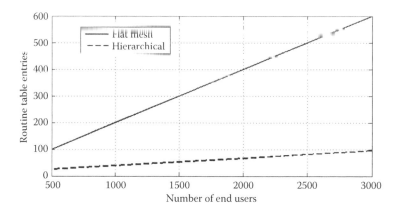

Figure 17.9 Routing table entries versus number of users.

Figure 17.9 illustrates the growth of the routing table as the number of the end users increases in both cases. The number of the users per mesh router follows the principles of the network design.

Furthermore, the hierarchical approach provides constant two-hop distance from every end user to the Internet, and it adds redundancy in the event of a gateway failure. The constant distance is expected to cause less delay variations which are very important to real-time jitter-sensitive applications. Due to the physical topology of the network, the two-hop constant distance to the Internet is a fact for flat topology, too. However, if one of the links to the Internet fails, then the number of hops toward the next gateway is increased significantly; in the alternative topology, cluster traffic is routed through the core network, and the number of hops is increased only by one.

As the entire cluster's traffic passes through the cluster head, the hierarchical architecture faces two disadvantages: single point-of-failure and bottlenecks at the mesh gateways. One way to avoid both would be redundant mesh gateways in every cluster. However, a new routing protocol with load-balancing and fail-safe capabilities is needed as none of the existing ones support such futures. Another way to avoid single point-of-failure would be the deviation of the cluster's traffic to another cluster through neighboring mesh routers of different clusters in the event of failure. Again, the limitations of the existing routing protocols would flatten the network hierarchy and its advantages.

17.6 Routing Protocol Selection

To support the network design goals and constraints (Section 17.1), the routing protocol selection was based on following criteria:

- Cooperation with existing wired routing protocols to support gateway functionality
- Ability to run in multiple wireless interfaces in the same node
- Implementation in test beds
- Availability in the OPNET modeler environment.

The two protocols that fulfill these criteria are AODV and OLSR. Theoretically, AODV, as a reactive protocol, is expected to produce higher routing traffic and introduce higher latency than the proactive OLSR, given the static environment of the WMN. Moreover, AODV and OLSR are the base of the proposed RM-AODV and RA-OLSR 802.11s standard.

17.7 Wireless Mesh Network Modeling and Simulation

The goal of this simulation is to evaluate whether our scalable and low-cost municipal WMN design using the existing 802.11 equipment is feasible and to reveal the strengths and weaknesses of the proposed hierarchical infrastructure as compared to the conventional flat one as discussed in Section 17.5.

17.7.1 Simulation Environment—OPNET Modeler

Simulation is the process of designing a model of a real system and conducting experiments with this model for the purpose of understanding the behavior of the system and evaluating various strategies for the operation of the system. Therefore, in our studies, OPNET modeler wireless suite 14.5 is selected as the network simulation tool such that it is used to build and evaluate our WMN design.

Figure 17.10 shows the OPNET network modeling environment. It is organized into three domains: network, node, and process. A node model specifies an object in the network domain, while a process model specifies an object in the node domain. Network models consist of nodes, links, subnets, and additional "nodes" that define traffic. One of the major advantages of modeler over other discrete event simulators (DES), such as ns-2, is its graphical, interactive network workspace.

Node models are built in the node domain using the basic blocks of processors, queues, and transceivers. Processors are fully programmable via their processing models in the process domain using state diagrams and blocks of C code. Packet streams and statistical wires provide the interface among the basic blocks of the node model.

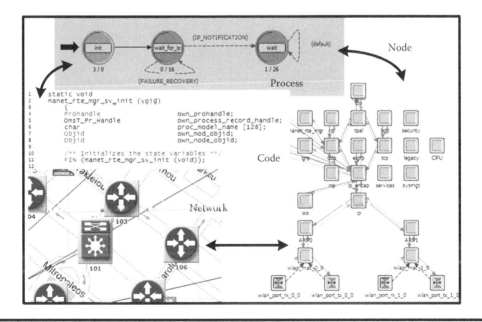

Figure 17.10   OPNET domains.

Project workflow in modeler includes some or all of the following basic steps:

- Create project
- Create basic scenario
  - Create or modify node models
  - Create topology
  - Create traffic
  - Choose results to be collected
  - Run simulation
  - View results
- Duplicate scenario (for comparison)
  - Modify nodes/topology/traffic
  - Re-run simulation
  - Compare results
- Iterate if necessary

17.7.2 WMN Model in OPNET

First of all, OPNET does not provide WMN features or WMN nodes. Multi-radio mesh routers and multi-interface mesh gateways have to be created. For the design needs, two custom models were created: a mesh router with two 802.11 wireless interfaces and a mesh gateway with three 802.11 wireless interfaces, one Ethernet and one point-to-point (PPP) interface.

The OPNET node model that was used as the foundation for both custom models was wlan_ethernet_router_adv. Additional interfaces were built using the appropriate processors and transceiver blocks. For example, an 802.11 interface can be created by connecting an ip_arp_v4 process model to the systems ip_dispatch as shown in Figure 17.11.

Interface MAC is integrated in wlan_dispatch model and PHY in one radio receiver and one transmitter block. Carrier sensing is possible via statistical wires (dashed lines) that report the received power level. One important parameter for a node to act as a gateway from other networks into WMN is manet_mgr.MANET gateway. This should be *enabled* in the attributes of the ip block, which is disabled by default and not visible in the nodes attributes.

After the custom nodes are available, the topology should be set up. The map was taken from Google Earth and the dimensions were accurately measured using Microsoft's MapPoint. The network's size was finally assigned to 2081.9 m by 1523.63 m. The free space propagation model was used throughout the simulation process.

Mesh routers were placed in the predefined locations described in Section 17.2 (see Figure 17.2). To form the flat architecture, all routers used two interfaces. The first (802.11a, 20 mW, IF0) is used to form the router-to-router backhaul and the second (802.11g, 5 mW, IF1) to serve the clients. All 802.11a interfaces participate in the same Service Set (SS) to form the ad hoc network. To form the hierarchical architecture, mesh gateways used the additional third wireless interface (802.11a, upper UNII band, 100 mW, IF2). The overall network consists of eleven ad hoc networks. The IF1 interfaces of the same cluster members participate in the same SS to form a total of ten independent ad hoc networks. Additionally, the IF2 interfaces of all cluster heads participate in a wide ad hoc network that spans the entire city center and interconnects the clusters.

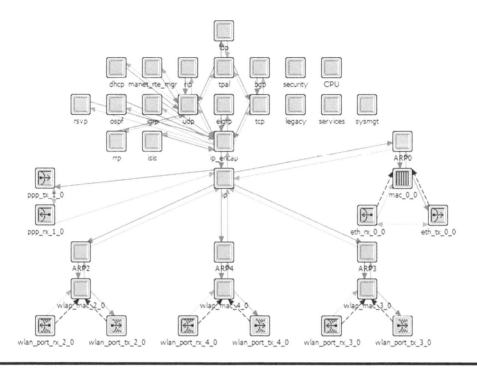

Figure 17.11   Mesh gateway/cluster head custom node.

One of the most difficult and time-consuming aspects of the simulation set up was the placement and configuration of the 700 end devices. Unfortunately, both OPNET's quick ways of massive node placement (rapid configuration and deploy wireless network wizard) were not adequate; in both cases, OPNET presumes that all nodes belong to the same service set. So, 700 nodes were placed in random but probable places and associated with 70 different mesh routers manually. A snapshot of OPNET's work space is illustrated in Figure 17.12.

OPNET-wise, the applications' traffic profile described in Section 17.3 can be modeled through the application config node using the following characteristics:

■ Web Browsing (Heavy HTTP1.1)
  – Page Interarrival Time: exponential 30 seconds
  – Page Size: constant 1000 bytes
  – Number of Repetitions: unlimited
  – Repetition Pattern: concurrent
■ File Transfer (Light)
  – Inter-repetition Time: exponential 60 seconds
  – File Size: exponential 1MB
  – Number of Repetitions: unlimited
  – Repetition Pattern: serial
■ Voice over IP Call (PCM Quality)
  – Encoder Scheme: G.711
  – Voice Frames per Packet: 1
  – Compression/Decompression Delay: 0.02 seconds

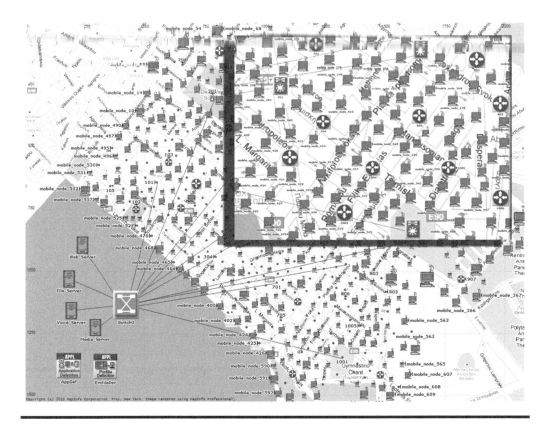

Figure 17.12 Municipal area mesh network.

- Inter-repetition Time: exponential 300 seconds
- Number of Repetitions: unlimited
- Repetition Pattern: serial
■ Video Conferencing (Light)
- Frame Interarrival Time: 10 frames/seconds
- Frame Size: 128 x 120 pixels
- Inter-repetition Time: exponential 300 seconds
- Number of Repetitions: 2
- Repetition Pattern: serial

17.7.3 Simulation Results and Analysis

Figure 17.13 shows the number of hops per route in the hierarchical and flat WMN architecture. It clearly indicates the superiority of the hierarchical design on large static WMNs. With the use of AODV in the model, from the figure it is clearly shown that the average number of wireless hops in the hierarchical architecture is two. Under the same traffic characteristics, the number of wireless hops in flat architecture could go up to six. Having fewer hops in the communication path, the hierarchical architecture has the advantage of achieving higher throughput over the flat architecture; [6] has shown how the throughput of 802.11 standards degrades exponentially in proportion as the number of hops increases. The throughput results in Figure 17.14 show a similar trend.

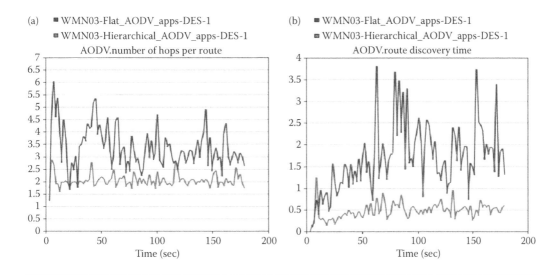

Figure 17.13 Number of hops per route.

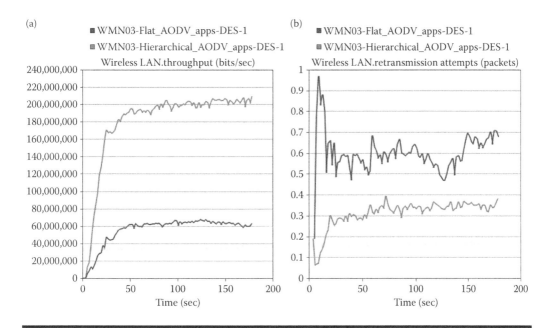

Figure 17.14 Throughput performance.

Besides the fewer number of hops, the hierarchical architecture has shown vast improvement on the route discovery time; it is 500% reduced as compared to the flat architecture seen in Figure 17.13(b).

Figure 17.14 shows that the hierarchical architecture achieved the throughput (or network capacity) by a factor of 2.5 which also has much reduced retransmission attempts as compared to the flat infrastructure. The vast improvement is not just the result of shorter paths (less number of hops) but also the attribution of route aggregation.

In the flat architecture, all the 70 mesh routers/gateways participate on the same service set and they all use the same channel, Intra-channel, mainly the inter channel Interference further degrades the performance. Moreover, the heavily congested flat links cause collisions and, as a result, increased retransmission attempts and data drops due to the retransmission threshold as seen in Figure 17.14(b).

On the other hand, the hierarchical approach of aggregating routes at different layers can lead to far smaller routing tables, less route advertisements, quicker look-up, and less processing and queuing delays. In the event the link between the mesh router of a subnet and the cluster head fails, the route discovery process will be activated within the cluster but the instability will not spread throughout the rest of the municipal network as the aggregated route is still active. These attributes have given much advantage to the hierarchical WMN architecture in dealing with both the file transfer (bursty and non-delay sensitive) and multimedia streaming (delay sensitive) applications; much improved traffics can be seen in Figures 17.15 and 17.16.

17.7.4 Limitations and Comments on the Simulation Process

During modeling and simulations' set up, a number of OPNET modeler limitations and deviations from the real-world implementations were aroused. These limitations considerably delayed the experimental process:

■ Every wireless interface with access point functionality must have a unique SSID in every logical subnet. In the real-world, all access points that belong to the same Extended Service

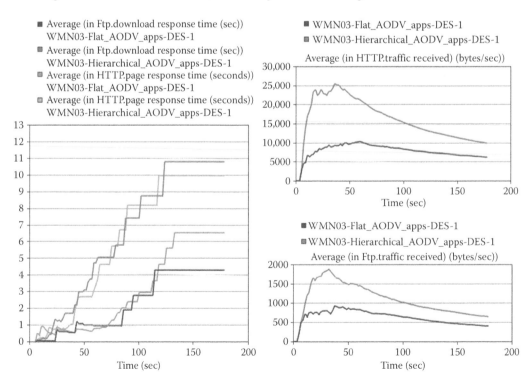

Figure 17.15   File transfer performance.

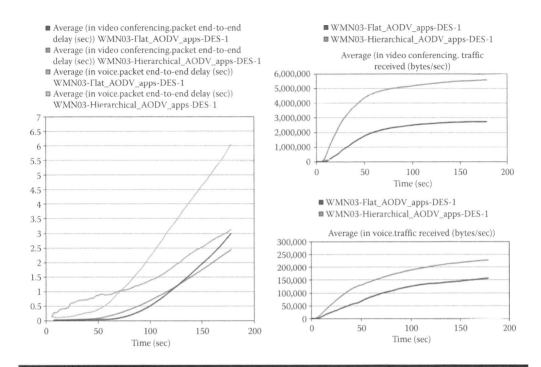

Figure 17.16   Multimedia applications performance.

set (ESS) have the same SSID. If a client is within reach of two or more, they select the one with the higher signal level or the one with the lower load. One side-effect of this limitation was the manual assignment of each one of the 700 clients of the large-scale simulation to one of the 70 access interfaces.

■ No auto rate functionality. Physical layer's data rate is statically assigned to each interface. In a network with more than 1000 wireless interfaces, assignment other than the maximum value is extremely time-consuming. As all interfaces have the same data rate, the minimum cost metric modeled in the third simulation has no effect on the large-scale simulation.

■ Massive assignment of variables that belong to the same process is not an option. For example, if a number of nodes that belong to different service sets are selected to change the PHY data rate, not only the selected value will change but also all the values of the wireless_lan_mac process as the SSID, physical characteristics, transmit power, etc. However, it works where all parameters are globally the same, e.g., routing protocol.

■ MANET gateway function works only with AODV and OLSR. In fact, only AODV can run simultaneously in a multi-radio node along with the MANET gateway function to enable a WMN where end users are unaware of the underlining mesh distribution system. OLSR was not an option in the large-scale scenario as clients should also run the routing protocol and in such a dense network would definitely route some other nodes traffic ruining the results.

17.8   Conclusions and Future Work

Based on their applications, WMNs can be defined as the backhaul technology that can disengage wireless local area networks from their limited range boundaries and expand them through large metropolitan areas. Network design and simulation results further support this statement and prove that even by using the existing PHY and MAC technology and MANET oriented routing protocols, large-scale WMNs are feasible and cost-effective.

The simulation results for the proposed hierarchical architecture show that it can enhance the performance and the scalability of WMNs. However, due to simulation environment limitations, only the AODV routing protocol was tested. These limitations can be overcome by routing modeling modification on the process level. Moreover, new architecture-specific routing protocols may be developed to further enhance the performance.

Another issue in the routing domain is the development of a sufficient mechanism of route redistribution among different protocols. For example, if a cluster is formed using AODV but the core-level network uses OLSR and is connected to an OSPF-based corporate network, then there is a need for the AODV routes to be advertised into the other domains using their routing protocols and vice versa.

Moreover, more link layer technologies could be put into test. Technologies such as 802.16 may have superior performance, especially in long-range core links.

Finally, other issues not directly connected to the performance of WMNs but crucial for their applications can be explored. For example, security is expected to be a huge issue, as never before was a backhaul network so exposed and vulnerable. Furthermore, as bandwidth is never expected to be plenty, end-to-end quality of service mechanisms should be developed to prioritize delay sensitive applications.

References

1. Silvester Kleinrock. Optimum transmission radio for packet radii networks or why six is a magic number. In *IEEE National Telecommunications Conference*, 1978, pp. 4.3.1–4.3.5.
2. Bob O'Hara and Al Petrick. *IEEE 802.11 Handbook: A Designer's Companion*, 2nd ed. John Wiley & Sons, 2011.
3. WiFi Alliance. (2010, Feb.) WiFi Alliance: Certification Programs. [Online]. http://www.wi-fi.org/certification_programs.php.
4. IEEE. 802.11 wireless LAN medium access control (MAC) and physical layer (PHY) specifications. IEEE Std 802.11™-2007, 2007.
5. Microsoft Support. Change the default maximum transmission unit (MTU) size settings for PPP connections or for VPN connections. Oct. 2006. Available at http://support.microsoft.com/kb/826159.
6. C. Zachos, S. Pouros, S. Kyrtopoulos, A. Kaloumenos, V. Oikonomou, and D. Kyrtopoulos. Effect of safety mechanisms, hardware, user and packet length on the operation of wireless networks LAN/MAN IEEE 802.11g. In *International Conference on Telecommunications and Multimedia (TEMU2008)*, Ierapetra, Crete, 2008.
7. Li Fuller. Classless inter-domain routing (CIDR): The Internet address assignment and aggregation plan. IETF, Request for Comments RFC4632, 2006.

Chapter 18

Adaptive Routing Provision by Using Bayesian Inference

Ilias Kiourktsidis, Jonathan Loo, and Grigorios Koulouras

Contents

18.1 Introduction

Many routing protocols have been proposed for Mobile AdHoc Networks (MANETs) by considering criteria for the routing selection mechanism such as hop count, end-to-end delay, bandwidth, mobility, and signal strength but there are not any solutions that successfully combine multiple criteria. Also, protocols that take into consideration the breakage lifetime of a path have been introduced. Instead a more realistic solution is proposed here, that targets the performance lifetime of a path and combines multiple criteria. This work presents a model that gives Bayesian inference about the performance lifetime of a routing path by utilizing efficiently cross-layer information. The proposed Bayesian network routing (BNR) protocol is designed to use this model in order to assist routing selection and route maintenance mechanisms for MANETs.

In fact, the aforementioned model represents a path in a MANET. The model can be used to give probability distribution of unknown parameters' states in the path, given the state of observed parameters. Thus, the model can be useful for evaluating the paths that are discovered and assist the route selection process. The evaluation of a path is supported by accurate statistical data collected from MANETs during operation. This statistical model can also be used for improving our knowledge about the MANETs and the elements that affect their performance. The BNR mechanism is a flexible cross-layer design that can adjust to the amount of available cross-layer information.

18.2 Routing in MANETs

A routing protocol for mobile ad hoc networks should minimize route setup and maintenance messages, in order to have a low communication overhead. The protocol should also converge quickly to select a new route before network changes make the route invalid. Furthermore, the routing protocol must be distributed, since no centralized host is available. Distributed nature makes also the protocol scalable. Finally, it should take advantage of the available technology and utilize information about the current state of the network by using cross-layer design. Most of the existing routing protocols focus solely on improving one or two aspects of the network and overlook the others. In addition, most of the protocols are relying only on specific metrics even though there are more metrics available that can give more and better information. Other solutions are assuming that some services are given, like global positioning system (GPS) devices and MAC layer capabilities. Consequently, they are not flexible enough to be used in every single situation. A system that can take advantage of the plethora of information about the network is needed while at the same time it can be tolerant to missing or inaccurate information in order to decide which route will be used. Cross-layer design between network and MAC layer is required in order to enhance the efficiency of the routing algorithm but at the same time the routing algorithm must not rely exclusively on the cross-layer information.

In spite of many path metrics for multi-hop networks that are available, new routing metrics are still needed to further improve the performance of routing protocols. In wireless mesh networks, many performance parameters and constraints are cross-related and make routing a more difficult problem than fixed cellular or wired networks.

18.2.1 QoS Aware Routing

One place in the stack of protocols where someone can look and try to improve the quality of service in a network is the network layer. The basic responsibility of the network layer is the routing of packets including routing through intermediate routers (in case of MANETs or other nodes). Hence, early research was focused on designing routing protocols that discover routes in the network that are able to support the communication. The routing protocols that are used now in MANETs are oriented on discovering paths in the network and using the ones with fewer hops. Some hop-count protocols are the on-demand Ad hoc On-Demand Distance Vector Routing (AODV) [1], Dynamic Source Routing (DSR) [2], the proactive Destination-Sequenced Distance Vector (DSDV) [3] and Optimized Link State Routing (OLSR) [4].

18.2.1.1 Additional Metrics

However, minimizing the number of hops maximizes the distance between two nodes in each hop. Usually, this approach results in low signal strength and high loss ratio. In addition, even if there are several minimum hop-count paths, there is not a way to find out which one is the best because there is not an additional metric to distinguish them. Douglas et al. [5] used a DSDV test-bed to prove experimentally and quantify the effect that minimum hop strategy usually delivers paths with low throughput and high loss ratio. They proposed the expected transmission count metric (ETX) as alternative metric. ETX metric predicts the number of retransmissions in a path by adding the packet loss ratios in each link in the path. Hence, the criteria for choosing a path will be the ETX and not the number of hops. ETX gives a good estimation of the link quality but cannot quickly track the changes in a link quality. Also in newly established links, ETX may not be available yet. Yarvis et al. [6] follow the same approach in sensor networks by observing the poor performance of the hop count metric and propose another loss aware metric that is based on the probability of a packet being lost in the path. Awerbuch et al. [7]also worked on an alternative metric—the medium time metric, that relies on the fact that the links in the network have different data rates. Their approach selects a route that the total medium time consumed sending a packet is minimized. This results in an increase in total network throughput. There are other proposed algorithms that take into consideration the mobility of the nodes which are based on the idea that nodes that have been stationary for a threshold period are less likely to move. Associativity based routing (ABR) uses stability and hop count as criteria to select a path [8]. In [9] a routing protocol that selects the most stable routes through the dynamic network is proposed. This on demand routing protocol, called *signal stability-based adaptive routing* (SSA) protocol, uses the signal strength and stability of each node in a path for routing decision. The simulation results show that location stability information can be a useful metric, although it must be used carefully since misinformation about stability patterns is considered quite costly and has negative impact on routing and performance. One of the reasons that a packet can be delayed is the queue waiting inside a node's buffer. The delay-sensitive adaptive routing protocol (DSARP) [10] utilizes the effect of buffer usage on delay. The route reply packet that follows the path from the destination back to the source gathers the number of packets waiting in the buffer of each node in the path. The path with the smallest number of packet waiting in the buffers is considered the one with the less delay. This also helps the network load balance which eventually helps to distribute evenly the energy usage in all nodes of the network. Again, it is a little optimistic to assume that the path with the less buffer usage is the faster one because delay also depends on other parameters.

18.2.1.2 Contention Aware Protocols

A non-flat hierarchy routing protocol is proposed by R. Sivakumar et al. [11] whichthat implements a core network for performing route computations (CEDAR). Each node in the core network keeps up-to-date information about its local dynamic topology and propagates the link state information about the more stable links that connect them with core nodes far away in the network. CEDAR reduces route maintenance overhead. The basic idea is that the information about stable high-bandwidth links can be made known to core nodes far away in the network, while information about dynamic links or low bandwidth links should remain local. Consequently, CEDAR approximates a minimalist local state algorithm in highly dynamic parts of the networks while it approaches the maximalist link state algorithm in highly stable parts of the networks. This technique assumes that MAC layer is capable of estimating the available link bandwidth. A heuristic interference-aware QoS routing algorithm (IQRouting) is proposed by [12] where the routing overhead is reduced by choosing routes based on localized information at the source nodes. In addition, it estimates a channel's residual capacity by measuring the channel usage from the links that interfere with each other. The same approach of using information from nodes inside collision domain range by overhearing neighbors' transmissions has been used in [13]. An admission control protocol is proposed that uses this information and predicts the available bandwidth in the channel. The requests for new flows through specific areas in the network are managed according to the available bandwidth.

18.2.1.3 Link Stability Based Routing

As it is already noted, QoS in MANETs in the sense of guaranteed resource reservation cannot exist. One way for soft QoS success is to select routes that are more likely to be stable by having long expected lifetimes. In [14] Rubin and Liu have created a link stability prediction mechanism. The probability distribution function (PDF) of link lifetimes under various mobility models is calculated and used to predict the residual link lifetime. By statistically predicting the lifetime of a link, it is easy to avoid paths that will fail soon. The difficulty with this solution is spotted by the fact that the node mobility pattern must be known and modeled accurately for correct predictions. Apart from that, it can be a waste of useful paths in case the protocol avoids links solely because of the mobility metric based prediction results, even if the signal to noise ratio, the capacity in the channel, or other metrics are in a very good state. Thus, the combination of this stability metric with other metrics can produce much better results. A similar approach is followed by Trivino-Cabrera et al. [15] and Tseng et al. [16]. A better link stability prediction mechanism has been proposed by Shen et al. [17] that defines the entropy of a link as a function of the relative positions and velocities of the end nodes in the link and the transmission range. The entropy of the path is the product of the link entropies along it. The more stable the path is the less entropy it has. The stability prediction has the potential to be more accurate as it involves more parameters, relative speed, and position, than just a general PDF. This approach does not depend on the need for an accurate mobility model. On the other hand, it assumes that nodes have a self-configuring localization mechanism. Another way to calculate the stability of a link and how stable the environment is around it, is to check the set of neighbor nodes and how they change in a period of time. The hybrid ad hoc routing protocol (HARP) [18] uses the quality of connectivity metric—which is a combination of buffer space and relative stability—by keeping a list of neighbors and checking how many of them are changed in a period of time. The less that changed, the greater the stability was. This method is more reliable as it measures directly the mobility in respect to the neighbors. As before, these protocols can also perform better if they are combined with other metrics for the routing decisions. In addition, these protocols target solely the link lifetime without considering the overall performance lifetime. A path is not very useful even when it is up if it cannot offer the required QoS.

18.2.1.4 Energy Aware Protocols

Mobile devices depend on their energy storage which is limited. If the power consumption is high, then the mobile nodes battery lifetime will be short. Several researchers are investigating methods to reduce the energy consumption in order to avoid node failures and link breakages. In [19], Yu et al. combine mobility prediction, link quality measurement by using ETX metric and energy consumption estimates to create a protocol that is QoS aware and at the same time makes energy efficient routing decisions. According to the results, the proposed protocol reduced link breakage and average delay in the network. But the protocol assumes that the nodes in the MANET are capable of measuring their mobility by using a GPS or other methods which, unfortunately, is not always true. Similarly, the residual energy in the nodes can be used as a routing metric to provide fair sharing of energy consumption in the network [20]. The distribution of the energy consumption in the network does not ensure that the traffic in the network will be distributed at every moment. In fact, it makes it more possible to create congestion in the network due to all traffic going through the nodes with more energy than others. In [21], the authors introduce five new power-aware metrics that can be integrated with other hop-count routing protocols to improve the overall network energy consumption. Energy aware protocols are useful in networks where the cost of maintenance in terms of recharging the batteries is high, and providing constant power is hard or impossible—like in sensor networks. In networks where nodes have enough power or there is no interest in saving power, most of the energy aware protocols are not useful; also, they cannot operate normally and provide the QoS improvements they may promise.

18.2.1.5 Position Based Protocols

Geographical position based routing algorithms have received significant attention since the progress of GPS and self-configuring localization mechanisms. One of the advantages of position based routing is that the overhead due to control messages and the size of routing tables is minimal. In geographical routing, the source node forwards a packet to the geographic location of the destination instead of using the network address. Taxonomy of position based routing algorithm can be found in [22]. One of the most popular techniques is the combined greedy forwarding (GF) and face routing. Greedy forwarding tries to bring the packet closer to the destination by forwarding it to the neighbor node that is geographically closest to the destination. GF can reach dead ends where no other neighbor node is closer to the destination. Face routing is then used to find a path to another node where greedy forwarding can be resumed. Nevertheless, geographic routing mechanisms assume that every device has a localization mechanism and the source node somehow knows the location of the destination node. These assumptions—especially the second one—are usually untrue. Furthermore, geographical routing has issues with mobility, and greedy forwarding fails when the network is not dense enough. In order to remove the need for localization and minimize the impact of localization errors, virtual coordinate systems have been proposed as an alternative to geographic location in geographical position based routing algorithms. Rao et al. [23] are using a relaxation algorithm that associates virtual coordinates to each node so geographic routing can be used. However, virtual coordination systems produce an overhead, cancelling one of the major advantages of geographical routing. Several studies also handle the issue that GF tends to forward the packets through lossy links because the route that is geographically closest to the destination node is not always the best choice in terms of link quality. Seada et al. [24] and Lee et al. [25] have proposed solutions that attempt a balance between proximity and link quality.

18.3 Cross-Layer Design

During the last several decades in the history of networking, the network protocols have been distributed into several independent layers. All the functionality is divided in layers and each layer is designed separately. The interaction between the protocols in the different layers is performed through a well-defined interface. This type of architecture follows the "divide and conquer" principle and has the advantage of complexity reduction and increase of the architectural flexibility. By defining exactly the functionality of each layer and the service access point interfaces in each layer, one implementation of a layer can be seamlessly replaced by another one. When a protocol is replaced by another one, there is no need to modify the rest of the network stack as long as the new protocol uses the same interface with the upper and lower layers.

However, the layered protocol design has been challenged for years for several reasons. One reason is that the need for quality of service is ever-increasing as multimedia and other time sensitive applications are becoming popular. Connectivity alone is not enough for a user anymore and features such as high speed transmission rate and quality of service (QoS) for time-critical applications are also desired. The conventional layered protocol design appears as a non-viable solution for these requirements. One hint is that most of the end-to-end QoS solutions that are already implemented for the Internet are using more than one protocol layers. Furthermore, the Internet consists of multiple heterogenic networks that are easier to be integrated all together with a layered architecture while, on the other hand, the performance is not optimized. One example that demonstrates the reason for that phenomenon is the routing. In order to decide which route is more applicable for a communication's needs, interaction with the MAC layer is needed to collect information about the links along the path. In addition, the fact that most of the networks today, and especially wireless networks, have no stable performance links between nodes increases the need for cross-layer design.

The cross-layer design helps the protocols to optimize their performance by enabling layers to exchange state information. The cross-layer information enables the different layers to have a better picture of the constraints and characteristics in the network and, as a result, allows them to make decisions that would jointly optimize the network and have better coordination. In order to achieve the desired coordination, the protocols must be developed in an integrated and hierarchical framework so as to take advantage of the inter-reliance between the protocols. The cross-layer information is related to the system constraints, several metrics, the requirements of the application, etc.

Even though the cross-layer design may assist drastically the performance optimization, it increases the cost of deployment, jeopardizes stability, and would hold back the proliferation and development of wireless networks, as industry standards are hard to be created. So, the aggressive use of cross-layer design must be avoided. A better approach would be to minimize its dependency on the information from other layers by careful placement of the various functionalities in the protocol stack. However, cross-layer design may be a suitable approach for stand-alone wireless networks that are dedicated to a single application, especially if the task is highly critical, and for cases where reliability and performance are more important than cost.

18.4 A Few Words about Bayesian Inference

A Bayesian network (BN) or belief network is a model. This model reflects the states of a system. The model can be anything; it can be a car, a human body, an ecosystem, a stock market, generally anything in the world. The objective is to create a BN that represents the communication between two peers in a MANET. Practically, a Bayesian network is a probabilistic graphical model that represents the joint probability distribution for a set of variables via a directed acyclic graph. The variables are some characteristics of the system that the Bayesian network models. For a car, the variables can be the car engine, the tires, the speed, the car's age, etc. A Bayesian network that models a car represents the probabilistic relations between the variables of the car we just mentioned. Given the state (condition) of the car engine and tire variables, the network can compute the probability distributions of the speed of the car and the age of the car variables or vice versa. Typically some states will tend to occur more frequently when other states are present. Thus, if you are sick, the chances of a runny nose are higher. If it is cloudy, the chances of rain are higher, and so on.

Bayesian networks are robust when there are missing input data. In this case, Bayesian networks will make the best possible prediction by using whatever information is presented. This is possible because BNs encode the correlation between the input variables. The more information you supply to them, the more accurate the results will be. This is one of the main advantages over other data analysis representations like rule bases, decision trees, and artificial neural networks, or over other data analysis techniques like regression, clustering, classification, and density estimation.

In order to construct a Bayesian network for modelling any system, a set of variables $X = \{X_1, \ldots, X_n\}$ is needed that are going to be the nodes in the BN. Those nodes need to be connected to create a structure S of the network that simply encodes a set of conditional independence assertions about variables in X. Suppose a node X_i whose condition directly affects the condition of another node X_j then there must be an arc from X_i to X_j to encode this impact. X_i is called the parent of X_j. Each variable X has a probability distribution table indicating how the probability of values of X depends on all possible combinations of parental values. For instance, if the parents are n Boolean variables then the probability distribution table will have 2^n entries, one entry for each of the 2^n possible combinations of its parents being true or false.

Let's consider that nodes are discrete random variables. The values should be both mutually exclusive and exhaustive, which means that the variable must take on exactly one of these values at a time. Common types of discrete nodes include:

- Boolean nodes, which represent propositions, taking the binary values true (T) and false (F).
- Ordered values—low, medium, high—can represent how full a buffer is.
- Integral values—for example, a node called age might represent a car's age and have possible values from 1 to 30.

The classical example model of rain, sprinkler, and wet grass is presented in Figure 18.1. By examining the BN, someone can find out which variables affect which others.

18.4.1 Bayesian Inference

Bayesian inference is an approach to statistical inference on model parameters quite different from classical methods, as all forms of uncertainty are expressed in terms of probability. It treats the unknown parameters of a model, θ, as random variables and not as fixed constants. The criteria of evaluating statistical procedures of estimation are conditional only on observed data and inference is based on the distribution of the parameters given the data, rather than the data given the parameters. Therefore, a prior probability distribution, $P(\theta)$, is assigned to the unknown parameters of the model, which represents our knowledge about these parameters. Clearly, it can be realized that Bayesian inference uses a subjective interpretation of probability, as prior distributions vary from

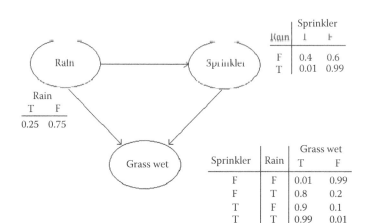

Figure 18.1 Bayesian network that models the relations between three variables—rain, sprinkler and wet grass. All of them are Boolean nodes. We can see that rain has a direct effect on grass condition and on the use of a sprinkler. We can see the joint distributions and the prior distributions for each variable (collected empirically or from statistical data).

person to person because they depend on people's belief and experience. The data supposed to be generated by the model provide information about the parameters.

$$P(\theta/E) = \frac{P(E/\theta) \times P(\theta)}{P(E)} \tag{18.1}$$

The prior distribution and the likelihood function, $P(E|\theta)$, are combined through Bayes theorem in Equation 18.1 in order to produce a posterior distribution, $P(\theta|E)$, from which we are able to update our prior beliefs about the parameters of the model, or compute predictive distributions for future observations.

Because a Bayesian network is a complete model for the variables and their relationships, it can be used to answer probabilistic queries about them. Thus, it can infer about variables that cannot be observed, by relying on the variables that can be observed. The more observations available, the more reliable answer it produces.

18.5 BNR Architecture

In this section we will describe by what means BNs can be useful for the selection of the routing path in a MANET. In a route discovery mechanism, several routes can be discovered. Usually the routing algorithm selects the route with the smaller cost, where cost can be the number of hops, smallest delay, the more throughput, most stable, etc. A BN can be asked to give probability of success for each route that has been discovered, like what is the probability that a specific path will provide the QoS that the application requires, or give the probability that this path will be stable long enough. The route that has the highest probability to succeed will be selected for the communication.

A Bayesian network (BN) is a statistical tool that can learn from previous events. Then, by observing current data, it can give inference about related events. Given the metrics for a discovered path in a MANET, the BN can be used to compute the probability of this route to be stable.

As it has been mentioned before, BNs are very tolerant to missing data; if a metric for a path is not available, the BN will make the best possible inference by using only the available metrics.

In order to use Bayesian inference for this purpose, three things are needed—a BN designed to model a path in a MANET, statistical data to create the joint distributions, and a routing protocol designed to collect the observed variables for each discovered route. In the following sections, we describe those three elements along with the variables that represent a BN's nodes.

18.5.1 Metrics Representing Nodes of Bayesian Network

The unique characteristics of MANETs invalidate existing solutions from both wired and wireless networks and impose unique requirements on designing routing metrics for mesh networks. More can be found in [26,27]. The routing metrics used in BNR design are presented later. The values of these metrics can give hints about the performance of the route the moment that they are collected and in the near future.

- **Hop-count** is the most classic routing metric and the simplest one that is used in existing routing protocols like DSR, AODV, and DSDV. The only information that it provides is how many links exist between source and destination, following a specific route. Sometimes hop-count information is enough for selecting a good routing path but most of the times there is the need for additional and more detailed information to determine whether a path is able to achieve good performance or not.
- **Path delay** is another metric; basically, it is the summation of the delays in each link that comprise the path. Path delay metric captures the packet loss ratio in the links, the queuing delays, and the load in the nodes, and the contention status in all neighboring nodes near the path area.
- **Data rate** for a path is the smallest data rate of all links in the path. Data rate is a concave metric because the link across the path that has the smaller data rate will be the bottleneck of the entire route.
- **Loss ratio** for the whole path is the product of the loss ratio in each link in the route.
- **Expected transmission count (ETX)** is the expected number of transmission attempts before a packet is successfully delivered on a link. The ETX or the whole path is the sum of the ETX on each link in the path. Packet loss ratio, collision probability, and link quality affect the value of ETX.
- **Frame transmission efficiency (FTE)** of a node is defined as FTE = 2/fACK + fRTS where fACK is the acknowledgment (ACK) failure count and fRTS is the request to send (RTS) failure count. The FTE for a path is the product of the FTEs of all nodes in the path.
- **Node density** is the number of nodes that can interfere with the channel in the area of a link. For the whole path, this metric is simply the summation for each link. Node density can affect the communication both in a positive and negative way. If the density is high, then it is easier to have local route repairs when the routing protocol supports this procedure but on the other hand high density means higher interference and lower channel capacity.
- **Relative mobility** or the similar **link life expectation (LLE)** is a metric that captures the trend of the link to stay up or not. For example, if the nodes are drawing away from each other, the link tends to become weak, and eventually the link will drop.
- **Signal to noise ratio (SNR)** for a link is a metric that can give a good estimation on the physical distance of the two nodes and the strength of the signal.
- **Node related information** like **buffer usage** and **processing usage** can give an indication on how much the delay will be. This metric will be additive for each node in the path.

There are also metrics that cannot be measured before the communication takes place and the path is used, like future delay, throughput, jitter, and route lifetime. All these metrics are combined in one parameter that denotes the route performance stability. A BN is called in to calculate the probability distribution of the routing performance stability for specific time duration and assist to the routing decision. Performance stability is a metric that denotes how stable the performance of a path is. Until now, some of the routing protocols targeted only the link lifetime without considering the overall performance lifetime. There is no point in using a path that is still alive but cannot offer the required QoS. Let us define as T_p the minimum time in seconds that is desirable to have stable performance in the path. Performance stability is a metric that includes the delay and throughput performance of the path in the next T_p seconds of the lifetime of the path. Performance stability of the path depends on how much the delay and the throughput will change along with the probability that the path will break or not after T_p seconds.

18.5.2 Protocol Mechanisms

The basic mechanism for a route discovery protocol that is usually used in protocols like DSR, involves the following steps.

1. A route discovery broadcast is sent by the source node. This broadcast contains the network address of the destination node.
2. All neighbor nodes receive the broadcast and each of them checks if it is the destination node.
3. If the node that receives the broadcast is not the one that the discovery is looking for, it stores their network address in the packet and broadcasts the packet again.
4. Eventually the destination node will receive the broadcast packet. Possibly it will be received more than one time from different paths. Then it sends back a reply addressed to the source node along with the path info that was stored and found inside the broadcast packet by using the exact reverse path.
5. The source node will receive the reply with the sequence of hops that is needed to reach the destination.

By using this mechanism, the only information that the source node has about the routes is the number of hops. The most direct route will be used although the links might have poor signal quality (and, hence, a greater chance that retries will be needed) or significant delays due to congestion. Our solution requires that the route discovery mechanism collect information about the metrics in each link that comprises the path that has been discovered. Every node keeps information about several metrics, and when the route reply packet passes each one of them, the node piggybacks this information. (The structure of the packet and the details of this protocol are not the main interest of this thesis and, thus, are not presented here.)

The measured metrics from each path, which has been discovered during route discovery, are fed to the BN model. The BN then estimates the probability for each one of the discovered paths to be stable after T_p seconds. Then the routing decision algorithm decides to use the path with the highest probability. During the communication, the packets give feedback about the path's condition by supplying the routing entity with the new metric values. The BN again is used to estimate the condition of the path for the next T_p seconds. If it realizes that the probability is less than a threshold value, a routing discovery is initiated in order to discover alternative paths that have higher probability of being stable.

18.5.3 Bayesian Model of a Routing Path

There are two major approaches in constructing a BN. One is the knowledge representation approach (KR) and the other is the machine learning (ML) approach. In KR approach, an expert on the subject area to be modeled constructs a BN by using his experience and knowledge in that area. In ML approach, the structure of the BN is learned from the data. In this hybrid approach, the experience from all the previous work that has been investigated about what way each metric affects the others in order to create the structure of the BN, and at the same time the creation of a correlation matrix, is used to justify the final design. Correlation matrix is a powerful statistical tool that can give the relation between many variables. For example, given a variable Y and a number of variables X_1 to X_p that may be related to Y, correlation matrix analysis can quantify the strength of the relationship between Y and every X_j. In that way, can be found which X_j may have no relationship with Y at all, and identify which subsets of the X_j contain redundant information about Y. Thus, once one is known, the others are no longer informative. Next, we present the procedure of collecting the required data, correlation matrix results, and the BN design.

18.5.3.1 Statistical Analysis

As we mentioned before, the BNR approach uses statistical evidence in order to design the BN and give successful inference. The data for the statistical analysis is produced in a simulated environment. A modified DSR routing protocol is used that discovers routes in a MANET, and then uses them in a random manner. For every route, the values of each routing metric are collected by the route reply mechanism and recorded along with the performance stability of the route after four seconds of use. The mean lifetime of a path in our simulated environment, as it is observed in Figure 18.2, is around four seconds. Thus, we choose to record the performance of the path after four seconds of use (T_p). By this way, thousands of path cases are collected.

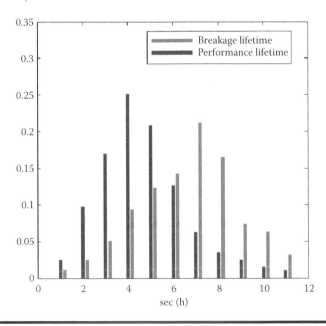

Figure 18.2 Probability distribution of a path's lifetime, considering path breakage and path performance degradation.

Table 18.1 is a correlation matrix for all routing metric variables in the model. The numbers in the table are Pearson correlation coefficients, from −1 to 1. Closer to 1 means strong correlation. A negative value indicates an inverse relationship (roughly, when one goes up the other goes down).

From the correlation matrix, information can be extracted in what way the routing metrics are related. The arcs between the BN nodes are placed according to this knowledge and the BN is constructed.

18.5.3.2 Graphical Causal Model

According to the general knowledge from the literature and research, and by using the correlation matrix that is presented earlier, the Bayesian network is constructed and is presented in Figure 18.3. In the graphical model, the only query variable that the protocol is interested in is the performance stability prediction.

There is a final step in order for the BN to be ready for operation—the creation of conditional probability tables in each BN node. The statistical data are used to train the network and create the required conditional probabilities.

18.6 Bayesian Network Queries

After successfully completing the BN construction, it is ready for evaluation by making queries and by observing the answers. The queries results are compared with the expected behavior that has been investigated in previous research. New conclusions about path performance in MANETs can emerge by experimenting with the BN. The initial probabilities for each state of every variable can be found in Figure 18.4. These probabilities correspond to the general belief of the condition of each metric in a path without observing any evidence yet.

It has been proven in literature that fewer hops usually results in longer distance hops, which in turn are usually not reliable links. While changing the hop count variable in the BN, it is observed in what way the performance stability probability distribution is changing. BN results are able to confirm the estimation that more hops give stronger links which in turn are more reliable. More interesting and complicated points also are coming up by experimenting with the model.

In a BNR model, delay is related to the number of hops; more hops result in bigger delay because of the additional delays in each hop. Furthermore delay is related to performance stability—longer delay means less performance stability. Hence, these two parameters conflict with each other. By increasing the number of hops, the performance stability increases negligibly and it even starts to decrease when the number of hops is at the maximum. But in case that the delay is measured and becomes fixed evidence, the hop count parameter cannot affect the delay—only the performance stability. With delay fixed, changes in the hop count make performance stability probability distribution change as expected. Table 18.2 depicts the values of performance stability when delay is unknown and when delay is fixed. This can also be observed in Figure 18.5 where the hop count evidence is nine-eleven hops, and the performance stability is increased along with SNR, data rate, and transmission efficiency (compared with initial values in Figure 18.4). These types of observations illustrate the complexity of metric relations and how each one depends on the others. For instance, enormous hop counts can denote either that the links are strong or not, depending on whether the destination node is close or far. At the same time, many hops can increase the delay because of the multiple retransmissions of the packet. On the other hand, if the frame transmission efficiency is high, then the delay is

Table 18.1 Correlation Matrix for All Routing Metric Variables in the Model

	Node Density	Relative Mobility	SNR	CPU & Buffer Usage	ETX	LOSS Ratio	FTE	Delay	Data Rate	Hop Count	Performance Stability
Node Density	1										
Relative Mobility	0.065	1									
SNR	0.079	0.005	1								
CPU & Buffer Usage	0.095	0.055	0.087	1							
ETX	0.657	0.153	−0.786	−0.028	1						
Loss Ratio	0.528	0.467	−0.845	0.158	0.737	1					
FTE	0.695	−0.225	0.683	0.085	−0.915	−0.896	1				
Delay	0.457	−0.115	−0.854	0.404	0.744	0.783	−0.853	1			
Data Rate	−0.278	−0.215	0.734	−0.383	−0.296	0.258	0.785	−0.150	1		
Hop Count	0.033	0.025	0.099	0.085	−0.076	−0.584	−0.392	−0.756	0.637	1	
Performance Stability	0.258	0.684	0.305	0.045	0.578	0.742	0.571	0.801	−0.589	0.872	1

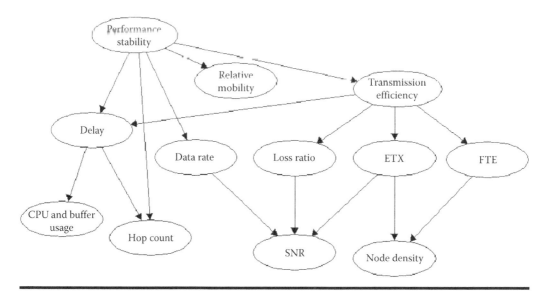

Figure 18.3 Bayesian Network structure that models a path in a MANET.

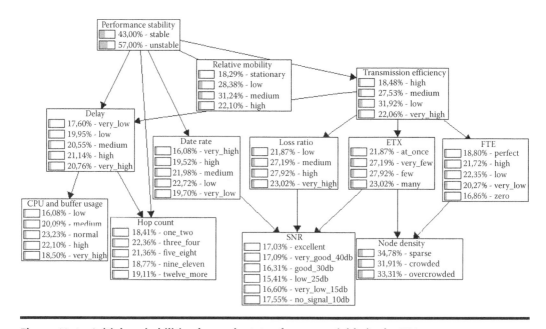

Figure 18.4 Initial probabilities for each state of every variable in the BN.

negligible even if there are many hops. Consequently, it is clear that a single hop count metric is not enough. The decision algorithm needs additional metrics in order to give a good estimation for the performance stability, such as delay, SNR, and more metrics as decision parameters. It is clear from the experiments that the use of the BN to quantify the effect of each metric to the others is beneficial.

SNR is another example in the BN that highlights both the complexity and the need for using multiple evidence for the conditions of a path in order to reach a better decision. Good SNR

Table 18.2   Hop Count Affect on Performance Stability and Delay

Hop Count	Performance Stability (Delay Not Fixed)	Delay (Very Low)	Performance Stability (Delay Fixed Low)
One – Two	40.72%	18.55%	54.72%
Three – Four	45%	23.05%	57.71%
Five – Eight	48.31%	23.82%	59.38%
Nine – Eleven	43.51%	14.32%	63.31%
Twelve – more	36.41%	6.59%	65.11%

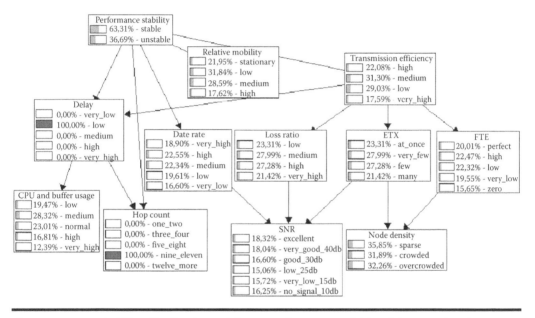

Figure 18.5   Probability distributions of the states in every variable of the BN when hop count is nine-eleven with fixed delay.

promises reliable links but if there is also high mobility then the path is not stable at all. Similarly, SNR increases the transmission efficiency but if the area is overcrowded, the transmission efficiency can be low.

In a case where the nature of a MANET is very specific—for example, if the mobility is very high due to the nature of the network—then the only choice for the protocol is to deal with the mobility and select a path where other metrics are in a state that counterbalances the high mobility effect in the performance.

By observing Table 18.3, it is clear that when only one good metric value is observed, it cannot be guaranteed that the path will have good performance even when it is evident that this value is in its best quality state. In this table, the probability of the performance stability is presented for each metric when it is observed in its best quality state. By looking only at the probabilities, someone cannot be certain that the path is stable. For example, even if the mobility is in a stationary state, the BN returns only 67.19% probability. So, there is a good chance that the path is not stable. In

Table 18.3 Stability Probabilities When Only One Metric is Observed in Its Best Quality State

Metric Name	Best Quality State	Probability That Path is Stable
Relative Mobility	Stationary	67.19%
Delay	Very Low	73.29%
CPU & Buffer Usage	Low	52.89%
Hop Count	Five-Eight	48.31%
Data Rate	Very High	64.19%
Loss Ratio	Low	50.91%
ETX	At Once	50.91%
FTE	Perfect	50.74%
SNR	Excellent	46.71%
Node Density	Sparse	52.13%

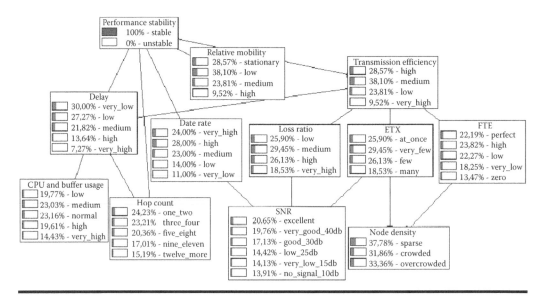

Figure 18.6 Probability distributions of the states in every variable of the BN when evidence of a stable path is presented.

order to increase the certainty, more metrics observations are needed. This is evidence that protocols which use very few metrics for their decisions are very likely to fail often.

On the other hand, the BN can also be used to find the most probable status of every variable in a path that is witnessed to have stable performance. In Figure 18.6, the BN gives the probability distributions when there is the evidence that the path has stable performance. In that case, as expected, it is more probably that all metrics are in a higher quality state. Figure 18.6 the states with the higher probability are the most common states in a path that has a stable performance.

This example has no practical value but it is presented here in order to demonstrate how the BN can be useful for improving our knowledge about the MANETs.

The number of the scenarios that can be examined can be as many as the combinations of each state in each variable node. The intention here is just to present the complexity of the system, rather than examining each different scenario.

18.7 Protocol Performance

To illustrate the effectiveness of the proposed routing algorithm, in this section the results of BNR application on MANETs in a simulated environment are presented. The performance of the BNR is presented in comparison to another established routing algorithm, the DSR.

The simulation environment consists of 100 mobile nodes with different flows of constant and variable bit rates between 20 of the nodes. DSR is used as the routing algorithm between these nodes; and throughput and delay, during the 100 seconds of simulation, are recorded for one of the flows. The simulation is repeated by using BNR as the routing protocol between the same two nodes. The results can be found in Figures 18.7 and 18.8 for throughput and delay, respectively. It is clear that BNR achieves better throughput and shorter delay than DSR during the simulation.

Next, end-to-end delay and throughput are presented for different mobility and different numbers of nodes in the network. These results are presented in Figures 18.9 and 18.10. BNR can achieve higher throughput and shorter delay than DSR due to higher quality links in the paths in different mobility and node density conditions. Also, from the figures it can be seen that BNR is more resilient to mobility; the rate of throughput decrease is smaller as the mobility increases when BNR is used. Results for delay are similar to the throughput results. Again BNR looks more resilient to mobility than DSR. Also, there are no significant BNR performance differences between different numbers of node scenarios in the network.

Routing overhead is defined as the sum of all the routing control packets like Route Request (RREQ), Route Reply (RREP), Route Reply Error (RRER), etc. In Figure 18.11, the number of control packets corresponds to number of packets in the 100-sec simulation duration. BNR uses

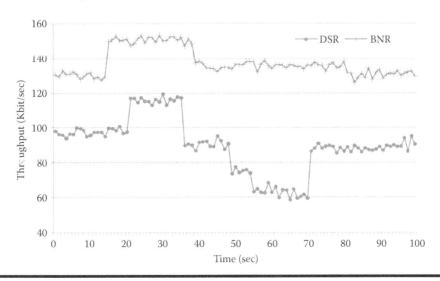

Figure 18.7   Throughput fluctuation between two nodes for DSR and BNR.

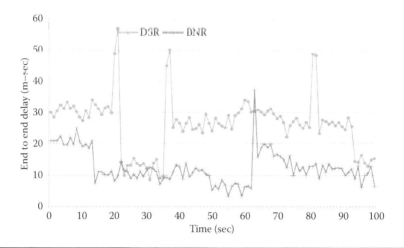

Figure 18.8 Delay fluctuation between two nodes for DSR and BNR.

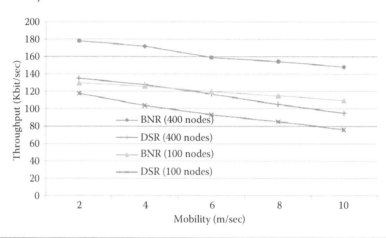

Figure 18.9 Throughput versus mobility for different number of nodes for DSR and BNR.

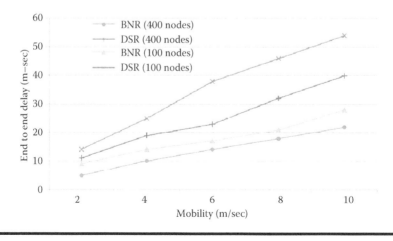

Figure 18.10 End to end delay versus mobility for different number of nodes for DSR and BNR.

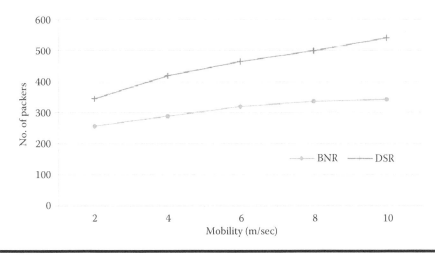

Figure 18.11   Routing overhead of DSR and BNR versus mobility.

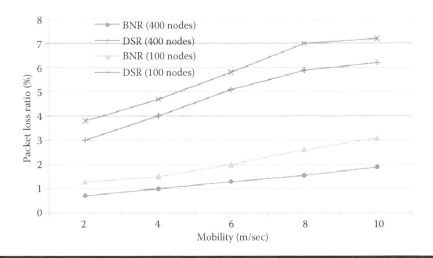

Figure 18.12   Packet loss ratio versus mobility for different number of nodes for DSR and BNR.

the same signaling as DSR, but from the results it can be observed that routing overhead is significantly reduced. These results are due to the fewer route discovery attempts when BNR is used because of the selection of paths that last longer.

Finally, the packet loss ratio is examined in Figure 18.12. Excepting the quality of the selected path, packet loss ratio reflects the amount of disruption in the communication, as a result of route breakages.

18.8 Protocol Performance When Metrics are Missing

One of the advantages of a BN as the routing decision algorithm is the flexibility it has with missing information. In this section, the performance of the routing protocol is investigated when various metrics are missing.

The first scenario that is presented assumes that the only available metric is the number of hops. Instead of using the path with the fewer hops, BNR will choose the path that is more likely to have stable performance according to the amount of hops. In Tables 18.4 and 18.5, the throughput and delay average performance is presented for DSR, BNR with hop count metric only, and BNR with all metrics available. BNR can achieve better throughput and delay than DSR even when the only available metric information is the hop count.

The delay metric can be measured easily but sometimes it is not possible to have a valid value quickly enough. When a delay measurement is not considered valid, it can be omitted in the BN. By observing the BN, we can see that transmission efficiency, CPU and buffer usage, and hop count are related to the delay value. Hence, when delay is not available, it can be estimated by the other three values. The results for this case can be found in Table 18.6.

In some cases, the cross-layer information can be inaccurate. For example, physical layer can return SNR measurements without using enough samples. Similar mobility measurements that are based in a GPS technology can be acquired when the GPS signal is low.

Here, we investigate the protocol performance when wrong cross-layer information is supplied. As can be seen in Tables 18.6 and 18.7, a wrong measurement of delay can produce worse results than a missing one.

Table 18.4   Average Throughput Comparison between DSR, BNR with Hop Count Metric Only, and BNR with All Metrics Available

Average Throughput (Kbit/Sec)		
DSR	*BNR (Hop Count)*	*BNR (Full)*
90.56	113.45	138.06

Table 18.5   Average Delay Comparison between DSR, BNR with Hop Count Metric Only, and BNR with All Metrics Available

Average Delay (mSec)		
DSR	*BNR (Hop Count)*	*BNR (Full)*
26.61	19.57	12.5

Table 18.6   Average Throughput Comparison When BNR is Operating with Full Information, without Mobility, and Full but Inaccurate Mobility

Average Throughput (Kbit/Sec)		
BNR (Full)	*BNR (No Mobility)*	*BNR (Inaccurate Mobility)*
138.06	130.89	121.424

Table 18.7 Average Delay Comparison when BNR is Operating with Full Information, without Mobility, and Full but Inaccurate Mobility

Average Delay (msec)		
BNR (Full)	*BNR (No Mobility)*	*BNR (Inaccurate Mobility)*
12.5	14.5	19.76

18.9 Discussion and Future Development

Many routing metrics have been proposed by researchers. It is difficult to combine these metrics in order to provide logical conclusions about the success of the path. An efficient and adaptive method to process the information from these numerous metrics for a routing path and give inference about the eligibility of the path is introduced here. Improved routing metrics in the future promise dramatic improvement of BNs' performance.

The BN was designed by using statistical data from the simulation environment. More accurate conclusions can be drawn by using real world statistical data. In that case, the BN must be redesigned using the new statistical data. The same can happen if this method is used for different types of ad hoc networks like mesh networks.

A further improvement that is worth examining is the use of soft evidence in the BN. Soft evidence is not conclusive. For example, we may get unreliable evidence that an event occurred, which may raise our belief in that event but not to the point where we would consider it certain. Thus, in the case of inaccurate cross-layer information, the performance of the protocol can be increased.

Considering the scalability of our proposed system, the protocol can suffer from the same scalability problems that DSR has. The focus of this work is to propose a decision algorithm solution and not to propose a mechanism for the rest of routing operations. However, the BN based routing decision mechanism does not have any scalability issues and can be integrated with other routing protocols which improve the scalability of DSR–like Modified Dynamic Source Routing (MDSR), as proposed in [28].

The proposed routing decision algorithm can be modified for heterogeneous networks. In this case, more parameters in the BN can be added. These parameters can be the technologies that are used in the links of a path assuming that different technologies provide different services and have different performance.

An interesting work for the future is how the integration in the protocol of a resource reservation mechanism can help to improve the performance. Also, modifications in the protocol can take advantage of the possibility of multiple route usage. The discovered paths can have a similar probability of stable performance. Every time more than one path can be used, from the ones that have high success estimation.

References

1. Charles E. Perkins, and Elizabeth M. Royer. 1999. Ad-hoc on-demand distance vector routing, *Second IEEE Workshop on Mobile Computer Systems and Applications,* New Orleans, LA, pp. 90–100.
2. D. B. Johnson, D. A. Maltz. 1996. Dynamic source routing in ad hoc wireless networks, *Mobile Computing.* T. Imielinski, H. Korth eds. Kluwer. ch. 5, pp. 153–181.

3. C. E. Perkins and P. Bhagwat. 1994. Highly dynamic destination-sequenced distance-vector routing (DSDV) for mobile computers. Conference on Communications Architectures, Protocols and Applications (SIGCOMM '94). In *Proceedings of the Conference on Communications Architectures, Protocols and Applications*, London, UK, Vol. 24, No. 4, pp. 234–244.

4. T. Clausen, and P. Jacquet. *Optimized Link State Routing Protocol (OLSR). Optimized Link State Routing Protocol (OLSR)*, RFC Editor, 2003.

5. S. J. Douglas, Daniel Aguayo, John Bicket, and Robert Morris. 2003. A high-throughput path metric for multi-hop wireless routing. *MobiCom '03: Proceedings of the 9th Annual International Conference on Mobile Computing and Networking*, pp. 134–146.

6. M. D. Yarvis, W. S. Conner, L. Krishnamurthy, A. Mainwaring, J. Chhabra, and B. Elliott. 2002. Real-world experiences with an interactive ad hoc sensor network. *Proceedings International Conference on Parallel Processing Workshops*, pp. 143–151.

7. Baruch Awerbuch, David Holmer, and Herbert Rubens. 2004. High throughput route selection in multi-rate ad hoc wireless networks. *Lecture Notes in Computer Science*, Springer, Berlin, Germany, Volume 2928, pp. 201–205.

8. C. -K. Toh. *Wireless ATM and Ad-Hoc Networks: Protocols and Architectures*. XXX: Kluwer Academic Publishers, 1996.

9. R. Dube, C. Rais, K. -Y. Wang, and S. Tripathi. 1997. Signal stability-based adaptive routing (SSA) for ad hoc mobile networks. *IEEE Personal Communications*, Vol. 4, No. 1, pp. 36–45.

10. M. Sheng, J. Li, and Y. Shi. 2003. Routing protocol with QoS guarantees for ad-hoc network. *Electronics Letters*, 39 (1), pp. 143–145.

11. R. Sivakumar, P. Sinha, and V. Bharghavan, V. 1999. CEDAR: a core-extraction distributed ad hocrouting algorithm. *IEEE Journal on Selected Areas in Communications*, Vol. 17, No. 8, pp. 1454–1465.

12. R. Gupta, Z. Jia, T. Tung, and J. Walrand. 2005. Interference-aware QoS routing (IQRouting) for ad-hoc. *IEEE GLOBECOM '05 Global Telecommunications Conference*, pp. 2599–2604.

13. Y. Yang, and R. Kravets. 2005a. Contention-aware admission control for ad hoc networks. *IEEE Transactions on Mobile Computing*, Vol. 4, No. 4, pp. 363–377.

14. I. Rubin, and Y. –C.Liu. Link stability models for QoS ad hoc routing algorithms. 2003. *Vehicular Technology Conference VTC 2003-Fall*. 2003 IEEE 58th, Vol. 5, pp. 3084–3088.

15. Alicia Trivino-Cabrera, Jorge Garcia-de-la-Nava, and Eduardo Casilari. 2010. Identification of stable links in MANETs. *IEEE 30th International Conference on Distributed Computing Systems Workshops (ICDCSW)*, pp. 276–281.

16. Y. –C. Tseng, Y. –F. Li, and Y. –C. Chang. 2003. On route lifetime in multihop mobile ad hoc networks. *IEEE Transactions on Mobile Computing*, Vol. 2, No. 4, pp. 366–376.

17. H. Shen, B. Shi, L. Zou, and H. Gong. 2003. A distributed entropy-based long-life QoS routing algorithm in ad hoc network. *IEEE CCECE 2003 Canadian Conference on Electrical and Computer Engineering*, Vol. 3, 2003, pp. 1535–1538.

18. N. Nikaein, C. Bonnet, and N. Nikaein. Harp-hybrid ad hoc routing protocol. Institut Eurecom Symposium Telecommunications, 2001.

19. M. Yu, A. Malvankar, W. Su, and S. Y. Foo. 2007. A link availability-based QoS-aware routing protocol for mobile ad hoc sensor networks. *Computer Communications*, Vol. 30, No. 18, pp. 3823–3831.

20. Z. Guo, and B. Malakooti.2007. Energy aware proactive MANET routing with prediction on energy consumption. *WASA 2007 International Conference on Wireless Algorithms, Systems and Applications 2007*, pp. 287–293.

21. M. Woo, S. Singh, and C. S. Raghavendra. 1998. Power-aware routing in mobile ad hoc networks. *MobiCom '98 Proceedings of the 4th Annual ACM/IEEE International Conference on Mobile Computing and Networking*, pp. 181–190.

22. I. Stojmenovic. 2002. Position-based routing in ad hoc networks. *IEEE Communications Magazine*, Vol. 40, No. 7, pp. 128–134.

23. A. Rao, S. Ratnasamy, C. Papadimitriou, S. Shenker, and I. Stoica. 2003. "Geographic routing without location information. *Proceedings of the 9th Annual International Conference on Mobile Computing and Networking*, pp. 96–108.

24. K. Seada, M. Zuniga, A. Helmy, and B. Krishnamachari. 2004. Energy-efficient forwarding strategies for geographic routing in lossy wireless sensor networks. In *ACM Sensor Systems (SenSys)*, pp. 108–121.
25. S. Lee, B. Bhattacharjee, and S. Banerjee. 2005. "Efficient geographic routing in multihop wireless networks. *Proceedings of the 6th ACM International Symposium on Mobile Ad Hoc Networking and Computing.*, pp. 230–241.
26. Y. Yang, J. Wang, and R. Kravets. 2005b. Designing routing metrics for mesh networks. *Proceedings of the IEEE Workshop on Wireless Mesh Networks (WiMesh)*.
27. I. Akyildiz, and X. Wang. *Wireless Mesh Networks*. Chichester, UK: Wiley, 2009.
28. M. Tamilarasi, V. S. Sunder, U. Haputhanthri, C. Somathilaka, N. Babu, and S. Chandramathi. 2007. Scalability improved DSR protocol for MANETs. *International Conference on Computational Intelligence and Multimedia Applications*, Vol. 4, pp. 283–287.

Chapter 19

Adaptive Flow Control in Transport Layer Using Genetic Algorithm

Ilias Kiourktsidis, Jonathan Loo, and Grigorios Koulouras

Contents

19.1 Introduction

The basic difference between MANETs and other ad hoc network types is the frequent topology changes caused by node mobility. An area of concern is guaranteeing quality of service in such a constantly changing communication environment. A known term in adaptive network protocol design called *self-modifying protocol* (SMP) is aiming to modify its network communication parameters at run time such that the protocol can adapt to the changing communication environment and user requirements on the fly. In this chapter, a self-evolvable protocol is presented using the principle of SMP with genetic algorithms. In essence, this approach applies the genetic algorithm in the design of an evolvable self-modifying protocol (ESMP) which controls the flow of the data in a way that it will improve QoS that applications experience. Initially, the protocol demonstrates how the size of the packets, combined with the time interval between the packets, can affect the performance of the protocol. How the genetic algorithm can discover the optimum values for these parameters is also discussed.

Genetic algorithms (GAs) have received increased interest in recent years, particularly with regard to the manner in which they may be applied for practical problem-solving. The main reason someone would adopt GA as an optimization method is due to its resilience in a changing environment. Traditional methods of optimization are not resistant to dynamic changes in the environment, and often require a complete restart to provide a sustainable solution. In contrast, GA can be used to adapt a system to find out a solution to dynamic changes. The available population of evolved solutions provides a basis for further improvement; in most cases, it is not necessary to reinitialize the population randomly. Furthermore, hardware implementations of GA are extremely fast and they are not very sensitive to network size [1].

19.2 The Need for an Adaptive Flow Control

Initially in network communications, flow control was the mechanism for assuring that a transmitting device does not overwhelm a receiving device with data. Without flow control, buffer overflow may occur at the receiver while it is processing old data resulting in dropped packets and poor communication. A flow control mechanism also proved useful in cases where congestion occurred in the network. In such cases, the flow control mechanism is able to provide relief to the network by reducing the amount of transmitted packets. In MANETs, the basic concern is the network capacity, which changes dynamically. The flow control mechanism is responsible for deciding when to reduce the flow and by how much. From the existing transport layer protocols, Transport Control Protocol (TCP) is the most widespread protocol that detects the need to reduce the flow by sensing the loss of packets. Unfortunately, it reduces the flow to a very low speed whereas a slighter reduction might be enough to solve the problem. Additionally, TCP only addresses the time interval between packets; adjusting other characteristics (like packet size) might be a better approach for each specific case.

19.2.1 How to Control the Flow

Most of the flow control designs on transport layers rely on controlling the rate of the transmitted packets on the network by the use of packets of a fixed size. Sometimes a flow control system might perform better if it transmits the same amount of data but in variable sizes of packets. For instance, the protocol is able to transmit 100,000 bytes of data by using 100 packets of 1000 bytes each, or 500 packets of 200 bytes each. More packets means more overhead but it implies less data lost if a packet is dropped. Which one will perform better always depends on the conditions of the network, and is relative to the needs of the application. In order to make this clear, this section explores some more examples.

Smaller data packets are susceptible to jitter, while larger packets are traversing the network through the same interfaces or occupying the media in neighbor nodes. On such occasions, the jitter possibly can be reduced by sending larger packets. In order to decide if this is the right action, the protocol must be aware of the reason that the application experiences deficient jitter. Perhaps at layer 2, a link fragmentation and interleaving (LFI) mechanism is operating which fragments large packets when there are signs of congestion in the link. Maybe in the next hop this kind of mechanism does not exist or is operating with different parameters and different thresholds. Finally, in cases where the protocol decides to send larger packets in a noisy environment, the probability to lose more data becomes significant at once, making the delay and throughput worse and reducing the quality in general.

The big questions are how can the protocol be aware of all these parameters and how can it evaluate the significance of each one with respect to the application requirements. In a MANET where the environment is changing all the time, things are becoming more complicated.

A genetic algorithm (GA) because of its nature is an optimization algorithm that does not need to collect information about the condition of the network environment but the information on how well the individuals performed in it. First, it randomly tries a population of individuals and evaluates them on how well they perform in the existing environment (calculating their fitness). Then, it selects the individuals that performed better and uses them to produce the next generation of individuals that will usually perform even better. In order to correlate the usual GA problems with the data flow control problem, we refer to the network itself and its condition as environment, and every set of parameters—like size of packets, time interval between packets, etc., as individuals. The evaluation of each individual is done by measuring the performance in relation to the application requirements.

19.3 Genetic Algorithm

The genetic algorithm (GA) is inspired from the natural process of evolution and natural selection, and belongs to the larger class of evolutionary algorithms. GA is a search heuristic, and is used in optimization and search problems. The basic operators of GA are the same as with the natural evolution—like selection, mutation, and crossover.

The basic element in a genetic algorithm is a population of strings which are called *chromosomes;* this population is the genotype of the genome. Each chromosome encodes candidate solutions for the optimization problem. Other names that are used for chromosomes are individuals or phenotypes. The GA usually starts with a random population of random individuals. Then, the fitness of each individual in this initial population is evaluated. Afterwards, some of the individuals are selected by using a method involving the fitness evaluation that has been modified by combining the individuals and mutating some of their genes to form a new population. The new population is then used in the next iteration of the same procedure. The individuals in each new population that is created can usually score higher fitness in the evaluation. In most cases, the algorithm terminates when a satisfactory fitness level has been reached for the population, or a maximum number of generations has been produced. In the latter case, a satisfactory solution may not have been found. A typical genetic algorithm requires a genetic encoding of the solution domain and a fitness function to evaluate the solution domain.

The structure of the representation of a solution is usually a group of bits. Groups of other types (like alphanumerical) can be used in essentially the same way. Each gene is represented by one or more bits (or alphanumerical) and all genes have the same fixed size in each individual. The fitness function is defined over the genetic representation, is a measurement of the quality of the represented solution, and is always problem dependent.

Once the genetic representation and the fitness function are defined, GA is ready to run. First, it randomly creates the initial population of solutions and then improves it through recurring application of mutation, crossover, inversion, and selection operators. Each operator is explained in more detail as follows:

1. **Initialization:** During initialization, many random individual solutions are generated. The size of the population depends on the nature of the problem and typically contains several hundred individuals in order to cover all the range of the search space. For a faster convergence, the initial individuals may be seeded in search space areas where optimal solutions are likely to be found.

2. **Selection:** In order to proceed into the next population, some of the individuals in the current population need to be selected. The fittest individuals are more likely to produce good quality offspring. Individual solutions are selected through a fitness-based process, where the fitter individuals are the most likely to be selected. Some more complicated selection schemes can be used but are always based on the fitness function.

3. **Crossover:** The next step is to use the individuals that are selected to produce the next generation. This procedure is called *crossover*. At the same time, mutation can be used to help the algorithm to avoid local maximal and "jump" into unexplored areas in the search space. Like selection, there are several ways to do the crossover. The basic idea is that two parents give a part of their genome and the combination of the two parts produces the child. Thus, the child shares many of the characteristics of its parents. Crossover is repeated again and again until a new population of solutions of appropriate size is generated. Although crossover methods that are using two parents are more compatible with nature, some proposals suggest that more than two parents can produce higher quality chromosomes. These processes will result in the next generation of individuals being different from the initial generation but with increased average fitness since the best individuals from the previous generation were used for breeding.

4. **Termination:** The repetitive generation production in the genetic algorithm has to come into an end. Usually, the termination condition that has to be reached for the GA to come into an end is the discovery of a solution that satisfies some criteria, when the successive iterations no longer produce better results, after a certain amount of time since the start is passed, or a fixed number of generations is reached. There are other times where the GA is running all the time as a system is functioning in order to fine-tune some system parameters; it then terminates along with the system.

5. **Genetic Algorithm Pseudo Code**
   Choose the initial population of individuals. Repeat until termination condition is true.
   　　Evaluate the fitness of each individual in that population.
   　　Select the individuals for reproduction.
   　　Crossover and mutation operations to give new offspring.
   The fittest in the last population is the solution.

19.4 GA Usage in Computer Network Designs

Most of the work in the area of computer networks using GA was about optimizing the routing in the networks. A genetic algorithm approach to the shortest path routing problem was used by Ahn and Ramakrishna [2] by working with variable-length chromosomes. Dengiz et al. [3] showed how a genetic algorithm with specialized encoding, initialization, and local search operators can optimize the design of communication network topologies. Another problem in the area of computer networks is the degree-constrained minimum spanning tree problem. Chou et al. [4] experimented with several different encoding crossover and mutation methods in different network sizes and found that encoding has the greatest impact on solution quality.

These researchers tried to experiment with GA in environments that are constant; however, in real-world situations, environments are not always constant. Thus, the experimental outcomes may not be relevant and accurate.

19.5 ESMP Architecture

The ESMP [5] like any other transport layer protocol is using the network layer service to communicate with any other node in the network and implements procedures like packet acknowledgement, etc., that assist the communication. Then it controls the size of the packets and the rate that transmits the packets according to the optimization algorithm decisions. Finally, it measures the performance of the communication and reports it back to the optimization algorithm (in our case, GA). This section will first describe the communication protocol and then the optimization algorithm using GA. Another concern in the design of the protocol was to avoid any stop-and-wait kind of behavior because this is a QoS aware protocol.

19.5.1 Description of the Transport Protocol

All network protocols are using a specific, designed header to accompany the data; this protocol is not an exception. The header can be found in Table 19.1 and consists of:

- IDs of the source and the destination ports like TCP and UDP.
- Sequence number: holds the number of the group of 16 packets that use the same configuration (belongs to the same individual).
- Delivery mask: a mask that has only one bit set to define the number of the packet in the group and the other bits are clear.
- Packet size: holds the size of the payload.
- Four flags: two of them are reserved and the other two are ACKp and ACKg flags that define if this is an acknowledgement for a packet or for a whole group. If ACKp is set, then the sequence and delivery mask fields define the packet to which the ACK belongs. When ACKg is set, then the delivery mask is all clear, and the sequence field holds the number of the group to which the ACK belongs. Data payload contains the measured values for throughput, end to end delay, jitter (latency variation), and packet loss (percentage) for the current group.

The flow charts of the protocol logic are presented in Figures 19.1 and 19.2. First, the initialization procedure is responsible for the creation of an initial population of 10 individuals.

Table 19.1 Packet Schematic Layout

Bit	0–15	16–31		
0	Source (16 bits)	Destination (16 bits)		
32	Sequence (32 bits)			
64	Delivery Mask (16 bits)	Packet Size (10 bits)		Flags
96	Data Payload			

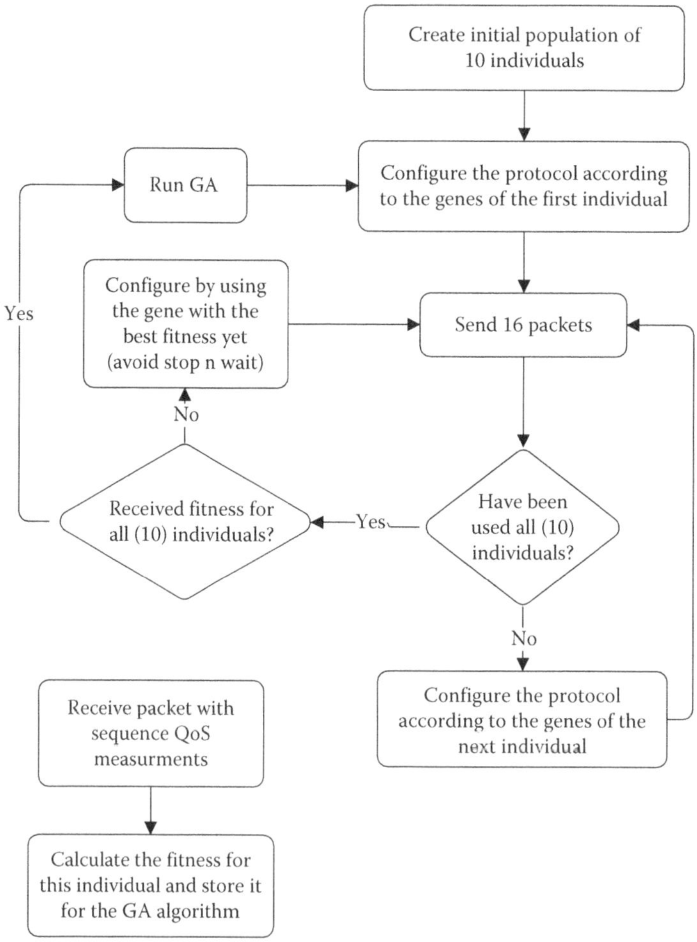

Figure 19.1 Flow diagram of ESMP instance in the sender node.

The first 16 packets are transmitted according to the genes of the first individual. The following groups of 16 packets are transmitted by using the configuration that is derived from the genes of the rest of the individuals until the first generation is completed. In the meantime, the receiver measures the delay, jitter, and throughput, and counts the lost packets for every group. These measurements are transmitted back to the sender. The sender receives these measurements and calculates the fitness for this individual (this corresponds to the measured group of 16 packets). If

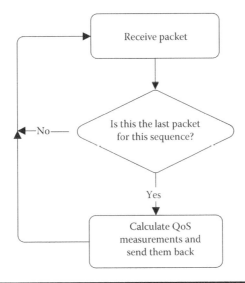

Figure 19.2   Flow diagram of ESMP instance in the receiver node.

all 10 individuals of the current population are used and the protocol has not calculated the next population yet, then in order to avoid a stop-and-wait behavior of the protocol, it continues the communication by using the configuration of the fittest individual that is available. When the fitness evaluation for all individuals is available, GA runs and produces the next generation.

19.5.2 Genetic Information Encoding

The genetic information is encoded in chromosomes that are a combination of genes. Genes are represented by using a binary system. Each chromosome represents a solution (different configuration in our problem).

ESMP problem solution uses a six-bit chromosome that is the concatenation of two three-bit genes, one for the packet size and the other for the time interval, as defined in Table 19.2 and Table 19.3, respectively. Note that the inter-packet gap is analogous to the Maximum Transfer Unit (MTU) packet transmission time. In the following example, one unit corresponds to 0.009s. When the chromosome equals 111000_2, the protocol will transmit packets of 1500 bytes at every 0.009s. In this case, the throughput is calculated as 1.33 Mbps, and that is the highest transmission rate for this configuration. In another example, when the chromosome equals 000111_2, the transmission rate would be 6.35 Kbps.

19.5.3 Population

After understanding how the chromosomes are represented, we have to deal with the population selection. There are two issues that need consideration: 1) The size of the population that is appropriate for the problem defined earlier, and 2) the actual individuals in the initial population. Intensive research has focused on how to calculate the population size [2], [6]. Large populations can spread more in the search space and produce better solutions, but on the other hand, they increase the computational cost. In the problem mentioned before, a larger population translates to longer periods of inadequate communication.

Table 19.2   Packet Size

Encoding Binary	Representative Value
000	100 bytes
001	300 bytes
010	500 bytes
011	700 bytes
100	900 bytes
101	1100 bytes
110	1300 bytes
111	1500 bytes

Table 19.3   Time Interval (0.009 sec for each unit)

Encoding Binary	Representative Value
000	1 unit or 0.009s
001	2 units or 0.018s
010	4 units or 0.036s
011	6 units or 0.054s
100	8 units or 0.072s
101	10 units or 0.09s
110	12 units or 0.108s
111	14 units or 0.126s

In [6], the author suggests that every point in the search space should be reachable from the initial population by crossover only. This means that there must be at least one instance of every allele at each locus in the whole population. When the initial population is generated by a random sample with replacement and the chromosomes are binary encoded, the probability, for this requirement to be satisfied, can be expressed as [6]:

$$P = \left(1 - \left(\frac{1}{2}\right)^{N-1}\right)^{L} \tag{19.1}$$

where N is the population size and L is the number of the encoding bits of a chromosome; in our example, L = 6 bits.

The graph of Equation 19.1 can be seen in Figure 19.3; for detail purposes, selected values can be found in Table 19.4.

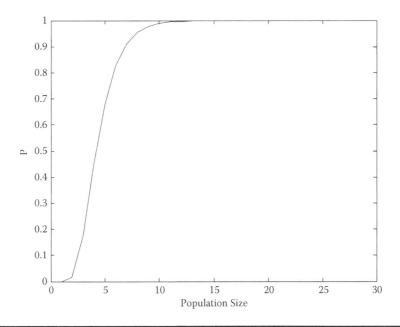

Figure 19.3 Population size versus probability that every point in search space is reachable from the initial population by crossover only.

Table 19.4 A Sample of Probability versus Population Size

Population Size (N)	Probability
6	0.8266
7	0.9098
8	0.9540
9	0.9768
10	0.9883
11	0.9942
12	0.9971
13	0.9985
14	0.9993

19.5.4 Fitness Evaluation

The fitness function computes a fitness value for each individual in the population. As mentioned before, the parameters that the protocol tries to reach a certain target value defined by the application are throughput, delay, jitter, and data loss. The more the measured values of these parameters are close to the target values, the better the fitness value will be.

The fitness for the delay when d is higher than D is calculated by:

$$F_d = 10 \cdot \frac{D}{d} \tag{19.2}$$

For jitter, when j is bigger than J

$$F_j = 10 \cdot \frac{J}{j} \tag{19.3}$$

For throughput, when t is smaller than T

$$F_d = 10 \cdot \frac{t}{T} \tag{19.4}$$

For lost bytes, when l is bigger than L

$$F_i = 10 \cdot \frac{L}{l} \tag{19.5}$$

Where D, J, T, and L are the target delay, jitter, throughput, and percentage of lost bytes, respectively, while d, j, t, and l are the measured values for the same parameters. The return fitness value would be higher when the measured values are close to the target values. In any other case that the measured value (e.g., d) is better than the targeted value (e.g., D), then the fitness value (e.g., F_d) is set to a maximum value of 10.

The total fitness for the individual is calculated by the summation of all partial fitness, multiplied first by a weight of significance w, as expressed

$$F = w_d \cdot F_d + w_j \cdot F_j + w_t \cdot F_t + w_i \cdot F_i \tag{19.6}$$

where the weight is used to regulate the priority for each metric and give a higher or lower priority over the other metrics.

19.5.5 Individual Selection for Reproduction

Selection algorithm is the procedure of the GA where individuals from the population are chosen for breeding. The goal is to improve the average quality of the population by transferring the fittest chromosome to the next generation. In that way, the GA focuses on areas of the search space that are more promising. According to the evolution theory of Darwin, the best ones should survive and create new offspring. The selection method that someone can adopt and use in a system can affect the performance of the GA in various ways. Hence, a selection of the fittest individuals helps to reach the equilibrium of the population faster. On the other hand, this leads to suboptimal solutions because the genetic diversity is sacrificed.

The ESMP protocol is ideal for use in environments that change continuously along with the search space. In such types of environments, the optimal solutions may last for only short periods; hence, the sub-optimal solutions sometimes are a better choice. On the other hand, because of the unsteady environment, GA needs genetic diversity to explore the new search space quickly and not stay locked to an area awaiting mutation to give the required genetic diversity.

There are many methods of selecting the best chromosomes. Some of them are:

- **Rank selector:** this is the simplest method as it simply picks the best member of the population every time.
- **Roulette wheel selector:** this method picks an individual based on the fitness score relative to the rest of the individuals. Individuals with higher scores are more likely to be selected. The probability for an individual to be chosen equals the fitness of the individual divided by the sum of all fitness values in the population.
- **Tournament selector:** this method uses the roulette wheel method for a few of the individuals in the population and then picks the one with the highest score. Then it runs a new tournament until it selects all individuals it needs. The tournament selector typically chooses higher valued individuals more often than the roulette wheel selector.
- **Deterministic sampling (DS) selector:** this method uses a two-staged selection procedure. In the first stage, each individual's expected representation is calculated. A temporary population is filled using the individuals with the highest expected numbers. Any remaining positions are filled by first sorting the original individuals according to the decimal part of their expected representation, then selecting those highest on the list. The second stage of selection is uniform random selection from the temporary population.
- **Stochastic remainder sampling (SRS) selector:** this method consists of two stages. In the first stage, a temporary population is created by using the individuals with the highest expected representation. Any fractional expected representations are used to give the individual more likelihood of filling a position in the population. For example, an individual with an expected representation of 1.2 will have the first position and then a 20% chance of a second position. The second stage of selection is uniform random selection from the temporary population.
- **Uniform selector:** any individual in the population has a probability, p of being chosen where p is equal to 1 divided by the population size.

Table 19.5 gives a comparison of various selection methods in terms of algorithm convergence speed and in what way and how much the genome diversity is affected. The results of the performance of each selection method are retrieved by using the ESMP in a MANET environment.

Table 19.5 Selection Methods Comparison for Mutation Rate Zero

Selection Method	Generation Number	Fitness	Diversity
Rank Selector	2	75	0.2
Roulette Wheel	8	61	0.08
Tournament	7	71	0.1
DS Selector	6	55	0.25
SRS Selector	10	65	0.07
Uniform	3	52	0.3

The diversity of the entire population is the average of all the individual diversities. If every individual is completely different from every other, the population diversity is higher than zero. If they are all the same, the diversity is zero. The probability of mutation is set to zero in order to make sure that the diversity is not caused by mutation.

The ideal selection algorithm is able to converge very fast, achieve the maximum fitness, and the diversity of the population must be high. Usually, in optimization problems that use GA, the diversity is not such an important parameter after GA execution. Without a doubt, though, it is crucial for the initial population. Once the optimum solution is found, the algorithm terminates and the problem is already solved. On the other hand, in MANETs, the environment always changes and the diversity of the population must be high enough for the GA to be efficient and continue to adapt repeatedly after any change in the environment. Since the rank selector has a superior performance, it is adopted by ESMP.

19.5.6 Crossover

After successfully completing the selection of individuals, crossover is used to create the next generation population by choosing a locus (or more than one locus in multipoint crossover) and then swapping the remaining alleles from one parent to the other.

In the previously mentioned example, there are two genes in each chromosome—so the only choice is to use single point crossover. In other words, when crossover takes place, the child will take one gene from one parent and the other gene from the other parent, or in real problem terms, will take the packet size from one parent and the time interval between packets from the other one. Another parameter for crossover is the probability for crossover to occur. Sometimes no crossover occurs and the parents are copied directly to the new population.

19.5.7 Mutation

Mutation is a simple operation in GA and, consequently, in ESMP protocol. Based on a low probability, a random change in the value of a gene in the chromosome will occur. The purpose of mutation in GAs is to allow the algorithm to avoid local minima by preventing the population of chromosomes from becoming too similar to each other, consequently slowing evolution. Mutation is, however, vital to ensure genetic diversity within the population. On the other hand, there is a risk that a gene will change drastically and degrade the performance of protocol.

19.5.8 Repair Function

In the populations that are generated after crossover and mutation, there is always a chance that some of the individuals are not able to meet some of the application requirements. Given the required throughput, it is possible to find out if an individual is going to reach it. It is possible to calculate the maximum throughput that each configuration can achieve by multiplying the size and the number of the packets that are transmitted per second. Figure 19.4 presents the theoretical throughput for the whole search space.

Usually there are two methods to deal with infeasible chromosomes. The first method uses a repair function to find and remove the infeasible chromosomes, and the second one uses another function to punish them by giving them a penalty. By giving a penalty, the probability that the infeasible chromosome reproduces will be lower; but, the current generation will suffer from the consequences and there is the risk that they will remain in the next population. By using a repair

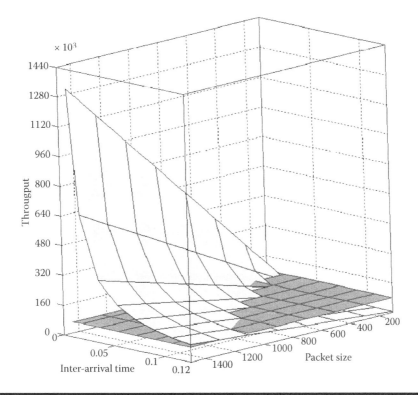

Figure 19.4   Theoretical throughput in the entire search space. The horizontal surface separates the throughput surface at 80Kbps as an example.

function, the infeasible chromosomes can be eliminated even before they will be used. By eliminating chromosomes, some of the genes possibly will be absent in future generations, except in the case where mutation recreates them. This will help the feasible genes dominate the population. This conceals a risk, though, because there might be some genes that give better results for delay and jitter but are eliminated because they were coexistent with other genes in infeasible chromosomes. The chance for those genes to appear in the population is very low.

In order for ESMP protocol to address this problem, it uses a mechanism that considers the importance of the throughput for the application. For instance, if the throughput is not crucial, the repair function will not remove a chromosome that is not too distant from the target.

19.6 ESMP Performance

The main goal of research and development of MANETs is to improve the performance of communication in different environments.

The protocols operating in a highly dynamic MANET environment must be designed to be self-configured, rapidly adapt as the network condition changes, and the target requirements must be posed from the applications. ESMP is a promising protocol for applications that need long-term connections in such types of environments.

In the following scenario, there are many nodes that are constantly in movement in an ad hoc network and some of them are producing traffic. This dynamic environment consists of 50 wireless nodes with one IEEE 802.11 interface. All the nodes are moving randomly in an area of 500 × 500 meters and 10 of them are generating Constant Bit Rate (CBR) or Variable Bit Rate (VBR) traffic. Because of the movement, the links are always changing along with the multi-hop communication paths. The different traffic connections sometimes cross each other and sometimes they do not. Sometimes the neighbor nodes are sending traffic and occupying a proportion of the communication medium. It is not difficult to understand that the factors that affect the availability of the network, the load at the nodes, and the quality of the communication in general (the network environment) are many and in constant change.

Two other nodes are using the ESMP with GA protocol to exchange data between them. In the 100-second duration of the experiment, there were 2785 total link changes for all nodes. Through the communication of the two ESMP nodes 178 route changes took place. Not all of the route changes affect the performance dramatically.

The purpose of this experiment is to investigate how the GA can follow the changes in the environment and adapt so the application requirements can be met.

Next, a comparison of communication's performance in terms of delay, when GA is used or not, is presented. In the case that GA is not used, fixed length and fixed interval of packets are used for the whole duration of the experiment and the metric delay observed. Throughput is always fixed and all three configurations are chosen to produce the same throughput. Specifically, the three configurations are:

- Configuration No.1: 900 bytes – 54 msec
- Configuration No.2: 1500 bytes – 90 msec
- Configuration No.3: 300 bytes – 18 msec

All three configurations produce the same 133 kbps throughput as an example. It was necessary to use the same throughput, in order to be able to compare more accurately the delay and other metrics. Finally, there is a comparison of these configurations with ESMP with target throughput of 133 kbps.

Figure 19.5 presents the delay of the three fixed configurations and the GA enabled ESMP. Table 19.6 shows that the GA has the lowest average delay. The delay outcomes from Configuration

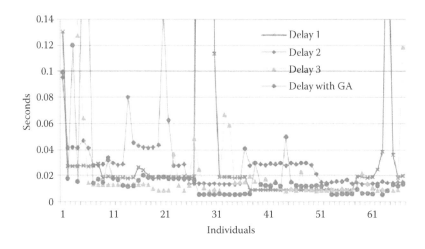

Figure 19.5 Delay for the three configurations and when GA is used.

Table 19.6 Average and Standard Deviation Values for Delay

	Average	*Standard Deviation*
Configuration No.1	36 msec	108 msec
Configuration No.2	27 msec	19 msec
Configuration No.3	19 msec	20 msec
GA Enabled	13 msec	9 msec

1 and 2 are the worst among all others. None of these three fixed configurations is the best solution. Clearer performance comparisons can be achieved from Table 19.6, which depicts the average and the standard deviation values of delay. It is observed that the GA self-configured packet size and interval achieves lower delay than any other fixed configuration.

19.7 Further BN Optimization and Integration with GA

In the previous chapter, a Bayesian network model for paths in MANETs was introduced. It has been shown how evidence is used to compute posterior probabilities for some hypotheses. The information for the evidence elements is cross layer information supplied mainly from lower layers. In this chapter, further investigation is made regarding the use of a BN to advise the other layers on how to configure some of their parameters, in order to improve the communication performance. The first part examines ways to support the transport layer and specifically the GA based self-modifying transport protocol. An extended version of the BN model that was introduced in the previous chapter is used to make the cross layer suggestions in order to improve the overall performance. The target again is to create a soft relationship between the layers without any critical dependency on the cross layer information. The next part illustrates the overall performance in QoS on MANETs by using both ESMP and BNR protocols that are introduced in this book. This section shows how each protocol adds up to the overall performance.

19.7.1 GA Initial Population Optimization

Genetic algorithm begins by creating a random initial population. As described previously, the size of the population plays an important role in the GA's performance. There is a trade-off between computational cost and search capabilities which affect the quality of the solution. To illustrate this phenomenon better, in Figure 19.6 a search space example is shown, along with ten random dots which represent ten random individuals in an initial population.

In Figure 19.6, the initial population contains ten individuals. This is the value of population size that, according to Section 0, is the minimum amount for the specific problem that satisfies the criteria that every point in the search space should be reachable from the initial population by crossover only. Note that all the individuals in the initial population are spread in the whole search space. The GA algorithm will select the individuals that display higher fitness, so the next

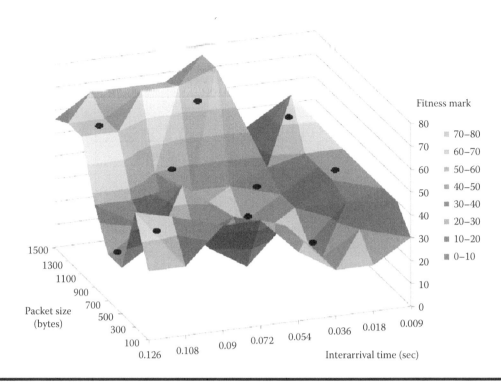

Figure 19.6 Search space example for the ESMP configuration, with ten random individuals (black dots).

population is expected to be gathered in a smaller portion of the search space. That portion of the search space will be in a higher fitness area.

It is clear that a higher population size can take up a larger area in the search space. Thus, more individuals result in a quicker and easier search. Usually, higher population size means computational cost but for ESMP it means a longer duration of inadequate communication. Notice that several of the initial individuals are placed in low fitness areas of the search space. Those individuals are translated to ESMP configurations that give bad performance. The experimental results show that ESMP is suffering from these bad quality individuals at the start of the communication.

If the algorithm knows approximately where the maximum fitness areas are, it can place the initial population on those areas. In that case, the start of the communication can be smoother.

19.7.2 BN Based Initial Population Selection

A modified version of the BN model that was introduced in Chapter 17 is used to give inference on which area of the search space the initial population should be deployed. The new BN model contains one additional node. The search space node has 16 states. Each state represents one area in the search space. The initial population will be created in a way that the state that receives the highest probability will have the most initial individuals. In other words, the probability distribution of the search space node will be used for the initial placement of the population.

For example, in Figure 19.7 the same search space can be seen from above divided in 16 areas. Areas like (A,1), (A,2), and (A,3) give the best fitness results; hence, it is preferable for most of the

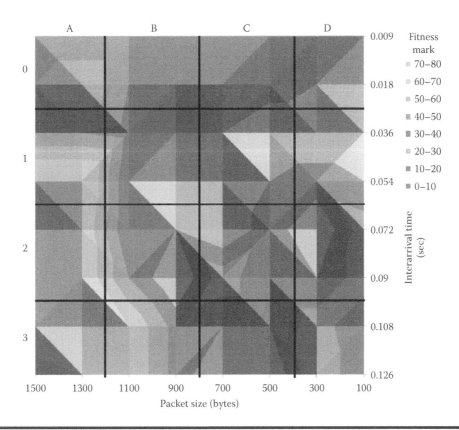

Figure 19.7   Search area divided by 16 smaller areas which are represented.

initial individuals to be placed on those parts of the search space. The opposite happens in areas like (C,1), (C,3), and (D,1) where individuals that may appear in those areas will result in bad performance.

Unfortunately, the search space is unknown. That is the reason, in the first place, why a search optimization algorithm like GA is used to explore it. BN can be used to give inference from statistical data on which of the 16 areas are more likely to have high fitness. The several metrics that the network layer collects for a routing decision can also be used for search area selection. For example, when the delay metric is observed to be high, then the selection of larger packets and a reduced rate might be beneficial. According to the statistical data, the BN will propose the initial configuration and GA will optimize and change it whenever it needs to adapt to network changes. As seen in Figure 19.8, the "initial population" node is directly related to the same metrics that, in turn, also directly affect the performance of a path.

19.7.3 Experimental Results

The use of the BN for the creation of the initial population in ESMP results in a smoother performance at the start of the protocol. In Figure 19.9, an example of the delay performance of the ESMP is shown, when a random population and when the BN suggested population are used. It is clear that the first 10 individuals of the first population have a higher performance.

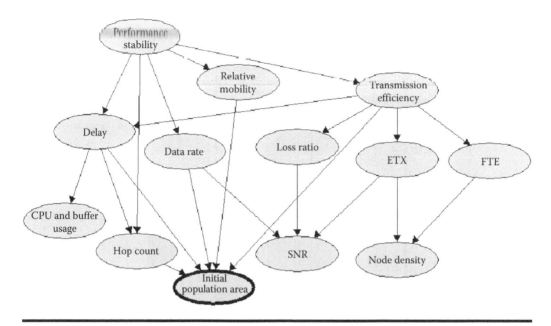

Figure 19.8 Bayesian network for initial individuals' placement selection.

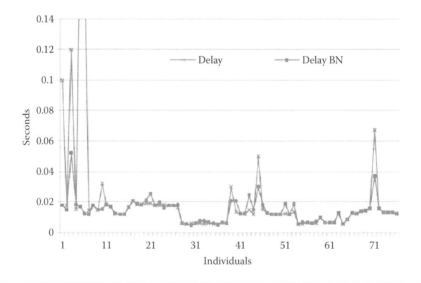

Figure 19.9 Comparison in the delay performance of ESMP with and without BN's initial population suggestion.

19.8 Overall Performance

As mentioned in the previous chapters, each of the protocol proposals targets an explicit area of improvement in the MANETs communications. The BNR is responsible for deciding which of the available paths is more stable to use for communication by using the best available cross layer information. In addition, ESMP utilizes the path, and is trying to improve the performance

Table 19.7 Average Throughput with and without ESMP over BNR

Average Throughput (Kbit/sec)		
DSR	BNR	with ESMP
90.56	138.06	145.86

Table 19.8 Average Delay with and without ESMP over BNR

Average Delay (msec)		
DSR	BNR	with ESMP
26.61	12.5	9.8

further, by discovering the best flow control configuration for the path selected. It adapts to any changes that happen in the path conditions during communication. Eventually, it becomes clear that the protocols complete each other. The performance increase is additive as the higher protocol builds on the achievements of the lower protocol.

19.8.1 Experimental Results

In this section, we investigate the improvement of the overall performance when ESMP is operating at the same time that BNR is making the routing decisions. The experimental results that show the performance improvements of BNR are already presented; based on these results, the integration with ESMP will be evaluated.

In Tables 19.7 and 19.8, throughput increase and delay decrease with ESMP usage are presented. Both tables show the average values. The performance increase is not so significant. Someone can expect greater performance increase because of the additive improvement that was already mentioned. This happens because ESMP performance is increased when the network conditions are critical. The benefit of using ESMP is greater when the environment is stricter and the good quality individuals are spotted more easily [5]. In other words, when the conditions in the network are not critical, any individual configuration will perform very well. Hence, the performance improvement of ESMP is varying and cannot be illustrated well enough by the average performance results presented in Tables 19.7 and 19.8.

19.9 Summary

In this chapter, we describe a system in which a flow control mechanism is used not only for congestion control but also to improve the QoS required by the applications. This system does not create the conditions for good quality services but takes advantage of the existing network capabilities for an optimized flow control, in order to improve service quality. As shown earlier, flow control can be achieved not only by changing the transmitting rate of the packets but also their length. The way the flow is controlled can affect not only the throughput but also the delay, jitter, and lost packets.

Additionally, we investigate how well GA adjusts the parameters of the flow control in order to exploit the underlying network conditions and produce the desired results for the applications.

Various parameters of GA have been analyzed, and the best configuration found and adopted in ESMP architecture. By exploring this problem domain, the unique characteristics are revealed. In MANET environments that are highly dynamic, we show that ESMP architecture is able to significantly improve throughput and delay performance.

GA can adapt quickly enough to the network conditions and re-adapt to any changes. This protocol is useful for long-term connections.

The effectiveness of a GA can nearly always be enhanced by hybridization with other heuristics. Another technique that can enhance the performance of the GA is annealing the mutation rate. In this method, the algorithm decreases the mutation rate by the time the individuals reach the optimum, and increases it again the moment it detects that the population is not good enough for the environment anymore.

The case of additional parameters, like battery consumption, can be investigated. If the network is in good condition and the application is not very demanding, then other secondary issues, like battery consumption, can be taken into consideration by the protocol. That can be added in the protocol by measuring the energy consumption for the time period of one individual. The parameter can be added in the fitness function with a smaller weight (priority) than the other parameters so it cannot have a significant effect on the selection of the individuals and "appear" only when the other parameters are satisfied.

Some may say that because evolution is a process that takes years to develop, it is a waste to throw away the results every time a new communication is needed. It is interesting to investigate how the performance can be increased if every node reuses the chromosomes that had the best performance as the starting population in a new communication session, or to give this information to a new node that joins the network. The protocol must make sure that this population must have enough diversity to be flexible enough and quickly adapt.

Furthermore, we utilized the cross layer design architecture so the Bayesian network can give advice to the other layers on how to achieve better performance. Network layer is chosen to host the BN for a good reason. BN already collects information for routing decision purposes; thus, it is easier to be extended a little and produce inference for additional issues.

BN is used to guess which part of the search space has the best configuration in order to give a "fast" start to the GA. BN is not meant to replace GA but only help it with where to start looking in the search space. GA will finally find the optimum solution and, more importantly, will adapt every time the network experiences any changes.

Finally, how the overall performance benefits by using both BNR and ESMP is examined. BNR finds the most performance stable path and ESMP finds the best configuration for the specific path and further increases the overall performance of the communication.

References

1. G. Tufte, and P. C. Haddow. 1999. Prototyping a GA pipeline for complete hardware evolution. *IEEE Proceedings of the First NASA/DoD Workshop on Evolvable Hardware*, pp. 76–84.
2. C. W. Ahn, and R. Ramakrishna. 2002. A genetic algorithm for shortest path routing problem and the sizing of populations. *IEEE Transactions on Evolutionary Computation*, 1 (3), pp. 566–579.
3. B. Dengiz, F. Altiparmak, and A. Smith. 1997. Local search genetic algorithm for optimal design of reliable networks. 1997. *IEEE Transactions on Evolutionary Computation*, 1 (3), pp. 179–188.

4. H. Chou, G. Premkumar, and C. –H. Chu. 2001. Genetic algorithms for communications network design—an empirical study of the factors that influence performance. *IEEE Transactions on Evolutionary Computation*, 5 (3), pp. 236–249.
5. Ilias Kiourktsidis. July 2011. Flexible cross layer design for improved quality of service in MANETs, PhD thesis, Brunel University, School of Engineering and Design.
6. C. R. Reeves, ed. *Modern Heuristic Techniques for Combinatorial Problems*. John Wiley & Sons Inc., New York, NY, 1993.

Index

C